Handbook
of Systems Analysis

Craft Issues and Procedural Choices

Edited by

Hugh J. Miser
Edward S. Quade

North-Holland
New York • Amsterdam • London

Elsevier Science Publishing Co., Inc.
52 Vanderbilt Avenue, New York, New York 10017

Sole distributors outside the United States:
John Wiley & Sons Ltd.

© 1988 by Elsevier Science Publishing Co., Inc.

First published in the United States in 1988 by Elsevier Science Publishing Co., Inc.
First published in the United Kingdom in 1988 by John Wiley & Sons Ltd.

This book has been registered with the Copyright Clearance Center, Inc.
For further information, please contact the Copyright Clearance Center, Inc.,
Salem, Massachusetts.

Library of Congress Cataloging in Publication Data
(Revised for volume 2)

Handbook of systems analysis.

    Bibliography: v.1, p.
    Includes indexes.
    Contents [1] Overview of uses, procedures, applications, and practice—[2] Craft
issues and procedural choices.
    1. System analysis. I. Miser, Hugh J. II. Quade, E.S. (Edward S.)
T57.6.H365 1985   003   84-8131
ISBN 0-444-00918-3 (v.1)
ISBN 0-444-00940-X (v.2)

Current printing (last digit):
10 9 8 7 6 5 4 3 2 1

Manufactured in the United States of America

Contents

Chapter 4. Using Operational Gaming   121

Ingolf Ståhl

Chapter 5. Social Experimentation: Some Whys and Hows   173

Rae W. Archibald and Joseph P. Newhouse

Chapter 8. Quantitative Methods: Uses and Limitations   283
Edward S. Quade

PART IV.   SOME COMPONENTS OF ANALYSIS   325

Chapter 9. Forecasting and Scenarios   327
Brita Schwarz

Chapter 13. Validation   527
Hugh J. Miser and Edward S. Quade

Chapter 14. A Framework for Evaluating Success in Systems Analysis   567
Bruce F. Goeller

Preface

Since the dawn of human history people have been facing problems arising from the operations of systems in which they themselves are important parts. However, the idea that systematic scrutiny in the spirit of science—and using its tools—can help solve them became widespread only in this century. Before 1935 the inquiries into such problems—and therefore into the systems from which the problems emerged—were few, scattered, and limited in scope. However, since 1935 the studies aimed at solving such problems by understanding the operations of their underlying systems have undergone rapid growth, encouraged by the successful work that took place just before and during the second World War.

This work has developed scientific knowledge of the behavior of man–machine operating systems, a technology to put such knowledge to work to solve their problems, a group of professionals with the relevant skills, and a body of successful problem-solving experience.

The success of this work—now often called operations research—led to a significant expansion of its scope, so that by the middle 1950s it was not uncommon for the problems of very large systems to be investigated, or for knowledge from such work to be employed to design new systems. This enlarged purview necessitated bringing into the work specialists from many disciplines; thus it was not unnatural for the enlarged community doing the work to adopt a name for it: systems analysis—to emphasize the fact that, associated with each of the problems then being investigated, there was an existing system, or one being planned. The ensuing quarter century has seen growth in both the scope and diversity of systems analysis, including its extension to problems arising in situations for which no system had yet been defined.

Systems analysis is now applied to a wide range of highly diverse problems, and the patterns of analysis exhibit a corresponding diversity, depending on the context, the possible courses of action, the information needed, the accompanying constraints and uncertainties, and the positions and responsibilities of the persons who may use its results. In a rare case, a problem may fall within the sphere of responsibility of a single policymaker; however, it is far more usual for the relevant responsibilities to be diffused among many persons, often with significant portions of the problem lying outside existing authorities.

This diversity of activity has spawned a variety of names other than systems analysis under which work that can be described this way is done, depending on the professional, institutional, and geographical backgrounds of the analysts involved: operational research, operations analysis, policy analysis, policy science, systems research, and others. Thus, choosing a title for this handbook presented some difficulties; as a compromise, we chose the name associated with a stream of thought and activity that began over four decades ago in the United States and that has had an important influence on developing the activity throughout the world.

For a field exhibiting such rapid growth and diversity—even in its basic nomenclature—is the concept of a handbook a meaningful one? With so many ramifications, and so many extensions going on in so many new directions, is it possible to take a snapshot of the state of the art that will have enduring value? If the answer is "yes," what should a handbook be?

The key to arriving at positive answers to these questions is twofold: to restrict the attention in the handbook to the central core of the field as it now exists based on extensive experience, avoiding the temptation to stray into the many attractive and promising new branches; and to provide a challenging perspective of the future opportunities as yet unexplored, while at the same time offering normative standards of quality and substance where experience warrants.

Thus, we have centered the attention in this *Handbook of Systems Analysis* on the type of problem that has mainly concerned systems analysts in the past: those relatively easy to structure and in which some important aspect is dominated by technology. Although there were early exceptions, it is only within the past decade or so that the people-dominated problems—such as those found in the influences of government and industrial operations on the public and the environment—began to be explored in a major way by systems analysts; thus, it is still too early in this experience to warrant a handbook synthesis.

In the United States much of the work spanning the field dealt with by systems analysis and one in which human behavior dominates is called policy analysis. While this handbook limits its reach toward people-dominated problems, it has much to say to the policy-analysis community.

We have found that, for our context, it is possible to assemble a summary of the main currents of knowledge and practice, as well as guidelines for future work. This handbook is the result.

Since systems analysis as it is currently practiced—and is likely to be for the foreseeable future—combines both old and new knowledge to evolve solutions to problems of operations, policies, and planning, the audience that needs information about it is extremely varied, ranging from technical specialists to intelligent nontechnical citizens with public concerns. Thus, we have divided the handbook into three volumes, depending on their purposes, contents, and audiences; although the three are intimately interconnected, each is a logical whole that is useful to its audience without reference to the others.

The first volume, *Overview of Uses, Procedures, Applications, and Practice*, is aimed at a widely varied audience of producers and users of systems analysis—government officials, legislators, business executives, scientists and technologists in other fields, public-interest groups, and concerned citizens, as well as students and practitioners; to keep systems analysis accessible to this broad audience, this book avoids technical details as much as possible. This volume provides a central description of the systems analysis process from the first appreciation that a problem situation exists through analysis and implementation to assessing the results. Several successful applications of the systems-analysis approach are described early in the book and used throughout to clarify the concepts and illustrate the nature of the work and its results.

Using the approach outlined in the first volume in a particular case confronts the systems analyst with many craft issues and procedural choices; they arise not only from aspects of the problem situation and its environment but also from matters interior to the analysis. In the problem situation and its environment the analyst must deal with the nature of the decisionmakers, the various forces at work, the internal characteristics of the situation, the time and resources available for the analysis, the potential forms that ameliorative steps might take, the likelihood of discovering feasible courses of action, the potential difficulties of taking these courses, the institutional frameworks that could be called into play, and so on. Internal to the analysis, the analyst must deal constructively with such matters as setting the problem to be investigated, choosing an analytic approach, reducing a large number of alternatives and their consequences to a manageable number, formulating findings so that they focus on the needs of the client, meeting demanding schedules, making the analysis team an effective unit, and so on. In sum, the work of the systems analyst must deal with many craft issues and procedural choices; it is faced with a host of "secondary decisions."

In making these secondary decisions the analyst will find that knowledge of science and the technology of systems analysis must be supplemented by judgment, intuition, and the wisdom gained from experience—in other words, knowledge gained from practicing the craft of systems analysis. This second volume of the handbook, *Craft Issues and Procedural Choices*, aims to support the work of the practicing analyst by gathering together and setting forth in systematic form much of the craft knowledge that has accumulated in this field.

Thus, the keynote of the book is the question: How does the systems analyst deal with _____ ? After a preliminary chapter on systems analysis and its craft, the book fills in this blank with discussions in four categories: selecting analysis approaches and procedures (choosing analytic strategies and their components, including expert judgment, operational gaming, and social experimentation), setting boundaries for the analysis (by generating and screening alternatives, facing uncertainties squarely, and recognizing the uses and limitations of quantitative methods), considering some important components of analysis (forecasting and scenarios, cost considerations, and program evaluation), and taking account of professional issues (such as foundational concepts, validation, evaluating success for analysis, and controlling quality).

The second volume of the handbook is not a treatise on techniques well described in the scientific literature, nor a compendium of models—nor does it summarize material that has been presented in well-developed and standardized form elsewhere. Rather, it focuses attention on aspects of the craft of systems analysis that may be useful to an analysis team as it plans and carries out a systems analysis study.

Consequently, the book's audience is primarily systems analysts. It will also be of interest, however, to students, specialists in other fields drawn into interdisciplinary systems analysis work, and other workers interested in how work in this field proceeds. In the United States it speaks also to the large and growing policy-analysis community.

The third volume, *Cases*, contains extended descriptions of actual cases in which systems analysis was used; thus, it illustrates the diversity of problems and approaches encountered in systems analysis. While some attention is paid to technical details, so that an analyst has a view of how the results were arrived at, as well as what they were, the descriptions are written so as to be accessible also to nontechnical readers who may want to skip the technical details. Insofar as possible, the descriptions view the cases from the point of view of the client as well as the analyst.

Although the third volume centers its attention on cases, both the first and second volumes make use of case descriptions to illustrate important points. Thus, the handbook as a whole weaves together generalities about systems analysis practice with particularities chosen from examples of good work.

The work of preparing this handbook has shown us that there is a great deal of existing knowledge, that there is a substantial body of practice, and that, in spite of its promise, the prospects for the field and its potential value for society are only dimly perceived in many quarters. Thus, at this relatively early stage in the development of systems analysis, a handbook has an important opportunity to bring the central core of knowledge and professional experience together in systematic form, to establish foundations for the further growth of systems analysis, to map its potentials, to indicate open questions and challenges, to offer guidelines for future development, and to provide a source of valuable information for actual and potential clients for systems analysis studies.

This handbook has been aimed at these purposes, and was prepared in the hope that it would serve them well enough to contribute significantly to the growth of systems analysis. If this goal is met, the Editors and contributors will be more than amply rewarded; they have worked in the belief that systems analysis, properly developed and skillfully pursued, can make important contributions to solving problems that the world will face in the future.

Hugh J. Miser
Edward S. Quade

Acknowledgments

The concept of a *Handbook of Systems Analysis* and a project aimed at creating such a work were originated at the International Institute for Applied Systems Analysis (IIASA), which owed its existence to the growth of systems analysis, particularly in the United States, and the perception that many problems of international and global scope could usefully be addressed by extending this approach appropriately.

IIASA is a nongovernmental interdisciplinary research institution located in Laxenburg, Austria, that was founded in October 1972 on the initiative of the academies of science or equivalent institutions of 12 nations (by 1978 these National Member Organizations numbered 17 and included countries from both East and West). In 1974 the National Member Organization from the USSR proposed that IIASA undertake to prepare a *Handbook of Systems Analysis*, and this project was supported by the Institute from 1974 through 1982. In 1974 and 1975, Roger Levien, later Director of the Institute from November 1975 until November 1981, conducted, together with Vil Z. Rakhmankulov, an initial survey of interests, needs, and possible contents. Edward S. Quade led the project from 1976 through 1978, Hugh J. Miser from 1979 on; Vil Z. Rakhmankulov, Władysław Findeisen, Giandomenico Majone, and Alexander P. Iastrebov participated in the project for various periods between 1974 and 1982. A conference held at IIASA in September 1978 devised the three-volume format in which the handbook now appears. The Institute's support for the project is hereby gratefully acknowledged.

In addition to the institutional support from IIASA and its 17 National Member Organizations, a great many individuals have contributed to the work of preparing the handbook, too many to make it feasible to recognize all of them in these acknowledgments. Some 160 persons responded to the survey that began the work leading to the handbook, 38 scientists submitted written comments on the manuscript of a preliminary version of the work as a whole, and 25 scientists from 14 National Member Organization countries attended the 1978 planning conference that gave the handbook the three-volume design that it has today. To all these people we are grateful for their support and for their contributions to the concept and details of the work.

With respect to the second volume, *Craft Issues and Procedural Choices*, there are many additional persons to whom we owe debts of gratitude:

Howard Raiffa, the founding Director of IIASA, and his successor Roger Levien, both gave the handbook project enthusiastic support, the former from 1974 through 1975, the latter from 1975 through 1981.

Ten scientists undertook to share their specialized expertise in aspects of systems analysis by preparing chapters in this book: Olaf Helmer of Carmel, California, wrote the chapter on using expert judgment; Ingolf Ståhl of the Stockholm School of Economics, Stockholm, Sweden, prepared the chapter on operational gaming; Rae W. Archibald and Joseph P. Newhouse of The Rand Corporation in Santa Monica, California, wrote the chapter on social experimentation; Warren E. Walker, also of The Rand Corporation, contributed the chapter on generating and screening alternatives; Yehezkel Dror of the Hebrew University of Jerusalem, Jerusalem, Israel, developed the chapter on uncertainty; Brita Schwarz of the Economic Research Institute at the Stockholm School of Economics, Stockholm, Sweden, wrote the chapter on forecasting and scenarios; H. G. Massey of The Rand Corporation prepared the chapter on cost considerations; Harry P. Hatry of The Urban Institute, Washington, D.C., contributed the chapter on program evaluation; and Bruce F. Goeller of The Rand Corporation prepared the chapter on evaluating success for analysis.

Five analysts gave the manuscript cover-to-cover reviews: Kenneth C. Bowen of the Royal Holloway and Bedford New College, University of London; Garry D. Brewer of the Yale School of Organization and Management; Peter deLeon of the Graduate School of Public Affairs, University of Colorado at Denver; G. D. Kaye of Ottawa, Ontario, Canada; and Kenneth L. Kraemer of the Public Policy Research Organization, University of California at Irvine. In addition, useful comments were received through three IIASA National Member Organizations: from Johannes Dathe of the Industrieanlagen-Betriebsgesellschaft mbH in Ottobrunn, Federal Republic of Germany; from Yoshikazu Sawaragi of The Japan Institute of Systems Research in Kyoto, Japan; and from Börje Johansson of the Umeå University, Umeå, Sweden. These persons contributed significantly by raising questions, challenging statements, pointing out omissions, and making suggestions. This is not to say that the Editors and contibutors accepted all of the suggestions and criticisms. Rather, they used the ideas that would, in their judgment, strengthen the arguments without making them unduly long and complicated, and that accorded with their own experience and the literature as they knew it. Thus, it cannot be said that the persons on this list of reviewers necessarily agree in detail with what is presented in this book.

The Editors are particularly grateful to the Elsevier Science Publishing Company, Inc., New York, for the subvention which has been used to defray the substantial administrative expense of carrying on the preparation of this handbook since the IIASA support was withdrawn from it at the end of 1982.

We are indebted to a number of organizations and individuals for permission to quote copyrighted material: to Addison-Wesley Publishing Company for the quotation from Frederick Mosteller and John W. Tukey in Section 11.4; to the *Air University Review* for the quotation from G. A. Kent in Section 15.8; to the American Academy of Arts and Sciences for the quotation from Harvey Brooks in Section 15.12; to Basic Books, Inc., and Gower Publishing Company, Ltd., for the extensive quotations from Donald A. Schön in Sections 1.3, 12.4, and 15.11; to Hylton Boothroyd for the extensive quotations from his *Articulate Intervention* (published originally by Taylor and Francis, Ltd.) in Section 12.3; to K. C. Bowen for quotations in Sections 2.4, 2.8, and 13.7; to Contemp Bks., Inc., for the quotation from Ludwig von Mises in Section 2.7; to the author and Elsevier Science Publishing Company, Inc., for the quotations from Alexander M. Mood in Sections 2.4, 2.7, and 2.8; to Greenwood Press, Inc., for the quotation from I. Wilson in Section 9.4; to the Harvard Business School, publisher of the *Harvard Business Review* (copyright by the President and Fellows of Harvard College), for the quotation from Peter Lorange and R. F. Vancil in Section 9.5; to The Institute of Management Sciences, publisher of *Management Science*, for the quotations from Olaf Helmer and N. Rescher in Section 3.2 and 3.5; to The Institute of Management Sciences and the Operations Research Society of America, publishers of *Interfaces*, for the quotations from B. F. Goeller and the PAWN Team in Sections 2.8 and 13.4 and from Saul I. Gass in Section 15.10; to the International Institute for Applied Systems Analysis for quotations from the Energy Systems Program Group (Figures 9.1, 9.2, and 9.3, Tables 9.1 and 9.2, and passages in Section 9.6), from C. S. Holling (in Sections 2.8, 13.4, and 13.5), from Giandomenico Majone (in Sections 1.5, 1.7, and 15.7), from Giandomenico Majone and E. S. Quade in Section 15.7, from D. H. Meadows and J. M. Robinson (in Sections 2.4, 2.8, and 13.7), from A. J. Meltsner in Section 13.7, from E. S. Quade (in Sections 8.3 and 13.8), from Howard Raiffa in Section 2.8, and from Ingolf Ståhl (Figures 4.2, 4.3, and 4.4); to the *Journal of the Operational Research Society* for the quotation from Samuel Eilon in Section 2.6; to Peter J. Kolesar for the quotations from Linda Green and Peter J. Kolesar in Section 13.7; to A. M. Lee for the quotation in Section 13.7; to Lexington Books and The Policy Studies Organization for the quotations from F. P. Scioli, Jr., and T. J. Cook in Section 8.4; to Giandomenico Majone for the quotations in Sections 15.1 and 15.7; to Martinus Nijhoff Publishers and Kluwer Academic Publishers, publishers of *Policy Sciences*, for the quotations from Wolf Haefele and H.-H. Rogner (in Section 9.6), from B. Keepin (in Section 9.6), from R. E. Strauch (in Sections 8.6 and 8.7), from V. Taylor in Section 8.4, and from Geoffrey Vickers in Section 8.7; to National Affairs, Inc., publisher of *The Public Interest*, for the quotation from J. Q. Wilson in Section 15.2; to Samuel Eilon, Chief Editor of *OMEGA*, for the quotation from Giandomenico Majone in Section 1.5; to the authors and the Operations Research Society of America, publisher of *Operations Research*, for the quotations from Saul I. Gass (in Section 13.3), from Saul I. Gass and B. W. Thompson (in Section 13.7), and from C. J. Hitch (in Section 2.6); to Pergamon Press, Ltd., for the quotation from Roy Amara and A. J. Lipinski in Section 9.5; to The RAND Corporation for the chapters written by members of its staff (Chapters 5, 6, 10, and 14) and for the material from RAND sources quoted therein, as well as quotations from Herman Kahn and I. Mann (in Section 15.11), from E. S. Quade (Figures 8.1, 8.2, and 8.3, and passages in Sections 8.3 and 8.4), and from R. E. Strauch (Figure 8.4 and passages in Sections 8.5 and 8.6); to Jerome R. Ravetz for the extensive quotations from his *Scientific Knowledge and its Social Problems* (published originally by Oxford University Press) in Sections 1.3, 12.4, 12.5, and 15.7; to the Russell Sage Foundation for the quotations from Martin Greenberger, M. A. Crenson, and B. L. Crissey in Sections 13.3 and 13.7; to Sage Publications, Inc., and the author for the quotation from J. E. McGrath in Section 1.3; to The Urban Institute for quotations from A. P. Miller, M. P. Koss, and H. P. Hatry (Table 11.3), H. P. Hatry, L. H. Blair, D. M. Fisk, J. M. Greiner, J. R. Hall, Jr., and P. S. Shaenman (Table 8.1 and a passage in Section 8.4), and H. P. Hatry, R. E. Winnie, and D. M. Fisk (Table 11.2); and to the authors and John Wiley and Sons, Inc., publishers (for Massachusetts Institute of Technology and the President and Fellows of Harvard College) of *Science, Technology, and Human Values*, for quotations from W. C. Clark and G. Majone (Table 15.2 and passages in Sections 15.1 and 15.2).

The Editors

PART I
SYSTEMS AND POLICY ANALYSIS

Chapter 1
Introduction
Craftsmanship in Analysis

Hugh J. Miser and Edward S. Quade

1.1. Background

Many of the problems of modern society emerge from interactions among people, the natural environment, and artifacts of man and his technology. Often such problems can be addressed by systems analysis, an approach that brings to bear the knowledge and methods of modern science and technology with appropriate consideration of social goals and equities, the larger contexts, and the inevitable uncertainties. The aim is to acquire deep understanding of the problems and to use it to help bring about improvements. In practice, analysis of this type clarifies and defines objectives; searches out alternative courses of action that are both feasible and promising; gathers data relevant to—and projects the nature of—the environments for which the actions are proposed; and generates information about the costs, benefits, and other consequences that might ensue from their adoption and implementation. In the usual case, the study's results and the evidence leading to them are presented to the responsible decision- and policymakers in order to help them adopt and follow an advantageous policy or course of action. Help at this last stage often requires the analysts to offer both advice and supporting arguments. A decision leading to an implementation can be followed, after a suitable period of time, by an evaluation to see if expected outcomes and benefits have actually accrued.

In this chapter—and throughout this book—we intend the term *systems analysis* to represent the portions of policy analysis, operations research, management science, and other professional fields that share the structure that has just been outlined, or some variant of it.

If analysis of the type considered here is to be fully helpful, it must not only provide sound information about problems and their solutions, it must also offer good advice and persuade its clients to adopt a reasonable belief in the information and to accept the advice appropriately.

> Systems analysis is concerned with theorizing, choosing, and acting. Hence, its character is threefold: descriptive (scientific), prescriptive (advisory), and persuasive (argumentative-interactive). In fact, if we look at the fine structure of analytic arguments we see a complex blend of factual statements, methodological choices, evaluations, recommendations, and persuasive definitions and communications. An even more complex structure emerges when we look at the interactions taking place between analysts and different audiences of sponsors, policymakers, evaluators, and interested publics. Moreover, descriptive propositions, prescriptions, and persuasion are intertwined in a way that rules out the possibility of applying a unique set of evaluative criteria, let alone proving or refuting an argument conclusively. [Majone, 1985, pp. 63–64; see also Majone, 1980a, p. 5.]

No infallible theory or sage philosophy exists to direct us safely through the maze Majone describes;[1] we have some theory and principles to guide us, but to make the host of decisions required we must also use information from the many disciplines involved, instinct, common sense, hunches, and especially craftsmanship gained from experience.

Much of the activity in systems analysis, particularly in its descriptive phase, can be guided by the knowledge and principles of the sciences, both natural and social, or by well-established scientific and professional canons, because systems analysis makes extensive use of information, tools, and methods borrowed from scientific disciplines. Nevertheless, many—and usually most—of the approaches, structures, and procedures used and the choices made in the course of systems-analysis work arise from its peculiar circumstances and properties, and thus depend on the knowledge the analyst has gained from his own experience, or learned from that of others. In other words, the analyst must be guided substantially by the knowledge of his craft. Almost every book on systems or policy analysis, operations research, or management science offers an introductory—but very partial—view of the craft knowledge available today; the purpose of this volume on craft issues and procedural choices is to extend these views significantly.

For the form of analysis discussed in this handbook we use its original name: systems analysis. As we have already noted, however, other names are also used for it. In the United States, where the term systems analysis has also become widely associated with an activity underlying the design

---

[1] The work of Giandomenico Majone, particularly that on craftsmanship in analysis carried out between 1974 and 1982 at the International Institute for Applied Systems Analysis, has provided much of the inspiration for the approach and structure of this volume.

of computer systems, the term policy analysis is often used instead, particularly when the analysis deals with public problems. When the problems are more technical and operational and do not involve demanding societal issues, it may be called operations research. But, no matter what name may be used, we are writing here about analysis to assist people responsible for policy or action to develop, understand, select, and implement what should be done in an uncertain environment to advance human welfare—and to evaluate the consequences of what was done.

1.2. The Purposes of This Volume

Although the knowledge and principles of science, together with its rules of professional conduct, plus the guidance found in the analysis literature (including, of course, the *Overview* volume of this handbook), offer assistance with many issues and choices, there still remain many others for the analyst to face that are important for successful work and for getting its findings accepted and used. Thus, the main purpose of this volume is to offer additional counsel on craftsmanship as it has been developed by leading workers in the field.

The discussions of craft issues in these two volumes, however, do not exhaust all that should be said on the subject. For some issues and choices the profession lacks the knowledge or experience to say more, as yet, than that they must be resolved on the basis of the analyst's judgment and intuition; for others there is not yet enough concordance of expert view to enable any sort of authoritative presentation. Nevertheless, enough experience on craft issues and procedural choices has accumulated to make it worthwhile to bring it together as a complement to other works in this field.

This book avoids, except where necessary for treating the craftsmanship matter at hand, all discussion of the largely mathematical and formal tools of operations research and other technical specialties that are so important for effective systems analysis work; such tools are well treated elsewhere.

When the findings of a systems-analysis study are ready and a potential user must decide what to do about them, he must proceed on the basis of his judgments and beliefs about what might ensue from any decision that he might make based on the analysis work; he does not have available to him the possibility of reviewing actual outcomes. His sense of the quality of the analysis and its pertinence to his concerns are important contributors to his judgments and decisions.

On the other hand, although it may seem to be the ultimately most desirable course, examining the actual results of an implemented program ensuing from an analysis is often not a practical way of evaluating the quality of analysis work; the results of some studies are not implemented, others are not implemented in accordance with the findings and recommendations, and the long delays that sometimes occur between study completion and implementation may make direct comparison for evaluation purposes either uninteresting or impossible. Even when the results of implementation are available for comparison, they sometimes do not offer a reliable guide to the quality of the work; too much depends on chance or changes in the context that may not have been anticipated during the course of the analysis.

Consequently, systems analysis needs standards of quality against which to compare systems-analysis work, both while it is going on and after it is complete. Unlike such well-established professions as law or medicine, systems analysis has not yet developed adequate and widely accepted professional standards to govern its work. Hence, the second purpose of this book is to contribute to the goal of setting standards by which the quality of systems-analysis work can be judged, not only after the work is finished but also during its progress—and, in particular, standards that depend on the procedures and craftsmanship used during the analysis process. Such standards cannot be confined to the descriptive aspect of systems analysis, but must extend to the advisory and persuasive aspects as well (as discussed further in Chapter 15).

After a brief discussion of the role of craftsmanship in scientific and professional work, the first chapter describes the important role of craft in systems analysis, explains the importance of craftsmanship to the quality and acceptance of systems-analysis work, and describes how craftsmanship forms an important basis for evaluating such work. It concludes with a preview of the later chapters and an appendix listing the steps in the systems analysis process and where they are discussed in the *Overview*.

1.3. Craftsmanship in Scientific and Professional Work

Various works on systems analysis, as well as the appendix to this chapter, sketch approaches to systems-analysis problems that, if carried out properly, can increase the chances that the suggestions emerging from the analysis will be adopted and implemented successfully, thus solving the problems—or at least bringing about improvements in them. Executing the details of the procedures in this approach, however, is seldom a straightforward or untroubled task. The analyst's path is strewn with issues and difficulties that force him to make choices, some of which will prove to be crucial. He must settle on the nature of the problem that is to be considered; he must choose among alternative assumptions, data sources, methods of analysis, and forms of presentation; he must make decisions that reflect the resources of time and money available, the nature of the decisionmaker, and the properties of the decisionmaking environment; he must choose how to limit the extent of the investigation, how to rank the alternatives realistically, and how to estimate the likelihood that they can be implemented successfully. Such choices—and many others—occur throughout any systems analysis, and how they are made will have important bearings on the quality and success of the work.

Some choices in a systems analysis can be guided by the knowledge and principles of social or physical science, and others by well established scientific and professional precepts, but many—and usually most—depend on the knowledge the systems analyst has gained from his own practical experience, or that of others. In other words, these choices will be guided by the analyst's knowledge of his craft.

In this regard systems analysis is like other scientific and professional work. Craft knowledge—less explicit than formal theoretical knowledge, but more objective than pure intuition—is important in every type of scientific and professional endeavor. Law, medicine, and engineering all depend heavily on craftsmanship, as does scientific research and the knowledge it produces. Even scientific research of the most theoretical kind requires extensive use of personal judgment and practical skill.

> [W]ithout an appreciation of the craft character of scientific work there is no possibility of resolving the paradox of radical difference between the subjective, intensely personal activity of creative science, and the objective, impersonal knowledge which results from it. [Ravetz, 1971, p. 75.]

Thus, no overall concept of the approach—or method—of science can, by any stretch of the imagination, describe the day-to-day work of scientists, nor, we now note, does it help them resolve most issues or make the practical choices they face in these activities. In these matters they depend largely on the craft knowledge they have gained from their teachers, their colleagues, and their own experience.

For example, McGrath (1982, pp. 13–14), reporting the views emerging from a meeting of an interdisciplinary group of scientists all involved with research on organizations, describes the choices that must be made in the course of a research project as "judgment calls," a term adopted from reporters describing the behavior of referees in American sports. He says that this term

> refers to all those decisions (some big, some small, but all necessary and consequential) that must be made without the benefit of a fixed, "objective" rule that one can apply, with precision, like a pair of calipers. . . .
> . . . such judgment calls accumulate in their effects; and, indeed, they quite literally determine the outcome of most games. Similarly, in research, there are many crucial decisions that must be made without the

benefit of a hard and fast, "objective" rule, or even a good rule of thumb. . . .

. . . It is our shared view . . . that one loses a great deal when one attempts to fashion sound research entirely on the basis of general decision rules routinely applied. Rather, we believe one should retain a very important place for the judgments of the skilled researcher who, at best, will be sensitive to nuances of both the substantive setting and the research impedimenta; and who, at least, can observe when something stunningly unexpected occurs, and can perhaps stop the churning of the research machine long enough to take a look at it.[2]

Similarly, in his searching inquiry into the professions, particularly those that are science-based, Schön (1983, pp. 49–50) finds that professionals are guided by knowledge gained from previous professional experience:

[T]he workaday life of the professional depends on tacit knowing-in-action. Every competent practitioner can recognize phenomena—families of symptoms associated with a particular disease, peculiarities of a certain kind of building site, irregularities of materials or structures—for which he cannot give a reasonably accurate or complete description. In his day-to-day practice he makes innumerable judgments of quality for which he cannot state adequate criteria, and he displays skills for which he cannot state the rules and procedures. Even when he makes conscious use of research-based theories and techniques, he is dependent on tacit recognitions, judgments, and skillful performances.

Schön advocates systematic inquiry into these craft skills as the basis for improving professional performance, and sharing it with clients, foreseeing a future in which such a "reflective practitioner" enters into a "reflective contract" with his client, as described in Section 12.4. In sum, Schön argues that a significant body of craft knowledge can be systematized into an "epistemology of practice" for each profession, although he does not envision that all elements of the craft associated with practice could be captured in such a doctrine.

1.4. Systems Analysis as Craft

Before defining what we mean by craftsmanship and craft knowledge in systems analysis, it will be useful to consider an illustration (from Majone, 1980a, pp. 12–14): handling data and information.

In systems analysis data are seldom produced, as often in the physical sciences, by planned experiment; they are usually taken from existing

---

[2] J. E. McGrath, "Introduction," pp. 13–16, edited by J. E. McGrath, J. Martin, and R. A. Kulka. Copyright © 1982 by Sage Publications, Inc. Reprinted by permission of Sage Publications, Inc.

sources. Thus, they are likely to have been compiled in categories that do not fit the objectives of the analysis. Moreover, since the process that obtained them was influenced by the methods and skills of the workers involved, they may or may not be representative of the general situation being contemplated in the analysis. Since perfection in data is unlikely in any case, and other sources may be hard to come by, the analyst may still want to use the available data, provided he judges that they are good enough for the functions that they must perform in the analysis.

Similar judgments of adequacy must be made in turning the raw data into usable information by means of such techniques as calculating averages or other statistical measures, fitting a curve to a set of points, or reducing the data by some multivariate statistical technique. This transformation of data into information involves at least three decisions for which the analyst has few formal rules to guide him: one is that the advantages achieved through a reduction make up for the information lost during the reduction process; another is that the model chosen to represent the data in the analysis is adequate; a third is that the model chosen from among the infinite number of possibilities is a significantly useful one for the problem under examination. Beyond these points there is the issue of how to supplement the data when they do not cover the ground needed in the analysis.

In the absence of theory, the success of the decisions involved in selecting data, transforming them into information, and supplementing them when they are not fully adequate depends on more than just intuition or common sense; it depends on knowledge of the craft of systems analysis.

Craft knowledge is common to all forms of disciplined intellectual inquiry. It is a repertoire of procedures and judgments that are partly personal and partly social, and are acquired mainly through practice and precept. They are derived in part from one's own experience, in part from the professional experiences and norms of one's teachers and colleagues, and in part from culturally determined criteria of adequacy and validity. Craftsmanship is the application of craft knowledge.

Some people speak of systems analysis as an art, but this term implies a lack of utility. Craft results may or may not be beautiful, but they are always intended to have utility. In a purely folk definition (Becker, 1978), a craft consists of a body of knowledge and skills that can be used to produce useful products; by extension, craftsmanship in analysis is the ability to perform in a useful way—say, to carry out a study that policymakers find helpful. Moreover, the organizational form associated with systems analysis is that of a craft, one in which the workers (the analysts) do their work for others (those responsible for policy and action), who define broadly what they want to achieve and to what purposes (the objectives).

Although systems analysis may not yet be a recognized profession in that it does not have qualifications for admission (such as an examination or a license) and widely accepted standards of quality (which we seek), its practitioners are usually professionals drawn from such established disciplines as economics or mathematics.

Success in both scientific research and professional practice depends heavily on craftsmanship meeting high standards.

> The importance of craft knowledge and experience is even greater in systems analysis [than in purely scientific research]. Because the conclusions of a systems study cannot be proved in the sense in which a theorem is proved, or even the manner in which propositions of natural science are established, they must satisfy generally accepted criteria of adequacy. Such criteria are derived not from abstract logical canons (the rules of the mythical standard scientific method) but from craft experience, depending as they do on the special features of the problem, on the quality of the data and limitations of available tools, on the time constraints imposed on the analysis, and on the requirements of the sponsor and/or decisionmaker.
>
> In short, craft knowledge—less explicit than formalized theoretical knowledge, but more objective than pure intuition—is essential for doing systems analysis as well as for evaluating it. [Majone, 1985, pp. 60–61.]

In every systems-analysis, operations-research, or policy-analysis study there are important points at which issues must be settled and craft choices made without adequate guidance from canons of science, from theory, or from widely accepted, "objective" ground rules. In these cases the analyst brings his craftsmanship to bear. It is probably impossible to define craftsmanship precisely for this context, and it may even be undesirable; in any case, we shall not attempt this task. It is possible, however, to illuminate the meaning of this concept by listing some typical issues and choices that an analyst may face in the course of his work, issues and choices that will call into play important elements of craftsmanship.

How to approach a mess of problems associated with a situation that needs improvement, and isolate one or more whose solutions will be relevant and useful to those with responsibilities for decision and action.

How to locate and collect data related to a problem and turn them into information relevant to its solution.

How to increase the decisionmaker's options by designing new and better alternatives.

How to select an analytic strategy that offers a reasonable hope of producing a useful result within the limitations of time and the available resources.

How to choose variables for investigation and how to quantify them in a useful way.

How to set the limits on the extent of the investigation, while balancing desire for comprehensive coverage against the limitations of time and resources.

How to choose models that represent the important variables in the problem, while balancing the competing factors of comprehensiveness and ease of calculation.

How to verify and how to establish the credibility of the chosen models.

How to formulate findings based on the results obtained from the analysis and its models so that these findings will focus squarely on the realities of the problem and the concerns of the client.

How to design and use forms of communication that will be effective and also facilitate interchanges of ideas with clients, their staffs, and other persons concerned.

How to relate to, and ensure the full cooperation of, the client and his staff throughout the inquiry.

How to interpret the findings of the work and help those responsible to formulate an action program.

How to participate effectively in programs aimed at implementing the decisions based on the study.

This list, in spite of its length, is far from complete; any experienced systems analyst could add to it. Its purpose is merely to suggest what we mean by the craft of systems analysis.

Sometimes the distinction between formalized textbook knowledge and what is, or originally was, craft knowledge can become blurred, or even eliminated. To illustrate, consider the task of helping a group of decisionmakers faced with the need to choose a course of action from a number of alternatives offering different advantages and disadvantages under various circumstances. One approach—sometimes found in operations research textbooks—is to apply decision theory. Here the analyst helps the decisionmakers to estimate probabilities and make calculations that eventually lead to an index that will rank the alternatives in order of preference according to a criterion chosen previously. It is a time-consuming process that is unlikely to work well with multiple decisionmakers who have widely differing objectives, values, or views of the future.

There is a contrasting craft approach based on the knowledge that people can agree on a course of action without necessarily agreeing on goals or criteria and without aggregating all the diverse consequences of each option into a single index. This alternative approach (very old, but further developed and first used extensively in systems studies by Goeller—see Chesler and Goeller, 1973) offers decisionmakers detailed information about the various possible options and their potential consequences on a series of carefully designed charts called scorecards (for an example, see Table 10.4, and, for further explanation, see Sections 5.7 and 10.3). Although these scorecards summarize, they do not suppress any information that could be important to anyone concerned with the problem that has been explored. Thus, a decisionmaker can balance positive and negative factors and choose his own preferred course of action without having to be explicit about his own criteria and goals. When several decisionmakers are involved, experience has shown that, against this sort of background, they can usually come together fairly easily to agree on a course of action.

This craft approach does not necessarily produce better results than other methods, and it does not avoid the need for each decisionmaker to estimate, at least implicitly, the probabilities associated with the outcomes or likelihoods that various potential situations will occur and to assess the weights that he should give them, but it does appear to ensure a reasonable level of satisfaction. Too, while speeding up the decision process, it eliminates much of the influence of the analyst's preferences and goals from the choice that is finally made. Moreover, in a craft process such as this one, it is possible to redesign the alternatives to take advantage of suggestions from the decisionmakers, and thus possibly offer more desirable options than before.

If the scorecard, or some other, approach becomes widely used—and is refined on the basis of broad experience—it may gradually cease to be regarded as a craft process and become another standard procedure in systems analysis associated with a certain class of situations to which it has come to be regarded as appropriate.

1.5. The Importance of Craftsmanship to the Quality of the Work

Unlike most problems in textbooks, few policy questions carry guarantees that there even exist solutions, let alone ones that can be established as correct. The relations between the outcomes predicted by analysis and the ones occurring in the real world are almost always indirect and uncertain.

Some systems analysts find themselves in the fortunate position of having their findings confirmed as the results of appropriate implementation programs. For example, the analysts who devised a new system for supplying blood for the Greater New York Blood Program were able to see within a year or two that the operations of the new system were bearing out the predictions of their analysis (for a brief summary of this work, see Section 2.2).

On the other hand, direct empirical confirmation of the outcomes of a systems analysis study is often impossible within a reasonable period of time, and sometimes is not possible at all. For example, the study of world energy demand and supply over the five decades from 1980 to 2030 described in Section 9.6 produced findings looking as much as 50 years into the future based on a complex combination of assumptions about conditions and possible administrative actions. The principal result implied that the world could have an adequate supply of energy at the end of the period studied, a finding that cannot be confirmed until this distant time is much nearer. If a shortage of energy should occur in 2030, this situation could not be construed as failing to support the analysis unless it could be shown that all of the assumed conditions had been met in reality. In any case, direct confirmation of the findings of this study, especially in the short term, seems remote at best.

Generally, where outcomes depend on chance effects or on actions of people who may or may not behave in accordance with study assumptions, actual results may differ substantially from those yielded by a systems analysis. This fact, however, does not vitiate the value of such work, since it can give early warning, based on early observation of variations from assumed inputs, of altered outcomes about to appear.

It is not only such factors as time, chance effects, and varying human behavior that make it difficult to evaluate the correctness, or even the usefulness, of systems analyses in terms of the results produced. Today it is not uncommon for a systems study to be sponsored by one organization, carried out by a second, used by a third, and evaluated by a fourth. What the original sponsor feels about the study and its results may differ from the judgments of the others in this chain. Actually, as Majone (1985, p. 62) notes,

> the situation is even more complex than this, for many policy studies in fields like energy, risk assessment, or education are "designed to influence congressional debates and to affect the climate of public opinion, not to guide decisions within individual corporations" (Manne, Richels, and Weyant, 1979, pp. 1–2). The effectiveness of such analyses can only be measured in terms of their impact on the ongoing policy debate: their success in clarifying issues, in introducing new concepts and viewpoints, even in modifying people's perceptions of the problem.

Craftsmanship factors are important to this uncertainty about results; a different conceptualization of the problem, different tools, a few different personal judgments made at crucial points in the work, can, separately or in combination, lead to differing conclusions.

Evaluation by results makes sense when the outcome can be clearly recognized as a success or failure, as in the blood-supply case cited above, or when the analysis product can be regarded as standard, as in the case of goods bought and sold in a market. What usually makes a difference in a market transaction is the nature and quality of the goods offered for sale, not by whom or how produced, except in the unusual circumstance when the buyer is willing to pay a premium for the reputation of the producer, perhaps because he has doubts about the quality of what is produced by others. For professional work, however, the reputation of the professional as a craftsman in his field is almost always important. In law and medicine, for instance, success in any particular case depends on many factors outside the control of the professional; the causal link between what the professional does and the outcome is complicated— and may be indirect and uncertain—so that the craft process and the decisions made in the course of this process are important. Thus, well developed professions pay considerable attention to the craft elements in their processes and develop canons of good practice to guide them; for example, physicians are expected to pursue the standards of good medical practice, and may even be vulnerable to suit for malpractice if these standards are not observed. Consequently, the criteria of evaluation in most professions depend importantly on the procedures used and the craft judgments of their relevance and adequacy.

Making a conclusion about the adequacy of a systems-analysis study based on the results of an implementation is complicated by the difficulty of verifying whether or not the decisions based on the analysis were correct; if the actual outcome is unfortunate, the failure may not lie in the analysis but in the failure of the client to take adequate decisions and actions, or in forces beyond the control of the client or the analyst that kept the implementation from going as planned. As Majone (1980a, p. 24) says,

> the conclusions of an analytic study can seldom be validated or refuted unambiguously. Hence, evaluation by results is either impossible or unfair (as when the quality of an analytic study is evaluated exclusively in terms of the actual success or failure of its conclusions and recommendations—too many factors outside the analyst's control determine the success or failure of a policy). Evaluation by process becomes unavoidable, and in this context the notion of craft and craft skills plays a crucial role.

As the evaluation of scientific work suffers from many of the same problems, we turn to it for guidance. Many early practitioners argued that operations research and systems analysis should be considered to be sciences, in spite of their concern with suggesting policy or action. The mistake, as Majone (1980a, p. 6) points out, consisted in the belief that the similarity was to be found in the outcome (at that time a "scientific explanation of the facts" and/or a "prediction of effects") rather than in the process—that is to say, in the activities required to carry out scientific research.

Since most systems analysts have come from backgrounds in the social, physical, or mathematical sciences, they borrow many methods from science, although their work is of necessity frequently less exact. But systems analysis work pursues goals beyond those usually motivating science; moreover, systems analysts aspire to prescribe or suggest courses of action, and sometimes even to urge their adoption, as well as to offer explanations and predictions. Nevertheless, owing to their intellectual heritage, systems analysts usually seek, insofar as possible, to use scientific methods in their work and to maintain scientific standards—and thus they often think of their work as scientific (see, for example, Helmer and Rescher, 1959).

Thus, establishing the correctness of the findings of a systems-analysis study may seem at least partially analogous to determining the truth of a scientific theory, although somewhat less likely to be successful. On the other hand, it is questionable that scientific knowledge can be proved to be correct in some abstract sense (see Section 12.2). Many scientists and philosophers of science agree with Toulmin (1974, p. 605) that

> if we wish to understand how actual sciences operate . . . we must abandon the assumption that the intellectual contents of natural sciences . . . have "logical" or "systematic" structures: we must instead consider how such sciences can succeed in fulfilling their actual explanatory missions, despite the fact that, at any chosen moment in time, their intellectual contents are marked by logical gaps, incoherences, and contradictions.

If scientific knowledge itself is always tentative and open to refutation, how then is the "correctness" of systems-analysis work to be established? Our most important clues come from what is done in the sciences: "If there is no demonstrative certainty for the conclusions of science, their 'truth' or, at any rate, their acceptability as scientific results, must be established by convention: through a consensus of experts in the field and the fulfillment of certain methodological and professional canons—the rules of the scientific game" (Majone, 1980b, p. 152). Thus, in science the existence of generally recognized rules of procedure, evaluation, and criticism is a necessary precondition for any reasonable claim to scientific status. In sum, good craftsmen know and follow these rules. In systems-analysis studies, as in science, the test of correctness is also adherence to the rules and procedures of the discipline.

The importance of relying on procedures to establish correctness is well recognized in some fields:

> In the law, in public administration and, to an increasing extent, also in business administration, decisions are accepted not because they can be shown to produce desired outcomes, but because of a generalized agreement on decision-making procedures. Reliance on detailed procedures, whether in environmental regulations, in the licensing of nuclear power plants, or in industrial quality control, greatly increases the costs of de-

cision-making, but it is also an unavoidable consequence of the cognitive
and social complexity of today's problems. [Majone, 1980b, p. 153.]

In both the fine and practical arts, the standards of quality are determined by craftsmanship. The influence of the master practitioner (helped
by patrons and critics) creates a standard of quality for lesser craftsmen.
These standards are hard to codify and are reached more by observation
and practice than in the lecture hall; nevertheless, they clearly serve to
guide workers—and especially beginners—to achieve work of high quality. Leading scientists provide similar guidance in craftsmanship for their
disciplines through professional societies, invited lectures, and publications, all of which establish exemplary standards for imitation. This handbook has a similar aim: to improve the craft skills of systems analysts,
and thus to improve the quality of systems analysis itself.

1.6. The Importance of Craftsmanship to the Acceptance of the Work

Systems analysis does not, and cannot, produce absolute proofs that its
results are in some sense correct; rather, it generates findings that are
more or less reasonable, depending on the complications of the situation
and the sophistication of the clients to whom they are presented. Thus,
it often happens that decisionmakers may not always be able to accept
the results without further clarifying arguments supporting their correctness. In some situations it would be necessary to supply additional persuasive arguments even if absolute proofs of validity could be provided,
for, as Majone (1980b, p. 160) remarks, "the psychological effect of purely
rational argument" may "not be strong enough to overcome the inertia
of long-established patterns of thinking—even after the need for change
has become clear." Furthermore, "the impact of rational arguments may
operate too slowly to bring about timely change."

Communication with the client is vitally important throughout a systems
analysis study. With other interested parties communication may be possible only in the final stages. However, "in our culture maximum effectiveness in communication is achieved neither by purely rational, nor by
purely persuasive means, but by a subtle blend of these two means of
redirecting attitudes" (Majone, 1980b, p. 160). Hence, elegance, visual
aids, humor, examples—anything that gets and holds the audience's attention and focuses it effectively on the subject—have a place. On the
other hand, none of these devices can substitute for a study that exhibits
a high form of systems-analysis craftsmanship; experience shows that this
important property shines through almost any form of presentation and
can usually be detected and judged even by persons with relatively little
experience in understanding analytic work. Good craftsmanship in de-

signing the presentation of the analyst's intentions, methods, and results will, of course, aid in effective communication; too, polish in the presentations may suggest craftsmanship in the work lying behind it.

It is sometimes argued that an analyst should be completely neutral with respect to the findings of his work, and that persuasion—or even presenting recommendations—is unethical. Experience does not support this view. Rather, experience tells us that a certain amount of persuasion is needed to get results understood and accepted for what they are. And most clients would regard an analyst who does not formulate the recommendations that his findings imply as shirking an essential form of assistance; many request recommendations explicitly. The analyst who has just completed an exhaustive study of a situation may have a far greater grasp of the portion of it included within the study's boundaries than others with responsibility for dealing with it. Majone (1980b, p. 160) argues "the question is not whether analysts should use persuasion in proposing new policy ideas, but what forms of persuasion may be used effectively and without violating basic principles of professional ethics."

The important distinction here is between the analyst acting as a persuader and the analyst acting as an advocate. The persuader acts to convince the client that, within the limits of what was considered in the study, the results are correct, relevant, and worth acting on; the advocate bases his urgings, not only on the results of his work, but also his own goals, values, and concerns (and possibly other matters not encompassed by the study). The two roles have areas of overlap, but the distinction between them is important, and the analyst should have it clearly in mind—and make abundantly clear to his client what the bases for his urgings are. As an analyst, he is entitled only to the role of persuader; in the role of advocate he is stepping out of that role and into a more general one.

As one of the most crucial steps in a systems analysis, its findings and their implications are presented to the client individual or group as well as, perhaps, other parties at interest. There is nothing intrinsically wrong or unethical for this process to select and emphasize, within the framework of the work, particular facts, goals, and logical arguments likely to appeal to these audiences. The desirability of doing so has long been recognized (for instance, by Kahn and Mann, 1956, p. 148):

> [M]ost analyses should (conceptually) be done in two stages: a first stage to find out what one wants to recommend, and a second stage that generates the kind of information that makes the recommendations convincing even to a hostile and disbelieving, but intelligent audience.

Thus, in presenting the findings and implications of a systems or policy analysis, the analyst selects the evidence likely to convince his audience; he does not present all of the available information (indeed, practical limitations preclude any such attempt). Evidence is not information in general; it is selected information introduced as part of an argument aimed at convincing the hearer that a factual proposition is either true or false. Selecting appropriate evidence is a craft skill. Introducing information at the wrong place in an argument, choosing inappropriate data, adopting a style of presentation inappropriate for the audience, all can destroy the effectiveness of evidence and undermine the persuasiveness of the study's findings.

There is a special difficulty in this regard about which systems analysts must exercise particular care. Mathematics is used throughout most systems studies and often acts to shield the quality of the evidence used as inputs from the outputs of its formulations. Thus, it is very important for the analyst to be sure that the quality of the input evidence is reflected appropriately in the interpretations placed on the outputs of mathematical and computer calculations.

In sum, there are many craft issues that must be dealt with properly in bridging the gap between a study's findings and their acceptance by the client.

1.7. Standards for Evaluating Systems Analysis Work

If, as we have argued, the systems analyst seldom is able to verify the correctness of an analysis by results, so that the quality of the work must be judged by looking at its process, what should the evaluator look for, and by what standards should he judge what he finds?

The first thing for the evaluator to look for is adequate documentation of everything that was done, together with the assumptions and arguments underlying the work. Not only is adequate documentation a hallmark of professionalism in systems analysis, it is also the absolutely necessary foundation for an adequate evaluation. To do his job, the evaluator must be able, not only to find out what was done, but also to discover why. Thus, he needs to answer such questions as these: What were the assumptions underlying the study as a whole, and what were all of the steps taken in it? What data were used and how they were processed? What factors were included in the investigation and what others that might have had a bearing were left out, and what were the arguments behind these choices? What models were used and how they were tested? What parameter values entered these models and what were the bases for their choice? And so on for scores of choices that may—or may not—have affected the results. Unless the documentation is thorough and complete, the ability of the evaluator to carry out his work will be seriously undermined, and he may be forced, *a fortiori*, to negative judgments.

Compared to other disciplines with well developed standards of quality for their work, systems analysis is relatively young; therefore, it is not surprising that widely accepted standards for controlling the quality of the work are still being evolved. Although the canons of scientific quality can be appealed to, the arguments and procedures in systems analysis can be judged by them only in part:

> [I]n fact, the nature of the testing process is more social than logical. This can be seen from the fact that the argument is never addressed to an abstract "universal" audience, as in the case of purely deductive proofs, but to a particular one (client, decision maker, special interest group, etc.) whose characteristics the analyst must keep constantly in mind if his argument is to carry conviction and affect the course of events. [Majone, 1980a, p. 20.]

Chapter 15 addresses the matter of criteria and standards of quality for systems-analysis work; it not only carries the argument further, it offers considerable detail about matters to be considered in controlling and evaluating quality in such work.

The important point to note here is that as these criteria and standards for evaluating the quality of systems-analysis work are evolved they will rest on an accepted concept of systems analysis (such as described in the *Overview* and developed further in this volume), standards of quality adapted to this concept, principles of choice and craftsmanship (such as discussed in this volume), and a body of work accepted as having high quality (such as the examples discussed in this handbook, especially its *Cases* volume).

1.8. Craft Issues, Procedural Choices, and Standards Discussed Herein

This volume is written for systems-analysis professionals and would-be professionals as well as others who are interested in important issues and choices that analysts face in the course of their work. It deals with craft issues and procedural choices for which professional experience has yielded useful advice enjoying a measure of acceptance based on practical use. This means, of course, that not all of the issues and choices an analyst may have to face in the course of his work are treated here. We have had, perforce, to restrict this book to material where there is at least some authority for its validity; where such authority does not exist, inclusion of an appropriate discussion awaits a future edition. The book does not, however, just look to the past; where it has seemed warranted, the contributors and editors have not hesitated to suggest prescriptions of what the subject's future should be.

After Part I, which consists of the introductory orientation of this chapter, Part II deals with selecting analysis approaches and procedures. Its first chapter discusses analytic strategies and their components: the need for an analytic strategy and some factors that should influence its choice,

problem formulation and techniques that can help with this process, analytic approaches available to analysts, the tactics that can be used in
analysis, and an overview of the tools that can be used. The other chapters
in Part II take up three somewhat unusual approaches that may be appropriate in some circumstances:

> Since a great deal of judgment is called on in any systems analysis study,
> it may be desirable to marshal such judgment systematically, espe-
> cially when experts outside the analysis team are called on; Chapter
> 3 takes up some ways that this can be done.

> Chapter 4 describes two-person operational gaming and how it can be
> used to contribute to systems analysis work.

> Chapter 5 discusses social experimentation in relation to systems anal-
> ysis: when to conduct such an experiment and when not to, how to
> manage a social experiment, and some practical advice for
> experimenters.

Thus, Part II deals with issues and choices facing the analysis team during
the research phase of its work, as listed at the left margin of Figure 1.1
in the Appendix (see p. 23).

Part III considers an important aspect of the research phase: setting
feasible and reasonable boundaries for the analysis so that it can be completed within the limitations of time and effort:

> Chapter 6 discusses generating and screening alternatives so as to have
> a reasonable number of attractive ones to deal with in detail in the
> analysis.

> Chapter 7 deals with the uncertainties surrounding the problems that
> systems analysis takes up and describes how they can be approached;
> it gives particular attention to political feasibility.

> Chapter 8 describes the uses and limitations of quantitative methods in
> systems analysis, while recognizing the many nonquantitative ele-
> ments that must enter any systems analysis inquiry.

Part IV offers discussions of three aspects of analysis that can pose
difficulties:

> Chapter 9 on forecasting and scenarios (see the right-hand side of Figure
> 1.1) deals with introducing a systematic view of the future into the
> analysis in ways that keep the work within feasible limits.

> Chapter 10 describes how cost considerations can be handled in systems
> analysis; to this end, it discusses the importance of cost, what it is,
> and the principles and methods of cost analysis.

> Program evaluation can help target candidates for further analysis by
> identifying programs that have not been working well; it can provide

baseline data on the performance of an ongoing program; and it can sometimes suggest ways to improve its performance. Chapter 11 takes up the objectives of program evaluation, discusses its various types, and describes actions that can be taken to improve its usefulness. The chapter also considers the conditions under which controlled experiments are likely to be appropriate for program evaluation.

Finally, Part V deals with professional issues underlying the entire systems analysis process.

Chapter 12 sets forth some important streams of thought bearing on the choice of an underlying conception for systems analysis: basic concepts of science useful for analysis, concepts underlying the applications of analysis, concepts of professionalism, and whether or not systems analysis can be considered mature.

Chapter 13 discusses issues relating to validation: verifying and validating models, validating analyses as a whole, and validation approaches and procedures.

Chapter 14 takes up matters closely related to making the analysis work effective, such as knowing the parties at interest in the problem situation and understanding the kinds of success that can be attributed to an analysis, so that these matters can be given appropriate consideration as the analysis team carries out its work and reports it.

Chapter 15 gathers together in reasonably systematic form the profession's scattered ideas about controlling the quality of systems analysis work as part of an ongoing research process, and being able to assess its quality at its completion as an input to the process of deciding what to do about its findings; thus, this chapter deals with the evaluation phase shown at the bottom and in the third box from the bottom of Figure 1.1. This chapter considers quality control in the descriptive aspect of systems analysis (including appraisal by input, output, and process) and evaluating the prescriptive and persuasive aspects of the work; it includes a number of checklists of matters to be considered in maintaining and assessing quality in systems analysis.

Since this volume of the handbook intends to deal primarily with technical issues and choices of main concern to analysts, it amplifies matters taken up very briefly—or not al all—in the *Overview*; on the other hand, matters adequately covered there are not taken up here. The earlier volume offers extended coverage of problem formulation in its Chapter 5 as well as discussions of values and criteria, objectives, and boundaries and constraints (see especially Chapter 6 there); Chapters 8 and 9 of the *Overview* take up matters relating to decision and implementation, and Chapter 10 discusses communicating results. Table 1.1 gives more detail on where matters outlined in Figure 1.1 are taken up in the *Overview*.

One of the important features of the *Overview* is a series of examples of systems analysis work that are described early and then used frequently throughout the text to illustrate important points (loc. cit., Section 1.2 and Chapter 3). These examples are also used from time to time in this volume, particularly in Chapters 2, 8, and 9 (Section 2.2 of this volume offers brief descriptions of these examples, and Section 9.6 describes the energy-study example in some more detail). Sections 6.3, 10.4, and 14.2 of this volume also contain extended descriptions of examples of systems analysis work. The reader may wish to review these examples as a point of departure for reading the rest of this book.

Appendix

Procedures and Craft Issues Discussed in the Overview

Because some readers of this volume may be unfamiliar with systems and policy analyses and how they are carried out, this appendix outlines the principal activities and stages in such work as presented in the *Overview* (see Figure 1.1). The discussion there is basically similar to what is found in other works on the same subject, although the activities may be grouped and labeled differently. Table 1.1 also lists the related craft issues and procedural choices discussed in the *Overview*; the plurality of references under each heading is an indicator of the interactive and iterative nature of systems analysis work.

The very brief description of systems analysis presented here—and its even more synoptic representation in Figure 1.1—is at best a sketchy first approximation to the full complexity of the systems analysis process. It portrays it as a linear one involving a set of interacting operations, whereas in actuality it almost always requires numerous iterations—and even re-starts—before the analysts can state conclusions in which they have a reasonable degree of confidence. Moreover, the process is full of potential pitfalls that take considerable craftsmanship to detect and avoid.

A systems-analysis study may begin with a problem, or more frequently, when someone, perhaps a policymaker, recognizes that a problem situation exists—an awareness that things are not as they should be—but without a clear idea of what is wrong or how it might be put right. The analysis starts by transforming the problem situation into a more clearly defined problem, a process known as problem setting or problem formulation. This process usually takes not only extensive communication between the analysts and those responsible for deciding what to do about the situation but also a great deal of disciplined effort by both parties; it

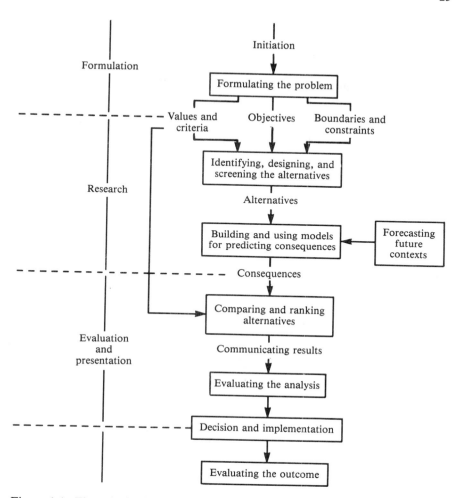

Figure 1.1. The principal activities in the systems-analysis process, as described and discussed in the *Overview* volume of this handbook.

also requires inquiry into, and agreement on, the goals, constraints, and limitations on what is to be investigated.

Next, the analysis identifies possible actions or alternatives that appear to offer some hope of improving the situation, collects data about them, and transforms these data into relevant information. At this time the analysts must forecast the environment or context in which each alternative is assumed to be implemented, and examine each alternative for political and economic feasibility and other desirable features. This preliminary analysis screens the inferior possibilities from the list of alternatives, leav-

Table 1.1. The Principal Craft Issues and Procedural Choices Discussed in the
    Overview Volume

Item	Where discussed in the *Overview*
Formulating the problem	Chapter 5 (pp. 151–170); see also pp. 127–132, 290–297
Values and criteria	Pages 119, 122–123, 130–131, 140–144, 220–239
Objectives	Chapter 6 (pp. 171–189); see also pp. 121–125, 128–129, 158–159, 223–229, 238
Boundaries and constraints	Chapter 6 (pp. 171–189); see also pp. 122–125, 131–132, 137, 159–161, 284
Identifying, designing, and screening the alternatives	Chapter 6 (pp. 171–189); see also pp. 119–126, 132–133, 222–223, 268, 298–299, 301–302
Forecasting future contexts	Pages 133–135, 137, 208–210
Building and using models for predicting consequences	Chapter 7 (pp. 191–218); see also pp. 18–19, 119–120, 136–140, 185, 219–220, 243, 299–301
Comparing and ranking alternatives	Pages 121–125, 140–145, 229–230
Communicating results	Chapter 10 (pp. 302–315); see also pp. 93–108, 122–127, 230–239
Evaluating the analysis	Pages 145, 316–317
Decision and implementation	Chapters 8 and 9 (pp. 219–280); see also pp. 75–79, 86–88, 306
Evaluating the program	Pages 17, 125, 145, 242, 277, 317

Note: For a guide to additional discussion of these items, see the subject index in the *Overview* volume.

ing the remainder to be analyzed further and in greater detail. Models of various sorts must be adapted or constructed to assist in this screening process as well as for turning the information about the remaining alternatives into evidence for comparing and ranking them with respect to cost, effectiveness, and other relevant measures of the consequences that would ensue from their implementation.

The analysis team then converts its results into findings bearing directly on the concerns of those responsible for deciding what to do about the problem situation and communicates these findings—and any recommendations that may arise from them—to all concerned.

As an important input to his consideration of the findings and recommendations of the analysis, the client may call for an evaluation of its quality (this process is indicated by the third box from the bottom of Figure 1.1).

If a decision is made to implement one of the alternatives dealt with in the analysis, the work may continue during the implementation process in order to keep the process properly focused and to deal with any new influences that may arise. A final step, not always carried out for a variety of reasons (some of which are suggested above), is to evaluate the program that results from the implementation effort.

Finally, it must be recognized that many systems-analysis studies do not carry through all of the steps outlined here, for a variety of reasons: the initial goal of the work may be merely to understand the problem situation more completely and not to propose changes, the underlying situation may change so much as to preclude an effective study of the sort being carried out, no prospective alternative may turn out to be feasible or acceptable, and so on.

References

Becker H. S. (1978). Arts and crafts. *American Journal of Sociology* 83(4), 865.

Cases. See Miser and Quade (to appear).

Chesler, L. G., and B. F. Goeller (1973). *The STAR Methodology for Short-Haul Transportation: Transportation System Impact Assessment*, R-1359-DOT. Santa Monica, California: The Rand Corporation.

Helmer, O. (1983). *Looking Forward: A Guide to Futures Research*. Beverly Hills, California: Sage.

————, and N. Rescher (1959). On the epistemology of the inexact sciences. *Management Science* 6, 25–52. Also reproduced in Helmer (1983), pp. 25–49.

Kahn, Herman, and I. Mann (1956). *Techniques of Systems Analysis*, RM-1829. Santa Monica, California: The Rand Corporation.

Majone, Giandomenico (1980a). *The Craft of Applied Systems Analysis*, WP-80-73. Laxenburg, Austria: International Institute for Applied Systems Analysis. This paper was first issued in 1974 as an internal document in the project at IIASA that was beginning work on this handbook; it was also published in 1984 in *Rethinking the Process of Operational Research and Systems Analysis* (R. Tomlinson and I. Kiss, eds.). Oxford, England: Pergamon, 1984, pp. 143–157.

———— (1980b). Policies as theories. *OMEGA* 8, 151–162.

———— (1985). Systems analysis: A genetic approach. In Miser and Quade (1985), pp. 33–66.

———— and E. S. Quade, eds. (1980). *Pitfalls of Analysis*. Chichester, England: Wiley.

Manne, A. S., R. G. Richels, and J. P. Weyant (1979). Energy policy modeling: A survey. *Operations Research* 27, 1–36.

McGrath, J. E. (1982). Introduction. In *Judgment Calls in Research* (J. E. McGrath, J. Martin, and Richard A. Kulka). Beverly Hills, California: Sage, pp. 13–16.

Miser, H. J., and E. S. Quade, eds. (1985). *Handbook of Systems Analysis: Overview of Uses, Procedures, Applications, and Practice*. New York: North-Holland. Cited in the text as the *Overview*.

————, and ————, eds. (to appear). *Handbook of Systems Analysis: Cases*. New York: North-Holland. Cited in the text as *Cases*.

Overview. See Miser and Quade (1985).

Ravetz, J. R. (1971). *Scientific Knowledge and its Social Problems*. Oxford, England: Oxford University Press.

Schön, D. A. (1983). *The Reflective Practitioner: How Professionals Think in Action*. New York: Basic Books.

Toulmin, S. (1974). The structure of scientific theories. In *The Structure of Scientific Theories* (F. Suppe, ed.). Urbana: University of Illinois Press.

PART II
SELECTING ANALYSIS APPROACHES AND PROCEDURES

Chapter 2
Analytic Strategies and Their Components

Hugh J. Miser and Edward S. Quade

2.1. Introduction

A systems or policy analysis begins when someone who feels a responsibility for matters related to a problem situation asks analysts for help. Their first activity is to explore the situation with a view to formulating a problem likely to be helpful and sufficiently well defined for systematic analysis. From this work the analysts gain an appreciation of the objectives that should be sought in investigating the problem, together with a concept of the constraints that must bound any potential solution—and at least preliminary views of some alternatives that should be considered, as well as the criteria that can be used to judge their value. Using their projection of the context in which a chosen alternative will become operational, and what they develop during the problem formulation, the analysts must evolve an appropriate strategy for guiding their work to findings useful to their client.

An analytic strategy is a plan for the intellectual activity that is the core of a systems or policy analysis study. It tells how the analysis team plans to structure and carry out the work. Its roots are in the formulation of the problem. Based on the objectives to be sought in its solution, the boundaries and constraints that must be observed, and the values and criteria that will help indicate the preferred outcome, this strategy describes how the analysts plan to explore and identify possible alternative solutions, to determine their consequences if they were to be implemented, and to help the decisionmakers involved select a preferred alternative and implement it.

Since the analytic strategy implies the analytic skills and support that will be needed, it is a necessary antecedent to developing a plan (sometimes called the analysis plan) that specifies the overt details of the work: the steps to be taken, the data and information to be gathered, the expected computer activities, the schedule, the desired products of the work, the resources needed, and so on.

In spite of the analytic strategy's place at the heart of any systems or policy analysis, detailed accounts of how such strategies were arrived at in individual cases are difficult to find in the literature, and there is seldom information on why a particular strategy was chosen, possibly because the original plan may so often have to be changed. Nevertheless, most successful analysts do have strategies and plans for their work. They frequently do not present them (for an exception, see Goeller et al., 1977, pp. 7–13), and when they do it is usually with no more than a flow diagram connecting models, not a listing of the concepts and procedures by which the analysis was carried out.

The purpose of this chapter is to present the various concepts and procedures that make up an analytic strategy, insofar as the existing literature and our experience allow us to do so. To this end, we have arbitrarily divided the various components of an analytic strategy into three parts: approaches, tactics, and tools. Few readers are likely to agree wholeheartedly with our choice of the category for each item, but it serves our purpose, which is primarily to organize the discussion.

The chapter begins with some example cases described in more detail in the *Overview*, with attention here to their analytic strategies. It then discusses the need for an analytic strategy and some factors that should influence its choice. After taking up problem formulation and mentioning techniques that can be particularly helpful in it, the chapter considers two broad approaches: optimizing and satisficing. Then it discusses a wide range of tactics that can be used to carry out various phases of the analysis. It closes with an overview of models, the essential tools of all analysis, paying special attention to a few types.

2.2. Some Example Analytic Strategies

The *Overview* volume's Chapter 3 describes a number of examples of good systems analysis work. Although none of these cases as originally presented (except for the estuary-flooding example) provided an overt discussion of the analytic strategy pursued, it is possible, from the descriptions and the material supporting them, to infer what the strategies were. The purpose of this section is to describe these strategies in terms of the procedures presented later in Sections 2.6 to 2.8 and indicate how they related to the work and its outcomes.

Improving Blood Availability and Utilization

Human blood has a unique medical value, but is perishable; in the United States it has a fixed legal lifetime during which it can be used for transfusion to a patient of the proper type, after which it has to be discarded. In the United States, blood is collected in units of one pint from volunteer donors at various collection sites and then shipped to a regional blood center, which acts as the distribution source for its region. After a series of typing and screening tests there, it is shipped to hospital blood banks, which store it so that it will be available to satisfy the random daily demands for transfusions to patients. In general, not all units demanded and assigned to a patient are used; therefore, a unit can be assigned several times during its lifetime before it is either transfused or outdated and discarded.

From the point of view of the regional blood center, efficient management of its perishable product is a difficult task, even if one neglects the uncertainty of having an adequate supply to meet all needs. In addition to blood's limited useful lifetime, there is the random nature of the demands and usages at the hospitals, the large variations in the sizes of the hospitals and their average demands, the relative frequencies of occurrence of the eight different blood groups, and the mixes of whole blood and blood fractions called for. Finally, there is the imperative of having blood available when and where needed (although additional deliveries can be made to meet shortages and elective surgery can be postponed, these acts incur additional costs to all concerned), whereas on the other hand there is the desire to operate efficiently and economically so as to make the available resources of blood and financial support stretch as far as possible.

There are two commonly used performance measures for a hospital blood bank that express the key management concerns: the shortage rate (that is, the proportion of days when supplementary unscheduled deliveries have to be made to satisfy the hospital's demand) and the outdate rate (that is, the proportion of the hospital's blood supply that is discarded owing to its becoming outdated). Suitable calculations convert these measures for individual hospitals to similar ones for a region.

In 1979 two systems analysts reported a study whose results had greatly improved the operations of the Greater New York Blood Program, which serves 262 hospitals in a region with a population of about 18 million. When the work began, the average number of deliveries per week to each hospital was 7.8 (most of which were unscheduled) and the average discard rate was 0.20; the hope was to find a way of bringing both of these indices down significantly, while maintaining or improving the standard of service, so that blood-use policies in the hospitals could be maintained.

Brodheim and Prastacos (1979), Prastacos and Brodheim (1980), and Prastacos (1984) give fuller discussions of the problem and how it was solved. In considering this problem situation, the analysis team noted that:

The procurement of the blood supplies was not considered part of the problem; rather, by reducing the discard rate, the goal was to make the existing supplies stretch as far as possible.

The hospital policies that produced the demands and usages were not considered to be part of the problem; rather, the intent was to continue to meet these requirements.

A delivery system existed and was expected to continue, although the way it was managed and operated could be changed.

The standard of blood availability at the hospitals had to be at least maintained—and improved if possible.

The goals, constraints, criteria, and measures of effectiveness were clear and acceptable to all of the interested parties.

The analytic strategy was to devise an alternative within these constraints that would be operationally acceptable to all of those concerned and that would reduce the average discard rate and reduce the number of daily deliveries needed, particularly those that were unscheduled. Thus, the strategy was not to try to discover the absolute optimum possibility, but to find a feasible and attractive one that would represent a substantial improvement.

Against this background, the tactical elements of this analytic strategy emerge fairly clearly: First, find a quantitative characterization for the hospital blood banks that combines the demand and usage patterns with the level of blood supply that the hospital blood bank managers are accustomed to accepting; second, devise revised supply alternatives that might be more economical of delivery effort and reduce the outdate rate; third, build a model that will allow the outcomes of these alternatives to be estimated; and fourth, convert the preferred alternative into a practical supply system. If this strategy is sufficiently successful to produce results adoptable by the regional-blood-center management, then a strategy of implementation will have to be devised and tested.

This approach might appear simple and straightforward to anyone not familiar with blood supply and distribution, but in actual fact the attendant complications not described here (see, for example, Prastacos, 1984) made previous research results infeasible of application. Thus, a fifth element of strategy was necessary if the approach was to be successful: Tactics of analysis would have to be followed that focused it on essential variables and rigidly excluded unessential complications in order to achieve a logical structure and models within it capable of practical use.

The availability rate for a blood bank is the fraction of days when the inventory of blood at the bank is sufficient to meet the demand. With a randomly fluctuating demand and a common desire to keep outdate rates as low as possible, hospital blood bank administrators had been accustomed to living with availability rates of 90 to 95 percent; when demand exceeded supply, the regional blood center would make supplementary unscheduled deliveries, or the hospital staff would adjust the schedulable portion of the demand. Similarly, the utilization rate for a blood bank is the fraction of the supply that is transfused.

These indices dictate another important goal for the regional blood center. In order to achieve fairness in its supply activities, its distribution policy should aim to equalize availability and utilization rates among the hospital blood banks it supplies so that the nonavailability and nonutilization risks are spread evenly among the hospitals, regardless of their size or location.

The analysis team was able to derive a fairly simple probability model relating the inventory level and mean daily demand to the availability rate for blood that was sufficiently close to the experience of blood banks to represent this relation in the later analysis, although in detail it was only an approximation to the experience of a given blood bank. Being able to represent all of the widely varying blood banks by this model was a significant tactical victory, as it enabled this model to represent the inputs to a model of the delivery system.

The next step was to achieve a model of the delivery system that included all of the essential relations but omitted enough complicating variables to enable it to be used practically to compute the outcomes of various supply policies and delivery procedures. This tactical goal was achieved by employing considerable skill in creating a finite-state Markov chain model, which the analysts used to explore a variety of possibilities in order to arrive at one with the desired properties.

Since one of the alternatives explored yielded significantly reduced weekly delivery and outdate rates while maintaining the previous standard of service to the hospital blood banks, the management of the regional blood center decided to introduce it, which led the analysis team to propose an implementation strategy that began in a modest exploratory way and expanded as favorable experience showed its worth. The tactics accompanying this implementation strategy emphasized the involvement and control of the existing managers, with only advisory services supplied by the analysts. To accompany this strategy the analysts developed and helped install a real-time information system to meet the needs of day-to-day operations as well as the periodic updating of basic parameters that the evolving health care system would demand.

The blood-supply approach evolved by the analysis team and adopted by the regional blood center, although it achieved very desirable results in practice, could not be claimed to be optimal in any global mathematical sense. Various practical considerations constrained the details of the supply policy and delivery procedures, and thus the alternatives considered could not be considered a complete set. Some computational explorations by the analysis team suggested, however, that the procedures adopted were yielding results very close to the best that could be expected if more complicated and comprehensive possibilities had been considered.

Improving Fire Protection

Fire protection is a basic and costly municipal service. In the face of significant increases in the demand for firefighting services, especially in the United States, city governments face important policy decisions on the sizes of firefighting forces and their distributions throughout the cities: how many fire companies to support, where to locate them, and how to dispatch them. A fire company is the basic firefighting unit; it consists of one or more motor vehicles, such as pumpers or ladder trucks, the associated equipment, and a complement of firefighters led by an officer. Such a unit usually has a firehouse, or fire station, of its own. A policy that leads to rapid response with appropriate firefighting resources can save lives and reduce property losses significantly as well as reduce threats to surroundings when a fire occurs.

In 1973 the Wilmington, Delaware, USA, fire department had eight firehouses, all but one antedating 1910. Since the city's growth and evolution suggested a fundamental reexamination of the firefighting deployment, a local project team was formed to carry it out, and systems analysts at The New York City–Rand Institute were asked to assist.

A desirable analytic strategy would have been to use descriptive models to estimate, for each deployment of firehouses and equipment that might be suggested, the cost to the city and the casualties and property damage that would result from fires of various types and intensities anywhere in the city. City officials could then decide, based on this information and taking into consideration other factors (such as the political consequences of closing a firehouse in a neighborhood), the deployment that they prefer. The strategy would thus be a satisficing one, for what would be sought was not a deployment that could be considered optimal in some sense but one good enough to be considered satisfactory and better than what had existed previously.

The desirable analytic strategy could not, however, be used. A fire department's primary objectives are to protect lives and safeguard property, measurable by the numbers of fatalities and injuries avoided and the value of property saved. Thus, the desirable research strategy demands that models be built relating deployment factors to these variables. This possibility, however, is denied by an unfortunate fact: there is as yet no reliable way to achieve these relations; for example, if the number of fire companies on duty were doubled, no one can say with a satisfactory degree of confidence what effects the change would have on the numbers of casualties and property losses. The directions of the effects may be predictable with some confidence for such a large change, but the amounts of the changes cannot be predicted with any reasonable accuracy based on what is known today. Moreover, for smaller—and thus more realistic—changes in deployment policy, it may be impossible to predict even the directions, let alone the sizes, of the changes in casualties and losses.

Because direct measurement of property saved and casualties avoided was not available, the analysis team had to settle for a less satisfactory strategy that was feasible—and acceptable to the sponsors of the work. (See the *Overview*, pp. 79–89, and Walker, Singleton, and Smith, 1975.)

This strategy began with three proxy measures, the first two of which are related to casualties and property damage (although the precise nature of the relation is not known): (1) travel time from the firehouse to individual fire locations, (2) average travel time in a region from its firehouse to possible fire locations in that region, and (3) fire company workload. It is possible to gather data related to these variables and to characterize them with approximate formulas. Then the analysis team used these data in two models:

A parametric allocation model determined fire-company allocations satisfying a wide range of objectives and permitted them to be evaluated in terms of the three proxy variables. This model contained a formula that specified the number of companies to be allocated to each region, given the total number of companies to be deployed in the city and a parameter reflecting the desired objective.

A firehouse site evaluation model was used to evaluate possible fire-company locations in terms of a variety of measures. This was not an optimization model, but rather a descriptive one that exhibited the properties of each location in terms of the chosen variables.

Thus, the strategy found its way past its fundamental handicap of not being able to use relations involving the ultimate variables of interest—casualties and property damage—by comparing historically acceptable values of related proxy variables with their values that would emerge from new deployments of fire stations.

Protecting an Estuary from Flooding

In 1953 a severe storm in the North Sea flooded much of the delta region of the Netherlands, killing several thousand people and causing tremendous damage. Determined not to allow this to happen again, the government started a program to increase protection from flooding by constructing a new system of dams and dikes. By the mid-1970s this system was complete except for the protection of the largest estuary, the Oosterschelde. Three engineering alternatives were under consideration: an impermeable dam to close off the estuary from the sea, a flow-through dam with gates that could be closed during a storm, and large new dikes around the estuary. To take no action would have been unacceptable to the Dutch people.

In 1975 the government agency responsible for water control and public works and The Rand Corporation of Santa Monica, California, began a joint systems-analysis project with a view to helping decide what should be done. Goeller et al. (1977) give a full description of the work.

Although security from flooding motivated the analysis, it was clear that the additional security had to be obtained without unacceptable damage to the environment. There were also seven other categories of concerns, with a large number and variety of impacts in each category: financial costs, ecology, fishing, shipping, recreation, national economy, and regional effects.

The problem situation was judged to involve too many varied political interests to obtain wide concurrence on an objective function, or any unitary criterion, such as that of benefit-cost analysis, that would be greatly influenced by the analysts' choices. Moreover, the constraints that would be acceptable could not be known until the final decision was made. The strategy, therefore, had to be a satisficing one, a search for one or more alternatives that would provide adequate protection, but in doing so would not cause unacceptable consequences in the other areas of concern.

Consequently, the analysis team took a satisficing approach. The primary objective of providing adequate protection from flooding was converted to a constraint. To do this, the team defined adequate as strong enough to protect the estuary from flooding in a storm so severe that its probability of occurrence was no more than once in 4000 years.

In addition to the task of designing the host of models required to estimate, for each of the alternatives, the many impacts in the categories of concern, the analysis team faced two other immediately discernible difficulties as it approached this problem:

Although three generic engineering possibilities had been specified by the client, there were a great many possible detailed variants that should be considered.

Each engineering possibility considered had dozens of consequences—called impacts by the analysts—that were of interest to various groups of citizens and officials.

These multiplicities yielded a centrally important problem for the analysis: How to reduce the number of cases to a feasible number for analysis, and to a number that could reasonably be considered by the many parties interested in the results.

The research strategy chosen by the analysis team had three parts: First, the number of engineering alternatives was severely reduced by screening, a tactic that identifies promising alternatives and rejects inferior ones on the basis of relatively simple analyses and criteria (for further discussion of screening, see Chapter 6). Second, owing to the large number and wide variety of decisionmakers and groups in the Netherlands who constituted parties concerned with the impacts produced by the way the problem was solved, the varied impacts were not reduced in number by some analytic device, but were retained in the analysis and its results. Third, the analysis team devoted a great deal of attention to devising a way to present the multiple results so as to make them clearly understood by the many parties at interest, who represented a wide variety of interests and backgrounds.

These choices of strategy dictated tactics that produced these results: The number of alternatives to be considered in detail was kept small. On the other hand, the number of impacts considered throughout was allowed to remain large; each was dealt with by the knowledge and models of the relevant scientific specialty, with the results presented in the units natural to the impact being considered. Finally, the collection of impacts for each alternative was set forth in a sufficiently standardized way by using scorecards (see Section 2.7) to make its relative importance among the other impacts readily visible. (For other examples of scorecard use, see Figures 10.4 and 10.8 and their associated discussions.)

This strategic approach depended on the parties at interest to produce mental syntheses of the results by relying on their own intuitive judgments about the relative importance of various factors. Each person viewing the results could thus choose an option that contained a preferred mix of consequences from his own point of view. This approach avoided the possibility that some analytic form of uniting impact measures would represent weighting judgments made by the analysts rather than by the parties at interest. Too, as the government worked toward a decision on what option to choose, the costs and benefits, as measured by the many impacts, could be kept continually before all concerned.

Providing Energy for the Future

In the early 1970s, during the so-called energy crisis, there was a rising concern about the world's ability to supply the energy (which then, and today, comes largely from fossil sources) that will be needed by its economies as they expand during the next century. An analysis team at the International Institute for Applied Systems Analysis in Laxenburg, Austria, undertook to explore the issues relating to this concern. For a brief description and discussion of this work, see Section 9.6; for fuller discussions, see Energy Systems Program Group (1981).

Since the team needed to look at least 50 years into the future, a major difficulty it faced was that there are so many possible—and even plausible—changes in consumption and growth patterns involved in looking so far ahead that it was clearly impossible to consider them all, or even a significant fraction of them, let alone develop models that would predict the outcomes with reasonable confidence. This difficulty was dealt with by restricting the scope of the study (an obvious and frequently used suboptimization tactic) by setting aside almost all political and social considerations. The aim of the study thus became to produce various perspectives on the technical possibilities of replacing the dependence on fossil fuels by energy from renewable sources, such as the sun. This approach was implemented by adopting a middle-of-the-road scenario containing some variants spanning likely possibilities for a surprise-free future, that is, one assuming no major catastrophes (such as a nuclear war) and no technological breakthroughs that cannot reasonably be anticipated today. (For a discussion of the scenario approach used, see Section 9.6.) The result was a framework with few enough variants to make the analysis feasible, while making it possible to relate the scenario view of the future to reasonable expectations.

On the other hand, the models used to represent various aspects of this future based on the scenario could not be claimed to be necessarily close to what the future would turn out to be; that is, they could not be claimed to be surrogates for this future. Rather, they represented enough possible properties of this future to offer useful perspectives on it. (For a further discussion contrasting models as surrogates with models as perspectives, see Sections 8.5 and 8.6.)

With the approach to this major strategic question settled, the analysis strategy was, first, to project worldwide energy demand over the 50 years of interest so as to bracket reasonable possibilities in the light of the future envisaged in the scenario, and second, to explore the various possible ways of meeting this demand within a reasonable envelope of constraints on natural resources, investment, manpower, risk, market substitutability, ecological effects, and so on.

Since the model of reality developed by the work was not a surrogate for a confidently known reality, but rather a perspective on a variety of possible realities, the analysts had to be careful to interpret the meanings of the findings. For example, the finding that the world would have enough natural resources to meet its energy needs for the next 50 years provided careful plans for their use are developed and implemented is considered fairly robust; that is, it applies to a wide spectrum of possible futures. On the other hand, the precise proportions of the demand that are likely to be met by various modes of energy generation can be expected to be quite sensitive to changing eventualities, whereas the conclusion that a variety of modes of generation will need to be used is believed to be robust.

Comparative Comments

These four examples exhibit a variety of analytic strategies. Although they cannot be claimed to show the full range of possibilities—they almost certainly do not—they nevertheless permit us to make some important points about analytic strategies for systems analysis, and the analysis tactics that they dictate.

Strategies differ. The four examples show wide differences. For example, although the strategy for the blood-supply case was to construct a good surrogate model of the reality represented by the chosen alternatives and to make it a close enough representation to be able to distinguish margins with confidence, the strategy for the energy-supply case was much more general: to enable valuable perspectives on future energy supply and demand to be displayed, even though the numerical values in them were quite uncertain. The flood-protection study strove to produce estimates of impacts that were as accurately realistic as possible, and used the best available knowledge to achieve this end; in contrast, the fire-protection work was forced to use proxy variables related to the true measures of effectiveness, but with these relations unknown beyond certain general impressions about trends. In spite of the wide variations in these four example strategies, each of the studies produced results useful to its parties at interest. The lesson of these examples is that there is no standard analytic strategy for systems analysis; rather, it exhibits a wide variety of approaches, each adapted to the problem situation being dealt with.

Strategies depend on a variety of factors. The choice of an analytic strategy emerges from the problem situation and the problem in it that is formulated for analysis. This choice must also consider the interests of those concerned, and how these interests can be helped, and it will be bounded by the available and obtainable knowledge, the possible forms and techniques of analysis, what can be done in the available time, and other factors. The reader may surmise—and probably correctly—how each of these factors influenced the choices of strategies in the examples. For example, the fire-protection analysis was limited to proxy variables by the absence of directly relevant knowledge of the relations between fire-protection efforts and casualties and economic losses. Since there were many significant groups of parties at interest in the solution to the flood-protection problem, and each group needed to know what impact a solution would have on its interests, the strategy included explicit consideration and display of results for each significant group. Since the energy work was aimed at a large class of policymakers with quite varied interests, it aimed at broadly applicable principles rather than highly focused and detailed recommendations, so that these principles could be applied in a variety of relevant circumstances. Since the blood-supply analysis was concerned with improving the margins of key variables in the blood-supply operations, its strategy aimed at modeling the situation closely enough to enable reasonably accurate estimates of these variables to be made for a variety of alternatives, thus enabling the analysis team to make solidly based recommendations for changing the pattern of operations.

Although the choice of an analytic strategy flows from all of these factors, plus perhaps others that we have not mentioned, it would be a mistake to represent that the analytic strategy is an easy and unambiguous choice after these factors have been marshaled. The appropriate strategy may sometimes be obvious at the beginning of the work, but it is far more likely to emerge later. Thus, the analysis team should hold its views of the possible analytic strategy to be tentative and subject to change and should keep it in this condition until there is a solid core of knowledge about the problem on which to base a final decision.

In particular, the analyst should be especially wary of choosing an analytic strategy with which he is comfortable because of past use, unless it clearly emerges as the most appropriate one to use.

It would be helpful if we had a taxonomy based on the diversity found in problems of systems and policy analysis that would influence the choice of an analytic strategy, but experience has not yet yielded such a guide. In fact, the analysis team may have to adopt an approach that experiments with various strategies until an appropriate one is identified.

The classic analytic strategy is to try to achieve a modeling approach that replicates the essential features of the problem situation as closely as possible. This heritage from operations research work served well in early systems analyses in which the structures were perceived as clear and the complexities and uncertainties modest. Later work, however, frequently dealt with matters made difficult to handle by the need to include socioeconomic factors, requiring the analysts to simplify the situation substantially in order to model it and to use descriptive, rather than optimizing, models. The success of the blood-supply work described at the beginning of this section was due in no small measure to the ability of the analysis team to cast aside complications studied by earlier blood-supply analysts and to concentrate on essential variables and effects in working toward an adequate and practical—but approximate—representation of the realities of the blood-supply operations. The other examples all employ the descriptive modeling approach that is typically used by systems and policy analysts.

Thus, another lesson of this discussion is that the analysis team should not drift into a strategy with a view to using a favorite tactical tool (such as linear programming, implying an optimizing strategy) unless this strategy is clearly appropriate to the situation being studied. In sum, the team should avoid the pitfall of allowing a method orientation to dictate the analytic strategy.

The analysis tactics flow from the chosen strategy. This statement may seem to be a truism, but experience shows that it can be ignored easily enough to make it worth emphasizing. Perhaps one of the best ways to view it is to see it as requiring that the analysis strategy and its tactics be in balance. For example, if, as in the blood-supply case, the strategy requires that the findings be able to give the client discriminating information about how to trim the margins of his operational parameters, then the analysis tactics will have to be chosen so that the actual operations are very closely replicated by the models adopted. On the other hand, if the strategy calls for the study to furnish a perspective on a highly uncertain and variable reality, then the analysis tactics need not try for so close a correspondence between model and reality; for example, in the energy study, the growth of the economies in the seven world regions considered in the work were represented quite successfully by a very simple model. Perhaps the worst pitfall is for the analytic tactics to represent great detail and precision in a situation where only a strategy calling for very uncertain and approximate results is feasible, owing to uncertainties not included in the analytic exercise. And we have already remarked that favorite—and therefore tempting—analytic tactics should not distort the consideration of an appropriate analytic strategy.

The nature of the desired results should affect strategy and tactics. Systems-analysis work has a three-fold character: descriptive, prescriptive, and persuasive (see Section 1.1). Although virtually all systems-analysis studies involve all three of these phases, the emphases may differ, and more importantly, the nature of the desired findings differ from case to case in these regards. For example, the blood-supply case sought results that would describe the operations accurately, use this description to discover and prescribe an improved system of supply, and provide a basis adequate to persuade the management to adopt the new system. On the other hand, the major emphasis in the energy study was descriptive; it portrayed the nature of the world's energy future under conservative assumptions with a view to suggesting broadly relevant findings that could act as a background for more detailed analyses of more local problems.

The fire-protection and flood-control examples both aimed to arrive at descriptive results that would enable decisionmakers to reach their own judgments and decisions; thus, in these studies the prescriptive and persuasive elements were more limited and were aimed largely at the understandings and beliefs of the decisionmakers. The differing emphases dictated by the desired results affected the choices of strategy and tactics for these studies.

The analytic strategies and tactics interact. Clearly an analytic strategy should prescribe analytic tactics that are feasible. For example, the strategy for the fire-protection example was constrained by the fact that tactics requiring explicit quantitative relations between fire-department operations and casualties and/or losses could not be used because the precise relations were unknown and could not be derived; thus, a strategy aimed at comparative results based on proxy variables was adopted, since it was feasible and offered useful guidance toward problem solution. Similar interactions can be traced in other examples.

2.3. The Need for an Analytic Strategy

There are a number of reasons why an analytic strategy is needed early in systems-analysis work, even though it may be revised as work proceeds.

It helps to organize the body of information, both quantitative and qualitative, encompassing the many factors related to the problem situation and the problem formulated from it.

It furnishes a logical structure for the intellectual work of solving the problem.

It suggests an appropriate balance among the various parts of the analysis and its follow-up activities.

Through synergy it suggests, at least crudely, forms of possible alternatives that may be useful to consider.

It guides the analytic tactics that will form the bridge between the problem and its solution and, in particular, the development of the models that will estimate the outcomes associated with the various alternatives.

It provides the initial basis for the analysis plan, which will include projections of the types of analytical skills needed, the manpower requirements, the support needed, and so on (see the *Overview*, pp. 294–297).

The analytic strategy, therefore, is an overview that constitutes a concept of how the systems analysis will bridge the gap between the problem formulation and what should be done about it.

2.4. Factors That Influence the Choice of an Analytic Strategy

The factors that can influence the choice of an analytic strategy are numerous and diverse. In addition to the nature of the problem and its context, there are the objectives, constraints, and criteria developed during the problem formulation. The analysis team must also consider the uses to which the findings of the analysis may be put, who may respond to these findings and what their concerns are, the generic nature of the possible alternatives and how they can be represented in the analysis, the length to which quantification can be carried without distortion, the techniques and tools that are available or that must be devised, the amount and quality of data and information that are available or must be collected, the time and resources allocated for conducting the analysis, the knowledge and skills of the possible analysis staff, the interest and cooperative posture of the sponsor, and so on.

The way an analyst perceives a problem is greatly affected by his past experiences with problems that seem to him to be similar, by his training, by the analytic tools he is accustomed to using, and by his values and preferences. Although all of these factors will bear on his responses to the problem as its formulation emerges, he should not, as far as he is able, allow his prejudices to interfere with the logical choice of an analytic strategy for the work. This is one reason that it is especially useful to have the analysis team as a whole participate in formulating the analytic strategy; since the members represent, or should represent, different professional backgrounds and experiences, they can act to avoid the adoption of a parochial course based on the experience of only one of their number.

In considering the organization for which the work will be done, Mood (1983, p. 2) suggests that it is useful to distinguish between task-oriented and behavior-oriented organizations:

> Organizations may be task-oriented or behavior-oriented; sometimes they may be both. Task-oriented organizations usually have some fairly narrowly defined specific purpose and often the members work full-time at accomplishing that purpose. Regularly employed persons are usually members of task-oriented organizations. . . . Members of behavior-oriented organizations are usually only occasionally concerned with the purpose, rules, and activities of the organization. . . . These organizations usually have multiple, often conflicting, goals. . . . [A] way of distinguishing between these two kinds of organizations is to note that a

task-oriented organization usually has a single primary goal, whereas a behavior-oriented organization usually has multiple goals.

In the blood-supply and fire-protection cases of Section 2.2, the organizations involved were task-oriented, whereas the Dutch government, for which the flood-protection analysis was done, was a behavior-oriented organization.

For the purposes of his discussion, Mood then distinguishes between systems analysis and policy analysis, the former done for task-oriented organizations, the latter for behavior-oriented organizations:

> The difference between the two kinds of analysis is partly a matter of the degree of complexity. Exploration of a possible change in a task-oriented organization usually comes down to assessing whether the proposed change will enable the organization to perform its task more efficiently, more rapidly, less expensively, less noisily, or whatever. That is, the organization has systems and procedures and facilities for carrying out its task, and the problem is confined to determining how the proposed change affects these instruments and thus affects the accomplishment of the task. A systems analysis is usually reasonably well circumscribed by the fact that the organization has a rather narrowly defined task.
>
> A policy analysis is usually not easy to circumscribe because a change in a political organization can have endless ramifications. In fact, any change is almost certain to confer benefits on some members and to harm others; sometimes benefits can be difficult to anticipate because ingenious members will invent a variety of devices and behaviors for turning the change to their advantage, or to the disadvantage of those with whom they are competing. Further, these benefits and disadvantages can be elusive, difficult to measure, and impossible to balance off in any sensible way in order to appraise the overall benefit or harm to the organization. There is no straightforward criterion such as efficient accomplishment of the task with which one can assess the various benefits and disadvantages; there is only some generality such as "welfare of the organization."
>
> Following from the above distinction is the fact that systems analysis can often be largely quantitative; that is, the analyst can frequently calculate numerical magnitudes for many of the benefits and disadvantages; he may even calculate an estimate of how much the efficiency of task performance will be changed by the proposed change. Policy analysis tends to be much less quantifiable, or if some aspects are quantifiable, there are nevertheless other important aspects that are not—thus requiring that the analysis rest at least in part on debatable value judgments. [Mood, 1983, p. 3.]

The blood-supply and fire-protection cases are examples of systems analyses in Mood's definition of the term, whereas the flood-protection case was a policy analysis in Mood's sense. The world energy demand and supply case is intermediate. Although Mood's distinctions are useful in considering the form of an analytic strategy, throughout this book we use the term systems analysis to refer to both systems analysis and policy analysis as he defines them.

Since the roots of systems analysis are in work that used quantitative models representing selected aspects of reality very closely, and many systems analysts have their basic scientific training in the quantitative sciences, there is often a strong preference for an analytic strategy including quantitative models as surrogates for the central parts of the problem (see Section 8.5). This preference is based in part on the advantages that such models offer, not the least of which are the many tools and techniques of manipulation and calculation that then become available. The difficulties arise when the preference leads the analysts to distort the problem unduly in order to make the use of quantitative models possible, or when unquantifiable aspects are simply ignored in order to make use of quantitative models. If these distortions are not recognized, the model may be falsely assumed to yield results that are reliable guides to the solution of the problem. The problem should dominate, rather than the mode of solution. Rather than simplifying the problem so far as to make it substantially unreal in order to use quantitative models, it is better to use the best model formulation that can be achieved for the true problem and then to obtain solutions from simpler nonquantitative or quasiquantitative methods that are adequate and feasible, but that may not show demonstrably optimal properties.

Time and resource constraints must be accepted as having important influences on what analytic strategy is chosen. The inexperienced observer may argue that any problem important enough to call for a major systems analysis effort to seek a solution deserves to have the decision on a solution postponed until the results of adequately supported work are complete. And in principle the observer is correct—but the real world seldom works ideally: many an analysis must be tailored to fit within time and resource constraints. Nor is this entirely bad; even an incomplete and somewhat hasty study may serve to illuminate some important relations, eliminate some seriously deficient options, and focus some light on the direction best taken.

Analytic strategies differ widely in the time and resources that they consume, and such differences must be compared with the prospective schedule and effort that is likely to be available. For example, operational gaming and computer simulation can easily be quite time-consuming and expensive, and can be used successfully only when time and resources make them feasible. Similarly, some problems are dominated by the data-collection and information-gathering efforts that are essential to them so that this phase of the work dominates resource and schedule concerns; if this work is essential, it must be accommodated, but any temptations to spend more time on this phase than really necessary should be sternly rejected. There have been other occasions on which the model building and analysis became so complex as to dominate; here again the strategy should seek a proper balance.

In general, considering what analytic strategy to adopt involves compromises between what it would be desirable to do in a world of infinite time and resources and what has to be done in the real world of serious limitations. Indeed, the limitations may be so severe as to force the analysts to undertake a "quick-and-dirty" analysis (see, for example, Patton and Sawicki, 1986), based on very simple or even purely mental models, or to resort to a strictly judgmental approach using structured committee discussions or the Delphi approach (see Chapter 3). With rare luck, an experienced analyst may be able to make a quick adaptation of an already developed model that was used elsewhere. In any case, the analysts must balance reality with what can be done:

> It is generally true that, as the model departs from "one-to-one correspondence" with the real world, the cost and time of analysis decreases. The skill in systems analysis is to depart without losing confidence in the model's ability to represent, through its solution, relationships which can aid the primary decision. In principle, models can always be solved, given enough money and time, but even then the solutions are not necessarily more useful than those of much simpler models. But, in some cases, it may be decided that it is not possible to get adequate solutions within the cost and time constraints. The analyst must then return to the problem formulation stage, and ask what sort of solutions can be obtained from simple models that relate to relevant, but less profound, aspects of the primary decision. [Bowen, 1978, p. 16.]

Thus, we see that formulating the analytic strategy must take account of the tactical implications of many prospective contributions to the work, ranging from the data and information to be collected through the forms of modeling to the kinds of findings that are appropriate to the problem and its context. Indeed, there may even be interactions between these tactical elements that must be taken into account. For example, econometric approaches and models that are based on statistical techniques, such as regression, require far more abundant and precise data than, say, system dynamics, a form of modeling that often accepts as adequate inputs that are quite approximate, perhaps even only of the right order of magnitude. Meadows and Robinson (1985, p. 81) offer this entertaining contrast between the attitudes of econometricians and system dynamicists with respect to data and information:

> System dynamicists recognize a spectrum of increasingly precise information, ranging from intuitions, hunches, and anecdotal observations at one end to controlled physical measurement at the other, with social statistics somewhere in the middle. They believe that this spectrum offers far more information than is currently used, and that the real need is not

for more data but for better use of the data already available. They point out that econometricians, by confining themselves to the narrow part of the spectrum consisting of social statistics, which contain no information about operating policies, goals, fears, or expectations in the system, are hopelessly restricted in learning about how social systems work.

These two approaches to the interpretation and use of various kinds of knowledge result in continuous, fruitless cross-paradigm discussions about the relative importance of structure versus parameters. Econometricians probably spend 5% of their time specifying model structure and 95% estimating parameters. System dynamicists reverse that emphasis. They find that their long-term feedback models are prone to wild excursions if even one small information link is left unclosed but are often maddeningly unresponsive to parameter changes. Having worked with such models, system dynamicists find it difficult to imagine why anyone would bother to estimate most coefficients precisely, especially when the coefficients are part of a model with what is to them an obviously open and linear structure.

The econometrician, on the other hand, may find that in his models a 6.5% growth rate produces a very different result from a 7.0% growth rate, and his client may care a great deal about that difference. To him the system dynamicists' cavalier attitude abut precise data seems both irresponsible and unsettling. Furthermore, since his paradigm provides no acceptable way of finding model parameters without statistical data, he cannot imagine how a system dynamics model becomes quantified. Since the numbers are not obtained by legitimate statistical methods, they must be illegitimate, made-up, suspect.

Although this contrast emphasizes the differences in the inputs required for the two kinds of models, a similar discussion could be written about the characters of the outputs, and what analysts could claim for them.

In considering an analytic strategy for solving a problem, a systems analyst will have to consider the models that will be part of his tactics, giving special attention to the inputs that they will need and the outputs that they can yield. Although a few overview discussions exist that will help get him started (see, for example, Greenberger, Crenson, and Crissey, 1976, pp. 85–127, and Meadows and Robinson, 1985, pp. 19–90), the analyst must depend on the acquaintance that he has with the models and the examples of their use that the literature contains.

It would be desirable to include here a conscientious catalogue of all models in common use in systems analysis, together with their key properties, for use in considering various strategies that might use them. Two important considerations, however, prevent this course: the limitations of space and the scattered literature, and the fact that each use presents a new demand that may render recognized properties only partially relevant. Thus, we must be content in this discussion to state the importance of these considerations.

Similarly, one might characterize the strategies used in the past in systems-analysis studies, but this would again be of only limited value, as each new problem brings new properties that the analytic strategy should respond to.

What this handbook does—and what seems preferable in view of the present state of the art—is to present a number of suggestive examples in all three of its volumes, the volume subtitled *Cases* being devoted entirely to this goal.

In any case, this section cannot stress too strongly the need for the analysis team to think carefully and in some detail about its analytic strategy as its work proceeds, beginning this consideration early in the problem formulation stage, and keeping the subject open for new influence and revision until the analysis is so far along that a full commitment to a course of action is demanded by the need to bring the work to a timely close. This chapter aims to help the analysis team with its view of key issues and choices to be considered, but each problem must be approached on its own merits from the analysis team's own unique vantage point.

2.5. Tactics for Formulating the Problem

The idea for a systems analysis often originates when someone, an administrator, a public official, a concerned citizen, or even an analyst, has the feeling that enough is wrong with a situation, an organization, or a program to warrant an effort to improve matters. He may or may not have an idea of where the difficulties lie, or how to formulate a problem to represent them, but he may realize a need for—and the advantages of—analyzing the problem situation with some penetrating care. Occasionally he may be able to do this himself, or he may seek analytic help from a systems analysis team.

The goal of the analysis may be limited to understanding the problem situation and where the problems in it lie or it may be more ambitious—but any more extensive analysis must start with this preliminary exploration.

Although setting the problem can sometimes be done unsystematically, the process almost always can benefit from using a penetrating and systematic approach, such as preparing an issue paper or a workbook. Even if the problem seems to be crystal clear, the real world is often more complex than it appears to be, whereas human perceptions are limited, so that it is well to verify very carefully that the original perception is correct—and in order to be able to solve the problem it may be necessary to simplify it to some extent.

Since Checkland covers the general approach to problem formulation well in the *Overview* (Chapter 5), the purpose of this section is to emphasize the importance of this step and the need to devote serious effort to it. It is also worth mentioning three slightly more formal tools that may help with this process in some cases: judgmental approaches, operational gaming, and system dynamics. These tools, however, are discussed later with other modeling devices, for, although they are particularly useful for problem formulation, they can also be used for other purposes and at other stages in the analysis.

Analysts in the United Kingdom are using analytic methods directed specifically toward problem formulation. (We are indebted to K. C. Bowen for the brief overview of these methods in the following three paragraphs.)

Within their overall processes, Eden, Jones, and Sims (1983) and Friend and Hickling (1987) approach problem definition or formulation through dialogues with the problem owners. The former team uses cognitive mapping, a procedure now well practiced: the problem situation develops in terms of the clients' verbal constructs, stored and easily retrievable in directed-graph form through associated software (COPE). The latter analysts take the problem-shaping process as an equal factor in an adaptive inquiry that, elsewhere, Hickling (1982) describes as *whirling*. This term aptly describes the movement across the phases shaping, describing, comparing, and choosing as understanding develops.

Bowen (1981, 1983a, 1983b) has developed a methodology based on his studies of conflict, a factor inherent in most problems, even though much of it is benevolent and, ultimately, constructive. He uses a simple diagrammatic notation; the process depends on the structural ability of the analyst to use it in interpreting the thinking of his clients. It is deliberately less detailed than cognitive mapping, although such mapping could be used as an extension.

All three of these methods are potentially compatible, although each tends to highlight different features of the problem situation. They could also, in theory, be adopted for the early stages of other methodologies, such as the Checkland (1981) systems practice, or soft systems methodology.

The central portion of a systems analysis study—and the analytic strategies, tactics, and models that we discuss below—come into play after the problem has been formulated (although there are occasional exceptions).

The problem that has been formulated, or isolated, from the problem situation is a model of an appropriate portion of that situation. In most cases, the formulated problem is expressed to a large extent in verbal terms, and it is the problem to which the chosen analytic strategy and its means of implementation will be applied.

In general, the solution derived from the formulated problem does not provide a complete response to the problem situation. This may be because the original problem situation was too complex to be captured in the single formulated problem, because the feasible solution is a suboptimization that does not extend to the larger problem situation, or because other difficulties exist (such as being unable to develop an adequate surrogate model). A suboptimization may have been occasioned by limited capabilities for analysis or by limitations on the sponsor's authority or interest. Nevertheless, the solution to the formulated problem usually offers at least a perspective on the original larger problem situation that will help further inquiry into the possibilities for ameliorating it.

Further, the problem-formulation process, which culminates with the formulated problem, enlightens the more general situation from which it was derived by (1) helping to organize the body of information related to the problem situation, (2) providing knowledge of some of the key mechanisms in that situation, (3) generating some understanding of how the varying inputs produce their varying outputs, and (4) giving some appreciation of their tolerable limits. Later, after gaining an understanding of how the various options and their combinations behave, the analysis team usually gets ideas that they can combine with their experience to yield expectations about how the problem situation behaves and even some clues about how each option might be brought to reality if it is chosen for implementation.

2.6. Possible Analytic Approaches

During problem formulation the analysis team reaches agreement with the sponsor about the problem and the objectives of the analysis, and both obtain some idea of the nature of the alternatives to be considered. These things, combined with the general knowledge of the context, enable the analysis team to choose, at least tentatively, one of two possibilities for the analytic approach: optimizing or satisficing.

Although we have not previously pointed it out, the typical framework used in systems and policy analysis (outlined in the appendix to Chapter 1 and described in more detail in Chapter 4 of the *Overview*) almost always leads to a satisficing approach. This state of affairs occurs because the list of alternatives—as generated through the ingenuity of the analysis team and the client—cannot be known *a priori* to contain an optimal one. When an alternative derived by this process is selected by the decision-makers for implementation, it is chosen if—and only if—it is judged to be "good enough" to satisfy their objectives and constraints; in their opinion, it is the "best" choice from among the set of available—and suitably analyzed—alternatives, but it is not necessarily the optimum among all conceivable possibilities. The exceptions to this general statement occur when the simplicity of the context and the similarity of the alternatives permit the analyst to use a mathematical model to investigate the *complete set* of possible alternatives, a circumstance relatively rare among systems-analysis problems and rarer still among policy-analysis problems.

With respect to public policy problems, Eilon (1972, p. 15) has remarked:

> My contention is that, as far as the application of O. R. is concerned, the major difference does not lie between the public sector and the private sector, but between the optimizing and satisficing approaches to decision-making. True enough, the optimizing philosophy is the one that prevails in the literature, but experience and observation suggest that satisficing is the approach that prevails in practice. There is far more to be gained from scrutinizing and ranking constraints than in constructing a super utility function to delight the heart of the optimizer. And where more so than in problems belonging to the public sector, where difficulties in defining quantitative measures and criteria are perhaps more readily understood and accepted than they are in the private sector, which has been subjected far too long to the single-minded and over-simplified notion of the profit motive.

Optimizing

In order for an optimization analysis to be feasible, there must exist a mathematical model—often called a prescriptive model—that will permit an optimal alternative to be derived, either by a mathematical algorithm or by a controlled trial-and-error procedure. Since any model is an approximation to reality, the optimum generated from it is also an approximation to reality: the better the model's approximation, the more likely the optimum generated from the model approximates the real optimum.

There is, however, a major difficulty with optimization: the need to select a single criterion or objective function. Some problem situations admit improvement on the basis of such a single criterion, but most involve multiple—and often conflicting—criteria. For example, in any attempt to improve a problem situation involving people, there is seldom a full agreement on the objectives or on ways of measuring their attainment. Even a single individual may have more than one objective and be uncertain as to how to measure or rank them. Operations-research workers have devoted considerable effort to finding ways to handle multiple preferences and utilities in order to reach sufficient agreement to allow analysis to be carried out, and this work has generated an extensive literature. We know, however, that without dictatorial powers there is no sure way in general for a group of individuals to reach the required agreement (Section 6.3 and Arrow, 1951). Hence, even if we could achieve true optimality with a single decisionmaker, there would be no way to do so with multiple decisionmakers, for any such concept depends on the individual decisionmakers' values, purposes, preferences, and needs. For a discussion of the tactics available when agreement on a single goal cannot be reached and the optimizer is faced with multiple goals, see the discussion of objectives in Section 2.7.

To facilitate discussion, we divide the optimizing approaches into three types: routine operations-research analysis, common optimization or programming analysis, and general optimization analysis.

Routine operations-research analysis. Very occasionally a systems-analysis problem may suggest the method of solution as soon as the problem is formulated, say, by means of a classical operations-research model, such as those for inventory control, scheduling, production planning, or queuing. Problems of this type are central to the operations-research literature, and clear approaches to their solution can be found in textbooks such as Wagner (1975) or a handbook such as Moder and Elmaghraby (1978). Such problems do not involve social, political, or other human behavior that cannot be handled by statistical means. Since they are well treated elsewhere—and are not central to the concerns of systems or policy analysis—they will not be considered further here.

Common optimization or programming analysis. If the problem is well structured, with both a clear objective function and precisely stated constraints, and with the relations among the variables well understood and involving mainly physical entities, an optimization strategy using a quantitative analytic model may become possible. Linear programming is the first choice if its structure is apposite, owing to the large body of theory and well-developed computational procedures. If linear programming is not appropriate, integer, quadratic, and geometric programming can be explored, as well as other procedures.

Optimization, particularly programming of one form or another, is an analytic strategy that is often used to aid decisionmaking in industry. It requires a clearly stated objective function, a knowledge of the control variables available to the decisionmaker, and a precise definition of the constraints. Because of the nature of the problems in this field, good surrogate models can usually be built. Meadows and Robinson (1985, pp. 70–73) discuss its typical uses, problems, and limitations.

General optimization analysis. Optimization models are attractive because, if they can be solved analytically, they identify the optimal alternative and its parameters, thus eliminating any additional problems of choice. If there are no optimization algorithms such as exist for linear programming, or some other analytic process for locating the optimum, the optimization process can sometimes be carried out numerically, using a purely computational approach. The result of such a process is, *a for-*

tiori, an approximation near the true optimum, but not necessarily exactly characteristic of the optimizing alternative.

The limitations of optimization in systems-analysis problems. As stated earlier, the sorts of problems for which optimization procedures of the types discussed so far are possible are relatively rare in systems analysis, even rarer in policy analysis, and are not characteristic of these fields. Here, when optimization procedures are used, they are more likely to be used in components of the problem, rather than as the overall approach.

Moreover, for the complex problems that are typical of systems and policy analysis, true optimization is a myth. If there were no other reason, the gaps that are always present between the real-world problem situation and the formulated problem and the models chosen to represent it preclude any true optimization of the real-world situation. An early leader in the development of systems analysis made these observations, which remain almost as true today as when he made them:

> There has been altogether too much obsession with optimizing . . . , and I include both grand optimizing and suboptimizing. Most of our relations are so unpredictable that we do well to get the right sign and order of magnitude of first differentials. In most of our attempted optimizations we are kidding our customers or ourselves or both. If we can show our customer how to make a better decision than he would otherwise have made, we are doing well, and all that can reasonably be expected of us. [Hitch, 1960, p. 444.]

In large-scale operating and policy problems there are also other factors to deny the analyst a true optimization in reality, of which two may be mentioned: there are always hidden constraints, often unknown until implementation is well started, but nonetheless important enough to bound what is possible; and the constraining rules that assure private stockholders and the public that administrators are fiscally responsible prevent analysts from trading off factors at the margin in a way that would assure a true optimum.

In sum, if the formulated problem has a surrogate mathematical model, results from this model are, with little or no modification, assumed to be results for the problem itself. The true situation is, however, that the solution to the model is an exact solution to an approximation to the problem, rather than an approximate solution to the real (exact) problem.

Against this background, the experienced and prudent systems analyst seeks from his model solutions that are on a preferred plateau in order to be reasonably sure that, when the perturbations of reality combine with the approximation of the model, they will not drive the solutions off this plateau.

Satisficing

After explaining what satisficing is, we shall discuss it as a deliberate strategic approach and then describe its relation to the typical systems-analysis approach based on descriptive models.

Since true optimization is impossible, and even a good approximation to an optimal solution may be beyond our capabilities (because, say, the computation is too difficult, the model does not correspond closely enough to the situation being represented, or we cannot select the best alternative because we have not been fortunate enough to have generated it), we must usually compromise and search for a satisfactory, rather than an optimal, alternative or policy. This process, which Simon (1961) calls *satisficing*, offers a practical way to provide an acceptable course of action in a world in which optimality is a myth and finding even a feasible alternative may signify success. In a political environment, it is often easier to obtain agreement on a proposed action than it is to get agreement on its purposes and on the meaning of effectiveness in achieving the purposes (Quade, 1982, p. 88). Fortunately, managers and politicians recognize the need for compromise, and decisionmakers are usually willing to accept a change that makes an obvious improvement.

To satisfice, goals are replaced by constraints. If disagreements on goals cannot be reconciled, the analyst tries to set a lower bound for each goal so that alternatives that attain this bound are good enough to be acceptable to those who regard the goal as important. After all the goals have been turned into constraints in ways satisfactory to their advocates, the analyst seeks an alternative bounded by the constraints, and, if one is found, it may be regarded as an acceptable solution. If no feasible solution can be found, it is necessary to relax some of the constraints until one becomes possible. As Simon (1964) has pointed out, the distinction between goals and constraints is a narrow one; they may to a large extent be viewed as interchangeable. In systems analysis, the goal is used to suggest the alternatives to investigate and the constraints to test them; if the alternatives fail these tests, the constraints suggest modifications to bring them into the area of feasibility. For success, both goals and constraints must be met. In fact, however, according to Simon (1964, p. 20):

> It is doubtful whether decisions are generally directed toward a goal. It is easier and clearer to view decisions as being concerned with discovering courses of action that satisfy a whole set of constraints. It is this set, and not any one of its members, that is most accurately viewed as the goal of the action.

Even for a decisionmaker who has a clear idea of his goals and priorities, a satisficing approach can have many advantages. The satisficer holds that, in most complex real-world situations, there are far too many uncertainties and conflicts in values for there to be any hope of working out anything approaching a true optimization, persuading those affected that it is desirable, and implementing the result. The systems analyst normally works under tight time restrictions, so that he may have barely enough time to design and test three or four promising alternatives. Hence, it is often more sensible—and possibly just as pleasing to his constituency—for the decisionmaker to set out with his analysis team merely to find a course of action that is demonstrably better than what has been done previously—and to have a reasonably good chance of achieving this goal.

Systems analysis uses a satisficing approach much more commonly than its literature would indicate, and so does policy analysis. For example, the study to protect the Oosterschelde estuary from flooding outlined in Section 2.2 used a satisficing approach, although only the major goal of flood protection was converted to a constraint early in the work. This goal was first used to generate three alternatives. Then it was converted to a constraint that these alternatives had to satisfy, namely, to protect against a storm so severe that its probability of occurrence was one time in 4000 years. Other goals for these alternatives—to limit damage to the ecology, to fishing, to water transportation and shipping, to recreation, and so on—were also treated as constraints, with the Netherlands government making the decision, after the impacts had been estimated, that they were acceptable.

Since it is usually impossible—or too difficult—to develop an analytic strategy based on prescriptive models that allow for easy optimization, the systems analyst typically turns to a strategy based on descriptive models. Indeed, this nonprescriptive approach is the one almost always used in systems and policy analysis—and it is often used in operations research. It is much more flexible than the prescriptive approach, whose models require a clear and unambiguous quantitative objective and a single measure of the performance characteristics of the alternatives. Thus, if a best alternative is to be determined, it must be among those tested in the calculations. On the other hand, descriptive models do not indicate whether an alternative A, which leads to an associated set of outcomes, is better than alternative B, which leads to a different set of outcomes; the models merely tell how A and B differ with respect to the multiple consequences they generate.

The fire-protection, flood-protection, and energy-supply examples discussed in Section 2.2 all used models in this descriptive manner. The flood-protection case's use of dozens of models from many sciences to predict, for each of the three principal alternatives, a large number of outcomes is a particularly instructive example of how models can exhibit comparisons of multiple consequences. Although the blood-supply example embodied surrogate models, and they were used to generate preferred courses of action, they were not used to calculate precise optima;

rather, they were used to show how a small number of feasible and attractive courses of action would reduce the discard rate from 20 percent to 4 percent and the average number of weekly deliveries per hospital from 7.8 to 4.2, while holding the availability rates for blood at the individual hospitals at the customarily acceptable high levels. The models were also used to show that the desirable courses of action that produced these improvements would reach levels of performance close to the possible absolute optima that could be achieved (provided, of course, that the models were very good surrogates for the real situation, and that the alternatives examined were close to the best conceivable ones).

In the descriptive approach the analyst uses the information acquired during the problem-formulation process about objectives and possible ways of achieving them to develop further alternatives, and possibly more refined objectives. Descriptive—or sometimes prescriptive—models are then used to supply information, both quantitative and qualitative, about what would happen if the alternatives were implemented. This information, refined in various ways and viewed through the lenses of the objectives, can then be used to help reach decisions about what to do. No overall prescriptive model that optimizes from among the full set of alternatives is used, nor is an attempt to find one made.

One important reason why prescriptive modeling may seldom be applicable to systems-analysis problems is that the alternatives one would like to compare may have radically different characteristics, and thus may require quite different modeling processes to estimate their effects. If all these processes are quantifiable, it may be possible to incorporate all of them into a single large model of some sort, but, as many analysts have learned, problems of cost, verification, and even feasibility may make contemplating such a course inadvisable. For example, if the objective is to reduce the losses of life and property from structural fires in an urban area, one alternative is to increase the number of fire stations in the area; another is to increase their efficiency through redeployment, training, and new equipment; still another is to make the fire-protection aspects of the building code more stringent by requiring such measures as more sprinklers, heavier electric wiring, or more use of fireproof building materials. Other alternatives that might be even more immediately effective would be for the municipality to furnish and maintain smoke alarms or to conduct intensive educational campaigns relating to smoking, especially in bed. In such a case, the variety of processes—and therefore the models of them—required to estimate the effectivenesses of the various measures rules out the use of a prescriptive model.

Another problem with using a prescriptive approach for this fire problem is that, as we saw in Section 2.2, there is no satisfactory way to measure the effectiveness of, say, a sum of money spent on improved fire response as opposed to the same sum spent on enforcing a stricter building code. It seems clear from some evidence (European cities, where building codes are much stricter than in the United States, have something like a fourth fewer structural fires than those of comparable population in the United States) that, as far as outlay by the city is concerned, the stricter building code would bring the greater relative payoff over the long term, but the additional costs of this alternative would largely be imposed on builders and property owners, which could make it politically infeasible. Prescriptive modeling is of no help in such a case.

A descriptive approach may use prescriptive modeling for some aspect of the work. For instance, after a type of alternative has been selected, a prescriptive model may be used to fine-tune its properties to get a better set for implementation. Any of the tactics or models discussed in the remainder of this chapter can form part of a strategy employing the descriptive approach.

2.7. Tactics

Having formulated the problem and arrived at an analytic strategy for tackling it, the analysis team must consider the tactics and tools that are needed to make up the complete strategy. The purpose of this section is to discuss some of these tactics; the next section considers tools. The various tactics are discussed in association with the main purposes for which they are used.

Not all of the tactics discussed under the various headings in this section are used in every analysis, although some of them are, nor are they equally useful with each approach. For instance, screening may not be needed with optimization, for the model itself selects the optimal alternative, but ways to handle the objectives and identify the constraints are essential in such a case.

Tactics for Selecting the Objectives

Selecting suitable objectives is crucial in systems analysis. Approaching a problem without the right objectives is equivalent to tackling the wrong problem, for the objectives determine the alternatives that emerge as preferred.

Almost every work on systems or policy analysis offers advice on the treatment of the problems of selecting and clarifying objectives (see, for instance, Quade, 1982, Hovey, 1968, and the *Overview*, Chapter 6) that need not be repeated here. Such works and Section 8.4 also discuss an associated difficulty: that of finding ways to measure the attainment of the objectives after they have been chosen.

For an optimization to be carried out there must be a single objective function. Unfortunately, in many problem situations—and possibly in most—there are multiple objectives that compete for resources and that may, indeed, conflict. Reconciling them, that is, replacing the set with a single measurable objective that represents the entire set satisfactorily, can present a serious problem. Although much attention has been paid to its solution, we will not discuss it further; the interested analyst can see Section 8.4 and numerous other sources (such as, for example, Raiffa, 1968, Eilon, 1972, Keeney and Raiffa, 1976, and Keeney, 1980). When agreement on a single objective cannot be reached and the optimizer is faced with multiple goals, he can sometimes resort to such methods as optimization in tandem, converting some goals into constraints, or target setting (Eilon, 1972).

In satisficing, the problem of multiple goals is less severe, for, in addition to the methods used in optimization, the analysts may deliberately turn the troublesome goals into constraints.

Tactics for Generating and Screening Alternatives

Generating a wide range and variety of alternatives is fundamentally important in systems and policy analysis. The "best" alternative—or even a very good one—cannot be determined unless it is among the ones investigated, and it may not be there unless significant effort has been made to search out the full spectrum of possibilities. But this process may produce so many candidate alternatives that it is too costly in time and resources to investigate all of them thoroughly. Consequently, it is desirable to have a logical process for screening out the inferior possibilities, thus reducing the remainder to a manageable number. Chapter 6 discusses generating alternatives and how to screen them.

Tactics for Ensuring a Feasible Solution

Not all alternatives that preliminary calculations show to be capable of achieving the objectives can, in practice, do so; political, cultural, legal, or other reasons—or even social prejudice—may interfere with implementation, preventing it entirely, or altering it sufficiently to vitiate the intended benefits. Thus, the analysis team must always consider whether an alternative, no matter how attractive it may look, is actually feasible. It makes little sense to designate a course of action or a policy as best, or preferred, if it is impossible to bring it to reality—unless, of course, there is some practical way to overcome or remove the obstacles that stand in its way. Thus, analysts should take particular care to avoid devoting energy to problems, particularly those involving political or societal factors, that have no implementable alternative. Similarly, analysis effort should not be wasted on infeasible alternatives. For many problems, the nature of public decisionmaking is such that, if any solution exists at all, it is likely to be suboptimal; when this is the case, systems analysis may help improve matters, even though a universally preferred solution may remain impossible.

Feasibility must be examined not only to determine what it is possible to do at a given time, but also to determine whether the obstacles that exist can be overcome and at what cost. In a review of feasibility, the analyst should even give some thought to whether it will be feasible to convince the client to adopt his findings and the solution that they will imply. Beyond this, the wider political feasibility poses some especially important problems for the analyst; Part 2 of Chapter 7 deals with them in some detail. Indeed, all issues that may affect potential implementation should be considered as early as possible. Some will come up during problem formulation. Others may arise later, and may well force the problem situation to be reconsidered; thus, the problem tackled may change.

Constraints can be both a help and a hindrance to analysis. They help by reducing the number and scope of alternatives but hinder by making it necessary to be sure that they are satisfied. The most troublesome constraints are probably the so-called political constraints, those that are generated by the people who are, or will be, affected by the course of action that may emerge from the analysis. Often these constraints are not discovered—and in some cases they cannot be discovered—until the process of implementation is far enough along for opposition to what is being done to develop.

The unwise analyst, overestimating the power of the decisionmaker, does not look for or try to anticipate such constraints. The wise analyst, however, avoids this pitfall; he knows that, although many decisionmakers appear powerful, their decisions, despite being announced at a high level, are carried out by the rank and file of their organization, which can modify what is done to considerable effect. The limits to power are many; in addition to the constraining effect of his own organization, the decisionmaker faces the power of other policymakers, other vested interests, poor or mistaken information, and competition with other programs.

If the formulated problem permits a prescriptive or optimizing approach (such as linear programming), the analysis team needs to begin by investigating the limits of feasibility, that is, to determine the constraints before seeking the optimum. On the other hand, if a nonprescriptive approach is in prospect, the team can examine individually each of the competing alternatives for feasibility. It may, however, be more efficient to investigate the set of constraints as a whole.

In considering the feasibility of a course of action, all possible constraints—social, political, administrative, institutional, technical, and economic—need to be searched out. When all of these constraints have been taken into account for public policy problems, the range of feasible alternatives is likely to be narrow. In fact, Majone (1975) argues that, because of the nature of government decisionmaking and the limitations of modeling, feasibility, not optimality, should be the realistic goal for public policy analysis.

In public decisionmaking the range of actual choice is much more limited than in the private sector. The public administrator is not free to trade off the different factors in production. No matter what the government, the administrator is restricted by the requirements for fiscal responsibility; if the budget prescribes particular combinations of inputs, as is almost always the case, then a search for optimal solutions becomes fruitless. On this point Majone (1975) quotes von Mises:

> In the conduct of government affairs, the discretion of the office holders and their subaltern aids is not restricted by considerations of profit and loss. If their supreme boss—no matter whether he is the sovereign people or a sovereign despot—were to leave them a free hand, he would renounce his own supremacy in their favor. They would do what pleased them, not what their bosses wanted them to do. To prevent this, it is necessary to give them detailed instructions regulating their conduct of affairs in every respect. Their freedom to adjust their acts to what seems to them the most appropriate solution of a concrete problem is limited by these norms. [von Mises, 1966, p. 310.]

Assuming we cannot determine the optimal alternative, should we be satisfied with just any feasible alternative, knowing that some feasible alternatives may be more efficient economically or technically than others? The term *efficient alternative* here means one that cannot be modified in such a way as to increase the utility for some member of the community without decreasing that for others. Should we seek an efficient alternative, or even the best in some sense, among the efficient ones? The answer is that we should not, for it is not clear that, say, an economically efficient alternative is superior to one that is merely feasible. Although normative models give precise conditions for an optimal solution, it has been shown that there is no corresponding set of prescriptions for achieving the second best, or even a better position (see Majone, 1975).

Moreover, to select the optimal from the efficient subset, the analyst needs a rule that aggregates the preferences of the different citizens into a utility index valid for the entire community. Arrow (1951) has shown that it is impossible to construct such a rule if the preferences of the community are sufficiently different unless some part of society is allowed to dictate to the rest. An alternative is to satisfice.

Constraints are not detrimental to analysis and decisionmaking in all aspects: if they are indeed real, they express very important boundaries that it would be foolish to try to cross, and they act to reduce the number of feasible alternatives, thus making analysis easier. Constraints that result from prior decisions, however, must be handled with considerable care, for they may have been set without careful analysis of their consequences and may rule out alternatives of great value. Thus, the basis for constraints deserves careful scrutiny.

Constraints are various: they range from global resources that are fixed and permanent or physical properties requiring an invention or a state-of-the-art improvement for removal to those that are man made, set possibly only by a decisionmaker's tastes. Thus some may be unchangeable, but others may not be. The latter can sometimes be removed—or at least widened—if analysis offers a good enough case for change. Hence, the opportunity cost of a possibly removable constraint should be considered carefully before it becomes firmly established in a study's analytic structure.

Sometimes it is more fruitful to ask what cannot be done than to focus on what can. The solvability of policy problems requires an investigation of all of the relevant constraints, including lack of knowledge. As discussed earlier, decisions are not directed toward a goal so much as toward discovering what can be done in the face of a host of constraints: If I know what cannot be done, then I may think of a way to get around the constraints and thus uncover new alternatives worth exploring. For additional discussion of political feasibility, see Chapter 7.

Tactics for Reducing the Problem

Problem solutions—or attempts at them—by systems and policy analysis are almost always suboptimizations (Hitch, 1953) in the sense that the criteria used to select the preferred alternatives are more appropriate to a lower level than the one with which the analyst and his client should be concerned. For example, an analyst serving the traffic manager of an amusement park might suggest a parking policy that maximizes the revenue from the parking lot. However, they should really be concerned with the effect of the parking policy on the revenues for the park as a whole. If they are and seek a criterion relating to this wider concern, they might explore a policy for free parking, or, at the other extreme, concealing the parking fee in the admission charge (Quade, 1982, pp. 211–212). But such policies would have to be adopted at a level higher than that of the traffic manager.

In many situations, however, suboptimization is both necessary and desirable. Not all decisions can be made at the highest level—and, even if they could be, it would be foolish to try to make them there. To be useful to a client, an alternative must be among his options. He will seldom want to support an analysis that is going to tell his superiors what they should do. Too, a solution that is beyond the client's capability to implement is likely to have no impact.

Another reason for suboptimization is that the full problem may be too difficult or too extensive for the analysis team to handle within the constraints of time and resources. In such cases, the problem may have to be broken into parts, with possible decisions about some aspects being postponed or even neglected and others assumed. The analyses of the remaining parts are then necessarily suboptimizations, because varying some or all of the aspects held fixed by assumption or neglect would almost always yield better results. Analyses made at a given decision level often take—or are forced to take—decisions made at higher levels as constraints.

Although suboptimization may make a solution easier by omitting elements or by treating as fixed elements that, in a broader perspective, would be treated as variable, it may not lead to improvement in the situation as a whole. In fact, improvement in a particular subsystem may even worsen the situation for the system as a whole. Thus, in the parking-lot example, a policy that maximizes profits from the parking lot may work to the detriment of the operations as a whole if the parking policy, owing to insufficient space or high cost, results in potential customers being turned away or going elsewhere, with subsequent loss of revenue for the park as a whole.

Nevertheless, it is undoubtedly an advantage of suboptimization that the narrower problem permits using greater detail in the analysis. Models can be smaller, with fewer factors, and, within their limitations, they can yield sharper estimates.

This discussion appears to present the analyst with a dilemma. On the one hand, any study must, to some extent, be a suboptimization, because it must, *a fortiori*, have limitations. On the other, a suboptimization leaves out important considerations. How then, is the analyst to suboptimize properly? The answer is that the analysis must make sure that the action that improves the part also improves the whole. The rule to keep the analysis on the right track, as Hitch (1956) states, is to keep the criteria in the lower-level problem consistent with the criteria at higher levels. In practice, this is largely a matter of judgment. Hitch's recommendation for obtaining consistency of criteria is the economist's "golden rule" of allocative effectiveness: Limited resources having interchangeable uses should be allocated so as to make each resource equally valuable at the margin in all uses. Such allocative efficiency can be achieved only if the resources can be freely combined and substituted for each other according to relative prices or scarcities. But this is not easy to discover without working out the higher-level optimization.

Suboptimization may enable the analyst to separate the original problem into component parts that may be investigated and solved separately—and thus more easily. How to make such a separation and whether it is possible depend on the nature of the problem. About all that can be said in general is that it seems desirable to make the separations at places where the interactions between parts are weakest.

Tactics for Exploring the Uncertainties

Sensitivity analysis is the most important tactic the analyst has for exploring the effects on the results of uncertainties in the assumptions, parameters, and input data used in the analysis. It is commonly carried out by changing the values used for one or more of the factors in the analysis and redoing the calculations. One aim is to identify the factors on which the results of the analysis are particularly dependent; another is to discover how great a change in the factors being investigated is required to affect the rankings of the alternatives.

In the usual sensitivity analysis, the uncertain assumptions or parameters are changed twice, once to see how increasing or strengthening them will affect the results, and once to see what lessening or weakening them will do. It is common to make these changes one factor at a time, a tactic that usually causes no serious computational or time problems, especially when the work is done on a computer. One-at-a-time changes, however, do not always constitute a fully satisfactory sensitivity exploration and cannot guarantee that simultaneous changes of two or more factors would not produce important changes when changes in the factors taken individually would not. If there are a large number of items to be investigated, however, examining all combinations, two, three, four, and so on at a time can become exceedingly time-consuming and expensive, even with a computer. In such cases, some form of Monte Carlo sampling of the parameters may be advisable (Emerson, 1969). In any case, the analyst will have to decide how far to take the process of sensitivity analysis in the light of the results and the context in which they fall. To neglect sensitivity analysis is a serious pitfall, as the analyst's findings will be seriously incomplete if he cannot accompany them with information about their sensitivities.

Bowen (1978) points out that there is a misconception about sensitivity analysis that sometimes arises to the effect that such analysis can compensate for simplifications that occur in a model when expected values replace stochastic processes. Sensitivity testing, however, varies assumed values, whereas a stochastic formulation allows a value to vary from event to event in some assumed manner. Sensitivity analysis cannot eliminate uncertainty from analysis and its results, nor can it deal with uncertainties in the context that lies around the analysis. It merely makes more explicit the types, degrees, and sources of some of the uncertainties that exist in the results from the models used in the analysis.

One aim in carrying out a sensitivity analysis is to discover a robust alternative, that is, one that, although possibly not the best for the situation considered most likely to occur, will do well if the actual situation is somewhat different. The future is full of surprises, many of them unpleasant, and an alternative that is too finely tailored to a particular anticipated situation may fail badly in other circumstances. Robustness, the degree to which the attainment of the objective will be sustained despite disturbances that might be encountered in normal operations, is almost an indispensable feature of an acceptable alternative. The farther into the future the decision extends its implications, the more desirable it is to have it based on a robust alternative.

Although uncertainty cannot be entirely eliminated from systems and policy analysis, it can be reduced. Delay, collecting more and better data, and more research can all help to do so—but at a cost. It is often a far more efficient use of mental and physical resources to devote analysis to finding ways to hedge against uncertainty, and to live with it, than it is to search for ways to lessen it.

Analysis results may be sensitive to the relations used in the model as well as to its parameters. If alternative relations seem plausible, this sensitivity can also be investigated by carrying out computations with an appropriately changed model. For further discussion of uncertainty, see Chapter 7.

Tactics for Helping the Decisionmaker Decide

An important aspect of the analytic strategy is to select a procedure for helping the client decisionmaker to reach a decision about what to do as a consequence of the analysis and its findings. He can, of course, reject the work and its results, take an action independent of them (including leaving things as they are), call for further analysis, or choose to implement one of the study's more promising alternatives, possibly with modifications. This subsection discusses the last of these options.

The commonest way for the analyst to help the client decisionmaker to judge the merits of various alternatives is to employ a criterion, or rule for ranking them, that is usually selected early in the study after discussion with and approval by the client. One commonly used criterion is the ratio of cost to effectiveness (for a discussion of cost-effectiveness analysis, see Section 10.3).

Since it is usually the case that there is no single measure that will give a satisfactory comparison of the alternatives, the analyst must consider other possibilities for helping the client.

Weighting approaches. If several measures are required to capture all the aspects of the comparison, they can be assigned weights and combined into a single number. This process can be carried out in several ways, including some that require extensive participation by the client and/or his staff. One is to use an index number; Mood (1983, pp. 59–61) gives an elementary explanation of how this might be done. Another is decision analysis; many works on operations research and systems analysis offer at least a brief insight into this approach (for an introduction to its theory, see Keeney and Raiffa, 1976, and for a view of some applications, see Keeney 1980). All of these weighting approaches suffer from two important drawbacks: they combine the large spectrum of information developed by the analysis into a single measure, therefore acting to conceal its details from any parties at interest who do not participate in the weighting process; the weights, although they may be acceptable to the analyst and the decisionmaker, may not represent the weights that might be assigned by the larger community interested in what is done.

Cost-benefit analysis. If the problem situation involves some aspect of the welfare of a community as a whole, the decisionmaker, in order to act in the broad public interest, may want the analysis and the criterion to take into account, insofar as possible, all the effects on the public, both favorable and unfavorable. A cost-benefit analysis is designed to meet this need, since it aims to establish a single measure (usually a monetary one) for all costs and benefits, thus allowing them to be summed; the *Overview* (pp. 224–227) gives a brief summary of this approach and lists references to supporting literature. This form of analysis, however, is fraught with both analysis and interpretation problems: the choice of a discount rate for future benefits and costs, the treatment of uncertainty, the estimation of the values of intangible benefits, the choice of a time horizon over which to consider the balance of costs and benefits, distribution effects, and so on; the choices made for many of these factors are often very hard for the analyst to defend (Merkhofer, 1987). In sum, cost-benefit analysis can be used naturally and effectively in a rather small proportion of cases, and great care should be taken to assure that it is, in fact, an appropriate form of analysis for the problem being considered. Although its unitary index of worth appears to make decisionmaking easy, this is often because many of the really important issues have been covered up.

Scorecards. When there are many impacts to be considered, they must be displayed and not concealed so that the decisionmakers can get a rounded picture of what impacts will ensue from adopting an alternative. This is particularly important when there are many parties at interest with varied concerns. Such a display avoids one of the problems of the cost-benefit approach, which requires the analyst to make many decisions about values and relative importance that are usually better evaluated by the decisionmakers and the other parties at interest.

The scorecard technique (as described and illustrated in Chapter 10 and, more completely, in the *Overview*, pp. 231–236) is a carefully designed and systematic way of presenting a large body of information in an attractive form easily assimilated by persons with varied interests; it is a matrix in which the columns are alternatives and the rows impacts, so that an impact in row 10 and column C shows the tenth impact of alternative C. Goeller has developed this form of presentation to a fine art, and others have copied it with considerable success (Chesler and Goeller, 1973; Goeller et al., 1977).

In a good scorecard, the analysis team focuses on presenting the most important and critical information that relates to potential decisions, keeping in reserve (but ready for presentation if requested or desirable) secondary considerations that, although significant and relevant, are of lesser importance. Too much information on a complex situation shown at one time may swamp the viewers, putting them into the unwelcome position of not being able to see the forest for the trees. Thus, the art of designing scorecards focuses on providing good coverage of important impacts while keeping the detail focused and reasonably spare.

General. Although the techniques for helping with decisionmaking discussed here may be useful in appropriate cases, the primary requisites for such helpfulness are carefully prepared oral and written presentations that deal candidly and effectively with what the analysis has found about the client's concerns. Portions of Sections 13.6, 13.7, 15.9, and Chapter 14 deal with how to approach this matter, but there is one point that is worth making here.

Presenting previously prepared material is never enough. The analyst must be prepared to answer a variety of "what if" questions, and must deal with them effectively. The parties at interest, in addition to asking about underlying details of the analysis, will almost surely inquire about the longer-range implications of its findings. With their imaginations stimulated by what they have seen and heard, they are likely to suggest new alternatives, either variants of those studied or entirely novel ones, and ask the analysis team's views on them. Although these alternatives will not have been studied, it is quite likely that the work that has been done will shed some light on them, and the analysts should not hesitate to share any reactions that they may have (being scrupulous, of course, to make clear the very partial basis for these reactions). These candid discussions among partners are an important ingredient in the process of moving toward decision. No matter how thoroughly the analysts may have studied the problem situation behind their work, the decisionmaker and his staff are likely to have a far deeper knowledge of its details—of what is going on and what can be done—than the analysts, but they lack the analytic skills to produce the sorts of syntheses that lead to well-supported decisions. Thus, these discussions constitute an important and significant sharing of knowledge and judgments that can move the situation along toward beneficent innovation.

Tactics for Making the Choice Implementable

Chapter 9 in the *Overview* discusses the things that can be done during the course of the analysis to ensure that the alternative selected for implementation can in fact be carried to reality as planned. However, it is worthwhile here to add some points about how to deal with losers and other opponents of the chosen course of action or policy, since they may be able to frustrate an otherwise desirable choice. They must be dealt with, but the ways are various, ranging from direct monetary payments to delaying the process of implementation until their objections are overcome, or at least until they have had time enough to accommodate their operations to the changed conditions (Cordes and Weisbrod, 1985).

Implementation problems are likely to be most acute in relatively loose organizations or public programs; they begin to appear once the decision is known, or sometimes even earlier in anticipation of what the new policy or program will be.

The great advantage that the decisionmaker and his analysis team have with respect to the special interests who may oppose implementation is that, if the work of analysis has been well done, the alternative selected for implementation is the one that brings the greatest total advantages to the community or the organization as a whole. Too, the analysis has presumably identified all the affected parties and taken their special interests into account, not only in the analysis, but also in the plans for implementation.

Although the analysis has developed arguments aimed at meeting the objections expected from special interest groups, it is too much to hope that this protective shield will be airtight. For one thing, every analysis is based on assumptions that, although reasonable and representing the best judgments of analysts and client, are nevertheless assumptions that cannot be proved correct; consequently, reasonable men may disagree about them. A thorough sensitivity analysis will help to defend the assumptions, but time and resources are always constraints on analysis and some assumption or some interest group may have been overlooked or judged to be insignificant, and thus not adequately investigated. The only defense in such a case is to admit the oversight and conduct an appropriate inquiry into the issue involved.

Any change in policy that affects the way an organization or a society operates will generate opposition, if for no other reason than that most people resist change. In some cases this response alone may turn a policy around, but those who resist implementation usually have a stronger motivation: their own special interests, whether selfish or unselfish. To overcome such resistance may require considerable persuasion and negotiation to convince the interest group that the program or policy is equitable, that other groups also suffer comparable losses to their special interests, and that all losses are more than offset by the gains for the community as a whole. Mood (1983, p. 285) goes so far as to suggest these possibilities:

> Sometimes it is good strategy in dealing with interest groups to give them, unobtrusively, some alternative compromises. That could be done in a report which would choose a particular compromise but, in showing its superiority over other compromises, would indicate that some of the others were not much inferior to the one chosen. Like everyone, leaders of interest groups dislike being herded to a position not of their own choosing and will sometimes go so far as to sacrifice a small piece of the pie in order to show their independence of the client and the analyst by agreeing upon a compromise different from the one promulgated in the report. Some sly analysts put the best compromise in second place in their reports so that interest groups can have the satisfaction of rejecting the solution put forward by the client and his analysts and yet arrive at the best compromise.

There are cases in which persuasion fails and it becomes necessary (and cost-effective) to compensate the losers. This may be done in several ways (Cordes and Weisbrod, 1985). Cash payments are often difficult to manage; among the other possibilities are tax breaks, postponement of the implementation process to allow time for adjustment to new conditions, or, as a last resort, modification of the decision itself.

Once implementation has started—but possibly not until it is well along—new interest groups may be created by people who formerly did not realize the full implications of the program until it actually began to impinge on their activities. As there are many ways to frustrate a program (see Chapter 9 in the *Overview* and especially Bardach, 1977), the implementation process must be monitored closely. It helps if a plan has been spelled out in detail, telling who does what and when, and if the participants have been carefully instructed as to how their tasks are to be carried out. The original analysts can be helpful during the implementation work, because they are familiar with what has to be done, what effects are expected, and what the interests of those affected are. Thus, the original analysis team is in perhaps the best position to spot aberrations early when they occur.

In designing alternatives, and in helping to implement them, a systems or policy analyst should keep in mind an important principle: he can fail even when his implemented recommendations fulfill the desired goals if the new program disrupts efforts to accomplish other equally important goals or hinders other desirable actions. The knowledge of this principle urges a wide study of ripple effects as part of the original systems-analysis study. Considering the wider problem situation at the formulation stage helps to decide what can reasonably be covered in this respect, and what should be part of the analytic strategy.

2.8. Tools

The central tools of systems and policy analysis—and, in fact, of all analysis—are models. Their importance to analysis cannot be overstated. Without models—that is, without simplified representations of reality to structure our thoughts—we would not even be able to think. Models come in infinite variety, ranging from simple mental models to elaborate mathematical structures that require the world's most modern and largest computers to keep track of their variables and relations. No matter how simple or how complicated, models are at best an imperfect representation of some aspect of reality.

Since models are important tools of systems analysis, choosing appropriate ones is one of the central tactical issues of systems analysis. We could bring in here a conscientious catalogue of models used in systems analysis, together with important information about each one, but this exercise would not be as useful as one might hope. Few systems analyses ever depend exclusively on models that have been used earlier; rather, the models are chosen so as to respond to the new problem and its characteristics as well as to produce the sort of findings that are likely to be useful to the client. Thus, model choice depends on having intimate knowledge of possibly relevant models represented on the analysis team—or readily available to it.

It is beyond the scope of this book to list the sort of expertise that such persons should have, and the technical literature that supports it.

Against this background, this section first makes some general remarks about model selection and then offers some comments about several classes of models often used but that call for careful consideration. The three following chapters deal with other classes of modeling: using expert judgment systematically, operational gaming, and social experimentation. Chapter 8 deals with the uses and limitations of quantitative modeling methods.

General Observations on Modeling

One thinks of models in systems analysis primarily as tools for estimating the consequences that will follow from implementing alternatives, but they have other important roles to play. For instance, they are used for education, to provide information, for forecasting, as instruments for exploring limited aspects of reality, to screen and rank alternatives, and for many other purposes, among them to reinforce or justify decisions that are not the result of analysis. In systems analysis, however, models are used as tools to advance the objectives of the analysis, never as ends in themselves.

A good deal has been said and written about models in the policy process; see, for instance, Greenberger, Crenson, and Crissey (1976), Gass and Sisson (1975), House and McLeod (1977), and Meadows and Robinson (1985). The shortcoming of much of this literature is that it focuses a great deal of attention on the models, and relatively little on the process in which they should be embedded if the work is to form a secure bridge between the formulated problem and the concerns of decisionmakers to ameliorate it. The pitfall well represented here is for the analyst to expect his client to take a strong interest in his modeling processes. In reality, decisionmakers are interested in problems and their solutions, and only occasionally in how analysis can help with the process of solving them; if their primary interest had been in modeling, they would possibly have become analysts rather than executives. In sum, modeling is not systems analysis. Nevertheless, models are the central tools of systems analysis, and many of the most important tactical choices that an analyst makes are the choices of models. Dutton and Kraemer (1985) and Kraemer et al. (1987) describe how several models were used in policy processes.

An analyst's background and training have strong influences on his choices of approach, and thus on the models he uses. The modeler's mental model of the situation or reference system underlies the formal models that he devises, whether quantitative or otherwise. Thus, if his mental understandings have been confined to a single discipline, his models are likely to be chosen from that discipline's models; if his mental understandings have been broadened by interdisciplinary work—as a systems analyst's should be—his models may well combine elements from several disciplines. For example, an economist working single-mindedly within his discipline on a program evaluation is likely to use an econometric model to assess costs and benefits for the entire community using market prices when available and, if not, willingness to pay, an approach confined entirely to economic variables, thus ignoring (or assuming as fixed) all other social and political variables. Similarly, an analyst working for a government agency or for a product-oriented industry may take the organization's goals and structure as given and then seek models that will enable him to calculate results showing how to come closer to these goals. Neither example is a horrible one provided the analyst's goals and claims for results are narrow enough to be appropriate for the work done. If the goals and claims are much broader, then, of course, the work may be too parochial and the examples those of bad work. The point here is that model choice must respond to the goals of the analysis and the claims that the analyst expects to be able to make about the findings. If the analyst expects widely applicable results for which broad claims can be made, he will undoubtedly not be able to stick to narrow model choices from a single discipline; he will have to combine elements from several disciplines, or perhaps have a set of models from several disciplines, each to cover a relevant aspect of the analysis. In sum, the nature of the problem and the expected character of the needed findings must dominate model choice. The choice of modeling tools can be helped by familiarity among members of the analysis team with a variety of modeling methods and with the history of both the problem situation and previous attempts to analyze and resolve it.

Most analysts do come from a disciplinary academic training and have previous experience with a narrow spectrum of analysis appropriate to the problems that they have helped solve. An academic training in a single discipline is usually an influence that tends to reject models from other modeling schools while trying to fit problems into the Procrustean bed of the discipline's modeling approach. A systems analyst must rise above such an influence; successes will arise from problems solved, not models elaborated within a discipline.

Some systems-analysis projects are dominated by the difficulties of gathering data and information, but it is more usual for building, verifying, validating, and exercising models to be the most time-consuming activity. Estimating consequences may require a single model, or, as in the flood-protection study of Section 2.2, it may require many models. If the analysts are in luck, the models will be independent, as they were in the flood-protection case. Frequently different models are needed for different portions of the process, so that some models depend on the outcomes of other models (as in the blood-supply and energy cases in Section 2.2). The analyst may be tempted to combine these models into one very large model that can deal with everything simultaneously (and in some cases he may have to), but it may be well to consider keeping them separate, so that the results emerging from each model can be inspected for difficulties and anomalies before the outputs enter the next stage as inputs (a course that was followed in the energy study).

Excessively large and complex models add so much to the difficulty and cost of analysis as to make it good advice to avoid them unless it is absolutely impossible to do so. Not the least of the difficulties of large models is the possibility that they may lead the analysts unwittingly into error (see, for example, Meadows and Robinson, 1985). Mood (1983, p. 78), offers this advice:

> There are two reasons why relatively simple models are usually quite sufficient for policy analysis. One is that there are inevitable unquantifiable aspects to almost all policy questions, hence it is impossible to deal with them on a purely quantitative basis; there is, therefore, bound

to be some fuzziness in the final policy recommendation no matter how precisely the quantitative aspects are handled; therefore, it is a waste of effort to seek great precision. The second reason is that reality is far more complicated than any model can be; in formulating a model one omits a great many factors that one judges to be relatively unimportant to the central issue of the analysis; it would be a waste of effort to try to make a highly elaborate model of a real situation when a judicious simplification will serve approximately as well. The marginal utility of increased accuracy in a model decreases with great speed once it begins to provide approximate results.

All systems analysts can well ponder the relevance of this advice as they proceed with their model-choosing and model-building activities. Raiffa (1982, p. 7) reinforces it:

> Beware of general-purpose, grandiose models that try to incorporate practically everything. Such models are difficult to validate, to interpret, to calibrate statistically, to manipulate, and most importantly to explain. You may be better off not with one big model but with a set of simpler models, starting off with simple deterministic ones and complicating the model in stages as sensitivity analysis shows the need for such complications. A model does not have to address all aspects of the problem. It should be designed to aid in understanding the dynamic interactions of some phase of your problem. Other models can address other phases.

It is frequently desirable to build simpler versions of very complex models. The complex model is first

> used to explore the problem space, educating the analyst about both the problem and the model—their sensitivities and insensitivities. With this knowledge, the analyst creates a simpler version of the same model, either an aggregated model or a "repro" model. (A repro model is constructed by systematically varying the inputs of the complex model, recording the outputs, and then fitting some functions or devising an interpolation scheme so as to "reproduce" the relevant outputs from the inputs with acceptable accuracy. The functions or scheme become the repro model.)
>   We use the simple model in most of the analysis, either as a component of a more comprehensive model, . . . or as a free-standing model for extensive trade-off analyses. We might use the original, complex model in the final rounds of the analysis or whenever greater accuracy of more detailed outputs are desired. [Goeller and the PAWN Team, 1985, p. 9.]

Even if a systems analyst has found a solution to a policy problem that it seems reasonable to call "optimal," he is likely to find that, when implemented, it is modified and "improved" in response to political, fiscal, and social expediency. What takes place is an accommodation among feasible variations of the chosen alternative in response to pressure from various affected parties. Some variations that may seem minor to the analyst may be much more acceptable to the affected parties than the original version of the alternative.

We have argued that any model represents only certain aspects of reality. Thus, the analyst must, as part of the model selection and construction process, choose how much of reality to include. He may decide to model the reality very closely with a surrogate model, or he may retreat to a limited but salient aspect of this reality with a model that offers only a limited perspective (see Sections 8.5 and 8.6). If he chooses the latter course, it is undoubtedly because he feels that the omitted aspects can be handled better in other ways, perhaps by the client's judgment, and the resulting limited model may still be valid for what it represents; however, the history of systems analysis shows that the pressure on many analysts, whether from themselves or others, has led them to pursue the opposite course of expanding their models to include as much as possible, perhaps in the mistaken belief that to make the models more realistic will lead to improved results. Giving in to this pressure can make models so large and complex as to be self-defeating (Lee, 1973).

This subsection has presented some of the generic issues that a systems-analysis team must face as it chooses its models, regardless of what part of reality they are meant to represent or what discipline they might be drawn from. The next four subsections add, partly by way of example and partly because of their intrinsic interest, further discussions of four classes of models: simulation models, econometric models, system dynamics models, and partially quantitative models.

Simulation Modeling

In the general sense of the term, every model is a simulation. In operations research and systems analysis, however, a more limited meaning is usually intended: a simulation model is a descriptive model implemented on a computer and used to explore the behavior of the reference system by experimenting with the model.

Simulation is employed in situations that are too complex or too detailed to be easily dealt with by analytical models, for instance, in situations where practical observations cannot be made or where it would take too long to understand and learn how to represent what is going on by mathematical equations. For example, simulation can be used to represent a queuing situation that involves nonexponential interarrival times. Most textbooks on operations research contain chapters on simulation, and there are treatises on this subject as well (for example, Tocher, 1963, or Bratley, Fox, and Schrage, 1987).

Simulation is often selected as the modeling technique because its relations to reality are fairly direct and developments in the field have made it simple and easy to use. Special computer languages, such as

SIMSCRIPT, DYNAMO, and GPSS, have been developed to make programming easy; using them also reduces the chance of errors in programming. Simple simulations are now so easy to construct that they open up the temptation to stretch this approach into overmodeling, to build simulations that are in almost one-to-one correspondence with a major segment of the real world. The results are models so large and complex that they become almost as difficult to understand and work with as the original real-world situation. As Bowen (1978, p. 4) remarks, "Some simulations are not appreciably less complicated than the real world. We need models to understand such models. This is no jest."

The standard advice is that one should pause to think carefully about the consequences before embarking on the development of a large-scale stochastic simulation. The analyst should be convinced not only that simulation is the only practical way to get the needed quantitative information about the problem situation, but also that simulation will in fact yield this information. The reason for stressing this second point is that, at the analysis stage when the simulation is run, the information one wants may be difficult and expensive to extract. All one has at the end of a single simulation run is a single answer to a completely specified numerical problem. But, if the analyst wants the best answer over a range of possibilities, he will have to run the simulation over and over again, possibly thousands of times with changed parameters.

> In principle, . . . a simulation is the least desirable of models. It has low insight, since it does not provide explanation of the observed outcomes, and it may involve an undesirably long, confusing, and expensive analysis phase. Nevertheless, it may be a correct choice of model, if only because no other choice is open. For example, the system to be modeled may not be well enough understood for mathematical relationships between variables, or some acceptable approximations, to be established. [Bowen, 1978, p. 4.]

An analytical model, even a highly simplified one, that will lead to some general information characterizing the situation under study, or some aspect of it, may be more enlightening than a collection of simulation runs, even though the simulation model may represent the problem situation much more closely.

Computer simulations represent a model world in which experiments related to the situation modeled can take place. To do this successfully, the analyst has to learn how to carry out such experiments (which can involve complicated statistical issues), which ones to undertake, and how to use the results to contribute to solving the problems occurring in the problem situation. Considerable usefulness can emerge, but the analyst must guard against the temptations to print out more information than is needed or to present a good deal that may be interesting to him but not very relevant to the problem.

The simulation model built by The New York City–Rand Institute for the fire department (Greenberger, Crenson, and Crissey, 1976, pp. 267–270; Walker, Chaiken, and Ignall, 1979, pp. 495–548) was not developed to determine an optimal deployment policy, but merely to be used (in a satisficing mode) to search for policies better than those in current use. This aspect may have contributed much to its success, for it allowed department officials, not just the analysts alone, to suggest the policies to be tried out. Another factor in its success was its obvious realism; it simulated sequentially all of the steps in the procedure followed by the department from the outbreak of the fire and the receipt of the alarm to the return of the engines to the station, taking account of such possibilities as that an engine might have an accident on the way to the fire or might be called to fight another fire; the most sophisticated version even allowed companies to be dispatched from other districts to occupy vacant fire-houses in case other fires should occur. By using the simulation, analysts and officials could test a range of policies under the same conditions (fire location, number of alarms, type of incident, and so on) to determine which deployment policy produced shorter response times or a lighter work load.

Later, when simpler analytic models had been developed, the simulation was used to verify the predictions of the simpler models so that they could be used confidently. These analytical models, which were much cheaper to operate than the simulation, were used in other jurisdictions to design better deployment policies; the fire-protection case described in Section 2.2 was a beneficiary of this work of simplification. The simulation was also used elsewhere: for example, it was used in Denver to verify results after other methods of analysis had suggested that a reduction from 44 to 39 fire companies, together with the construction of six new fire stations, could provide almost the same level of protection as the then-current configuration. The simulation served here and elsewhere to give fire officials evidence they could believe that proposed policy improvements already evaluated by using simpler models were workable and effective.

Econometric Modeling

Econometrics, the commonest form of economic modeling, is based on the use of economic theory to specify the structure of the model and on statistical methods to estimate the parameters used in it. It is a modeling tool frequently used in problems involving economic variables, for instance, when the concern is something like tax equalization, welfare reform, land use, or housing or when a precise forecast for the short-term economic future is desired, particularly for the performance of regional or national economies.[1]

The phase of econometric modeling that occupies most of the modeler's time and attention and is the center of his concern, is estimating parameters from historical data by statistical techniques, primarily regression analysis. This method puts heavy reliance on data availability, particularly time-series data.

> Econometricians can also represent an unobservable concept by means of a closely-correlated tangible substitute or proxy. Literacy may suffice as a stand-in for degree of modernization, rainfall may be a proxy for all the effects of weather on crop production, or advertising expenditures may be used to represent some of the perceived-utility assumptions underlying a consumer demand equation. In other words, an econometrician can transform a direct causal hypothesis ("in early stages of modernization people's material aspirations rise, and so they consume less and save more for investment to increase their future consumption") into an indirect hypothesis about correlated observables ("at low income levels literacy is inversely correlated with consumption and positively correlated with savings"). The use of correlated rather than direct causally-related variables allows econometricians to proceed in spite of the requirement of empirical validation, but it also reinforces that requirement because a double set of assumptions must be made. A relationship must be hypothesized not only between modernization and consumption, but also between modernization and its proxy, literacy. Both relationships must be demonstrated to have existed over the historic period from which data are taken, and to exist in the present and into any forecasting period. Since correlational relationships are tenuous and subject to change, they must be rechecked continuously against the latest available real-world data. [Meadows and Robinson, 1985, p. 49.]

The requirements for statistical estimation, especially those of least-squares regression, sometimes impose restrictions on the structure of econometric models, forcing the analysis to compromise the form of the model in order to make use of existing data when the analyst would prefer, for theoretical reasons, to use a different functional form. What variables are included may, in fact, depend on the existing data. As a consequence, econometric models sometimes

> represent a world quite different from either the world of economic theory or the world seen by other modeling schools. Econometric models tend to consist of linear, simultaneous relationships connecting the observable economic variables by means of coefficients derived from the

---

[1] What we say in this subsection is largely drawn from Meadows and Robinson (1985, pp. 42–54, 75–86.)

historic operation of the system. Economic theory, however, recognizes that economic relationships may be nonlinear, greatly delayed, dependent on unobservable goals and desires, and capable of behavior patterns not contained in the historical data. [Meadows and Robinson, 1985, p. 51.]

One criticism that even econometricians often voice about their field is that econometric modeling is sometimes done badly, particularly with respect to its use of statistical techniques. Correlation is frequently confused with causation (see Chapter 11, Action 11, and Strauch, 1970).

Input-output analysis is another important form of economic modeling. It is a way of representing flows of money, products, or resources among the producers and consumers in an economy. The main use has been to forecast the states of various national and regional economies. For an overview discussion and references to key literature, see Meadows and Robinson (1985, pp. 55–64).

System Dynamics Modeling

System dynamics is a well-developed and well-documented technique for representing complex, nonlinear, multiloop, feedback systems.

As its name implies, system dynamics is concerned with questions about the dynamic tendencies of complex systems—what kinds of behavioral patterns they generate over time. System dynamicists are not primarily concerned with forecasting specific values of system variables in specific years. They are much more interested in general dynamic tendencies; under what conditions the system as a whole is stable or unstable, oscillating, growing, declining, self-correcting, or in equilibrium. To explore these dynamic tendencies they include in their models the concepts from any discipline or field of thought, with special emphasis on the physical and biological sciences and some tendency to discount (or rediscover and rename) theories from social sciences. [Meadows and Robinson, 1985, p. 34.]

In view of this brief description of system dynamics by an early pioneer and one of her students, it is perhaps not surprising that it has not proved to be a widely useful tool of systems analysis for identifying preferred alternatives and urging their adoption. Since it is well developed and easily used, however, it has a useful place in exploring the properties of systems for which it is an appropriate model, and therefore it can be a useful aid in problem formulation.

Meadows and Robinson (1985, pp. 27–42 and 75–86) give an authoritative and discriminating overview of the properties, strengths, and limitations of this technique as well as a guide to the literature that will be useful to someone desiring to become familiar with its practices with a view to using them in a problem-definition exercise. Greenberger, Crenson, and Crissey (1976, pp. 120–182) offer an early historical and critical review of system dynamics from the systems-analysis point of view. The basic references in system dynamics are Forrester (1968), Goodman (1974), Richardson and Pugh (1981), and Roberts et al. (1983).

Gaming and Other Partially Quantitative Modeling

No matter what systems or policy problem the analyst investigates, there will almost always be aspects for which quantitative tools are not fully adequate. Often this is of little consequence, for the quantitative model being used provides enough insight to enable the analyst, or his client, to bridge the gap between the inadequacies of the model results and the real world judgmentally, that is, by using mental models. In other cases, however, a tool such as operational gaming, role-playing, Delphi, or a structured conference may be used to replace a quantitative model in which there might be little or no confidence (see, for example, Chapter 3). We call these tools partially quantitative rather than qualitative, for they always seem to have some quantitative aspects, even if, as often in Delphi, nothing more may be involved than using the median and the quartiles to indicate trends.

In systems analysis, partially quantitative models are seldom used as the central model to estimate the consequences of alternatives and to compare them, although they have sometimes served advisors well in these roles elsewhere. They deserve mention here because they are less well known in the systems-analysis community than they might be, and their advantages and difficulties need discussion. They offer a way of taking modeling into the domain of the behavioral sciences, where measurement and quantitative analysis are notoriously difficult. Despite the importance of human behavior in almost all issues investigated by systems analysis, these tools have been little used.

Among these tools, gaming and role-playing models are the ones used most frequently—but not really often—as the main model to assist a primary decision. For instance, ways to improve a logistic system have been found by having people act out the roles of decisionmakers in various sections of interest throughout the system by supplying them with computer-generated data similar to data generated in the real system and observing how they handle the problems that arise (Geisler and Ginsburg, 1965). Operational gaming is widely used throughout military establishments and has been an important tool in developing military policies; there has also been some use in industry and civil government (Ståhl, 1983, and Chapter 4).

In a general sense, an operational game is a simulation model for determining or representing the behavior of people who are faced with decisions under uncertainty. Unlike the computer simulations discussed earlier in this section, the course of the game simulation is largely determined by the decisions of the participants (with perhaps some interference from the game director). In contrast, in the computer simulations (sometimes called programmable simulations) the course is determined by rules prescribed in advance, that is, by decisions made by the analyst who constructed the simulation.

The key to a successful model is usually considered to be whether its output can be analyzed and interpreted with respect to the real-world segment that it represents. Large-scale games cannot be evaluated this way, because the game's outcome is too dependent on chance and the decisions of the individual players. Such games are time-consuming, almost impossible to replicate, and usually require expensive facilities and large numbers of man-hours from highly qualified personnel. Unlike the smaller games Ståhl describes in Chapter 4, an unambiguous outcome cannot be determined statistically after repeated plays to guide a general understanding or decision. However, the classical, minimum-rule, "free" game—often called a war game because of its similarity to the traditional military war game with its red, blue, and control teams housed in separate rooms—does offer its players an intense learning experience. The participants, presumably chosen for their expertise and experience related to the issue being investigated, acquire further knowledge of behavior and issues of the problem situation. To gain additional insight into the problem, the game may even be played three or four times—seldom more because of the expense—with different scenarios and the same participants, possibly in changed roles. Advice to a decisionmaker from such a game, or a small series of games, clearly cannot be based on the game outcomes alone, for there is seldom any consistent pattern in the results. Although the outcomes of the games, or the comparative status of the two sides if not played to a conclusion, are a consideration that the players take into account, it is their judgments that are important—and any advice given to a decisionmaker should be based on this judgment. The judgment can be assembled after the game(s) by means of a group discussion, or a Delphi inquiry can be used.

Games of the type just discussed are particularly valuable when the decision under consideration is one that is important, but which is not of a type that takes place frequently enough for a body of real-world experience to have been built up on which an analysis could be based.

The type of problem suited to investigation by gaming has characteristics such as these: complexity; the interests of a large number of people; some elements of competition; and neither a completely unstructured plethora of data, beliefs, and issues nor a situation with generally well understood internal interactions. The greatest difficulty in applying gaming is to find a way to represent the original problem situation or issue in the form of a game (unless, like a combat or bargaining situation, it already has some of the characteristics of a game). Nevertheless, some unlikely problems have been formulated as games; for example, gaming has been proposed as a way to forecast the future of the United States (Helmer, 1983, pp. 300–302) and to study the global economy (ibid., pp. 348–358).

A free-play, minimum-rule game can also be a useful device for the preliminary investigation of a problem situation looking toward a problem definition. Helmer (1983, pp. 337–348) gives an example showing how such a game might be used to initiate a study of a developing economy.

An operational game that engages experts and specialists in the various aspects of the problem situation excels as an educational device. It quickly makes the participants aware of the intricacies of the situation and challenges them to contribute ideas and insights. It is a particularly useful device for generating possible alternatives and subjecting them to preliminary evaluations and comparisons. On the other hand, the extent to which results of games of this type can be used to predict real-world outcomes or to make policy recommendations is much more limited. Nevertheless, gaming is an excellent device for organizing a large group and focusing their attention on the details of a problem situation; it puts more pressure on the participants to work together than a workshop or a Delphi exercise.

Gaming helps to organize a problem by exploring assumptions—in particular, by bringing the implicit ones to light—and by drawing out divided opinion, thus examining the feasibility of a concept or plan. It tends to draw attention to areas that are particularly sensitive, or where more information is needed. Because the game assembles a great deal of expertise, its participants exchange facts, ideas, possibilities, and arguments in a highly stimulating environment. Thus, bits of information about the problem area, known at the outset to only one participant—or perhaps only a few—become common information very rapidly. Similarly, ideas about possible problem solutions become common property very soon, with the varied reactions to them shared. Thus, this somewhat informal form of operational gaming allows individual team members to achieve a broad understanding of—although not necessarily an agreement with—the perceptions, opinions, and feelings of their colleagues about the problem situation being investigated.

This interactive process may even converge to a commonly agreed problem definition. If it does not, however, it can give the systems-analysis team a very valuable background for its own work of problem formulation.

In addition to the frequent and necessary use of individual judgment that permeates every analytic process, systems analysts may make systematic use of the pooled judgments of knowledgeable people at various stages in their work; they do this by means of such activities as conference workshops, contextual mapping, the Delphi procedure, and cross-impact gaming.

Chapter 3 discusses the Delphi procedure, cross-impact gaming, and the systematic use of experts; Helmer (1983) covers these subjects more extensively, in addition to scenario writing and other related topics. These activities, although useful in the later stages of systems analysis, are particularly appropriate for the problem-definition stage.

A group of environmentalists used a series of conference-workshops as a tool of analysis in an approach that they called adaptive environmental assessment and management (Holling, 1978). This device is used to define the problems to be treated:

> The workshops that define the sequence of steps are the heart of the approach. . . . They provide a series of sequential targets, maintain integration while minimizing organizational and emotional overhead, and allow involvement of a wider spectrum of key actors than is normally possible. The policymaker, busy as he is, is involved at key points for short intervals.
>
> Each workshop draws upon up to twenty specialists, the choice depending on the particular stage of the process. The first workshop is critical, for it is then that the problem is defined and focused. It is essential to have all prime "actors" present at that time—scientists, managers, and policy people. The policy people and the managers provide a balance to the scientist's penchant for exquisite detail and excessive resolution. The scientists provide the rigor and understanding of the fundamental physical, ecological, and economic forces. During such a workshop, impact categories are classified, key information needs defined, alternative actions described, and the framework and crude working version of a computer model developed. Even if, through lack of expertise, facilities, or time, a model is not developed, the techniques of organizing elements in preparation for a formal modeling effort are themselves of fundamental value. The point is that, at the very beginning of a study, all elements—variables, management acts, objectives, indicators, time horizon, and spatial extent—are jointly considered and integrated. Even the crude model that is developed at this stage can be a powerful device to explore the significance of unknown relationships. By testing alternative extremes, priorities can be established for data and for scientific and policy analysis. [Holling, 1978, p. 13.]

A continuing decision seminar, as described by Brewer (1972) and by Brewer and deLeon (1983), is much the same as the series of conference-workshops that Holling describes.

The contextual map, a rather obvious device first described in the operations research literature by Kennedy (1956), is likely to be useful for a committee or panel—or even for the analysis team as the work progresses. It is simply a chart room with a large matrix on the walls with, say, time on the ordinate and places along the abscissa for each of the variables that the analysis team is interested in; it can include other exhibits, such as maps or charts showing progress toward key milestones.

Experimenting

Experimentation is an attractive tool for systems analysis when knowledge central to the problem is not available, or when key relations are so poorly known as to require some form of field inquiry. Scientific experiments, both in the laboratory and in the field, are frequently used to gather data needed in analysis. Large-scale experiments to help a policymaker choose a course of action (see Chapter 5), however, are relatively infrequent, although they may be necessary when there is no way of getting the required information about the consequences of implementing certain alternatives without actually putting them into effect.

When people are involved, particularly in large numbers, experimenting is attended by many difficulties; such social experiments are very costly in resources and time, and usually involve important administrative and political problems. Since the basic function of systems analysis is to estimate the consequences that will follow from implementing possible alternatives and to achieve these estimates before implementation, there is a sense in which the experiment defeats the purpose of systems analysis with its more limited menu of choices. Nevertheless, experimentation is used when centrally important data are needed and can only be obtained in the field. The approach makes the data very expensive to get, but there are times when the costs are warranted.

Since Chapters 5 and 11 discuss experimentation in considerable detail, there is no need to expand the remarks here.

2.9. Conclusion

The universal advice to systems and policy analysts is to take a systematic and comprehensive approach to the client's problems—which implies that the analysts should take the same approach to their own analysis concerns. Mostly they do, although they are sometimes unduly influenced by work on a previous problem that looks at first like a strong analogue of the new one, or by fascination with some modeling paradigm. This is not to say that guidance from previous problems or disciplinary allegiances is valueless, but merely to suggest that it should be accepted with caution.

Although systems analysis has reached a sufficiently mature state to warrant a handbook synthesis, it is still working toward the stable criteria of quality and standards of procedure that are the hallmarks of mature professionalism, as discussed throughout this handbook. The craft of designing an analytic strategy appropriate to the form and substance of a problem, however, remains only partially developed; it will be subject to change and refinement as further experience accumulates. Thus, while this chapter offers an overview on the basis of the modest literature and our experience, the systems analyst facing a problem situation, although he may find some guidance here, will also have to think carefully about what approach may be appropriate.

In well-developed applied fields, such as engineering, the problem-solver has a large body of established knowledge and standards of procedure to guide his approach. The systems analyst's similar background is not only more modest, he also is usually involved in exploring situations involving factors from economic, political, and social fields in which models such as those of the natural sciences and engineering are lacking. Consequently, he must rely heavily on his judgment, complemented by that of others involved, to an important extent. He must design his approach as best he can on such bases as he can bring to bear. Thus, often his strategy, tactics, and models "have an *ad hoc* quality, representing merely the best insight and information that he happens to have available" (Helmer, 1983, p. 52).

The material in this chapter thus serves two purposes. It reminds the analyst of some of the things that are known about appropriate strategies, and suggests a framework for thinking about his approach. It also serves as a reminder of the importance of working out an analytic strategy as part of the plan for approaching a problem situation.

Finally, the developmental state of the art of strategy formulation for systems-analysis work reminds all practicing analysts of their responsibility for sharing in the development of this art by reporting their experiences to the larger community of professionals in this field.

References

Arrow, K. J. (1951). *Social Choice and Individual Values*. New York: Wiley.

Bardach, Eugene (1977). *The Implementation Game: What Happens after a Bill Becomes Law*. Cambridge, Massachusetts: The MIT Press.

Bowen, K. C. (1978). *Analysis of Models*. Unpublished paper. Laxenburg, Austria: International Institute for Applied Systems Analysis.

———— (1981). A conflict approach to the modelling of problems of and in organizations. In *Operational Research '81* (J. P. Brans, ed.). Amsterdam: North-Holland.

———— (1983a). An experiment in problem formulation. *Journal of the Operational Research Society* 34, 685–694.

———— (1983b). A Methodology for Problem Formulation. Ph.D. Thesis, University of London.

Bratley, Paul, B. L. Fox, and L. E. Schrage (1987). *A Guide to Simulation*, 2nd ed. New York: Springer-Verlag.

Brewer, G. D. (1972). *Dealing with Complex Social Problems: The Potential of the Decision Seminar*, P-4894. Santa Monica, California: The Rand Corporation.

——, and P. deLeon (1983). *The Foundations of Policy Analysis.* Homewood, Illinois: Dorsey Press.

Brodheim, E., and G. Prastacos (1979). The Long Island blood distribution system as a prototype for regional blood management. *Interfaces* 9, 3–20.

Cases. See Miser and Quade (to appear).

Checkland, P. B. (1981). *Systems Thinking, Systems Practice.* Chichester, England: Wiley.

Chesler, L. G., and B. F. Goeller (1973). *The STAR Methodology for Short-Haul Transportation: Transportation System Impact Assessment*, R-1359-DOT. Santa Monica, California: The Rand Corporation.

Cordes, J. J., and B. A. Weisbrod (1985). When government programs create inequities: A guide to compensation policies. *Journal of Policy Analysis and Management* 4(2), 178–195.

Dutton, W., and K. L. Kraemer (1985). *Modeling as Negotiating.* Norwood, New Jersey: Ablex Publishing.

Eden, C., S. Jones, and D. Sims (1983). *Messing about in Problems: An Informal Structured Approach to their Identification and Management.* Oxford, England: Pergamon.

Eilon, Samuel (1972). Goals and constraints in decisionmaking. *Operational Research Quarterly* 23, 3–15.

Emerson, D. E. (1969). *UNCLE—A New Force Exchange Model for Analyzing Strategic Uncertainty*, R-480. Santa Monica, California: The Rand Corporation.

Energy Systems Program Group of the International Institute for Applied Systems Analysis, Wolf Haefele, Program Leader (1981). *Energy in a Finite World: Vol. 1. Paths to a Sustainable Future.* Cambridge, Massachusetts: Ballinger. (See also *Energy in a Finite World: Executive Summary.* Laxenburg, Austria: International Institute for Applied Systems Analysis.)

Forrester, J. W. (1968). *Principles of Systems.* Cambridge, Massachusetts: Wright-Allen Press.

Friend, J., and A. Hickling (1987). *Planning under Pressure: The Strategic Choice Approach.* Oxford, England: Pergamon.

Gass, S. I., and R. L. Sisson, eds. (1975). *A Guide to Models in Governmental Planning and Operations.* Potomac, Maryland: Sauger Books.

Geisler, M. A., and A. S. Ginsberg (1965). *Man-Machine Simulation Experience*, P-3214. Santa Monica, California: The Rand Corporation.

Goeller, B. F., A. F. Abrahamse, J. H. Bigelow, J. G. Bolten, D. M. de Ferranti, J. C. DeHaven, T. F. Kirkwood, and R. L. Petruschell (1977). *Protecting an Estuary from Floods—A Policy Analysis of the Oosterschelde: Vol. 1. Summary Report*, R-2121/1-NETH. Santa Monica, California: The Rand Corporation.

Goeller, B. F., and the PAWN Team (1985). Planning the Netherlands' water resources. *Interfaces* 15(1), 3–33.

Goodman, M. (1974). *Study Notes in System Dynamics.* Cambridge, Massachusetts: Wright-Allen Press.

Greenberger, M., M. A. Crenson, and B. L. Crissey (1976). *Models in the Policy*

Process: Public Decision Making in the Computer Era. New York: Russell Sage Foundation.

Helmer, Olaf (1983). *Looking Forward: A Guide to Futures Research*. Beverly Hills, California: Sage.

Hickling, A. (1982). Beyond a linear iterative process. In *Changing Design* (R. J. Talbot et al., eds.). Chichester, England: Wiley.

Hitch, C. J. (1953). Sub-optimization in operations problems. *Operations Research* 1, 87–99.

———— (1956). Comments. *Operations Research* 4, 426–430.

———— (1960). Uncertainties in operations research. *Operations Research* 8, 437–445.

Holling, C. S., ed. (1978). *Adaptive Environmental Assessment and Management*. Chichester, England: Wiley.

House, P. W., and John McLeod (1977). *Large Scale Models for Policy Evaluation*. New York: Wiley Interscience.

Hovey, H. A. (1968). *The Planning-Programming-Budgeting Approach to Government Decision-Making*. New York: Praeger.

Keeney, R. L. (1980). *Siting Energy Facilities*. New York: Academic Press.

————, and H. Raiffa (1976). *Decisions with Multiple Objectives: Preference and Value Tradeoffs*. New York: Wiley.

Kennedy, J. B. (1956). *A Display Technique for Planning*, P-965. Santa Monica, California: The Rand Corporation.

Kraemer, K. L., S. Dickhoven, S. F. Tierney, and J. L. King (1987). *Datawars: Computer Models in Federal Policy Making*. New York: Columbia University Press.

Lee, D. B., Jr. (1973). Requiem for large-scale models. *Journal of the American Institute of Planners* 39(5), 163–178.

Majone, G. (1975). The feasibility of social policies. *Policy Sciences* 6, 49–69.

Meadows, D. H., and J. M. Robinson (1985). *The Electronic Oracle: Computer Models and Social Decisions*. Chichester, England: Wiley.

Merkhofer, M. W. (1987). *Decision Science and Social Risk Management: A Comparative Evaluation of Cost-Benefit Analysis, Decisions Analysis, and Other Formal Decision-Aiding Approaches*. Dortrecht, Holland: D. Riedel.

Miser, H. J., and E. S. Quade (1985). *Handbook of Systems Analysis: Overview of Uses, Procedures, Applications, and Practice*. New York: North-Holland. Cited in the text as the *Overview*.

————, and ———— (to appear). *Handbook of Systems Analysis: Cases*. New York: North-Holland. Cited in the text as *Cases*.

Moder, J. J., and S. E. Elmaghraby, eds. (1978). *Handbook of Operations Research: Vols. 1 and 2*. New York: Van Nostrand Reinhold.

Mood, A. M. (1983). *Introduction to Policy Analysis*. New York: North-Holland.

Overview. See Miser and Quade (1985).

Patton, C. V., and D. S. Sawicki (1986). *Basic Methods of Policy Analysis and Planning*. Englewood Cliffs, New Jersey: Prentice-Hall.

Prastacos, G. (1984). Blood inventory management: An overview of theory and practice. *Management Science* 30, 777–800.

————, and E. Brodheim (1980). PBDS: A decision support system for regional blood management. *Management Science* 26, 451–463.

Quade, E. S. (1982). *Analysis for Public Decisions*, 2nd ed. New York: North-Holland.

Raiffa, Howard (1968). *Decision Analysis: Introductory Lectures on Choice under Uncertainty*. Reading, Massachusetts: Addison-Wesley.

———— (1982). *Policy Analysis: A Checklist of Concerns*, PP-82-2. Laxenburg, Austria: International Institute for Applied Systems Analysis.

Richardson, G. P., and A. L. Pugh, III (1981). *Introduction to System Dynamics Modeling with DYNAMO*. Cambridge, Massachusetts: The MIT Press.

Roberts, N., D. F. Andersen, R. M. Deal, M. S. Garet, and W. H. Shaffer (1983). *The System Dynamics Approach*. Reading, Massachusetts: Addison-Wesley.

Simon, H. A. (1961). *Administrative Behavior*. New York: Macmillan.

———— (1964). On the concept of organizational goal. *Adminstrative Science Quarterly* 9(1), 1–22.

Ståhl, Ingolf (1983). *Operational Gaming: An International Approach*. Oxford, England: Pergamon.

Strauch, R. E. (1970). Some thoughts on the use and misuse of statistical inference. *Policy Sciences* 1(1), 87–96.

Tocher, K. D. (1963). *The Art of Simulation*. Princeton, New Jersey: Van Nostrand.

von Mises, Ludwig (1966). *Human Action*, 3rd ed. Chicago: Henry Regnery.

Wagner, H. M. (1975). *Principles of Operations Research with Applications to Managerial Decisions*, 2nd ed. Englewood Cliffs, New Jersey: Prentice-Hall.

Walker, W. E., J. M. Chaiken, and E. J. Ignall, eds. (1979). *Fire Department Deployment Analysis: A Public Policy Case Study*. New York: North-Holland.

————, D. W. Singleton, and Bruce Smith (1975). *An Analysis of the Deployment of Fire-Fighting Resources in Wilmington, Delaware*, R-1566/5-HUD. Santa Monica, California: The Rand Corporation.

Chapter 3
Using Expert Judgment

Olaf Helmer

3.1. Introduction

Systems analysis is characteristically team research, with members of the team drawn from a number of disciplines. Its activities use not only the scientific and other formal techniques of the contributing disciplines but also the craft knowledge and experience that the analysts possess (see Chapter 1). Throughout the analysis, those engaged in it contribute both their expertise and their experiential judgments in ways that arise naturally from their professional backgrounds.

Bringing together these contributions may sometimes be a reasonably straightforward process, especially because, as the team works together, its members share a great deal of their individual knowledge. But concocting a consensus from the separate judgments in an *ad hoc* manner is not always sufficient. In particular, if outside experts or specialists need to be consulted, a more systematic procedure may be needed, one that does not rely on the team's shared picture: When projections are being made into an uncertain future, or into areas where scientific knowledge is limited, individual professional judgments become less reliable and need more careful scrutiny. Thus, as systems analysis deals with increasingly complex problems, there is an increasing need for methods that can marshal judgments, not only to see the issues involved more clearly, but also to enable more confidence to be placed in the judgments—and therefore the analyses that depend on them as inputs.

The purpose of this chapter is to discuss some of the ways in which judgments can be developed systematically for use in systems-analysis studies. To this end, it first explains the need for, and the objectives in, making use of expert and specialist judgments. Then it discusses ways of selecting experts and aiding their performance as individuals, after which it takes up methods of making effective use of them in groups: in particular, by operational gaming, cross-impact gaming, and computer-assisted conferencing.

3.2. The Need for Systematically Developed Judgments

The principal sources of the need in systems analysis for developing judgments by systematic means are the interdisciplinary character of the work and the future (often distant) context of its findings and recommendations. Both sources emerge from the social systems to which the work is generally directed and in which it is always embedded. Judgments beyond what are the logical consequences of established knowledge may be needed, and they must be open to critical appraisal.

Interdisciplinarity in Systems Analysis

The problems that systems analysis addresses invariably possess certain societal features, as in the cases of a firm, a city, a transportation system, or a nation. But even if they do not, as illustrated by a totally inanimate object such as a machine tool, there are inevitably certain social considerations involved in deciding how such a system ought to be designed and in judging how well it is performing. In fact, the features of the system itself and the aspirations and values of the system's users are generally so intricately interwoven that the systems analyst needs to include the users as part of the overall system being investigated. Thus, the very distinction between social and nonsocial systems becomes extremely tenuous.

Systems analysis, therefore, inevitably involves interdisciplinary considerations. Often, numerous disciplines come into play, such as physics, chemistry, engineering, economics, sociology, political science, human values, or ecology. In particular, it should be noted that the relevant disciplines always include some that belong in the realm of the social sciences.

Since systems analysis is driven by the problems it undertakes to understand and solve, rather than by a desire to extend disciplinary knowledge, it happens not infrequently that the analysis team, in spite of having representatives of several disciplines on it, must respond to the need for information, estimates, or assumptions that lie outside its domains of knowledge. Sometimes these gaps can be filled by consulting other specialists, sometimes by using judgments from members of the analysis team; but there are, too, occasions on which neither of these expedients will serve. Such cases can also arise when disciplinary knowledge is spotty and incomplete, or perhaps not yet extensive enough to cover the issue in question, or when the issue involves *trans-science*, that area of potential knowledge involving "questions that can be stated in scientific terms but that are in principle beyond the proficiency of science to answer" (Weinberg, 1972, p. 211). In these cases, as well as others in which authoritative interpretations from single individuals may be lacking, it may be desirable to arrive at judgments by systematic processes involving a number of persons with relevant authoritative expertise.

The Future-Directedness of Systems Analysis

Since systems analysis usually addresses itself to problem situations with a view to finding ways to ameliorate them, and thus to activities in the future or with future implications, it is clearly a future-directed planning activity.

It is, however, useful to distinguish two kinds of planning activities: those based on the assumption that the operating conditions at the time the plans might be implemented will be essentially the same as those presently prevailing, and those where the operating conditions are expected to be substantially different from the conditions existing at the time the plans are being made. The former case, where the relevant features of the world are assumed invariant over time, is only nominally future-directed; it is as though plans were being made for the present. Many short-term commercial operations are of this kind, and in times past, when the general pace of societal and technological change was much slower, most planning operations would have fallen into this category. The latter case, on the other hand, where no such invariance is implied, may properly be called future-oriented; it requires forecasts of future operating conditions as a prerequisite of adequate planning. The relatively new field of futures research (see, for example, Helmer, 1983, Schwarz et al., 1982, or Fowles, 1978), which has come into being in recognition of the importance of future-oriented planning, concerns itself with developing methods of forecasting and ways of using the forecasts to improve planning.

In the hard sciences we have the luxury of having at our disposal laws that, for the purposes of analysis, can be assumed to be time-invariant; those of thermodynamics, for instance, can be expected to apply in the twenty-first century as they did in the nineteenth. It may be reasonable in a rare case for a systems analysis to be only nominally future-oriented, but such a stipulation should be recognized as requiring justification.

Such justification is harder and harder to attain as the maturation of systems analysis continues. The reason for this is two-fold. First, many of the problems addressed by systems analysis concern increasingly complex situations where the time horizon is farther in the future, as illustrated by current efforts to solve the energy crisis, to plan new cities, or to develop underdeveloped countries. Second, even in the case of studies having a short time horizon, the systems analyst cannot afford to forget that we live in an age of ubiquitous future shock, in which changes in our environment occur with increasing rapidity, due mostly to technological advances and their societal implications. The time is past when such changes, even from one generation to the next, were barely noticeable and, consequently, planning "for the present" was entirely proper. Now, even in the short run, the assumption of environmental time-invariance as a basis for systems planning is usually found to be unwarranted.

Thus, it is evident that systems analysis generally is in an essential way, rather than merely nominally, future-directed.

Implications for the Use of Judgment

The interdisciplinary and future-directed character of systems analysis makes some reliance on judgment indispensable. To see that this is so, it is only necessary to reflect on the conditions under which judgment-free conclusions, applicable to real-world planning, could be drawn: the presence of data derived from observations; the existence of a coherent, well-confirmed, and objectively (that is, intersubjectively) testable theory establishing correlational and, if possible, causal connections among the observed phenomena; and the application, through extrapolation, of such a theory to the future consequences of contemplated actions. This is the pattern followed, by and large, in the case of straightforward engineering applications of the laws of physics.

These conditions are not satisfiable in the context of a typical systems analysis. First of all, well-confirmed theories are not presently available in any of the social sciences, and all we have are, at best, mildly confirmed, *ceteris paribus* regularities. Although these are an enormously valuable substitute for the kind of well-confirmed theories to which the hard-science tradition has accustomed us (Helmer and Rescher, 1959), they are generally strictly unidisciplinary, and established regularities connecting several disciplines—especially the all-important case of a social-science discipline and engineering technology—are virtually absent. Second, even if such well-confirmed interdisciplinary theories were available, their applicability to determining the future consequences of planned actions would be questionable because the *ceteris paribus* condition (the invariance of the operating environment) would generally be violated.

Consider, for example, the case of the government of a developing nation contemplating measures intended to curb the future growth of its population. (Such measures might include birth-control propaganda or tax incentives encouraging smaller families.) The obstacles to an objective, preferably theory-based, approach to planning would here be formidable. The first thing the planner would need, and could not obtain from extrapolation from past time series, would be a baseline in the form of a population forecast on the assumption of no government intervention. Routine extrapolation from past observations would, no doubt, imply something close to an exponential growth pattern; but, since common sense suggests that this could not continue indefinitely, the population projection would have to reflect influences other than those in the past, such as changes in people's values, general economic conditions, and possibly even unprecedented famine. Moreover, the effectiveness of governmental intervention, even if it could be foreseen clearly under *ceteris paribus* conditions, would be altered by future changes in the operating environment, such as (again) the economy and, especially, technological advances and institutional modifications in regard to birth control.

This example illustrates the futility of trying to proceed in a complex planning context by formal theoretical means; but it also points the way to the solution. The answer, of course, lies in the proper utilization of expert judgment. Fortunately, in the social sciences we are able to call on a vast reservoir of expertise of many kinds. Although in some cases the expert judgment that is available may be derived from the explicit application of rudimentary (and almost invariably intradisciplinary) theories, more often it may be highly intuitive in character and based on insights that, although no less reliable, may have thus far defied articulation within a theoretical framework.

For instance, in the population-policy example, experienced demographers would, on the basis of a mixture of theoretical considerations and intuitive insights, be quite capable of correcting naive exponential extrapolations and of thereby providing a more acceptable baseline forecast. Similarly, specialists in the disciplines relevant to birth-control techniques (such as social psychology, biomedicine, pharmaceutical manufacturing methods, and so on) could supply estimates of the time of occurrence of certain technological advances in this area and thereby provide inputs valuable in assessing the effects of governmental birth-control propaganda. The combined result of utilizing such cross-disciplinary expertise, although of course falling short of the comparative certainty associated with hard-science predictions, is nevertheless likely to impart to forecasts a considerable improvement in reliability over those made without the benefit of such consultation.

Emphasis on the intuitive character of much of expert judgment is not intended to imply that all judgment of this kind is necessarily purely subjective and unsupportable by objective fact or intersubjectively accepted theory. Very often, an expert serves primarily as an efficient transmitter of knowledge commonly accepted within his specialist community. But the point to be borne in mind is that an expert, through extensive experience within his field of specialization, often will be able to augment his objectively based information by purely intuitive, hence subjective, insights that will make his judgments more authoritative than those of persons without his background.

Experience clearly supports the view that in general the quality of decisions based on intuitive forecasts improves with the reliability of such forecasts, and that this reliability, in turn, depends greatly on the level of expertise of the forecasters. Therefore the need for care in selecting forecasters and in eliciting and using their forecasts cannot be overemphasized.

Justification for Using Expertise

The admission of intuitive expert judgment, however indispensable, as a source of information amounts to using subjective procedures rather than objective ones, and the question may well be raised about the epistemological justification of such an abandonment of standard scientific principles. Briefly, the answer consists of two parts: (1) systems-analytical applications are broader than those of science interpreted narrowly and (2) subjective judgment is capable of a measure of objectivization. Both of these statements require elaboration.

Systems-analytical applications are broader. The principal function of systems analysis is not to gain scientific insights but to provide assistance to planners and decisionmakers. This does not mean that the value of scientific insights is questioned or unappreciated, or that systems analysis is an "art" without regard for objectivity. As Chapter 1 brings out, to designate systems analysis as a craft is more appropriate, one in which the quality of the results—that is, the advice it is able to offer decisionmakers—is capable of objective evaluation, regardless of whether the source of the advice is objective or subjective.

> Compared to a traditional scientific investigator, there is a crucial difference in emphasis on the part of [a systems] analyst who is, of necessity, a pragmatist, interested primarily in effective control of his surroundings and only secondarily in detailed understanding of all the underlying phenomena. Thus, both the exact scientist and the [systems] analyst tend to make use of what is called a "mathematical model" of the subject matter; but, though in the case of the exact scientist such a model is apt to be part of a body of well-confirmed scientific knowledge, [a systems-analysis] model is usually of a much more tentative character. In other words, even if the current status of science provides no well-established theory for the phenomena to be dealt with by the [systems] analyst, he must nevertheless construct a model as best he can, where both the structure of the model and its numerical inputs have an *ad hoc* quality, representing merely the best insight and information that the

analyst happens to have available. As further insights accrue and more experimental data become available, the [systems] analyst has to be prepared to discard his first model and replace it with an improved one. This tentative procedure, dictated by pragmatic considerations, is thus essentially one of successive approximation. [Helmer, 1966, pp. 5–6.]

In short, the pragmatic spirit in which a systems analyst has to operate may compel him to abandon purist scientific principles and accept a certain amount of subjective insight in order to fulfill his primary, advisory mission.

Subjective judgment is capable of some objectivization. The prescientific status of systems analysis may warrant the admission of subjective factors in the form of judgmental input data. On the other hand, an objective basis for a study's output remains desirable, even though it may not always be attainable. When using the subjective opinions of experts, it is conceptually possible to introduce some objectivity at a higher level by thinking of the experts as equivalents of measuring instruments:

Epistemologically speaking, the use of an expert as an objective indicator . . . amounts to considering the expert's predictive pronouncement as an integral, intrinsic part of the subject matter, and treating his reliability as a part of the theory about the subject matter. Our information about the expert is conjoined to our other knowledge about the field, and we proceed with the application of precisely the same inductive methods that we would apply in cases where no use of expertise is made. Our "data" are supplemented by the expert's personal probability valuations . . . and our "theory" is supplemented by information regarding the performance of experts. In this manner the incorporation of expert judgment into the structure of our investigation is made subject to the same safeguards that are used to assure objectivity in other scientific investigations. The use of expertise is therefore no retreat from objectivity or reversion to a reliance on subjective taste. [Helmer and Rescher, 1959, p. 43.]

In sum, the use of expertise need not amount to a retreat from objectivity, provided that one is prepared to take the step to a metalevel where the experts can be treated as part of the subject matter under consideration. This step involves a broad effort to establish a theoretical structure in the part of applied psychology that deals with the articulation and utilization of the existing expertise potential.

Admittedly, we remain today in the same position with regard to this metadiscipline as the one in which we find ourselves in relation to most of the social sciences; that is, we rely largely on intuitive common sense rather than on well-confirmed theory when we believe that expert opinions can make a significant contribution to improved planning. When such opinions are articulated and utilized through systematic methods, such as the ones discussed later in this chapter, we still operate at the intuitive level, but have structured the opinions given so that they can be judged more readily by those who may ultimately have to rely on them.

3.3. The Objectives in Using Expert Judgment

Expert opinions are of potential utility to many aspects of systems analysis. The most important forms in which expertise enters into the analytic process can be subsumed under four headings: conceptual inventiveness, factual estimation, factual forecasts, and normative forecasts.

Conceptual Inventiveness

It should go almost without saying that progress in any kind of intellectual pursuit—be it a strictly scientific activity or an enterprise such as a systems analysis—relies heavily on the occasional intuitive spark of an imaginative mind. New discoveries or new insights into relations among observed phenomena usually come about as the result of some person's conceptual inventiveness.

There are many types of such inventiveness. Leaving aside strictly technological inventions, they include innovative concepts, such as the differential quotient or the social indicator; novel experiments, such as a test of the improved learning ability of worms after having eaten some of their own previously trained colleagues; and new strategies, such as the creation of an international food bank to stabilize agricultural prices and reduce the risk of famine.

The influx of intuition in the form of such inventiveness does not, incidentally, represent a favoring of subjective over objective influences. Clearly, even mathematics, the most objective of all disciplines, needs, as much as any other, an occasional brilliant inspiration to arrive at new discoveries. However, once a new fact or a new idea has been conjectured, no matter on how intuitive a foundation, it must be capable of objective test and confirmation by anyone. Thus, the standards of scientific objectivity are not in any way put in question by the use of intuition in this form, although test and confirmation may, as in systems analysis, often be achievable only in part (see Chapter 13).

Novel ideas are valuable but rare; unfortunately, there are no ways to generate them by mass production. It may be worthwhile, however, to give some thought to systematic ways of encouraging their invention. Informal brainstorming sessions have sometimes been productive. There are also more structured procedures for using groups of experts. Among them analysts have found both operational gaming and Delphi inquiries to be expedient tools, since they combine a structured approach with some of the stimulative characteristics of brainstorming. Both of these techniques are discussed further later in this chapter.

Factual Estimation

The informational starting point for a systems analysis is often a description of the current status of the situation plus past time series of relevant indicators. This assumes that such data are readily available from reliable sources. In practice this is often not the case, either because adequate statistical information has not been recorded (as may occur in the case of the per capita income in a developing country) or because the desired information (say, the quality of a certain educational system) is too vaguely identified to permit anything but subjective evaluation.

In such cases, the systems analyst may be tempted to restrict himself to using whatever hard data are accessible, ignoring judgmental information, however potentially important it might be to the analysis.

Instead of thus omitting vital factors from the analysis, the analyst would generally do far better to obtain judgmental inputs from qualified experts. In the examples cited, an area specialist is likely to supply estimates of per capita income that, although probably imprecise, would be accurate enough for many applications; similarly, an educational expert, given a suitable evaluation scale, could give useful comparative estimates of the quality of education, either for the same educational system at different times, or for different systems at the same time.

Thus, in the absence of hard data, expert opinion can have an important role in supplying "soft" judgmental inputs.

Factual Forecasts

The term *factual forecasts*, as used here refers to (hypothetical) forecasts of what is expected to be factual at some future date.

In practice, the explicit resort to using expertise occurs most frequently in connection with factual forecasting. One reason for this, as pointed out earlier, is that extrapolation from past observations is generally insufficient to establish a theoretical foundation for valid forecasts and that it needs to be supplemented with the intuitive, not yet theoretically articulated, insights and interpretations of appropriate specialists.

Factual forecasts fall into two important subcategories: absolute and conditional forecasts; both are properly made in probabilistic terms.

There have been many instances in recent years of absolute forecasts, particularly of technological developments, but also including societal elements. Such forecasts, each stating that the occurrence of some development by a certain date has such-and-such a probability, serve to establish the likely operating environment, or the range of likely scenarios, upon which a planner has to set his sights. They thus fulfill the indispensable function of providing the all-important background information without which a plan might be inappropriate or obsolete before it is considered for implementation.

By contrast, conditional forecasts state that if some condition were to arise, specifically if some action were taken, then the occurrence of a particular development by a certain date would have such-and-such a probability. A conditional forecast may have one of two significantly different purposes. One is the case where an absolute forecast, say of an event E, is to be made, but the forecaster knows, or senses, that the probability of E occurring is quite different depending on whether some condition C does or does not prevail. In this situation, conditional forecasts of E, given either C or non-C, can be combined into an absolute forecast of E, provided an absolute forecast of C is available. Typically, in such a case, the conditional forecasts of E and the absolute forecast of C are made by different experts, because event E and condition C may well belong to different disciplines (as in the case of forecasting the state of the economy, depending on whether or not a technological breakthrough regarding a new source of energy will occur).

The other application of a conditional forecast is concerned directly with the planning process. A planner may wish to know about the consequences of a contemplated action A. If B is one of the potential consequences, the decision on whether or not to carry out A may depend on the conditional forecasts of B, given either A or not-A. More generally, if a decisionmaker is faced with a choice among actions $A_1, A_2, \cdots, A_n$ (where doing nothing may be one of the options), the conditional probabilities of various potential consequences $B_1, B_2, \cdots, B_m$, given either A_1 or A_2 or $\cdots$ or A_n, may have a bearing on the decision and thus may have to be estimated.

It is worth noting that conditional forecasts often represent the first step toward theory formation. In the simplest case, if A and B are properties attributable to objects $i, j, \cdots$, then a series of conditional forecasts such as "if $A(i)$ then $B(i)$, with probability p_i," "if $A(j)$ then $B(j)$, with probability p_j," and so on, may lead to a general hypothesis of the form "whenever $A(x)$ then $B(x)$, with probability $p(x)$," where $p(x)$ is a function obtained by curve-fitting from the individual values $p_i, p_j, \cdots$.

Tentative mathematical models built by systems analysts are not uncommonly put together on the basis of experts' estimates of the kind just described. The resulting "theories" are rarely intended to be more than first approximations of reality, to be discarded when new insights provide better approximations, and always treated with caution appropriate to the experts' view of the accuracy of their estimates.

Normative Forecasts

Ordinary forecasts, whether absolute or conditional, represent estimates of what the future will, in fact, be like. At times they are, for emphasis, called factual forecasts.

Normative forecasts, by contrast, are opinions about what the future ought to be like, or more precisely, which among the possible futures is considered preferable. They are thus related directly to the goal- and objective-setting part of the planning process. It is in this area that the virtually total absence of a comprehensive theoretical approach is felt very keenly, and reliance on expert opinion is here even more indispensable than in the case of factual forecasting.

There are as many aspects to normative forecasting as there are to the planning process; in fact, there is little difference between the two, except perhaps that the term *planning* sometimes implies an operationally systematic and practical procedure, whereas what is designated as *normative forecasting* is less impeded by considerations of feasibility and resource constraints.

A planning process generally involves a number of logically distinct steps, which, however, are not always clearly separated from one another in practice. They are discussed in the following five paragraphs.

Assessing the future operating environment. This process amounts to estimating the characteristics of the so-called exogenous changes that will have taken place in technology and society by the time any plans are expected to be implemented—exogenous in the sense that they are not under the control of either the planner or the decisionmaker. The previous subsection on factual forecasts discussed the potential role of the expert in this process.

Setting overall goals. Such goals are usually stated in very general terms, such as "enhancing the welfare" of the organization (for example, the nation, the state, the city, the firm, the educational system) for which plans are being made. The systems analyst's task here is primarily one of interpretation, preferably in close consultation with the organization's goal setters. That is, the systems analyst has to establish what goals, if any, are clearly defined and help to cast others, possibly stated merely as vaguely expressed desires, into such a form that it becomes possible to argue meaningfully whether the attainment of specific objectives would indeed, in view of the expected future environment, contribute to reaching the stated goals.

Usually what is at first expressed vaguely as "welfare" turns out to be a multidimensional utility, and an important part of normative forecasting consists in assigning preferred weights to the individual components of this utility. Since it is rare that explicit and universally-agreed-upon exchange ratios among the components can be defined, the intuitive insight of experts may be required to estimate how different attainment mixtures with respect to the overall utility's components would compare in the eyes of various interested groups, taking into account future changes in the operating environment. It should be mentioned in passing that the utility dimensions include not only substantive differentiations (such as health, education, income, security, and the like in the case of national planning, or market share, return on investment, and the like in the case of corporate planning), but also matters of preference between lower risk and higher expectations and between short-term and long-term gain.

Specifying objectives. Once the goals have been identified clearly, objectives may sometimes be specifiable in terms of the degree to which one ought to strive to attain the various goal dimensions—that is, if these are in fact measurable so that a quantitative objective can be stated. Otherwise it is necessary first to define appropriate numerical indicators (social indicators in the case of societal planning) in terms of which objectives can then be formulated. (Examples of such objectives are: raising the literacy level to 98 percent, lowering unemployment to 4 percent, or expanding production facilities by 20 percent.) Specifying objectives may be relatively straightforward, except that both choosing appropriate indicators and selecting objectives that are likely to be feasible may require an understanding of the subtleties of the subject matter not available to anyone except a highly trained expert.

Formulating action programs. Translating objectives into action programs, under budgetary and other resource constraints, is seldom a routine matter leading to a unique prescription. To arrive at an efficient recommendation generally takes a great deal of ingenuity and imagination as well as cognizance of likely changes in operating conditions. Here again, the role of intuitive judgment on the part of trained specialists is evident.

Evaluating alternative programs comparatively. Choosing between two or more proposed action programs requires an analysis of the candidate programs with regard to cost, effectiveness, and possibly other aspects (all of which are usually multidimensional) that may affect the degree to which various stated objectives can be expected to be met. Such assessments involve all of the difficulties mentioned above in the paragraph on setting overall goals, such as the exchange ratios among utility components and—in view of the uncertainties regarding exogenous environmental changes—the risk-versus-expectation preferences. Although standard decision-theoretical analyses may be of assistance in guiding such a choice, there are inevitably so many intangible aspects present as to make some resort to intuitive insight virtually unavoidable.

In summary, the need for intuitive judgmental inputs is evident and multifold throughout many aspects of systems analysis. Indeed, in practice, whether we look upon it approvingly or with a jaundiced eye, the informal use of intuition on the part of experts is already implicit in many activities in this field. The questions remain whether anything can be gained by giving explicit recognition to the role of expertise and whether systematic methods are in existence, or can be invented, providing experts with enhanced effectiveness in contributing their intuitive insight to systems analysis. The answer to the second question, I contend, is in the affirmative, and therefore by implication, so is the answer to the first.

3.4. Methods of Using Expert Judgment

As we have seen, expert judgment enters the systems-analysis process in many forms. It plays an essential role in

inventing new concepts;

designing the structure of models, including modifying existing but inadequate models;

supplying judgmental input data, especially in the form of forecasts;

identifying specific components of stated goals and providing the moral perception necessary to apportion weights of emphasis to these components; and

offering normative visions of the future and inventing strategies for their attainment.

Once the fact has been accepted that using intuitive judgment in systems analysis is not a temporary expedient but a permanent and integral part of its normal procedures, it becomes imperative for orderly methods to be employed so that judgmental contributions can be made, and can be seen to be made, as effectively as possible.

In discussing such methods, it is useful to distinguish between those relating to individual experts and those concerned with employing groups of experts.

Individual Experts

If an analysis relies in part on judgmental inputs, the quality of its results clearly depends significantly on the quality of these inputs. Therefore, it is important for the persons selected to supply such judgments to be highly qualified and for the conditions under which their expert opinions are solicited to be as conducive as possible to proficient performance.

Selecting experts. Experts for consultation are usually selected on the basis of what may vaguely be called their reputations. This process, which in itself requires a certain amount of expertise, breaks down logically into two parts: determining which categories of expertise are needed, and selecting among the available persons those most expert in each category.

It is often obvious which types of expertise will be needed, but this is not always the case. What is sometimes overlooked is that a first-rate specialist may not always be good at taking a sufficiently general systems view, whereas a generalist with a flair for model building may lack crucial specialized information, and either or both may not have a sense of the ethical implications of alternative courses of action. Also, ascertaining precisely what kinds of specialists need to be enlisted may require a certain amount of preliminary analysis of the problem area. For example, a common failing in carrying out analyses of national economies aimed at long-range forecasting of economic indices is neglecting to consult sociological area experts in addition to economic specialists.

Regarding the second part of the selection process, the actual choice of individual experts, there is an opportunity for numerous empirical studies that may be highly rewarding. There is, first of all, the question of determining expertness on the basis of past performance (a problem of considerable logical intricacy) and of studying stability in personal expertise, that is, the correlation between future and past performance as an expert. Too, one can make an analytical approach to the questions of an expert's performance in terms of reliability and accuracy (Helmer, 1983, pp. 40–41).

Aside from such *a posteriori* assessment of performance, there is the question of *a priori* recognition of a person's qualifications in terms of performance other than specific expert pronouncements, such as years of professional experience, number of publications, and status among peers. The relation between such objective indicators and the ability to make reliable predictions is clearly susceptible to empirical investigation. For example, a 1965 experiment has suggested the tentative conclusion that an expert's self-appraisal of his relative competence in different areas of inquiry may be well enough correlated with his actual relative performance in these areas to be of significant aid in selecting experts for specific tasks (Helmer, 1983, pp. 150–152; Brown and Helmer, 1964).

In practice, the selection of experts often proceeds along the following lines:

Having listed the areas where expertise is needed, the analysis team surveys the recent literature to identify some of the more prominent persons with publications in these fields.

Some of these persons are contacted for suggestions of suitable experts who may be available to contribute to the subject matter under investigation.

Of those nominated by two or more persons a subset is invited to serve as expert consultants and, regardless of whether or not they accept, to propose additional experts for inclusion in the panel.

In selecting such a subset, it is wise to give representation, if applicable, to opposing schools of thought within the subject area; this is especially important when the experts are to serve as a panel or respondents in a Delphi inquiry (see Section 3.7).

Incidentally, although appropriate compensation of experts for their services is generally a matter of course, some Delphi studies have been conducted without offering any honoraria. Such a practice may be acceptable when the respondents are members of an organization on whose behalf the Delphi study is being conducted, or when the subject matter of the inquiry is of great and unquestioned public interest. Otherwise, even a relatively small compensation will be helpful, if only to convince the participants of the earnestness of the investigation and thus of the need to contribute a full measure of their thought and to maintain their effort through several rounds of inquiry.

Aiding the performance of an expert. In this context it is well to note the difference between generalists and specialists.

If an expert who is primarily a generalist is asked for advice, his contribution may be, or should be encouraged to be, in the form of constructing a model. Even if such a model is tentative, which is particularly likely when a firm theoretical foundation is still absent, it may serve to clarify concepts and aid communication. If a generalist's opinion is sought when a preliminary model is already in existence, it is of the utmost importance to assure him of the desirability of any modifications of the model that he regards as imperative, lest the entire effort be devoted to the "wrong problem."

If the expert is a specialist, his contribution can be aided by facilitating communication. First, the prior formulation of a suitable model serves both to communicate the problem to the specialist with clarity and to receive an answer without risk of misinterpretation. Second, the expert is greatly aided if he has ready access to relevant information that may exist elsewhere; in this regard, rapid progress in data processing is opening up new possibilities by which the swamping with irrelevancies is gradually being replaced through automated libraries with push-button availability of pertinent data. Third, in order to provide access to intuitive knowledge as yet unrecorded, an expert's performance can be enhanced most significantly by placing him in a situation where he may profitably interact with other experts in the same field or in different fields related to other aspects of the same problem. Traditionally, a simple mode of this type of communication has been in the form of round-table discussions; other modes, possibly more systematic, are discussed below.

In all these selection and aiding processes, the systems analyst should have, both professionally and experientially, some important expertise. He will have some, but not necessarily the sole, responsibility for planning the progress of the overall study and for integrating its various findings. In such a broad role he will, as an expert, be a generalist, although he may also be a specialist in some of the methodological processes he adopts.

Groups of Experts

In a context such as systems analysis, where some reliance on intuitive judgment is unavoidable, there are several reasons why it may be preferable to employ more than one expert.

First, intuition is not infallible; to speak only statistically, enlisting more than one expert should lead to significant improvement. Hence there is some justification in applying a principle enunciated by Norman Dalkey: "*n* heads are better than one." Applied in particular to numerical estimates—which include current-status assessments as well as probabilistic forecasts—the median of *n* estimates of a value can be seen to compare favorably with at least half and often substantially more than half of the original estimates.

Second, whenever the use of imagination is called upon, the increase in the number of ideas generated from several experts, as compared to the number from a single respondent, affords a much better chance that a really good idea will not be omitted from consideration. Such ideas may concern the pertinence of past observations, the design of experiments, the structures of models to be built, the descriptions of desirable futures, and the invention of methods and strategies for their attainment.

These two reasons for using groups of experts rather than individuals apply even in cases where the required idea or information lies strictly within a single intellectual discipline.

Third, systems analysis is interdisciplinary, which makes it difficult if not impossible for the intuitive insights of a single person to do justice to all relevant aspects of such an inquiry. Hence, it is important to place proper emphasis on the need, not only to tap the expertise of a number of specialists from different disciplines, but also to seek suitable means of interdisciplinary communication that mesh the contributions from different fields that do not normally interact effectively with one another.

The traditional—and simplest—way of drawing out a group of experts is to have a round-table discussion in which ideas can be exchanged freely. The stimulus of open conversation is particularly helpful when the objective is to generate innovative concepts; this objective can be enhanced by declaring explicitly that the conference is a brainstorming session in which criticism of ideas—even farfetched ones—is to be suppressed, at least initially.

On the other hand, if the objective is factual—such as assessing the current situation, making conditional forecasts, or evaluating the feasibility of proposed action plans—a free, face-to-face discussion may not be the best mode of operation; a more structured procedure is generally preferable.

There are a number of reasons for this preference. In an open discussion of ideas among several persons, the noise level (not only in the literal sense, but more importantly also in the intellectual sense) may outweigh the actual information passed along, particularly in view of the known difficulties of cross-disciplinary communication. There are likely to be all kinds of psychological undercurrents that may affect the quality of the end product: the undue influence of a domineering personality, the deference to authority (real or supposed), the reluctance to contradict opinions voiced by persons of higher status (because of age or organizational rank), the bandwagon effect of majority opinion and, conversely, an unwillingness to abandon one's previously taken public stand on some issue.

Taken together, these potentially detrimental effects of an unstructured meeting may negate the undoubted assets of the intellectual stimulation it provides.

Most of the remainder of this chapter will be devoted to a discussion of two methodological approaches that seek to find a way out of this problem: operational gaming and the Delphi technique. Both replace open discussion with a controlled form of dialogue, both discourage the kind of psychological undercurrents just enumerated, and both lend themselves well to interdisciplinary discourse.

3.5. Operational Gaming

The term *operational gaming* refers to the use of a particular kind of simulation model that includes role-playing by persons who participate in the operation of the model.

This section first discusses simulation models and then turns to interactive simulation and how to elicit expert opinions by means of gaming procedures.

Simulation Models

The complexity of the real world is so great that any attempt to construct a model of some segment of it involves, of necessity, some simplification in the sense of abstracting from many features of reality that have only marginal relevance and incorporating in the model only the few that seem to be indispensable to the situation of interest to the model builder. Such simplification is carried out in the hope that the simplified picture of the real world will be sufficiently transparent to facilitate analysis and understanding, without being so overly simplistic as to lead to erroneous conclusions.

In addition to simplification by abstraction, model building may involve conceptual transference. Instead of the situation being described directly (itself a move away from complex reality), it is often the case that each element of the real situation is simulated by a mathematical entity or a physical object, with its relevant properties and relations to other elements mirrored by corresponding simulative properties and relations. For example, any geographical map may be considered a physical model of some sector of the world; the planetary system can be simulated mathematically by a set of mass points moving according to Kepler's laws. A model that, in addition to simplifying assumptions, involves such transference is called a *simulation model*: Any results obtained from an analysis of a simulation model,

> to the extent that it truly simulates the real world, can then later be translated back into the corresponding statements about the latter. This injection of a [simulation] model has the advantage that it admits of what may be called "pseudo-experimentation" ("pseudo" because the experiments are carried out in the model, not in reality). . . . Pseudo-experimentation is nothing but the systematic use of the classical idea of a hypothetical experiment; it is applied when true experimentation is too costly, or physically or morally impossible, or . . . when the real-world situation is too complex

to be analyzed directly. (Helmer and Rescher, 1959, p. 48.)

Past experience with simulation models, especially when they are multidisciplinary, suggests that they can be highly instrumental in motivating the participating research personnel to communicate effectively with one another, to learn more about the subject matter by viewing it through the eyes of persons with backgrounds and skills different from their own, and thereby, above all, to acquire an integrated overview of the problem area. This catalytic effect of a simulation model is associated, not only with the employment of the completed model, but equally with the process of constructing it. In fact, the two activities usually go hand in hand. The application of the model almost invariably suggests amendments, so that it is not uncommon to have an alternation of construction and simulation.

The heuristic effect of collaborating on the construction and use of a simulation model is particularly powerful when the simulation takes the form of an operational game, where the participants act out the roles of decision- and policymaking entities (individuals or organizations). By being exposed within a simulated environment to a conflict situation involving an intelligent opposition, the "player" is compelled, no matter how narrow his specialty, to consider many aspects of the scene that might not normally weigh heavily in his mind when he works in isolation. Thus, the game laboratory induces an integrating effect amounting to a systems approach to the subject matter.

There are many kinds of simulation games, which will be discussed briefly in the next subsection. (This is a first introduction to their uses specifically for aiding expert judgment. Chapter 4 discusses games more fully, although with definitional restrictions that are not insisted on here; in particular, this chapter accepts "one-player" simulations as games, whereas Chapter 4 does not. As Section 4.2 points out, it should be noted that games terminology has not been standardized, and care must be taken to examine definitions before taking statements from one writer and translating them into the language of another.)

Interactive Simulation

The terms *game* and *gaming* tend to convey a flavor of frivolity, which is especially inappropriate when the subject, as is so often the case, involves matters of grave human concern. It is for this reason that the alternative designation of *interactive simulation* is coming into more frequent use. In this context, the word *interactive* refers, not only to the obvious interaction among several players participating in a game, but also (and, in fact, particularly) to the interaction between a player and the game model.

Interactive simulations, or games, can be classified from several points of view, of which only a few are discussed here. A first, obvious classification could be by subject matter, of which there is no limit. The areas to which gaming has been thus far most frequently applied include military, corporate, and economic problems.

Second, games may be distinguished according to the number of players. (It should be understood, incidentally, that when we speak, say, of a three-player game, this means that the game involves three separate decisionmaking entities that are being simulated, where the actual simulation of each of these may well be carried out by a team rather than by an individual.) Of special importance among these are the one-player simulations, which are planning games in which a single player (or playing team) plays "against nature." Thus, there is, in these cases, no live opposition, and the problem is only one of coping with the vicissitudes that have been designed into the game model, which are often controlled by random distributions. The one-player game against an automaton is the most controllable type of game, thus particularly valuable for research purposes (Bowen, 1978, 1986).

We note in passing that, regardless of the number of players, a player's assignment in an operational game may be either to optimize something or to simulate someone. In the first case, he is to attempt, within the constraints of the game rules, to maximize a personal score (in game-theoretical terminology, his "payoff"). This tends to put the verisimilitude of the game model, which after all is only a simplified and conceptualized form of reality, to a severe test and to suggest the need for amendments in the underlying assumptions. The second mode in which a player may function—as a simulant—is more likely to utilize his expertise properly; for, in this role, he is required to contribute constructively to the developing analysis by feeding in such simulated decisions as, in his estimate, will most faithfully reflect the decisions that his actual counterpart would make in the corresponding real situation.

A third classification might divide games on the basis of the rules governing them. A *rigid-rule game* is one whose rules are complete enough that at each stage the strategic options at the players' disposal are wholly specified, and also that the consequences resulting from the joint exercise of these options are entirely determined. This means that the model represents a complete miniature theory of the phenomena simulated in the game. In an open or *free-form game*, by contrast, at least one of these factors is not completely determined by the rules, in which case it is the responsibility of an umpiring staff to allow or disallow proposed strategies and to assess their consequences. Clearly, umpiring in this sense represents yet another important device for using expertise within the framework of an interactive-simulation model.

Finally, a fourth classification of games is according to their intended purpose. Gaming has been variously described as

a teaching device (providing conceptual understanding and substantive indoctrination),

a training device (for coping with specific problem situations),

an experimental approach (for preliminary sifting and comparative evaluation of alternative strategies),

a research tool (for theory development, sensitivity analysis, and the like),

a policy-formulation method (serving to identify overall goals, both by normative intent and by practical feasibility),

a planning method (providing a means of constructing scenarios and action plans), and a heuristic technique (aiding the intuition and stimulating the inventive imagination).

In practice, these distinctions of purpose usually become blurred, and in any case the intended applications tend to overlap. To the systems analyst, the second-named purpose is likely to be of least interest, whereas the last-named corresponds most closely to his needs. Indeed, when used in heuristic fashion, the gaming activity—both playing the game and, even more importantly, designing it in the first place—serves as a self-educational device for the planner or the analyst acting as a planning advisor. Such heuristic gaming represents a trial-by-error process that combines some of the features of the other six uses of gaming and involves a complex feedback relationship between invention and testing. The inventive portion of the effort consists in selecting appropriate elements to be included in the game model and promising strategies to be tested for their closeness to optimality. Preliminary tests, through actual gaming, may yield information on relative sensitivities, thus suggesting the elimination, or addition, of conceptual elements from, or to, the model as well as the formulation of other strategies to be subjected to examination.

Using Expert Judgment through Gaming

The preceding discussion has evidenced the role of gaming as a medium for eliciting the ideas and opinions of experts. In fact, the process of interactive simulation constitutes an ordered structure that encourages cross-disciplinary communication, stimulates both the production of ideas and their critical review, furthers a clearer intuitive understanding of the subject matter, and thereby provides the basis for at least tentative theory formation in areas where no established theory preexists.

For reemphasis it should be stated once more that there is no guarantee that expert opinions will in every instance, or even most of the time, be correct. Thus the results of gaming, since it relies heavily on judgmental inputs, must always be taken with a grain of salt; yet it can be expected that the lessons learned from gaming activities are, by and large, conducive to more effective planning. Fortunately, experience confirms this, and it is fair to say that, as a heuristic aid to intuition and a tool for interdisciplinary cooperation, operational gaming appears to have no equal at this time.

3.6. Cross-Impact Gaming

A special kind of interactive simulation known as *cross-impact gaming* deserves discussion here because it strongly encourages its players to employ a systems-analysis approach, and because it lends itself well to constructing scenarios systematically.

The Cross-Impact Concept

The cross-impact idea first arose in the context of conducting forecasting studies; it was an effort to increase the degree of sophistication in estimating the probabilities of occurrence of potential future events, such as scientific or technological breakthroughs, political events, or acts of legislation (Helmer, 1976). The thought was that a set of potential events $E_1, E_2, \cdots, E_n$, which might be relevant to an area of inquiry, would not necessarily occur independently, and hence ought to be considered in relation to each other.

Specifically, the occurrence of one of the events E_i might affect the probability of the subsequent occurrence of the event E_j. These mutual effects can be displayed in a cross-impact matrix (Figure 3.1), where the cell at the intersection of the E_i row and the E_j column carries information x_{ij} about how the probability of occurrence of E_j would be affected if E_i were to take place. (There is an implied balancing effect in the opposite direction in case E_i does not occur.)

The cross-impact concept has been extended from an events-only model to one that, in addition, includes a representation of trends, such as fluctuations in population, gross national product (GNP), air pollution, quality of life, and so on. (It may be noted that many traditional models, such as in econometrics or system dynamics, are of the trends-only type.) In the corresponding cross-impact matrix (see Figure 3.2), the events-on-trends impacts refer to raising or lowering trend levels as the result either of an event or its nonoccurrence. Conversely, a deviation of a trend from its anticipated course may cause event probabilities to change or other trend levels to rise or fall; the corresponding entries in the cross-impact matrix reflect these effects numerically.

To construct a cross-impact model requires that these items be selected:

The time period to be covered. In a planning context, this is usually the time from the present to some future time horizon.

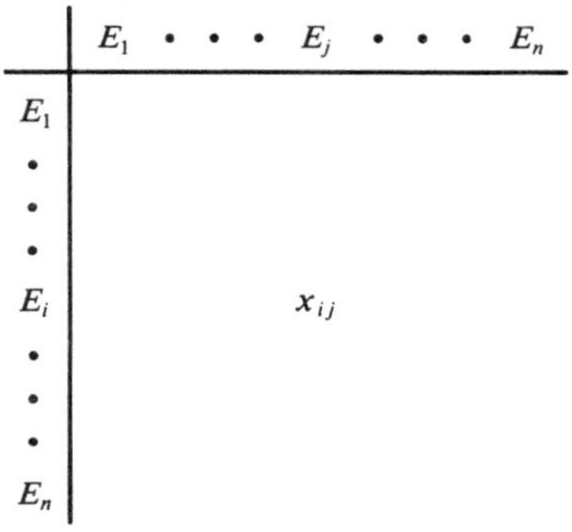

Figure 3.1. The cross-impact matrix for a set of n events $E_1, \cdots, E_n$.

Figure 3.2. The cross-impact matrix for a set of events $E_1, \cdots, E_n$ and trends $T_1, \cdots, T_m$.

The number of *scenes* into which the total time period is to be subdivided. The scene length indicates the fineness of the temporal grid, that is, the frequency with which the developing scenario is to be monitored.

The potential future events that have a more than negligible chance of occurring and that would, if they occurred, have an important effect on the situation under consideration.

The probabilities of occurrence, within each scene, of these selected events.

The trends that ought to be monitored, either because they indicate the system's performance (as, for example, social indicators), or because any unexpected deviations of their values from their anticipated courses would affect the situation significantly.

The anticipated values of these trends at the beginning of each scene.

The volatility v of each trend, that is, the precision (or lack of it) to be attached to scene-by-scene trend forecasts. For example, if a trend's value in some scene is t and its anticipated value for the next scene is t', then the interval from $t' - v$ to $t' + v$ represents the 50 percent confidence interval. Thus, the volatility concept serves to incorporate uncertainties regarding future trend values. (Incidentally, it also acts as a convenient measure for the magnitudes of cross impacts on the trends.)[1]

The cross-impact matrix, that is, a square array as shown in Figure 3.2

---

[1] For further details on this and other computational aspects of cross-impact analysis, see Helmer (1977). What is here called the volatility v is there called the surprise threshold s.

of coefficients representing the cross impacts among the selected events and trends.

With all of this information in hand, the analyst runs the model in this way: For each event, he decides (by a standard Monte Carlo procedure) whether or not, according to its stipulated probability, it occurs in Scene 1; then he adjusts the probabilities of the other events in subsequent scenes, as well as the future trend values, in accordance with the coefficients in the cross-impact matrix; for each trend, the initial values for Scene 2 are adjusted further by adding to them random deviates drawn from normal distributions having the appropriate volatilities. Proceeding to Scene 2, the analyst again decides randomly which events occur, and observes the deviations of trend values from their anticipated values at the beginning of Scene 2. He adjusts the event probabilities and trend values for subsequent scenes according to the cross impacts caused by event occurrences and nonoccurrences and trend-value deviations, repeating the procedure for all remaining scenes.

The result of such a run is called a *scenario*,[2] that is, a record by scenes of event occurrences and trend-value adjustments. Repeated runs produce different scenarios owing to the randomness of event occurrences and trend-value fluctuations. For a large number of runs, however, the averages of the number of occurrences of a given event and of the trend values should closely replicate the corresponding input values.

This model lends itself well to sensitivity studies: any one of the inputs (a probability, a trend value, a cross impact, a volatility) can be changed; a rerun of the model then allows one to observe how sensitively the resulting scenario depends on the change. In particular, sensitivities to event and trend changes yield important clues to appropriate strategy selection when the model is used as part of a game, as discussed below.

Action Impacts

To convert a cross-impact model to an interactive, game-playing mode, it is necessary to specify the following additional inputs:

The identities of the actors (that is, of the decisionmaking agencies being simulated).

The actions that may be taken, by which actors, and under which conditions, if any (for example, what prior technological breakthrough is a prerequisite). Note that actions can often be taken at various

---

[2] In systems analysis the set of conditions defining the context or environment under which the reality is assumed to operate is usually called a scenario without regard to how the set was generated (see Section 9.4).

levels of intensity (for example, constructing n new plants, or increasing minimum wages by m monetary units).

The impacts of actions (as functions of their intensities, where appropriate) on event probabilities and trend values.

An overall budget for each scene and each player.

The cost of each action (as a function of its intensity, where appropriate).

When the cross-impact model is used in a game, the players are informed at the end of each scene about the current state of affairs, as depicted in the events and trends scenario up to that point. They are then given an opportunity to decide on further actions within their budgetary and other limitations—and the game continues. Such play yields a scenario, which now, in addition to event occurrences and trend-value adjustments, includes a scene-by-scene record of player actions.

In view of the budgetary constraints imposed on each player, the need for his action plan to fare adequately in the face of random vicissitudes and potential counteractions by other players, it is virtually mandatory for each player to consider his options in a systems sense. Thus, the structure of the model and the context of interaction with other players provide an excellent opportunity for an expert, simulating the role of a real-world actor, to apply his intuitive judgment and reconsider his opinions, if necessary, in the light of unforeseen interventions by nature or by other actors.

This is by no means to say that cross-impact simulation is the systems-analyst's panacea, or even that it is an impeccable tool for extracting judgmental information. In its present, highly imperfect state, it is still encumbered with too many conceptual simplifications and procedural difficulties to make such a claim.[3] It has nevertheless been explained here briefly because it represents a promising prototype for a style of modeling that deserves further exploration in view of its interdisciplinary flexibility, its encouragement of thinking in systems terms, and its attention to the effects of uncertainty on the planning process. (See also Helmer, 1983, pp. 159–193.)

---

[3] Work in cross-impact modeling so far has typically involved such simplifications and limitations as these: cross impacts caused by trend deviations proportional to the sizes of these deviations; constant scene lengths; cross-impact coefficients constant over time; interactions considered only among pairs of developments; inadequacies in the mathematical treatment of the superimposition of multiple impacts, as a result of which the required balance between the effects of event occurrences and nonoccurrences is only approximately satisfied; and the absence of safeguards against double counting cross impacts (for example, when A affects B, which in turn affects C, while A is also assumed to affect C directly).

Scenarios

The term *scenario*, as used here, refers to a potential future sequence of developments, a "history of the future," as it were. In the terminology introduced in connection with cross-impact analysis, a scenario consists of a description of event occurrences and of trend deviations from their anticipated courses. Scenarios, in this sense, may constitute merely the possible backdrops against which plans for the future have to be composed, or they may include interventive actions. (In fact, whether a prospective occurrence should be classified as an event or an action may often depend on the identity of the decisionmaking agency on whose behalf a planning model is being built; hence a legislative act should be interpreted either as an action or an event, depending on whether or not the legislature is among the game's players.)

Occasionally the term *scenario* has also been applied to normative prescriptions for desirable future developments or to descriptions of the world at some particular moment in time (for a discussion of these and other uses of the term, see Chapter 9). For clarity, however, this term will here be reserved for what we have called factual forecasts of sequential developments.

In our present context of exploring judgmental techniques, the notion of scenarios is of some importance because of their role in the planning process. Plans for the future clearly need to stand up under a variety of likely contingencies, such as those represented by different scenarios. Although it is not feasible to conduct a complete survey of all possible future scenarios and test contemplated action programs against each of them, an accepted procedure has been to select a few scenarios, thought to be reasonably representative of all possible contingencies, and to confine oneself to examining how well the plans would do in each of these selected cases.

The role of expert judgment here is twofold: to select, in the first place, scenarios that are reasonably representative of all possibilities; and to evaluate alternative action programs with respect to these representative scenarios.

In supplying this kind of judgment, an expert may find a cross-impact model to be of considerable help. For one thing, each basic run (that is, a run without player interventions) produces a scenario, and a number of such runs represent a random sample of scenarios from which an expert may be able to select intuitively a set of representative scenarios. This is perhaps most easily done by the simple procedure of omitting scenarios that are close to others in the sample. An alternative, more systematic procedure consists in determining, for each event included in the model, (1) its overall probability p of occurring within the planning period under consideration and (2) its overall impact i on trends and other events, where

i can be defined as the sum of the coefficients in the cross-impact matrix representing the event's impacts on events and trends. For these two parameters a planner would be interested in the dichotomy created by events for which p is close to $\frac{1}{2}$ (because of the marginal uncertainty attached to the occurrence or nonoccurrence of such events) and for which i is large. One possible criterion is to compute for each event a score

$$s = (1 - |1 - 2p|) \cdot i,$$

which has the required property of being the largest when $p = \frac{1}{2}$ and i is maximal, and to order the events according to decreasing values of s. If the events taken in this order are E_1, E_2, E_3, $\cdots$, then a set of four "representative" scenarios might be selected by choosing the first two in this series of events and defining scenarios generated by new events $E_1 + E_2, E_1 + \overline{E}_2, \overline{E}_1 + E_2, \overline{E}_1 + \overline{E}_2$, where the plus sign indicates the logical union and the overbar the nonoccurrence. Similarly, if one wishes to have eight "representative" scenarios and is willing to accept the greater complexity associated with the comparative analysis of strategies with respect to that many reference scenarios, the four cases can be subdivided by distinguishing subcases characterized by E_3 and $\overline{E}_3$. And so on for larger numbers of scenarios.

As for the actual evaluation of alternative action programs or, more generally, of alternative strategies for selecting action programs, the availability of a cross-impact model again could be an aid in systematizing and simplifying the required work. Taking as an illustration the case of the four representative scenarios just described, the procedure for each strategy would be to stipulate, in turn, that $E_1 + E_2, E_1 + \overline{E}_2, \overline{E}_1 + E_2$, and $\overline{E}_1 + \overline{E}_2$ be satisfied, leaving to chance the occurrence of other events as well as the fluctuations of trends, and carrying out the given strategy in sufficiently many runs to obtain a clear evaluation of how well it would fare, on the average, in each of the four alternative scenarios.

3.7. Delphi

The Delphi technique, mentioned earlier as another approach to eliciting opinions from a group of experts, is not an alternative to operational gaming. It can serve as an integrative device whenever quantitative judgments from more than one expert are sought; as such, it can be employed in a gaming context as well as in the course of inquiries unrelated to such interactive simulation.

The prime motivation for inventing the Delphi technique was the desire to avoid the psychological problems, noted earlier, of face-to-face discussions that can be detrimental to establishing a reliable group consensus. The method was first designed by Dalkey and Helmer to be used in a military estimation problem that arose at The Rand Corporation in the early 1950s (Dalkey and Helmer, 1963). Later, in 1963, it was used in a major technological forecasting study (Gordon and Helmer, 1964). Since then it has been employed in several thousand cases all over the world, covering such widely divergent subjects as educational reform, long-range corporate planning, the future of medicine, assessments regarding the quality of life, and public-sector planning at the highest levels (for a view of this broad experience, see Linstone and Turoff, 1975).

The Structure of a Delphi Inquiry

A Delphi inquiry proceeds by using a series of successive questionnaires, where, in each questionnaire after the first, the respondents receive feedback information about the outcome of the preceding round (without learning, however, which opinion was contributed by which particular respondent). Some of the questions directed to the participating experts may, for instance, inquire into the reasons for previously expressed opinions; and a collection of such reasons may then be presented to each respondent in the group, together with an invitation to reconsider and possibly revise the earlier estimate or, conversely, to state counterarguments explaining why the reasons presented are unconvincing. This inquiry into the reasons for stated opinions and the subsequent feedback of the reasons adduced by others constitutes the elements of what may be thought of as an anonymous debate serving to stimulate the experts into considering factors they might inadvertently have overlooked, and to get them to give due weight to considerations they may at first have been inclined to dismiss as unimportant.

The procedure most commonly applied involves four successive rounds: a first estimate; a second, possibly revised estimate, accompanied by reasons for deviating from the first-round median; a reestimate in the light of these reasons, accompanied by counterarguments where appropriate; and a final estimate made in the light of the counterarguments. The median of the fourth-round responses is then accepted as the nearest thing to a consensus among the respondents. The residual divergence of the opinions is usually indicated by stating the interquartile range of the fourth-round responses. This range is the length of the interval containing the middle half of the responses.

Statistically speaking, successive rounds of a Delphi inquiry exhibit some convergence of views, indicated by a shrinkage of the interquartile range from each round to the next. To be sure, such convergence does not always occur. Even in cases where the views of the experts do not converge but, as sometimes happens, cluster around two distinct values, the Delphi inquiry nevertheless serves the purpose of identifying the reasons for disagreement among the experts and of ascertaining whether the nature of such disagreement is factual or purely semantic.

It goes without saying that convergence *per se*, even when it occurs, is not sufficient to validate the method, because what matters is convergence toward the correct value. A number of experiments have been conducted to ascertain whether, on the average, such convergence in the right direction does indeed take place. The results have generally confirmed that this is the case (see, for instance, Dalkey, 1969), although admittedly the amount of such experimentation to date has been insufficient to satisfy the severest critics (see, for instance, Sackman, 1975).

Of course, the quality of the results of a Delphi inquiry depends largely on the expertise of the respondents, and the proper selection of experts is, as always, of great importance. It should be noted, however, that the Delphi process as such ought to be regarded as a communication device among a panel of experts, regardless of how they have been selected. In particular, Delphi is not an opinion poll, and the criticism that it does not include random sampling from ''the population of experts'' (a rather dubious notion at best) simply does not apply.

Another caveat, to be generally observed in selecting experts and especially pertinent to the composition of a Delphi panel, applies to the case where the subject of inquiry concerns the future of some societal group (say, a racial minority, the elderly, a group of political radicals, or the citizens of a developing country). In such a case, it is claimed by some, care must be taken to have members of that group properly (and perhaps predominantly) included as respondents, on the ground that they are the ''experts'' with respect to that group. The error in this argument is that, although members of a group certainly are experts as to the group's present dissatisfactions and future aspirations, their emotional involvement is likely to bias their view of the future and cause them to substitute wishful thinking for objective forecasts. More reliable forecasting judgments are obtainable in this case from experts once removed, that is, from social psychologists and other specialists who have made a detailed study of the circumstances, the potentialities, and the desires of the group in question.

Applications

The Delphi technique can be employed in any situation where expert judgment is required. Its most straightforward application is to cases of numerical estimation, such as forecasts (of the future value of a trend, of the probability of occurrence of an event within a given time period, or of the time by which its occurrence attains a given probability), estimates of past or current indicator values, or even normative determinations (such as desired expenditure levels, pollution reduction goals, and the like).

Even if, say, a forecast is at first not formulated in numerical terms, it can usually be rephrased so as to permit quantitative estimation through a Delphi procedure. For instance, if it is a question of forecasting the identity of the next president of a country, the Delphi inquiry proper can be preceded by an opening round in which the respondents are asked to list all candidates who, in their opinion, have a nonnegligible chance of being elected; then, in subsequent rounds, numerical probability estimates for the election of each candidate on the combined list can be elicited in standard Delphi fashion. Or, similarly, if the question is one of choosing among alternative action plans, it is generally not difficult to quantify the inquiry by defining one or more measures of merit for each alternative and asking for assessments of merit, which in turn will lead to determining the preferred choice.

Simulation also offers opportunities for using the Delphi technique. Since expert opinion enters the gaming activity at several levels (designing the game structure, estimating judgmental inputs, choosing moves and strategies, evaluating outcomes, and amending the game in order to improve it as play proceeds), there is ample opportunity for group opinions to be applied, as may have been elicited by Delphi inquiries. An abbreviated, conference version of the technique, known as mini-Delphi, is particularly useful in such a context. It consists in (1) each participant independently writing down his choice (or estimate) of the required number; (2) the set of all choices being revealed to the entire group, without attributing the opinions to their authors; (3) a brief discussion of divergences; (4) each participant again independently writing down his (possibly revised) choice; and (5) the median of these second choices being accepted as the group's decision about what the number should be. This procedure abandons complete anonymity, but preserves voting independence, as well as determining the final consensus by computing the median.

3.8. Computer-Assisted Conferencing

The evolution of computer networks extending both nationally and internationally is bringing into existence a powerful new tool that can benefit many aspects of society. Computer conferencing is one of the far-reaching applications that offers a new way to utilize expertise. Some early experiments have been carried out in which each member of a group of experts is equipped with his own computer terminal, and communication among them, or between them and the conductor of the conferencing exercise, is carried out via a computer network.

There are still some problems regarding the man-computer interface, since many potential users still feel uncomfortable about certain facets of the process of interacting with a computer terminal. Moreover, the orderly conduct of a conference requires a set of procedural rules, such as Robert's *Rules of Order*. The latter were designed to apply to face-to-face situations, and considerable changes have to be made in order to adapt them to computer conferencing. These difficulties will no doubt soon be resolved satisfactorily, and once they have been, a number of applications of this new mode of utilizing experts will become available.

These applications, of course, include both Delphi and interactive simulation, but are not confined to these two modes.

In the case of Delphi, accessibility to experts via a computer network (sometimes referred to as a *D-net* when used for this purpose) immensely shortens the time required for conducting such an inquiry. The usual time span of six to nine months needed to carry out a Delphi study through mailed questionnaires can thus be reduced to hours or, at most, a few days. Moreover, it can easily be foreseen that, once D-nets become an accepted institution, they will serve to bring about a more systematic utilization of what is sometimes called "the advice community," that is, the collection of specialists and scholars the world over whose counsel is regularly sought by high-level decisionmakers in both the public and private sectors. D-nets, together with well-categorized rosters of experts willing to act as occasional (and presumably adequately compensated) respondents, might thus assume the nature of a new kind of public utility, through which appropriate expertise could be tapped rapidly in order to improve the decisionmaking process.

As for simulation gaming, it is clearly possible to conduct games in which the players are located remotely from one another and enter their moves through a computer network. This mode of computer conferencing may be particularly useful when it is a question of simulating the socio-economic development of a developing country. A gaming laboratory set up in a developed country, possibly under the auspices of an international organization, might construct suitable simulation games, tapping the expertise of area specialists to obtain the required judgmental inputs, and then play such a game with the remote participation, via computer link, of planners in the developing countries for whom such planning games were designed.

A computer-conferencing facility could be used more generally for conducting structured conferences of experts, where the conference chairman might compose an agenda by suitably intermingling various types of conference modules, some of which might consist of Delphi inquiries, some of gaming sessions, and some of still other modes of controlled interaction. Inputs could be kept anonymous when expedient, and participation in a particular submodule could be restricted to the conferees judged (by themselves or by their peers) most competent to contribute to that phase of the proceedings. Throughout such an exercise the computer could be used, not just as a communication medium, but as a reference source, data processor, and record keeper, thus greatly facilitating the performance and enhancing the productivity of the human participants.

Although this description of the organization of many brains, both human and artificial, into a single operation may smack of the notion of a superbrain of the kind favored by science-fiction writers, it is well to divest such a concept of its romantic aspects and to take sober cognizance of the current trends that in a very few years will noticeably advance in this direction. By anticipating developments of this kind and by pursuing techniques and procedures that will aid the exploitation of such new capabilities, the discipline of systems analysis may be able to derive substantial benefits.

It may be worth pointing out that computer-assisted conferencing, while offering the important possibility of collaboration at a distance, may also have distinct advantages when applied in situations where the participants are colocated. For instance, staff members of a firm, a government agency, or a research institution may find it convenient to participate intermittently in an inquiry or a gaming effort without having to leave their offices and without being exposed to the psychological disadvantages enumerated before of face-to-face conferences.

Similarly, computer conferencing offers the potential of extending—or even replacing—some of the activities of professional society meetings. It is possible to imagine rededicating them to their original purposes of exchanging ideas and information and engaging in professional collaboration by organizing them in the form of highly prestructured computer conferences. That this form of interaction would no longer require travel to a common meeting place might, under many circumstances, be regarded as an additional benefit.

References

Bowen, K. C. (1978). *Research Games*. London: Taylor and Francis.

——— (1986). Games in research. An invited paper at a conference on "Spel som Metod," sponsored by the Swenska Operationsanalysföreningens Årlige, 6 November 1986.

Brown, B., and O. Helmer (1964). *Improving the Reliability of Estimates Obtained from a Consensus of Experts*, P-2986. Santa Monica, California: The Rand Corporation.

Dalkey, N. (1969). An experimental study of group opinion: The Delphic method. *Futures* 1(5), 408–426.

———, and O. Helmer (1963). An experimental application of the Delphi method to the use of experts. *Management Science* 9, 458–467. Also appears in Helmer (1983), pp. 134–145.

Fowles, J., ed. (1978). *Handbook of Futures Research*. Westport, Connecticut: Greenwood Press.

Gordon, T., and O. Helmer (1964). *Report on a Long-Range Forecasting Study*, P-2982. Santa Monica, California: The Rand Corporation. Also appears as an Appendix in Helmer (1966); a portion appears in Helmer (1983), pp. 220–251.

Helmer, O. (1966). *Social Technology*. New York: Basic Books.

——— (1976). Interdisciplinary modeling. In *World Modeling: A Dialogue* (C. W. Churchman and R. O. Mason, eds.). Amsterdam: North-Holland, pp. 73–80.

——— (1977). Problems in futures research. *Futures* 9(1), 17–31.

——— (1983). *Looking Forward: A Guide to Futures Research*. Beverly Hills, California: Sage.

———, and N. Rescher (1959). On the epistemology of the inexact sciences. *Management Science* 6, 25–52. Also appears in Helmer (1983), pp. 25–49.

Linstone, H. A., and M. Turoff, eds. (1975). *The Delphi Method: Techniques and Applications*. Reading, Massachusetts: Addison-Wesley.

Sackman, H. (1975). *Delphi Critique: Expert Opinion, Forecasting, and Group Process*. Lexington, Massachusetts: D. C. Heath.

Schwarz, B., U. Svedin, and B. Wittrock (1982). *Methods in Futures Studies: Problems and Applications*. Boulder, Colorado: Westview Press.

Weinberg, A. M. (1972). Science and trans-science. *Science* 177 (21 July 1972), 211.

Chapter 4
Using Operational Gaming

Ingolf Ståhl

4.1. Introduction

How to model human behavior in a problem situation is a major issue in
systems analysis. Among other difficulties, there is often a conflict between achieving realism and a formal logical structure. Operational gaming is a modeling approach that tries to bridge the gap between human
intuitive behavior and mathematical rigor; within the framework of carefully defined rules of a game it uses human players to replicate each of
the key actors in the problem situation.

Operational gaming, however, has some properties—problematic in
some situations—that should be well understood before making a decision
to use it as part of an analysis program: the scale of the activity and its
cost, the variability of the results, and the problems of generalizing on
the basis of the outcomes.

The purpose of this chapter is to provide an overview of operational
gaming in order to assist analysts in deciding whether or not to employ
this approach. In serving this purpose, the chapter also gives an overview
of how analysts can go about designing certain types of games, organizing
their play, and interpreting their results. In the process, this discussion
takes up many of the issues and decisions that analysts must face as they
consider operational gaming as an approach in systems analysis.

To set the stage for what will follow, the chapter first describes gaming
and the major types of games, together with clients for games and how
they can use games to obtain information for the decision process. After
an overview of the main elements in the gaming research process, the
chapter then takes up each element in more detail: the question of whether
or not to use a game, whether an already existing game can be adapted

for use or a new one needs to be constructed, developing an operational game and testing it, and using it as a tool of analysis. The chapter then closes with discussions of five important supporting subjects: gaming for testing models, the size of the game, developing robot players, using interactive man-computer dialogues for determining game parameters, and how to proceed toward gaming. Finally, in view of the special need to have a focus on the portion of the large literature on gaming that is related to operational gaming, the chapter concludes with additional references providing listings of games and bibliographies of the relevant literature.

4.2. What Is Gaming?

Before proceeding to discuss operational gaming, it is important to describe what this term and its related concepts mean in this chapter. This step is needed because no standardized terminology has as yet emerged in either literature or practice. Not only is the term *gaming* variously interpreted, but also more specialized terms such as *operational gaming* and *research gaming* are given different meanings in different settings. The definitions adopted for this chapter have been chosen to fit its purposes; they do not necessarily represent a wide agreement in the literature (in fact, they differ to some extent from those used in Chapter 3).

This chapter is not an appropriate place for an extensive discussion of gaming definitions (see, for example, Ståhl, 1983, pp. 25–59, and Gibbs, 1978). Some discussion of definitions, however, is needed for two purposes: to lay the foundation for discussions of other issues in later sections, and to make clear the limitations on what is taken up in this chapter (this chapter does not cover everything that could be regarded as gaming under other definitions).

Of the two words *operational* and *gaming*, the second offers the greatest difficulty in reaching a consensus on meaning, while at the same time representing the more fundamental concept. Thus, this section deals with gaming and its associated concepts; the next section takes up the term *operational* and the context in which it is used.

Since the concept of gaming emerges from a setting that is not perfectly general but has some important restrictive properties, it is important to start toward a definition of the term with some prior concepts.

The word *game* is used in the literature (for example, in game and decision theory) in connection with two types of decision situations:

Game situations with two or more decisionmakers. (This chapter uses the term decisionmaker to cover both individuals and groups of people—decisioncenters—acting with a joint purpose, that is, as one person.)

So-called games against nature, that is, decision situations with only one clearly defined decisionmaker.

This chapter uses the term *game situation* to represent only the first of these two situations, that is, one with the property that there are several decisionmakers and the decisions made by each will noticeably influence the payoffs to some of the others. Thus, in a game situation there is an interdependence between at least two decisionmakers in the sense that neither can make any kind of optimal decision without first considering the decision the other is likely to make.

On the other hand, in a game against nature there is no other explicit decisionmaker whose actions have to be taken into account, but only the stochastic response of nature (which may, however, include many anonymous decisionmakers).

This chapter confines its attention to games with several decisionmakers; thus, it excludes games against nature, as well as other one-person games where the thoughts and actions of the other side(s) are either pre-planned or discussed and assumed through some structured process by the playing group (the "one-person"). The main reason for this choice is that including such games would make our purview far too wide. The definition of game situation adopted here is somewhat narrower than much usage elsewhere, since some regard some simulations of the one-person game type as gaming. Nevertheless, it seems confusing to have some cases of simulation referred to as gaming whereas other very similar ones have completely different names, such as stochastic simulation, interactive simulation, interactive heuristic programming, and so on. Some of these simulations may have roles in operational studies that need to consider human behavior; many of the aspects of gaming discussed here are relevant to these simulations also (for example, see the later comments in this section relating to the treatment of robots in Section 4.16).

To make the choice of definition more concrete, game situations are represented in such games as chess and Monopoly, but not in such games as roulette or solitaire.

It is also desirable to distinguish clearly between the real situation and its model. *Game situation* is the name for the real situation with at least two interdependent decisionmakers, and *game* is the name for the *model* of the game situation. A game is thus a model—that is, a simplified representation—of a game situation. For example, the game of chess can be thought of as a crude representation of a battle in Persia in the first millennium B.C., and the game of Monopoly as a very simplified representation of real estate dealings in Atlantic City in the 1930s.

It should be stressed, however, that a game is not just any representation of a game situation, but rather a special type of model. If a model is considered to consist of a set of assumptions, two types can be distinguished, institutional and behavioral:

Institutional assumptions deal with the physical properties of the game situation, such as how many players there are, how and when action may be taken, what physical payoffs are possible, what information is available, what time spans are involved, and so on. In an experimental replication of the situation, the experimenter controls the factors covered by these institutional assumptions.

Behavioral assumptions concern the properties of the players, their thought patterns, and their patterns of behavior. In an experimental replication of the game situation, these assumptions are not controlled by the experimenter.

A *complete model* of the game situation is one that contains the totality of both institutional and behavioral assumptions. A complete model can therefore describe how the game is played, and thus assign a solution to the game situation. An *institutional model* of a game situation contains only institutional assumptions; since the behavioral assumptions are missing, it cannot assign a solution to the game situation.

For the purposes of this chapter, a *game* is defined as the institutional model of a game situation, and *gaming* is the process by which the game is used by human players replicating the roles of the interdependent decisionmakers, thereby adding their specific individual behavioral assumptions to the model. Thus the game constructor—or game director— provides only the institutional elements and their structure, mainly in the form of rules, scenarios, and physical paraphernalia such as boards, cards, formulae relating key variables, and so on; he does not specify the behavioral assumptions, because the players will be allowed to play in whatever ways they choose, as long as they are within the limits of the rules of the game.

It should be noted that the definition of gaming used here includes the condition that the roles be played by different human players. This is one factor that distinguishes gaming from some types of simulations of game situations.

Using any model of reality can, of course, be seen as a simulation of some aspect of it. In the more specialized sense in which the term is usually used, simulation can be defined as a form of systematic manipulation of a model of a problem situation for the purpose of observing its behavior for specific sets of inputs. In this sense gaming is also a simulation—but a special kind of simulation. In particular, simulation—and therefore gaming—is not the manipulation of a closed mathematical model to achieve optimization.

First of all, gaming is an interactive simulation. In contrast to most other forms of simulation, where all of the input is made at the start of a run, in gaming inputs are made several times, often after feedback is received and responded to. Second, unlike ordinary simulation, in which one person, the experimenter, conducts all the experimentation with the model, gaming is a multiperson simulation: several decisionmakers are involved, and the role of each important one is played by a different person. Thus, from our definition of gaming we exclude simulations in which one person, such as the research worker, plays the roles of all the decisionmakers by, for example, supplying behavioral equations for all of them.

To sum up, this chapter defines *gaming* as an *interactive simulation, involving more than one player, of a game situation.*

With these definitions in mind, it is possible to clear up some common confusion about the difference between game theory and gaming. Unlike gaming, game theory provides a complete model of the game situation constructed by one person, the game theoretician. In gaming the constructor supplies at most the institutional assumptions for the game situation; in game theory the game theoretician also supplies the behavioral assumptions. In gaming, the participants in the play embody the behavioral assumptions, consciously or unconsciously, in their decisions during the play.

Game theory, however, is not the only type of complete model for a game situation. Although game theory assumes both rational behavior and rational expectations, there are also complete models of game situations—for example, of a simulation type—in which such rationality is not assumed, but where one person provides both the institutional and behavioral assumptions. The best known example of such simulations of behavioral rules that are not necessarily rational in the game-theory sense is that of Axelrod (1984).

Figure 4.1 shows the relations among the terms and concepts discussed this far.

The distinction between a one-person simulation (in which the behavioral assumptions are defined as part of the computer program) and gaming (in which behavior is governed by the players) has important implications. Since the players in gaming manipulate the model, the researcher who constructs the game must make its institutional assumptions completely explicit to the players in an understandable way if they are to be expected to know how to use them in their play. On the other hand, in a one-person simulation the analyst can keep many of the assumptions of the model implicit—and he often does so. The distinction between gaming and ordinary simulation—that is, one-man simulation—is also important for differentiating the way to develop an ordinary simulation from how to construct a game, as discussed later in Section 4.9.

4.3. The Purposes of Gaming

With a view to reaching a definition of operational gaming, it is appropriate to start by distinguishing various purposes a gaming activity can have. This approach also focuses on the kinds of benefits that can be obtained,

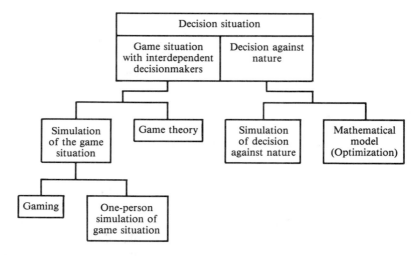

Figure 4.1. The relations among the terms and concepts discussed in Section 4.2.

the time scale they relate to, and who obtains them. This section considers five purposes:

1. Entertainment
2. Education
3. Experiment
4. Research
5. Operational aims

Figure 4.2 summarizes information on these types of purposes, the usages for them, and their relations.

Additional comments, albeit incomplete, are in order:

Entertainment games have the purpose suggested by their name; all the positive results of the game are obtained during its play.

Educational games have teaching, learning, and attitude-changing goals; all the direct benefits of the game are obtained by the players, but these benefits are usually fairly general and intended to endure over some period of time.

Experimental games are aimed at investigating hypotheses or theories; they are without specific situational content or context and without any direct intent to apply the results to a practical situation. The main planned benefit of the game lies in reporting its findings to an outside audience of persons interested in the hypotheses or theories examined by the game.

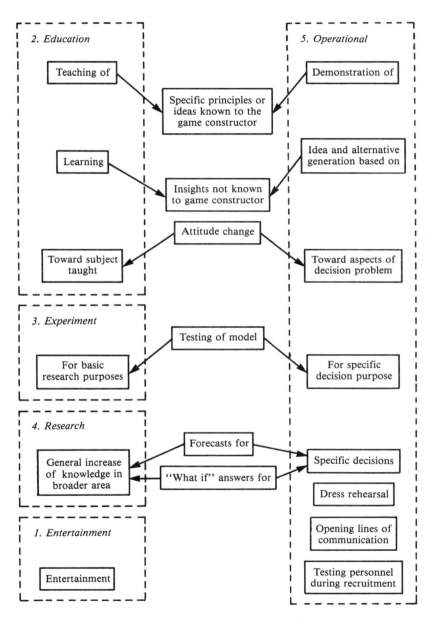

Figure 4.2. The purposes and uses of games. (*From Ståhl, 1983, p. 33.*)

Research games are played to obtain empirical findings (such as forecasts) concerning fairly broad areas; the practical application of these
results to immediate decision and policy problems is not usually
apparent.

Operational games are played to aid decisionmaking, planning, and
implementation in specific and fairly immediate situations. The findings are reported to the decisionmakers concerned, and the main
benefits accrue shortly after the play of the game is over.

As Figure 4.2 shows, operational games can exhibit a variety of different
ways of helping a decisionmaker, of which nine are worth listing here:

1. Demonstrating principles
2. Generating ideas
3. Changing attitudes (for example, by motivating people)
4. Testing models
5. Forecasting
6. Answering "what if" questions
7. Providing dress rehearsals for future new operations or policies
8. Establishing communication
9. Testing personnel during recruitment

It may also be inferred that for each of these uses the benefits of a game
are obtained in a different way and aimed at a different audience.

Several of the uses listed as operational aims have, as Figure 4.2 shows,
counterparts in the other four purposes, the differences lying in the ultimate uses of the results of the gaming activities. It is also evident that
there are no clear-cut boundaries among games designed for different
purposes, the differences being largely of emphasis and degree. This is
particularly the case with games designed for operational and educational
purposes: many games lie somewhere between these two categories.
Sometimes the same game can be used for both. Games designed with
an educational purpose, for example, although focused primarily on teaching specific lessons, can function as operational games if they bring specific issues to the attention of management, or if they act as dress rehearsals for future implementation activities. The main difference
between the educational and operational game is that in the latter the
findings are expected to be put to more immediate management use than
in the former.

There are also similarities between certain games for research and operational purposes (see, for example, Bowen, 1978). Where such similarities exist, the main difference is that the research game deals with a
more general subject area and is not focused on a particular decision,
whereas the operational game deals with a specific problem and the decision that has to be made to cope with it. Since there is a middle ground

in which these differences fade together, it appears appropriate to define a category of *operational research games* lying between research and operational games. Several of the games to be mentioned later in this chapter are of this intermediate category.

Like an operational game, an operational research game has the purpose of being an aid to decision, planning, and implementation, but, unlike the operational game, it does not focus on a single problem situation but rather on a class of similar problem situations. Therefore, the operational research game tries to produce findings worth communicating to analysts and managers who may later face problem situations from this class. Its players behave like real decisionmakers, although they are not usually actual executives with appropriate responsibilities. In contrast, the results of an operational game are aimed directly at the specific problem and concerns of a designated decisionmaker, often acting as one of the players.

The operational research game also differs from research games and experimental games in its proximity to potential applications of the knowledge that it generates. The operational research game deals with a real—albeit generic—decision situation.

This chapter deals with both operational and operational research games, with the term *operational game* usually intended to cover both types.

Finally, it is useful to distinguish between rigid-rule gaming and free-form gaming:

> In *rigid-rule gaming* all of the institutional assumptions in the model of the game situation are supplied by the game constructor. Hence, all of the rules of the game are defined before the game starts, often in the form of a computer program. Thus, the outcome of every possible combination of players' decisions is defined precisely.

> In *free-form gaming* at least some of the institutional assumptions are supplied by the game players as the play proceeds. Thus, the players to some extent invent the rules as the game goes on. As a consequence, the outcome of a particular decision might, for example, emerge from discussions among the participants. In some cases, the game constructor supplies only a small portion of the institutional assumptions in the form of a scenario; in others, some of the rules are supplied by a special player called a referee or by a control team. (For discussions of free-form games, see, for example: Brewer, 1978; deLeon, 1981; Jones, 1985.)

Usually, management games in which the players decide on quantities such as price, production, and so on are rigid-rule games, whereas games involving the exchange of verbal messages, such as international diplomacy games, are free-form games. We shall see in Section 4.16, however, that computerized games can also be of free-form type.

4.4. Clients for Operational Gaming

Operational gaming has a long tradition in military affairs, but its use in the civilian sector is much more recent. This chapter, however, focuses entirely on civilian gaming, a limitation that has implications for the methodological issues it discusses. Military gaming has often been of free-form type, with teams of players, including a control team, and involving a long playing time and large resources. The games for civilian use that are the focus here are generally smaller in scope than military games, require shorter times to play, are more likely to involve more than two players, and are more often rigid-rule games (Brewer and Shubik, 1979, and McHugh, 1966).

A comprehensive overview of the extent to which operational gaming is used in practice is difficult to find in the existing literature. Accordingly, in 1981–1982, in connection with a gaming project at the International Institute for Applied Systems Analysis, Laxenburg, Austria, Richard Duke and I conducted a questionnaire survey of the operational uses of gaming in practice. Although we were able to reach only a modest portion of all of the analysts in the gaming field, the approximately 50 usable answers we received indicated that operational gaming is used fairly widely in the civilian sector, not least in socialist countries (Assa, 1983, and Marshev, 1983), and that many interesting applications have not been documented for an audience outside the community of immediate users of the findings. For our survey we arrived at partial—but nonetheless useful—answers to three key questions:

Who are the clients for operational gaming? International organizations like the United Nations Educational, Scientific, and Cultural Organization, the United Nations Environment Program, and the World Health Organization; in national governments, central government departments like the U.S. Department of Energy; national health services; regional administrative bodies; corporations and industrial enterprises (in the latter case particularly in the socialist countries); and many others.

What areas are covered? Agriculture, construction, disaster planning, educational systems, energy, environment, financial policy, garbage disposal, health care, location policy, management information systems, marketing strategy, tourism, and so on.

For what purposes is operational gaming used? Changing attitudes, dress rehearsals, forecasting, improving communications, generating new alternatives, planning, testing decision aids, testing personnel, training for specific tasks, and so on; indeed, the list covers much the same areas as the list indicated in Figure 4.2.

Since this chapter emerges to some extent from personal experience, I should mention that during the last several years I have been involved with operational gaming dealing with three issues:

What kind of a cost-allocation method would be acceptable to planners for intermunicipal cooperation on water projects in southern Sweden? (Ståhl, 1983, pp. 211–230.)

Could the use of negotiable emission permits lead to a cheaper way of reducing the amounts of hydrocarbons in the Gothenburg area of western Sweden? (Ståhl, 1986b.)

Is a drastic change in price formation in the coal market likely within the next two decades owing to the environmental problems caused by carbon emissions? (Ståhl and Ausubel, 1981.)

All three games focused on getting specific information, and all three used players who were as much like actual decisionmakers as possible.

4.5. How Clients Obtain Information from Gaming

Since the purpose of using operational gaming is to aid decisionmaking, planning, and implementation, this section focuses on the different ways that clients can get new knowledge from it, be it hard facts or intuitive insights, to use in their decision processes.

A client, or a representative of a client, can obtain information from gaming in one of these four roles:

As a player of the game

As an observer of the play of the game

As the game's constructor

As an outsider who reads a report on the game's play and its results

The knowledge gained from these four roles differs considerably in quantity and quality.

The outsider who reads a report on the game is getting knowledge of it at second hand, as he would from a report on a simulation or ordinary survey research. Such a report may provide statistics on how the participants played, and whether or not their behavior confirmed or denied some hypothesis, but it will not usually convey subtle interactions that they may have been involved in during the play.

On the other hand, a player in the game obtains knowledge of quite a different sort. For example, many users of gaming claim that through its activities and interactions the players gain valuable intuitive insights, and one can think of a gaming experience transmitting a gestalt to its players. Whereas the reader of a report on a game gains information in textual form, a participant gathers experiences that include interactions, impressions, and mental images. More importantly, there is a qualitative difference in the experiences of readers and players. The reader's activity is relatively passive, whereas the player must be active throughout the game, thinking in some detail about the stimuli that come his way; thus, the player's experience is more detailed, although the reader may gain a better perspective and overview, depending on the quality of the report that he reads.

The rapid feedback that most games supply the players gives them an incentive to make their decisions as good as possible. The participants are thereby forced to make their own mental models of the game. On the other hand, the reader of a report on the game usually accepts the mental model of the game transmitted by the writer of the report. Building a mental model of a game, based on cumulative observations of the behavior of other players and the results of all the decisions, often leads a player to change his thinking about the essential aspects of the situation being modeled by the game.

The experience of an observer of the game lies somewhere between those of the players and the report reader. Although the observer does not have the same incentives as the players to form his own mental model, his perception of the game is based on more detailed stimuli than those reaching the report reader, who is viewing a distilled version of the mental model developed by the researcher who prepared the report.

Finally, the knowledge that the game constructor gains through the process of developing the game is usually especially deep and comprehensive—and for many games it has proved to be the most important product of the gaming activity.

Later, Section 4.9 discusses why the information-gathering activity for an operational game is often more intensive and searching than that for an ordinary simulation. Here, it is worth mentioning that this process may itself confer important benefits on those who are involved with the problem situation as they seek to answer questions from the game's constructor, and as they produce information for him that sheds new light on the problem situation. In fact, this effect may be strong enough to force those involved to revise their mental models of the problem situation, both as a result of the new information and the need to consult widely to obtain it. Duke and Duke (1983) describe a case in which the information-gathering part of the game-construction process had a substantial effect on the large corporation being served.

4.6. The Main Elements of the Gaming Research Process

Systems analysis begins when those involved in a problem situation decide to seek analytic help. One of the techniques available to systems analysts is operational gaming. A decision to use this approach will set in motion a series of steps that is here called the *gaming research process*. This process, of which playing the game is only one part, yields important information, some of which has been described in the preceding section. In fact, the gaming research process can be thought of as a total information process related to the problem situation of concern.

In the gaming research process, it is possible to distinguish five steps.

1. The first step is a decision about whether or not to use a game, either as the central tool of the analysis, or as a complement to other methods. Can a game substitute for other possible analysis procedures? If it complements other procedures, how can an appropriate match be arrived at so as to achieve a tightly knit research program addressed to the problem that emerges from the problem situation? Section 4.7 addresses these issues.

2. Once a decision has been reached to use a gaming process, the next question is whether to use an existing game or to construct a new one. If the decision is to use an existing game, will it have to be modified to fit the new situation? Section 4.8 discusses these questions, as well as the process of modifying existing games to fit new situations.

3. If a new game is decided on, it must be constructed. This process, as well as various issues associated with it, is discussed in Sections 4.9, 4.10, and 4.11. However, it is worth mentioning here that game development includes three important components: information gathering, constructing the game (often including computer programming), and testing the game (for example, by playing the game several times for the sole purpose of getting feedback that will enable the design of the game to be improved).

4. The next step is to plan and carry out the game-playing program so that it will yield information about the problem that prompted building the game. It is usually not enough to play the game only once. There are, as Section 4.12 discusses further, usually strong reasons for playing the game a number of times, one of which is to strengthen the conclusions and comparisons that have emerged.

The essential part of the fourth step of the gaming research process, playing the game, can be considered to have three parts: the pre-game activity, the actual play of the game, and the postgame activity.

The pregame activity consists, among others, of the following items:

Choosing players for each run of the game.

Instructing the game director on how to manage each run.

Making the physical arrangements (space, schedule, personnel, computer time, administrative support, and so on).

Determining the parameter values to be used for each run of the game. Although this is usually specified in the game plan, and often flows from the way the game is constructed, it is sometimes desirable for the players to determine some of the parameter values just prior to their play of the game. This last-minute cooperative setting of parameter values can be carried out by a computer dialogue method, as discussed in Section 4.16.

The actual play of the game has two parts: (1) the start of the game (mainly consisting of briefing the players on the structure of the game and its rules, usually with the aid of player manuals that often contain background information on the problem situation and sometimes more detailed descriptions of the decisionmaking units that the players are trying to replicate) and (2) playing the game through a number of *rounds*, or *runs*. Each round can be seen to contain these important elements:

Decisions by the players, that is, their inputs into the game model.

Calculating the outputs from these inputs, that is, the feedbacks to the players from their decisions.

Presenting these feedbacks to the players.

Where deemed necessary by the game director, introducing changes to the rules of the game.

The postgame activity consists mainly of the postgame debriefing. This colloquy serves different purposes, depending on the purposes of the game's play. For example, if the game was played in order to assist in the process of developing it, the debriefing focuses on how the game experience can be used to improve the game's design, to make it more playable, to make the instructions more understandable, to determine whether or not the game was too complex, to discover whether or not the players had enough time to contemplate their decisions, and so on. When the play of the game is part of the problem-investigation process, the debriefing focuses on extracting information from the players that bears on the problem that the game was designed to investigate, and that therefore can be used as part of analysis for the report on the gaming activity. The debriefing procedure can also help the players to understand what they have learned from the game-playing process, thus contributing to their learning experience. (See also Lederman, 1984.)

5. The final step in the gaming research process is the analysis of the results of the game's play and the preparation of a report on what was done and what it implies about the problem to which the analysis is addressed. Preparing this report, and the communication process of which it is a key part, follows the usual practices of good systems analysis work (see the *Overview*, pp. 303–315, as well as Chapter 15 in this book). If the clients of the work have been players in the game, or close observers of it, the knowledge they have gained may obviate the need to prepare a report. Good professional practice, however, urges that, wherever pos-

sible, a report should be prepared to help future users and gaming professionals generally.

4.7. To Game or Not to Game?

The first step in the gaming research process is the decision about whether or not to conduct a gaming procedure as part of the problem analysis. This decision must hinge on the answer to the question: Can gaming be a useful device for assisting in the analysis? It is, of course, impossible to give any absolutely general advice on when gaming is suitable and when it is not. It is important, however, to distinguish between the cases when gaming is used as the only analytical device (perhaps as a substitute for other systems-analysis methods) and when it is used to complement other approaches.

Let us first look at the case when gaming is of interest on its own. It often makes sense to use gaming by itself when it is desirable to increase meaningful communication among key persons in the organization with the problem, or when the goal is to convey to the participants and observers the gestalt of the problem and thus deepen their intuitive understanding of it—sometimes referred to as an "aha!" experience (Duke, 1974, and Rhyne, 1974). In such cases free-form games are generally more suitable than more highly structured types. Gaming can also be seen as one of the ways of using and extending expert judgment as discussed in Chapter 3. Since it is very difficult to measure the relative efficiencies of alternative methods for such purposes, the choice of gaming is necessarily fairly subjective. The only general comment possible is that the more formalized the game the more likely it is that the game will be more expensive to develop—but not necessarily to use—than the less structured interactive methods. However, these less structured types of gaming—some of which have been covered in Chapter 3—are not the main focus of this chapter.

We next turn to the case where gaming is considered as a complement to other methods. In particular, we want to consider the additional information that gaming can offer when it has been decided that an ordinary simulation model or a standard analytic model is also to be used.

In this case, one of the benefits of gaming is the way it forces all concerned to make all of the assumptions explicit, including the choice of variables and other simplifications of the problem situation, and to test the relevance of these assumptions. (It is all too easy to neglect doing this rigorously for simulation and analytic models.) Thus, the decision about whether or not to use gaming as one of the techniques of analysis sometimes rests partly on the extent to which the analysis team wants to force all of the relevant assumptions into the open and explore their consequences.

One special use of gaming is to test the behavioral assumptions of formal models, such as game theory or simulation. This special use of gaming as a complement deserves particular attention, and Section 4.13 discusses it further.

The role of gaming in making assumptions more explicit is not confined to formal models; it can also apply to mental models. Gaming can bring out the assumptions of mental models in several ways. For example, when an executive says that he feels strongly uncomfortable with the game he has participated in because "it doesn't portray my company in a realistic way," the analyst knows that the game does not correspond to the executive's mental model of his firm. The same executive faced with an "executive summary" of the results of research based on an ordinary simulation or an analytical model will generally have little possibility of assessing the assumptions of the model, and his response to this report, if any, will seldom reveal his own mental model.

Similarly, changes suggested by the players during debriefing sessions can bring out the mental models of the participants. Suggestions that do not add to the complexity of the game are particularly interesting. On the other hand, suggestions that only add to the game's complications may be less interesting, particularly if they do not reflect the player's mental model but rather an unwillingness to work with a simplified model of the problem situation.

Making the institutional assumptions of the game explicit allows analysts to improve the other institutional models that are being used in the analysis. Through playing the game and sharing debriefings, the players also reveal their mental models of the situation to each other, thus improving their mutual communication about the problem situation. This shared experience can help to create revised and realistic institutional models in the participants' mental models.

There are, then, many potential benefits from gaming exercises, but they can only be achieved at a cost, the subject to which we now turn.

Estimating Gaming Costs

The costs of gaming can vary widely (Shubik, 1968, and Shubik, 1975b, pp. 59–97). As a basis for estimating them, it is convenient to look at four phases of the gaming process and the man-months of effort required for each: concept development, information gathering, game development, and the game runs.

Concept development. This phase is usually a one-man activity, and, though of critical importance, of small cost consequence. The more previous experience the game developer has, the smaller this effort is likely to be. My experience suggests that this effort will seldom exceed three man-months.

Information gathering. The cost of this activity can vary greatly. For some games where the constructor (for example, as part of an in-house team) already knows most of the relevant facts, only a few hours or days may be needed to assemble the information in good order. In other cases, it may constitute the largest share of the gaming cost. For example, in the Emissions Permits Trading Game (by the standards of Section 4.14 a small game), information gathering regarding emissions-reduction costs took more than half of the total effort of around one man-year.

Game development. This activity transfers the game concept into a playable game. Its cost may depend importantly on whether or not a manual or computer game is being developed.

In my experience, to develop a manual game—such as a board game—seldom takes more than a man-year; most board games probably take less effort, unless the game contains an exceedingly complicated manual or takes many days to play. The game-development effort should allow enough test runs of the game to verify its relevance and stability. For example, to develop a board game on the CO_2 issue (Ausubel et al., 1980) required less than half a man-year, including the time for a dozen test runs, but excluding the time spent on information gathering.

For computerized games the variations are much greater. A game that requires its own dedicated computer system can incur very high costs. For games run on computers also used for other purposes, the total computer costs (development, testing, and game runs) are generally of minor importance. If the game can be run on a microcomputer, which is becoming increasingly possible as they increase in power and size, costs can generally be measured in only tens of dollars per day. Only for games that have to be run on a number of terminals in a time-sharing system, or for games requiring special peripherals (such as special graphic displays), do the computer costs become much higher. However, the costs would be even higher for games played over a computer network with the terminals located in different cities.

On the other hand, the programming costs for computer games may be more significant. As a very rough guide, let us assume a programming productivity of 5 to 10 lines of code (for example, in BASIC or FORTRAN) an hour. Some of the small games I have developed, running around 1000 lines of code, have required 100 to 200 man-hours of programming time. Most larger games, however, involve much larger programs, and may involve at least ten times this effort.

Game runs. In many games the largest cost item—yet one that is frequently neglected—is the direct cost of the time spent by the players. In a company whose employees play the game this cost must be taken into account; for a highly paid executive, this cost can be quite high, perhaps on the order of $100 per hour. These costs may be even higher if we look at the opportunity costs, that is, the values of the functions forgone owing to the persons playing the game rather than attending to their usual duties, but there is no general way of estimating these costs. Thus, a game with many qualified players can incur costs that are sufficiently high to call for games requiring only short times of play (see also Section 4.14).

The cost of the time that the game director devotes to it is likely to be much smaller, although my experience suggests that it will be at least twice the actual playing time, even in cases where the game is repeated many times. The cost of rewards, if any, and logistical support (even including entertainment expense) is usually of minor importance.

Since the cost of the gaming effort is mainly dependent on the size of the game, a question that should be asked is: What is the smallest game that will make a meaningful contribution to the analytic effort? Section 4.14 deals with the question of whether or not the analyst should also consider a small game. Here I can say that the games from my experience mentioned above (that is, the games on carbon dioxide emissions, on emissions permits, and on water-project cost allocation) were all small games with a total effort for the four phases amounting to something between 6 and 12 man-months. My guess is that in most cases one cannot do a meaningful operational game in significantly less time. Rather, it seems that many operational gaming activities will require substantially more effort to be meaningful.

In line with classical economic principles, an analyst should consider the cost of a game in comparison with the costs of alternative methods of analysis, together with the benefits captured and foregone in each case. In reality, it is almost always impossible to make such estimates with any precision, but even crude cost projections can be useful, particularly if—as is often the case—the research effort has a fixed budget and a preassigned time horizon. In spite of the many uncertainties in estimating costs, it is usually possible to answer the question: Can a meaningful gaming activity take place within the cost and schedule limitations?

Finally, it must be pointed out that in most cases it is likely that considerations other than costs and benefits determine whether or not a gaming approach is attractive and will be used. Questions like these are likely to be more important: Do we already have experience in gaming? Has this experience been successful? Or, if there has not been prior gaming experience in the parent organization: Is gaming an area in which we want to build competence over the long run? Do we have the staff that would like to build this competence? Is the topic of the game of real interest to at least some key people? Will management want to participate in the game? In sum, a gaming project must be looked at within the total context of the organization's concerns and its style of working.

4.8. Using an Already Existing Game as a Basis for a New Game

Before deciding to develop a costly new game, an analyst should spend some effort on investigating whether it is possible to use an already existing game with appropriate modifications. Such an investigation has three phases: surveying existing games to identify ones possibly suitable for use or adaptation, investigating in detail the suitability of the games selected for consideration, and modifying the chosen game to fit the purposes in view for the prospective gaming activity.

Surveying Existing Games

There are four main sources of games: books, journals on gaming, personal game banks, and gaming organizations.

Books containing game descriptions. These books, which are listed in the references under the heading "listings of games," contain game descriptions: Bierstein and Timofievsky (1980), Elgood (1984), Gibbs (1974), Horn and Cleaves (1980), Rohn (1984), and Stadsklev (1975).

Journals on gaming. Since the gaming field is developing rapidly, books tend to become obsolete, so that the journal literature is a more important source of up-to-date information. The leading journal in English is *Simulation and Games* (Beverly Hills, California: Sage); it is published quarterly and contains, among other things, reviews of recently published games. (The December 1984 issue contains a listing of all the games reviewed up to that time; see Basinger, 1984.)

Personal game banks. Some people keep personal game banks, often as computer files, and they are generally kept up to date. Among such persons are Walter Rohn (Wuppertal, Federal Republic of Germany), Richard Duke (University of Michigan, Ann Arbor, Michigan, USA), and Dennis Meadows (Dartmouth College, Hanover, New Hampshire, USA). Hans Gernert (Humboldt University, Berlin, German Democratic Republic) keeps a data bank on games in the socialist countries.

Gaming organizations. One of the pleasanter ways to investigate the existence of suitable games is to participate in the meetings of professional gaming organizations, such as the International Simulation and Gaming

Association (ISAGA), the North American Simulation and Gaming Association (NASAGA), and the Association for Business Simulation and Experimental Learning (ABSEL). The journal *Simulation and Games* publishes details of the future meetings of these organizations.

Investigating the Suitability of the Selected Games

If the survey of existing games has identified some that appear to be suitable, they must then be investigated more thoroughly. It is my conviction that such a thorough investigation demands actually playing in the game for at least a couple of runs. One can generally not get an adequate— or even correct—opinion of a game from only reading about it. Obviously, in at least one of these test runs the participants should be persons like those who are going to play in the game that is finally used.

Modifying the Chosen Game to Fit the Purpose

Playing a game a couple of times will almost surely give an analyst a perspective on the extent of its suitability for the new purpose, and how it will have to be modified. The question then is how easy it will be to make the needed modifications.

Unfortunately, even a small modification of a game is generally more difficult to make than one might expect. In a computerized game, for example, modifying the game probably requires changes in the computer program, which demand knowledge of the programming language used in the game. Even with such knowledge, changes may be difficult to make unless the program is very well structured and documented. Changes are impossible if the program is available only in binary object code rather than source code.

It should be stressed that modifying a game is generally easiest with games that are constructed from the beginning so as to be easy to modify, and hence usable for a variety of purposes. One type of such games is the *frame game*, in which one part is left undefined, allowing different users the possibility of defining this part to suit their varied purposes. (For further information on frame games, see, for example, Duke and Greenblat, 1979, and Thiagarajan and Stolovitch, 1979.)

4.9. The Difference between Gaming and Ordinary Simulation

Before the next section takes up the methods for developing operational games, it is useful to look at the differences between what I shall call ordinary simulation (that is, one-person simulation, in which just the researcher—or some other expert—interacts with the computer) and gaming (that is, multiperson simulation). The reason for discussing these differences here is to enable the next section to focus its attention on the aspects of game development that are peculiar to gaming and not covered by the literature on developing simulation models.

It should be stressed that this section focuses on the differences due to the numbers of users, and the ensuing degree of user familiarity with the model. In ordinary simulation, the model is usually run on a computer, with all of the data input done by one person well acquainted with the model (usually its constructor). Although this input process normally occurs once at the start of the simulation, this discussion also includes ordinary simulation cases in which the researcher works interactively with the simulation model (for example, by changing parameters during the simulation run). It should also be noted that an ordinary simulation can cover decisions against nature as well as game situations. According to the definition in Section 4.2, gaming exists when the simulation of the game situation is handled by several interacting players.

Many of the differences between ordinary (one-person) simulation and gaming discussed below are therefore mainly associated with the fact that the behavioral assumptions in gaming are dealt with by the players—persons other than the game constructor—and with the ensuing requirement that the players play the game in a serious way (that is, as though it were a slice of real life), so that the game will yield information that has meaning for reality. Thus, there are special requirements on the game and the entire gaming program.

1. The game must contain incentives for the players to make efforts, not only to play the game to the best of their abilities, but also in a way that imitates actual decisionmakers (see also Section 3.5). To achieve these goals the game should first of all be interesting to play, or at least not boring; the players must not let their attention wander and thus play erratically out of inattention or boredom. It is not always enough that the research problem behind the game is of interest; the game itself must be so well constructed that it does not diminish the initial interest of the players. To induce the players to play as professionally as possible, it may in some cases be desirable to have some special physical incentives, such as money or other rewards.

2. If the players are to play so as to resemble real-life counterparts as realistically as possible, it is important for the game's time compression not to be so restrictive as to force erratic or poorly thought-through play. A way of alleviating this problem is to facilitate decisionmaking; one way to do so is to provide decision aids, as discussed further in the next section. Another way is to make the way to input the player decisions as simple as possible, as well as to make the outputs as easy for the players to read as possible. In simulation, where the constructor is thoroughly acquainted

with his model, these forms of simplicity may play a minor role, but in gaming, input and output simplicity may distinguish a good from a bad game.

3. To get players in an operational game to act seriously, it is of prime importance for them to regard the game's institutional assumptions as realistic. There is always the problem that essential simplifications introduced to make the game tractable bring some unrealism with them, but they are not likely to be misunderstood. Rather, what is of concern here is unreality that accompanies data that strike the players as incorrect. For example, if players take part in a game that is presumed to represent the operations of their own company and they observe data that seems to them to be erroneous either in magnitudes or internal relationships, their interest in playing seriously has been observed to decrease markedly; some have refused to continue playing on these grounds.

In an ordinary simulation, there is no comparable immediate punishment for having incorrect institutional assumptions or data. Operating more or less by himself, the researcher can work completely under cover, so to speak, until the results are presented either to a client or for publication. Thus, in simulation there is sometimes more focus on results than on the assumptions and inputs, although in a simulation done to support a client-sponsored, problem-oriented systems-analysis study this should not be the case.

This difference between simulations and games as regards the incentives to collect valid data is important. In my experience, a game may require a much more thorough data search than a simulation of the same size. This explains, for example, the relatively high data-collection effort in the Emissions Permit Trading Game that has already been mentioned. A game's need for a thorough data base has the consequence that one of the most important and valuable byproducts of game-program research is the data collected in connection with developing the game.

The fact that games demand realistic institutional assumptions implies that a single data-gathering effort is unlikely to suffice; rather, it is nearly always the case that the data-gathering effort has to be recycled, often after getting feedback from the players about the realism of early test runs of the game.

4. If players are not to lose their interest in the game, it is absolutely essential for the game to run smoothly. If a simulation breaks down, the researcher can fix it and then run the program again from the beginning. On the other hand, if a computer program for a game breaks down during the play, it is generally not possible to rerun the game from the beginning, owing to two factors: the risk of boring the players, and the fact that the players now have information that they otherwise would not have, enabling them to play the rest of the game differently from the way that they might have played it if the breakdown had not occurred.

To avoid this last difficulty, it is desirable to have the program designed to enable a rapid restart from the point of the breakdown, so that there is no need to replay any part of the previous period. But this requirement is easier to state than to fulfill. It is particularly difficult if the problem is a program error resulting, for example, in an erroneous output. In this case the game must be interrupted until the program is debugged. The time lost may well make it impossible to continue the game. Thus, it is much more important to have an error-free program for a game than it is to have one for a simulation.

5. It is important to run a game several times, as discussed in Section 4.12, and it is also important to run a simulation several times. There is, however, an important difference: The same experimenter can repeat a simulation many times, but operational gaming may in many cases require different players for each run, especially in operational research games. This requirement is prompted by two desiderata: a desire to have the game results represent the variations in play that are brought to it by various players, thus reproducing a variability reflecting that of reality, and the desire to avoid having players who have learned the game, and thus who play it differently. The need for different players for each run of a game sometimes dictates that it be played in different locations. This is particularly true when the players need to have certain specific backgrounds: for example, a game that needed municipal water planners had to move from city to city because most cities had only enough water planners for a single game.

The need for repeated plays of a game, sometimes in different locations, may require more than one person to act as the game director. Thus, in contrast with simulation, in which the experimenter can easily repeat the simulation as many times as he likes, the gaming research program may require explicit special instructions for the game director, perhaps in the form of a game director's manual. Furthermore, the need to play the game in more than one location within the framework of one research program may make it essential to make the game and its associated program easily portable, a need that is felt less often in connection with a comparable project involving simulation.

6. Another fundamental difference between gaming and simulation concerns knowledge of the rules. In simulation the details of the model, and hence the rules for what input parameters to use from run to run, are well known to the experimenter; in gaming the corresponding rules of the game are generally not known to the players before the game run, since they differ from run to run. Hence, the description of the rules of the game, that is, its institutional assumptions, is a very important part of the game. Owing to the limited amount of time available to the players, it is often important to present the rules as concisely and precisely as possible in player's manuals.

4.10. Developing an Operational Game

This section assumes that a decision has been reached to use an operational game as one element in the systems-analysis effort addressed to a specific problem situation, and that a search of existing games has ruled out the possibility of adapting one of them. Thus, a new game must be constructed—and the purpose of this section is to describe the phases of this process.

Many of the steps in developing a game are the same as steps in developing a computer simulation model. Since much has been written about simulation development, this section focuses on the aspects, as brought out in the previous section, that are different for game development.

The development of a game should follow a well-thought-out, orderly process. A good start is to look at some of the extensive literature concerned with this development process, such as Duke (1980), Easterly (1978), Ellington, Addinall, and Percival (1982 and 1984), Greenblat and Duke (1974), and Yefimov and Komarov (1982). This process, as this section discusses it, has four phases: a conceptual phase, a data-gathering phase, a problem-model phase, and a final phase in which the problem model is transformed into a game.

It should be stressed that, although our discussion of the development process is divided into four phases, the actual process is seldom a well-defined, step-by-step process. Rather, it is more like proceeding from the large to the narrow end of a funnel, during which the choices are gradually narrowed. Nor can the process usually be carried out in a straightforward manner; rather, as in systems analysis generally, the process is iterative: As more work is done it becomes clear that earlier decisions and estimates need revision, so that the actual process emerges more like a cycle with frequent feedback.

The Conceptual Phase

Developing an operational game begins, as in systems analysis generally, with work aimed at defining the problem (see the *Overview*, especially Chapter 5). With this definition in hand, the analysts must evolve the goals that the game should seek, that is, what the gaming research process should produce. At the same time, there should be an appreciation of the constraints of time and resources within which the effort must be conducted.

Even at this early stage the analysts should make at least a crude preliminary estimate of some aspects of how the ultimately developed game will be played: how many runs may be needed to explore various options or establish some results, what types and numbers of players will be needed, and how long each game will last. These matters will be contem-

plated in greater detail in later phases (see Section 4.14), but early estimates are essential, as they will be needed to support requests for people
and resources for the gaming project as part of the larger analysis effort.
In many cases, these results from the conceptual phase form important
inputs to the decisions to proceed with the analysis and what resources
to allocate to it.

At the end of the conceptual phase, it is often desirable to write a formal
concept report including a description of the simulated system, a rough
estimate of the resources needed for the gaming activity, and a justification for an activity of this size. Such a concept report is often used to
obtain top management support and funding for the game.

Data Gathering

As in other systems-analysis approaches, the conceptual phase of the work
includes gathering a good deal of information and data in order to provide
a foundation for the concept. The concept report then describes the additional information that has to be gathered, an activity that may be extensive enough to take a considerable amount of planning, time, and effort. This information is essential for both the modeling effort and the
work on defining the institutional assumptions of the game. No matter
how much care and effort are put into this phase, it usually happens that
the work of developing the game will generate further information needs,
and thereby the effort to satisfy them; this must be allowed for.

The Problem Model

The next step is for the constructor, using the information that has been
gathered so far, to make his mental model of the situation explicit by
putting it on paper. His model may be purely verbal, or it may be a
mathematical model; it is more likely to be a combination of both. It
should be noted carefully that this model, which is called the *problem
model*, is not yet a game. When the game is finally developed, it will also
be a model of the situation. In fact, one can see the game as a simplified
model of this problem model.

The main differences between these two models are these: Although
the game, as has already been pointed out, is always a partial model, since
it contains only the institutional and not the behavioral assumptions, the
problem model may be a complete model that includes the constructor's
perceptions of how the participants will behave (that is, it may also contain
behavioral assumptions). The problem model may in some ways be closer
to reality than the game because the necessary simplifications that must
accompany any model are not focused on a need to make the game play-

able, and there has not yet been the need to make the cruel choices necessary to bring the analysis exercise down to the size that will make practical gaming work feasible. The simplifications in the constructor's problem model are only a reflection of the content of his mental model and his desire to write down on paper the essential features of the problem situation. It is my view, however, that work on developing the game benefits from having the constructor's mental model formulated explicitly in a problem model document.

Transforming the Problem Model into a Game

The next step is to transform the institutional assumptions of the problem model into an operational game. In this process, the behavioral assumptions in the problem model can be useful in suggesting the kinds of players that will be needed (see Section 4.12) and also in running a general check on the validity of the game in pregame tests (see Section 4.11), but the institutional assumptions are at the core of the transformation of the problem model into a game.

In this transformation step, it is necessary to return to the fundamental question of the size of the game (to be discussed in more detail in Section 4.14). This question deals with the length of the game (that is, the number of hours needed to play it), the number of players in each game, the number of periods of play, and the number of decisions to be made in each period. The size of the game is importantly related to the number of times the game is to be played; within a given budget, the more times the game is to be played, the smaller it has to be. Section 4.12 discusses the issue of determining the number of times to repeat the game.

Suppose that the analyst has decided on a small game playable in three to four hours, that is, in an evening's time, in order to be able to get real decisionmakers on the executive level to take part (see further discussion in Section 4.14). The next question is: How many periods of play and how many decisions in each period can be handled in such a short time? The more decisions per period, the fewer periods can be run. The number of periods needed depends in part on how much feedback is needed to bring out the fundamental pattern of the game. In some games, where a steady-state level is reached fairly soon, as few as four periods of play may suffice. In other games, where one goal may be to bring out cyclical patterns, a much larger number of rounds may be needed. For example, a management game might be aimed at exhibiting what happens when companies increase investments, and thereby outputs, until prices fall enough to make investments, and therefore outputs, decline enough to make prices rise again, thus inducing higher investments and outputs. For a game to exhibit such cyclical patterns might require at least six to ten rounds of play.

A fundamental issue, to be decided, if possible, early in the transformation stage, is whether to have a computer or manual game. It is usually the case that the general issue being addressed and the purposes of the game have already made it clear which choice should be made. However, there are cases where there exist strong arguments for either approach, so that it is desirable to postpone the commitment to later in the development process when more light can be shed on the issue.

In a manual game, the transformation phase deals with translating the various components (such as decision variables) into the most effective symbols, so that the symbols themselves help communicate the essential rules of the game. In particular, the payoff system must be given an effective symbolism; if it involves money, for example, chips or special paper money can be used. A well planned system of symbols may also help to speed the decisionmaking.

In a computer game, the symbols of the manual game (board, pieces, cards, chips, paper money, and so on) are replaced by inputs into and outputs from the computer. In some recent games, each team has its own computer terminal, and its players make their inputs directly on the keyboard; the output then appears on the terminal's screen. This arrangement presupposes a time-sharing computer system with a number of terminals, one for each team, usually located in different rooms. In such a game system, it is not absolutely necessary to divide the game into periods, since in principle the teams are able to change their decision variables continuously.

In spite of the technical flexibility offered by games played on such a time-sharing system, most computer games are run in a more traditional fashion, with the teams filling out a *decision form* for each period and then obtaining feedbacks on their decisions in the form of printed reports generated by the computer. In fact, even in situations where input and output over individual terminals would have been possible, games have been run with decision forms and printed feedback reports. There are, in fact, some advantages to the older approach. If the decision sheet is carefully designed, it can help the teams to plan their decisions more efficiently, thereby speeding up their decisionmaking. Too, input errors are less frequent with decision forms than with terminals, particularly when the players have inadequate prior experience with using terminals. The printed output also helps planning for the next round.

In a computerized game that uses a decision form, the design of this form is very important; the clearer it is, the better the understanding of the rules. This document should be evolved through repeated tests and should be given a final polish through feedback from the early test runs of the game.

Besides making the decision form suitable for fast and easy decision planning, the analyst must also decide during this phase whether or not other decision aids should be provided to facilitate and speed up the decisionmaking process (in order to achieve a given level of performance). For example, such aids could include simple calculators, microcomputers (or terminals) with spreadsheet packages, like Lotus 1-2-3, simple LP packages, and so on. Without decision aids, experience has shown that the players will often spend a considerable amount of time making manual calculations, thus reducing the time available for strategic planning and problem solving (Schuenemann, 1983).

The *output report* also requires serious development effort. The output must be presented in a suitable and effective way in order for the game to be easy to play. In particular, the constructor must decide on how much and what type of output to return to the players as feedback. Too large an amount of information tends to slow down the play, whereas too little information makes the game appear unrealistic and may tend to make decisionmaking less serious and more random. It may be suitable to contemplate presenting different amounts of information in different periods, a situation that is not without its counterpart in real life. For example, in one of my management games, only a simple profit-and-loss account is provided in each period, whereas a market report is presented every second period and a set of balance sheets every third period (Ståhl, 1986a).

In connection with the output, the constructor has to decide on the proper *payoff structure* for the game, a factor that is closely connected to its basic purpose. The constructor must ask questions like these: Do I want the players to focus on a single specific factor, such as maximizing a single variable like the net worth of the corporation? If so, what is a good way to get the players to focus on this target? Is it enough to state this goal clearly in the player's manual? Do I want also to offer some special incentives to ensure that the players pursue such maximizing behavior? Incentives can involve a system for translating the scores obtained in the game's reports into real prizes at the end of the game. For example, the players could be offered a real dollar for each million dollars of net worth in their corporation at the end of the game.

On the other hand, there has been some debate about such a procedure and about how much real money is needed to influence players significantly. In the experimental games that I know of, rewards have been at most $100 per team. Some scientists have raised ethical objections to monetary rewards related proportionally to the scores, particularly in cases where the teams have started with widely differing assets (Ståhl, Wasniowski, and Assa, 1981). Another approach used in some games is to reward only the team that has been the most efficient, as measured by some criterion (which may, for example, take into account differences in starting positions). The problem with such an approach is that the players may become unduly competitive with each other; instead of focusing only on their own payoffs, they may be more interested in the difference between their own score and that of their competitors. In game-theory terms, a game that in reality represents a nonzero-sum situation may become a zero-sum game. A third approach, which I used recently, is to relate the monetary reward for a player to his results set in relation to how he fared in earlier games (Ståhl, 1986b).

In connection with the discussion of scores and incentives for playing well, there is an issue of *end effects*. If players focus their attention, perhaps owing to a reward system, on the value of some variable (such as the net worth of the company) to be obtained at the end of the last period of play, their behavior may differ from what it would have been if the game had continued for more periods. In reality, most corporations and organizations are going concerns, never influenced by such end effects. One way to avoid end effects in games is not to declare in advance how many periods are going to be played or how long the game is going to take. Another, more sophisticated method, which I used in a recent game, is to use one microcomputer as a clock that stops the game at random after a certain time, but before a fixed limit. For example, the game may stop at the earliest after 80 minutes, but at not later than 90 minutes, with the probability of a stop by the end of the eighty-fourth minute being $(84 - 80)/(90 - 80) = 0.4$ (Ståhl, 1986b).

Whether computerized or manual, most games require some kind of *player's manual* describing the game and listing its rules. This manual thus contains a summary of the game's institutional assumptions and becomes the basis for each player's mental model of these assumptions. The clarity of the manual will have an important influence on how well and how quickly the players make their decisions.

In designing a manual for an operational game, one important decision deals with how specific it should be in presenting the details of the game situation. For example, suppose a company in a game wants to test its competitors' reactions to a change in its marketing strategy. Should the player's manual then specify the names of all the companies in the game, or should the manual be less specific, saying that the game depicts a situation "largely similar to our branch of industry?" If the manual is very specific, giving the players for example the names of known corporations, some participants in the game might become excessively preoccupied with the fact that the model leaves out various aspects of the real situation; on the other hand, more detail might in some cases influence the players to be more intensely involved in the game.

All computerized games, as well as some manual games, contain a mathematical model that translates the decisions (the inputs) into the scores (the outputs). Designing this model and coding it are often the most time-consuming parts of the game design. In designing the mathematical equations that relate the decisions to the scores, it is important to strike a balance among realism, viability (or robustness), and, to some extent, simplicity.

It should be noted that simplicity in the computer program is not as essential as simplicity of the inputs and outputs. Owing to the power of today's computers, highly complex calculations can be carried out in a second or less, even on a small personal computer, although it is hardly ever reasonable to allow less than a couple of minutes for each decision by a player.

It is more important for the relations in the game to be *viable*, or *robust*, terms implying that the game should not reach extreme conditions (such as bankruptcy or extremely high monopoly profits) when all the values of the decision variables appear to be normal. In order to avoid such extreme outcomes, the constructor may have to make additions to the original model, such as, for example, temporary government subsidies to avoid bankruptcies or new competition from robot players in case of excess profits from collusion. (See, for example, Ståhl, 1986a.)

A game may need a special kind of robustness that protects it from "kamikaze" players (see, for example, Hutchings and Robertson, 1983). Extreme decisions by one player should not lead to extreme outcomes for all the other players, in, say, a game with a half dozen players. One way to achieve such robustness is to be careful when defining average decision values against which the decisions of each player are measured. As an example, suppose that the sales of each player are partly dependent on how much his price differs from the average price of all the similar products in the marketplace. Company A's sales should not then change very much in case Company B, which already has the highest price, doubles its price. To reflect this fact, the average value should not be too seriously affected by the extreme prices quoted by Company B. In this case the geometric mean might be a better measure than the arithmetic mean. Gold and Pray (1984) discuss the proper forms of mathematical equations defining the relations between decisions and scores, such as the forms for demand and cost functions in management games.

The constructor must also consider how the computer is to handle the input decision values. For example, it is suitable for the computer to carry out forms of *feasibility control* to avoid read-in errors and inappropriate decisions (such as prices outside a preset feasible range).

As regards coding the program, the rules of *structured programming* that are recommended for ordinary simulation models are also recommended for game models (see, for example, Payne, 1982, and Solomon, 1984). There are, however, some features specific to gaming—as the preceding section suggests—that are important. Since the cost of having a game fail owing to programming errors is very high, the principles of "safe programs" free of logical errors are of even greater importance.

The computer code should also include a provision for continuous storing of the decision values on a nonvolatile medium, such as a disk. This provision ensures that a computer crash will not stop the game, and that the computer program can restore the game completely up to the point it had reached when interrupted on the basis of these historical decision data.

If the game has to be played in several locations, it should preferably be *portable*. In this case, it is especially important for its supporting computer programs to be written in a highly standardized language. Many operational games that have been widely transferred (Gernert et al., 1983; Uretsky, Mozes, and MacWilliams 1983) have been written in ANSI Standard FORTRAN IV or 77. In the future, other standardized languages, such as ADA, may be used. Writing the computer programs in a highly modular way and then documenting them thoroughly are also very important if the game is to be reasonably easy to adapt to local conditions, to expand, or possibly to combine with other games (Hutchings and Robertson, 1983).

In this connection, it should be mentioned that recent developments allow games to be played over computer networks, with the players at various locations at any distances from each other. It appears that this kind of gaming is most appropriate for fairly large games. However, to assure that the rules are properly understood, it is usually desirable to have an initial session face to face with the game director.

As a rule, the design phase of an operational game involves writing a *game director's manual*—especially important for games to be played in different locations. This manual describes how to inform the players orally about the rules of the game (as a complement to the player's manual description), how to set initial parameters (such as those depending on local conditions, like the number of players, the time available, and so on), how to recreate the game in case it breaks down, and so on. It may also include an administrator's section on how to select players and staff, how to conduct debriefings, and how to make and report observations as part of the research program.

4.11. Testing the Game

After the game and its supporting models have been constructed, a very important part of game development begins: testing.

The lesson of experience is this: It is very likely that the first version of the game can be improved considerably. Duke and Duke (1983, p. 250) have suggested *the rule of 10*: A game should be tested at least ten times, with revisions to improve it after each test. Although this rule is not intended to apply to games that take a very long time to play, and which will only be used a few times, my experience is that it is suitable for games of smaller size intended for repeated use. This rule is hardly at odds with this further suggestion: After several games have been played without the results suggesting significant changes, it is reasonable to put the game to its intended use with players of the sort that should be involved.

If the game is a computer game, the computer program can be tested by itself before it is used in actual game tests; these tests check whether or not it performs correctly for a suitable variety of inputs.

With the tests of the computer program out of the way, the tests of a computer game can focus on the game manuals, the decision forms, and the output formats to be sure that they are precise, clear, effective, and fully understandable. In a manual game, the focus is also on the efficiency of the corresponding decision and payoff symbols. The test also checks, for any game, to be sure that enough—but not too much—time is given for decisions and how many periods can be played within a specified time. It is not infrequent for the game to have to be simplified in order for the desired time scheme to hold. Finally, it is important to observe how seriously the players act in the game. This can be done by studying the decisions and their results, and, in games involving teams as players, by observing directly the deliberations that lead to the team decisions. At the conclusion of the test games, the debriefing sessions should be as frank and openly critical as it is possible to make them.

The test runs of a game should use different sets of players for most runs in order to increase the variety of the feedbacks. These test players do not have to have backgrounds and experiences like those of actual decisionmakers. For cost reasons, students or other young persons with relevant interests are often used, with satisfactory results.

4.12. The Operational Phase

With the testing completed, the operational phase of the game is ready to begin. The first decisions to be made are to determine the number of runs to be made and to select the players for them.

The Number of Games to Be Played

This chapter has already indicated that, to obtain useful information from a game, particularly an operational research game, it has to be played repeatedly. Before these runs begin, the leader of the gaming project must consider how many runs should be planned for, at least in a contingent manner. In this planning he is likely to consider at least these four topics:

The number of runs likely to be needed to support significant
  conclusions

An experimental design for answering "what if" questions

The desirability of introducing one complication at a time

Developing robot players

Each of these topics will be discussed in turn.

The number of runs needed to support significant conclusions. Although operational research games are usually played repeatedly in order
to buttress the significance of the conclusions to be drawn from them, it
should be stressed that one cannot, in general, say that these conclusions
are drawn in the statistical sense of confirming or denying a null hypothesis to a certain degree of confidence. To do so would require that the
game participants be drawn completely at random from the predetermined
universe of players with the desired properties, a possibility clearly practically remote in most real situations. It is possible, however, to ask this
question: If the player-selection process had been a random process, and
if the choices had been made from a large universe of players with the
desired properties (such as, for a water-planning game, the universe of
all water planners), could appropriately significant conclusions have been
drawn from the results of the game runs? From this point of view, each
game run can be regarded as a sample from a large universe of many
potential game runs, and we know that significant conclusions will require
a reasonable number of runs.

If a game is played three times with the same result each time, this
result is not usually enough to infer, with what is commonly regarded as
statistical significance, that the result is not due solely to random causes.
In the simplest case, with result A as likely as result not-A in the population of all games, three A results in a row can occur one time in eight;
for a result significant at the 5 percent level at least five runs with the
result A would be needed. It is rare, however, to encounter five A results
in a row, so, even in so simple a case, it would be very unlikely for only
five runs to suffice to establish A as the result for the analysis to adopt.
Most games are, of course, more complicated than this trivially simple
example suggests, and one cannot be absolutely sure that some group of
players will not play the game quite differently from the rest of the players.
Thus, the number of runs planned for must not be very small; seldom will
less than ten runs suffice (see, for example, Siegel and Fouraker, 1960).

It may not be suitable to determine the number of runs exactly in advance, but rather to reestimate the number needed after the conclusion
of some runs, provided the number already played is not sufficient to
establish a significant result in this special sense.

An experimental design for answering "what if" questions. Gaming can be used to answer "what if" questions such as this: What happens to factor *C* if factor *A* is changed to factor *B*? In such an experiment, one can have two groups of runs, with an important parameter (or important part of the scenario) fixed at one value in one group, and at another value in the other group. With everything else being the same (except for the players), it is possible for the results to shed some light on the question of what behavior ensues from the preassigned difference in input.

As an example, in the Emission Permits Trading Game, in which four Swedish corporations were allowed to trade emission permits, and thus reduce emissions either more or less than prescribed by the local environmental agency, we worked with two information conditions: in some of the runs, the companies did not know each other's costs; in the other runs, each player had a cost table showing the other players' costs. One question investigated was whether total emission-reduction costs would be lower in the complete-information case than when the information was hidden. (Ståhl, 1986b.)

Even in such a simple situation, two groups of runs are needed to explore the two different inputs, and within each group a number of runs is required to stabilize the random effects, as argued in the previous subsection. Thus, the total number of runs needed if a game is to address "what if" questions is likely to be fairly large, even if the nature of the question is fairly simple.

The desirability of introducing one complication at a time. One way to approach a complex reality is to construct a very complex game, which, owing to its size, can only be played a few times. Another, sometimes better, approach is to play a smaller game many times and to include and vary different key factors in different groups of runs in order to determine which of these factors are important for the larger situation. One advantage of playing smaller games in order to study the variations of one key factor at a time, rather than dealing with a large game with many varying factors, is that this simpler approach is sometimes better related to the mental models of the real situation held by the players (as discussed further in Section 4.14). On the other hand, this simpler approach risks losing the influence of correlated effects between factors that might appear in the larger game, but be lost in the simpler ones.

Developing robot players. This subject is dealt with later in Section 4.15, but it is necessary here to point out that developing reasonable equations to govern the behavior of robot players requires that the game be run with real players a substantial number of times, so that their typical behavior can be characterized.

Selecting the Players

The discussion so far has frequently referred to the desirability of using real decisionmakers, or persons whose behavior can be expected to be similar, as players in operational games. Thus, it is worthwhile here to describe what these terms mean in this context.

When dealing with a specific decision or planning problem, one can usually identify the real decisionmaker; he is the person with the ultimate responsibility for what is decided. It is clearly desirable to involve him in a game dealing with his responsibility. On the other hand, in many cases it is as undesirable as it is impossible to have all the players be actual decisionmakers with relevant responsibilities. For example, consider a marketing game for the purpose of studying the reactions of competitors to changes in the sponsoring firm's sales strategies. Here it is obviously impossible—or at least inadvisable—to ask the executives in the competing firms to participate in the game. However, it is very important to use people with backgrounds similar to those of the executives in the competing firms, insofar as possible, since one wants to simulate the behavior of these competing executives. Thus, with this example in mind (as well as many similar situations), this chapter defines a real decisionmaker broadly as a person who, through background and experience as well as knowledge of the role and behavior of real decisionmakers in situations like the one under study, is likely to be able to play the role of a real decisionmaker very much like the way a real decisionmaker would behave.

The main alternative to using real decisionmakers in this sense has been to use university students. One advantage of using students as players is that they can often be ordered by their professor to participate, particularly if he is the game constructor and this experience is part of their training—and they will usually participate willingly, even if the amount of time needed is considerable. The disadvantage is that students, owing to their ages, inexperience, and rather incomplete knowledge of the problem situation and its context, are often likely to find difficulty in playing their assigned roles in operational games.

In my own experience with games played by both professional decisionmakers and students, I have noted a considerable discrepancy between their modes of behavior. Students behaved more randomly than the professionals, whereas the professionals appeared to be more aware of certain social norms such as the principle of good-faith bargaining, and observed them in the game (Ståhl, 1983, p. 275). Against this background, it is my personal opinion that it is generally quite worthwhile to make the extra efforts to get real decisionmakers to play operational games rather than settle for students as players. For a broader discussion of this issue, see Alpert (1967) and Khera and Benson (1970).

4.13. Gaming for Testing Models

This section and the three that follow discuss some special aspects of gaming in more detail than is included in what precedes. This section looks at how gaming can be used to test another model, such as game theory or an ordinary simulation model.

The Purpose of Gaming Tests

Before going into more detail about such testing, it is important to look at the purpose of the test. Broadly speaking, the question being addressed is this: Is the theoretical model valid for decisions in reality? Gaming cannot give a full and general answer, but it can offer certain critical steps toward an answer. To explain this point we begin with Figure 4.3, which shows the relationship between theory and reality in the context of this chapter. Within the larger relation of the model, or theory, and reality, we are interested in three relations: R_1, the relation between the institutional assumptions in the model and the actual decision environment; R_2, the relation between the behavioral assumptions in the model and the decisionmakers; and R_3, the relation between the model outcomes and outcomes in reality.

Of the three relations within the larger relation, we are ultimately most interested in R_3, particularly as regards future applications of the model. As it relates to gaming, the relation R_3 can involve three main sorts of

Figure 4.3. The relationship of a model, or theory, and reality. (*From Ståhl, 1983, p. 215.*)

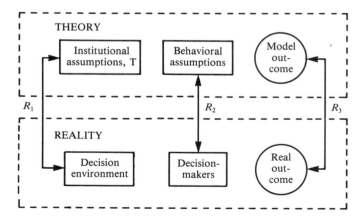

correspondence: predictive, descriptive, and normative. In the predictive and descriptive cases, the model outcome has direct relevance without the decisionmakers knowing about the theoretical model. In the predictive case, the model will, prior to the actual event, forecast the outcome; in the descriptive case, it generates a similar outcome after the event. In the normative case, the model outcome is relevant to the real outcomes by influencing the decisionmakers to reach a solution close to the model outcome. When applied to a specific decision situation, the main question is whether the results suggested by a theoretical model on the basis of some specific behavioral assumptions correspond to the observed outcomes, will soon become true, or are acceptable to the decisionmakers.

Since we are generally interested, however, in a broader application of the model to situations with differing arrangements of the institutional assumptions, the focus of attention shifts from relation R_3 to R_2, that is, the relation between the behavioral assumptions of the theory and the way actual decisionmakers will (or want to) behave. The important questions for the predictive and descriptive aims then become: Will the parties behave, or have they behaved, in line with the behavioral assumptions of the model? In the case of the normative aim, this question can be restated as: Will the decisionmakers want to behave in line with the behavioral assumptions of the model?

It is in answering this type of question that gaming comes into the picture as a useful tool. Figure 4.4 shows the relationship between a model and reality (as depicted in Figure 4.3) with operational gaming introduced at an intermediate level. In line with this chapter's earlier definitions, gaming activity consists of the players and the game, which includes only institutional assumptions; the behavioral assumptions of the model have as their counterparts in the gaming activity the behavior of the players.

What gaming can do is to test the relevance of the theory for game playing. As mentioned, the main interest is in estimating the validity of R_2, which depends on R_5 and R_8. By choosing suitable players for the game, one can get a high degree of correspondence between the players and real decisionmakers and can hope for R_8 to be valid. Since, as mentioned earlier, one can make the institutional assumptions of the game very similar to the institutional assumptions of the theory, that is, make R_4 valid, R_6 can be used to check the validity of R_5.

In view of this argument, one examines the degree of similarity between model outcomes and game outcomes to determine the similarity between the behavioral assumptions of the model and the way the players behave—or want to behave—in the game. Thus, in order to test the relevance of the theory for decisions in reality, one first checks the relevance of the theory for the play of the game.

The model's predictive ability for reality is thus tested by testing its

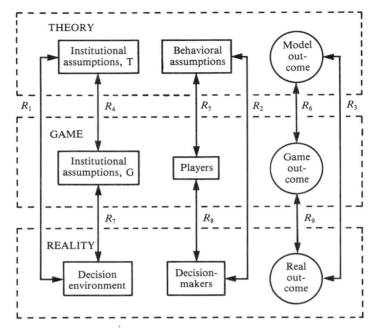

Figure 4.4. The relationship between a model and reality with operational gaming introduced as an intermediate model. (*From Ståhl, 1983, p. 217.*)

predictive ability for the game. It is appropriate to construct the model before running the game; otherwise building the model could become merely an exercise in curve fitting.

Testing a model's normative relevance for reality is more complicated. According to the traditional idea of a normative model, its main contribution is in helping the decisionmaker deduce (and possibly compute) his solution to the problem at hand. According to this approach, testing the validity of the model for normative purposes consists only of finding out whether the decisionmaker says that he wants to follow the stated behavioral assumptions. There are, however, two significant problems with such an approach:

Behavioral assumptions often cannot be tested in an abstract situation that is not connected to a concrete decision situation. Behavioral assumptions in an abstract setting often seem to decisionmakers to be meaningless generalities.

Even if a decisionmaker is able to establish that he wishes to behave according to the behavioral assumptions, it is by no means certain that he therefore wants to act in accordance with the model. He may distrust the deductions of the model if he believes—sometimes on justifiable grounds—that some of its deductions are erroneous. Too, he may fear that deducing the solution involves other implicit assumptions to which he may not subscribe. Or, he may just feel uncomfortable about relying on a procedure that he does not understand completely.

Because of these problems, it appears that methods of testing the normative aim of a model should be focused on this question: Would decisionmakers want to use the game model if they had been properly and thoroughly introduced to it, and would they, after having used it, want to use it again?

Therefore, one way of testing a normative model experimentally is for the game director to instruct the players about it and then let them play the game to see if they use it. There is, however, the problem that the desire to conform to higher authority may override other considerations in the experiment, particularly if the players do not gain or lose considerable amounts. In order to avoid such effects, one can present the players several normative theories and observe which ones they choose, but presenting them in an adequately detailed way would be time-consuming for the players.

An analyst may instead choose to run only one normative model in each experiment, with a new experiment and new subjects for each new model, or to run two different models at a time in a kind of tournament. Another approach is to give each player information on one model in the form of a consultant's report containing arguments for using the model presented to him (Ståhl, 1980). Another way to test the normative value of a model is to use a robot player, that is, a computer program that behaves according to the model's prescriptions (this is discussed in Section 4.15).

A completely different approach to normative testing is to run the test mainly in a predictive way without any normative influences of the types described above. This procedure avoids any authority effects. In this approach, differences between the normative and the predictive aims are differences of degree and not of kind. In the normative test, the analyst takes measures to increase the likelihood that the players will behave according to the behavioral assumptions; he selects players carefully, increases the time allowed for play, supplies the players calculation aids, increases monetary incentives, and so on.

Conclusions from Gaming Tests

Suppose that an analyst has tested a model by game playing, for example, in the way described above. What conclusions can he draw? There are two main cases: the positive case, in which the game outcome is similar to that of the model, and the negative case, in which the game outcome is different from that of the model.

The positive case: the model outcome supported by the game. Using the notations of Figure 4.4, we now have that R_6 holds. Even though R_6 and R_4 hold, it is not certain that R_5 holds, that is, that the players behave in accordance with the model's behavioral assumptions, since it is quite possible for another set of behavioral assumptions to lead to the same model outcome. In order to be certain that R_5 holds, we must either analyze the model structure to see whether other behavioral assumptions could possibly lead to this outcome, or study the whole game process to see whether the players seemed to behave in line with the model. Both of these tasks are, of course, very difficult. One cannot study all possible behavioral assumptions, and models often deal with implicit thought processes.

Let us assume, however, that through such an analysis we have become reasonably certain that R_5 holds, that is, that the players behave in the same way as in the model. We have already assumed that R_8 holds. Then, only in the very special case where R_1 also holds, implying that the institutional assumptions of the game map very closely onto reality, can we be reasonably confident in applying the model in the real world.

In the more probable case that R_1 does not hold, there is still the problem that the decisionmakers may behave differently in the more complex reality than they behaved in the game. Several factors suggest that this is the case (Ståhl, 1983, p. 226).

To sum up, a positive result in a game very seldom allows one to draw the conclusion that the model can be applied confidently to reality. At best, one may revise the subjective assessment of the model's probable applicability in the real world. Depending on the cost of gaming relative to the cost of making a mistake in applying the model, in this situation one may proceed to further, more elaborate, tests of the model.

The negative case: the model outcome refuted by the game. The conclusion that R_6 does not hold implies that decisionmakers do not follow the behavioral assumptions of the model. To analyze the implications of this apparent conclusion further, however, two cases may be distinguished on the basis of how well the institutional assumptions of the game represent the real decision environment, that is, whether R_7 holds or does not hold.

Institutional assumptions correspond to reality: R_7 holds. If R_8 holds, that is, if the players are "decisionmaker similar," one can conclude that the behavioral assumptions of the model are not valid in reality. This is the clearest case for using gaming as a diagnostic test: If gaming refutes the model, there is no reason to try to apply it to the real situation.

Institutional assumptions differ from reality: R_7 *does not hold.* This case does not allow such strong conclusions to be drawn. Even if the players are decisionmaker similar, their behavior in reality may be different, owing to the fact that R_7 does not hold. Without further investigation, one cannot rule out the possibility that decisionmakers will behave differently in the generally more complex real environment. Indeed, one may note that the chance cannot be ruled out that they may, in reality, act in accordance with the behavioral assumptions of the model. Consequently, R_7 must be looked at closely to see how the institutional assumptions of the game differ from the real environment and to what extent the differences cause changes in behavior. Ståhl (1983, p. 227) discusses institutional factors such as time compression and incentives for decisions that could affect behavior. On balance, it appears in this case that a negative outcome in the gaming experiment offers evidence to support the hypothesis that the model is not valid for use as a representation of reality. This suggests that the theoretical model should not be applied.

Conclusion. In sum, testing a model by gaming mainly functions as a test to screen models that one wants to use for real decisions. Models unsuccessfully verified in gaming experiments can be rejected as invalid for real applications. On the other hand, success in gaming experiments is far from sufficient to ensure successful real-world model applications. Rather, success in gaming experiments is equivalent to encouraging further tests of the model.

Finally, it should be stressed that rejecting a model on the basis of a negative gaming outcome need not mean that the model should be scrapped; more often it means that a redesign should be considered, making use of the experience gained from the gaming experiments. This remark indicates another important function of gaming: a generator of ideas for constructing new or revised theoretical models. When a model is revised after gaming tests, it will have to be tested again. Hence, to use gaming in testing is only one link in an iterative process of model construction, testing, revision, and further testing leading step by step to a model that is more valid for real applications.

4.14. The Size of the Game

This chapter has noted frequently that the size of the game is one of its most important properties, and one about which a good decision should be made. A key question is whether or not it is enough to have a small game.

To make the discussion about size meaningful, it is essential to define what is meant by the term *size* when it refers to a game. The size of a game can be measured in a variety of ways, such as:

1. The time required to play the game
2. The number of players involved
3. The resources used for constructing the game
4. The total size of the game model on the computer (when the game is computerized)
5. The total size of the game's physical paraphernalia (the size of its board, the sizes of its manuals, and so on)
6. The total size of the entire gaming activity measured, for example, in time and money

This section has points 1 and 2—and to a lesser extent point 3—in mind, and regards the others as somewhat less immediately relevant. In particular, point 6 is not necessarily directly related to the size of the game; in some situations with a given total amount to be spent on gaming, the choice is often between using a small game many times or a large game just a few times.

A small game, therefore, is one with a relatively small number of players that can be played in a modest length of time. With this definition in mind, it is possible to discuss the advantages of keeping games small, the point of departure being a situation in which there is a given fixed budget for gaming. In this situation, should a small game be designed to be played many times, or should a large game be chosen that can be played only once, or at most a few times?

The advantages of choosing a small game will be presented under five headings: the advantages of repeated play, the greater likelihood that real decisionmakers will play small games, testing small models, the motivations of the players, and start small—get started fast.

In the early eighties, when professional people were still buying microcomputers with only 64K bytes, I included in a similar list the possibility of using small and inexpensive computers. Today, however, when it is possible to buy a relatively small computer with 640K bytes for less than a thousand U. S. dollars, even large games can be run on relatively inexpensive computers.

The Advantages of Repeated Play

The substantive advantages of repeated play have been dealt with in Section 4.12. Here, it suffices to point out that the smaller the game the more times it can be repeated within the framework of a given budget.

The Greater Likelihood that Real Decisionmakers Will Play Small Games

The main reason that real decisionmakers are reluctant to take part in a large game is undoubtedly their hesitation, in the face of many other pressing demands, to commit a major block of time to what must inevitably appear to be a somewhat speculative enterprise. Busy executives will find it hard to spend several days participating in a game even if they are absolutely sure that it has something substantial to offer.

Another factor that makes businessmen and executives reluctant to take part in games during business hours is that the name seems to connote a lack of seriousness. On the other hand, persons unwilling to play during office hours are often prepared to play in the evening—for example, for three hours after dinner, when it is not regarded as improper to be involved in such an ostensibly frivolous activity as game playing. My experience is that it is difficult to get qualified decisionmakers to play a game during office hours, even for just a day, but I have been quite successful in getting highly qualified people to play games during the evening.

Based on this experience I can give a quite concrete definition of a small game: one that is playable in an evening of at most three to four hours.

Testing Small Models

Most real problems are complex, but the models people have of them are much simpler (see, for example, Miller, 1956). In order to grasp a problem mentally, a person has to make a mental model of it, focusing on a small number of factors that he regards as significant. This mental model is generally verbal, and only seldom does it contain any mathematical elements. Just as for a formal and overtly explicit model, a mental model can be divided into two parts: institutional and behavioral assumptions. Thus, the mental model includes a small set of institutional assumptions dealing with what the person perceives to be the most important aspects of the complex situation. Similarly, it also includes a few behavioral assumptions about how the most important actors will behave.

This chapter has argued that gaming can be used to test the behavioral assumptions associated with a given set of institutional assumptions. Thus, to test the mental hypotheses that a person has about how people will behave in the face of the institutional assumptions of his mental model, one can construct a game that replicates these institutional assumptions. Hence, if the person has a small and simple set of assumptions in his mental model of the situation, the game that replicates its institutional assumptions should contain only a comparably small and simple set of them.

Consider an example. A corporation is involved in a competitive market and wants to test some hypotheses about the responses of its competitors to a change in its own marketing strategy. Although there are a score of competing corporations in this market, according to the corporation's management only 4 of them really matter. In other words, in the mental model of the corporation's managers the institutional assumptions contain only four competitors. Since the management's hypotheses regarding its competition are based on this small and simple set of competitors, an appropriate game to test the hypotheses about the behavior of the competition in the face of a new marketing strategy should also deal with this small set of 4 competitors and not with the larger set of 20.

The Motivations of the Players

The following hypothesis, arising from experience with educational gaming, can be extended to operational games: Players are generally more highly motivated to play well in a small game of a few hours' duration than in a large game taking several days. In a game played over several days—perhaps stretched out over several weeks—it is fairly common to observe that players get bored with the game, and some will drop out of it, or at least minimize their involvement. Small games also have the motivational advantage of rules that are easy to learn, thus allowing the participants to start playing after only a short introduction.

Start Small—Get Started Fast

There are many situations in which small games have the problem of lack of realism, and therefore there are good reasons to aim for a larger game. However, even if the ultimate aim is a large game, there can be advantages in starting with a small game and building it into a large game through a series of steps. Starting with a small game requires less development effort before it is possible to get feedback on the initial and central hypotheses, and it is sometimes even possible to draw early conclusions on the basis of the smaller initial game. If these conclusions are strong enough and cover the main questions that prompted the gaming effort in the first place, then there is no reason to extend the effort. On the other hand, if the analyst decides to build a larger game, the feedback from playing the smaller game can often be useful in indicating the directions in which the game needs to be extended, such as, for example, those in which the players have regarded the smaller game as simplistic.

4.15. Developing Robot Players

Robot players are artificial players that have exact instructions on how to play in every conceivable situation. Today almost all robots in games are in the computer programs associated with the game, but in principle a human could play the role of a robot by following a comprehensive set of instructions exactly. A computer program playing the role of a robot in a game should preferably pass the so-called Turing test: A human shall not be able to say whether the role played by the robot is played by another human or by a robot (Turing, 1963; Hoggatt et al., 1978).

There are several reasons for introducing robots into a game:

Robots provide a practical way of making a game that is small with regard to the number of human players involved into a larger game with more players. For example, in situations where it is difficult to find enough suitable players, robot programs can play some roles, especially those that are not critically important to the outcome of the game (Shubik, 1975a, p. 238).

Robots provide a way of minimizing random variations due to differences in players and their playing styles. Since robots can be made to play in exactly the same way in, for example, both of two experimental runs, the observed variability in the two runs must then be due either to the difference in the run designs or to the behavior of the remaining human players.

To test the normative value of a model, one can proceed by having robots act in accordance with the normative model. The game can then investigate whether or not their behavior has enough impact to induce the human players to also act in accordance with the normative model.

Developing robots requires repeated game playing. Experience with experimental gaming and computer simulations to replicate the outcomes of gaming experiments has shown that it is often virtually impossible to specify any reasonable behavioral equations for the players, even in very simple games, without first having studied a number of actual runs of the game. After running the game a large number of times, it is possible to derive these behavioral equations from the behavior of the actual players.

As an example, consider determining a robot's equation for setting a price in a simple management game (Ståhl, 1982, Chapter 3). The robot's decision on price in a period t can be expressed as a linear function of his own price in period $t - 1$, the change in his inventory during this preceding period, and the difference between his own price and the average market price in the preceding period. To establish the coefficients of this equation, we used linear regression on data from a number of games with real players.

In this connection, it should be noted that, by using a game repeatedly, it may be possible to replace more and more players with robots, perhaps even making it possible ultimately to reach an ordinary one-person simulation of a game situation. In this way, it is possible to view gaming, in some circumstances, as a constructive device for developing ordinary simulation models; or it may be said that one-person games are simulations that are limiting forms of gaming as I have defined it in this chapter. (See Bowen, 1978.)

4.16. Interactive Man-Computer Dialogues for Determining Game Parameters

It is particularly important in an operational game to have data and forecasts that are as valid as possible. It is, however, often difficult for just one or two analysts to gather all of the data necessary to specify the correct parameters of the payoff functions, such as those for demand and cost. It is also important for the players in the operational game to regard the game as reasonably realistic, particularly with respect to the aspects with which they are most familiar.

One way of making operational gaming activity more efficient in these respects is for the players to use an interactive man-computer dialogue procedure in a pregame session to estimate the most essential parameters of the game (Ståhl, 1983, pp. 309–321). The procedure is that the computer asks a player sitting at a terminal a number of questions about various data; for example, for a marketing game, the computer might ask about the quantities demanded at prices A and B. Using the responses supplied by the player, the computer then calculates a demand function based on a preassigned functional form and uses this function to present a table or curve, including extrapolated values. The player can then accept or reject this function. If he accepts it, it goes into the computer to be used in the game; if he rejects it, he gets a chance to supply additional point estimates, which the computer then uses to prepare a revised demand function, which is presented for decision as in the first step. This process can continue until the player is satisfied that the game includes a demand function that accords with his knowledge and experience.

Sometimes the computer will have relevant data in its data base before this dialogue process starts. If so, it can present these data and the estimates based on them to the player, a step helpful to him if he does not start with a very good mental data base of his own.

This dialogue procedure helps to make the assumptions of the game more transparent to a player than they might be otherwise. Even if the player does not change any of the parameters and just accepts the computer's suggested data and functions, going through the dialogue can still be very worthwhile, since it makes the player fully aware of the data assumptions behind the functions used in the game as well as the functional forms of the relations being used.

There may be cases where the players are unwilling to take the game seriously because it seems to be a black box that hides the assumptions behind the payoff functions. Players with this view tend to regard the responses of the payoff function to their decisions as arbitrary, and possibly unrealistic. The dialogue procedure can help to overcome such a problem, since it gives the players a clear view of the payoff functions and how they behave, thus making them appear less arbitrary. Such enhanced transparency also facilitates constructive criticism of the game structure, thus contributing to fruitful discussion at the end of the game.

Transparency alone does not always act to increase a player's confidence in the game, and thus induce him to take it seriously. If a player regards some data or an assumption, now made transparent by a dialogue, to be unrealistic, his view of the whole will be seriously discolored by this fact. On the other hand, he has before him an avenue that allows him to make changes to bring the game toward his view of realism. Care must be taken, however, to guard against players changing the data merely in order to benefit in the game.

4.17. How to Proceed toward Gaming

The purpose of this chapter, in line with the purpose of this volume, is not to describe how to design a game or how to play it and report its findings, but rather to discuss the issues and choices facing a systems-analysis team that is considering using operational gaming as a tool of analysis. A secondary purpose, however, is to refer the reader who wants information on such operational details to literature that will give him entry to the professional experience in this field.

In the course of the chapter suitable references have been made to the literature when it was appropriate. For example, Section 4.8 gave some sources of information on available games.

As a whole, the literature on gaming is striking for both its breadth and the variety of the backgrounds of game constructors. Thus, the first problem facing an explorer of this literature is to locate the portion of it that is apposite to his interests. It is, in fact, quite difficult to get a good overview of the field in order to guide such a search, so that a beginner is probably better off to search for bibliographies, and the lists at the end of this chapter are aimed at helping him begin. Owing to the great variations in the gaming literature, there are no bibliographies that are generally accepted as standard. Hence, the reader may or may not find the ones referred to here useful, depending on his interests. It is hoped, however, that, together with the references used in this chapter, these bibliographies will open the door to appropriate literature.

The gaming literature is growing rapidly, whereas much of the earlier literature, particularly that on computer-supported gaming, is rapidly becoming outdated. Therefore, it would appear that the ideal way to keep an up-to-date listing of the most relevant current literature would be on a computer, where the listing can be updated easily. Section 4.8 referred to some computer-stored game banks, but the need for a more general up-to-date computer-stored literature listing has yet to be met. Thus, the serious worker must rely on the classical approach in science and technology, keeping in close touch with the journal literature and the activities of professional associations in the field (some of which are listed in Section 4.8). There are also university programs in gaming (for example, at the University of Michigan).

Although the literature of this field is valuable, it is my strong conviction that it cannot take the place of active participation in games, both in their construction and play; this is the only really effective and efficient way to learn the methodology of this field. Gaming associations organize meetings where attendees can take part in many different kinds of games; besides being very instructive, participation in such games, which are usually of very high quality, can often be very enjoyable.

By ending on this note I hope to convey an important lesson of my experience: that gaming, besides being a useful tool of analysis, is in general a highly enjoyable approach to systems analysis.

References

Items Referred to in the Text of the Chapter

Alpert, B. (1967). Non-businessmen as surrogates for businessmen in behavioral experiments. *Journal of Business* 40(1), 203–207.

Assa, I. (1983). Management simulation games: A comparative study of gaming in the socialist countries. In Ståhl (1983), pp. 63–81.

Ausubel, J., J. Lathrop, J. Robinson, and I. Ståhl (1980). *Carbon and Climate Gaming*, WP-80-152. Laxenburg, Austria: International Institute for Applied Systems Analysis.

Axelrod, R. (1984). *The Evolution of Cooperation*. New York: Basic Books.

Bowen, K. C. (1978). *Research Games*. London: Taylor and Francis.

Brewer, G. D., (1978). Scientific gaming: The development of free-form scenarios. *Simulation and Games* 9(3), 309–338.

——, and M. Shubik (1979). *The War Game—A Critique of Military Problem Solving*. Cambridge, Massachusetts: Harvard University Press.

deLeon, P. (1981). The analytic requirements for free-form gaming. *Simulation and Games* 12(2), 201–231.

Duke, R. D. (1974). *Gaming: The Future's Language*. Beverly Hills, California: Sage.

—— (1980). A paradigm for game design. *Simulation and Games* 11(3), 364–377.

——, and K. M. Duke (1983). Development of the Conrail game. In Ståhl (1983), pp. 245–252.

————, and C. S. Greenblat (1979). *Game-Generating Games*. Beverly Hills, California: Sage.

Easterly, J. L. (1978): Simulation game design: A philosophic dilemma. *Simulation and Games* 9(1), 23–28.

Ellington, H., E. Addinall, and F. Percival (1982). *A Handbook of Game Design*. London: Kogan Page.

————, ————, and ———— (1984). *Case Studies in Game Design*. London: Kogan Page.

Gernert, H. R., I. Assa, M. Habedank, and W. Wagner (1983). The transfer of games between the socialist countries. In Ståhl (1983), pp. 117–128.

Gibbs, G. I. (1978). *Dictionary of Gaming, Modeling, and Simulation*. Beverly Hills, California: Sage.

Gold, S. C., and T. F. Pray (1984). Modeling market- and firm-level demand functions in computerized business simulations. *Simulation and Games* 15(3), 346–363.

Greenblat, C. S., and R. D. Duke, eds. (1974). *Gaming-Simulation: Rationale, Design, and Applications*. New York: Halstead.

Hoggatt, A. C., H. Brandstätter, and P. Blatman (1978). Robots as instruments in the study of bargaining behavior. In *Bargaining Behavior: Contributions to Experimental Economics 7* (H. Sauerman, ed.). Tuebingen: J. C. B. Mohr.

Hutchings, D. J., and W. C. Robertson (1983). Transferring a computer-based management game between capitalist countries. In Ståhl (1983), pp. 107–116.

Jones, W. H. (1985). *On Free-Form Gaming*, N2322-RC. Santa Monica, California: The Rand Corporation.

Khera, I. P., and J. D. Benson (1970). Are students really poor substitutes for businessmen in behavioral research? *Journal of Marketing Research* 7, 529–532.

Lederman, L. C. (1984). Debriefing: A critical reexamination of the post-experience analytic process with implications for its effective use. *Simulation and Games* 15(4), 415–431.

Marshev, V. (1983). Gaming in the USSR. In Ståhl (1983), pp. 83–96.

McHugh, F. A. (1966). *Fundamentals of War Gaming*. Newport, Rhode Island: U.S. Naval War College.

Miller, G. A. (1956). The magical number seven, plus or minus two: Some limits on our capacity for processing information. *Psychological Review* 63(2), 81–97.

Miser, H. J., and E. S. Quade, eds. (1985). *Handbook of Systems Analysis: Overview of Uses, Procedures, Applications, and Practice*. New York: North-Holland. Cited in the text as the *Overview*.

Overview. See Miser and Quade (1985).

Payne, J. A. (1982). *Introduction to Simulation Programming Techniques and Methods of Analysis*. New York: McGraw-Hill.

Rhyne, R. F. (1974). Communicating holistic insight. In Greenblat and Duke (1974), pp. 25–28.

Schuenemann, T. M. (1983). Operational research methods as decision aids in gaming. In Ståhl (1983), pp. 295–308.

Shubik, M. (1968). Gaming: Cost and facilities. *Management Science* 14(11), 629–660.

——— (1975a). *Games for Society, Business, and War: Towards a Theory of Gaming*. New York: Elsevier.

——— (1975b). *The Uses and Methods of Gaming*. New York: Elsevier.

Siegel, S., and L. Fouraker (1960). *Bargaining and Group Decision Making*. New York: McGraw-Hill.

Solomon, E. (1984). *Games Programming*. Cambridge, England: Cambridge University Press.

Ståhl, I. (1980). *Cost Allocation in Water Resources—Two Gaming Experiments with Doctoral Students*, WP-80-134. Laxenburg, Austria: International Institute for Applied Systems Analysis.

——— (1982). *Six Small Games for Research and Education*, internal discussion paper. Laxenburg, Austria: International Institute for Applied Systems Analysis.

———, ed. (1983). *Operational Gaming: An International Approach*. Oxford, England: Pergamon.

——— (1986a). The development of a small business game. *Simulation and Gaming* 17(1), 104–107.

——— (1986b). *The Emission Permits Trade Game*, EFI Working Paper. Stockholm, Sweden: Stockholm School of Economics.

———, and J. Ausubel (1981). *Gaming Approaches to World Coal Trade and the CO_2 Issue*. Paper presented at the 12th annual International Simulation and Gaming Association conference, Haifa, Israel.

———, R. Wasniowski, and I. Assa (1981). *Cost Allocation in Water Resources—Six Gaming Experiments in Poland and Bulgaria*, WP-81-83. Laxenburg, Austria: International Institute for Applied Systems Analysis.

Thiagarajan, S., and H. D. Stolovitch (1979). Frame games: An evaluation. *Simulation and Games* 10(3), 287–314.

Turing, A. M. (1963). Computing machinery and intelligence. In *Computers and Thought* (E. E. Feigenbaum and J. Feldman, eds.). New York: McGraw-Hill, pp. 11–38.

Uretsky, M., L. Mozes, and H. MacWilliams (1983). Transfer of gaming technology: A U. S.—Hungarian case story. In Ståhl (1983), pp. 129–142.

Yefimov, V. M., and V. F. Komarov (1982). Developing management simulation games. *Simulation and Games* 13(2), 145–163.

Listings of Games

Basinger, A. M. (1984). A bibliography of *Simulation and Games* game reviews. *Simulation and Games* 15(4), 500–504.

Bierstein, M., and A. Timofievsky (1980). *Catalog of Soviet Games*. Leningrad. In Russian; contains a list of 136 Soviet games.

Elgood, C. (1984). *Handbook of Management Games*. Aldershot, Hampshire, England: Gower.

Gibbs, G. I. (1974). *Handbook of Games and Simulation Exercises*. Beverly Hills, California: Sage. Lists 2000 games and simulations.

Horn, R. E., and A. Cleaves (1980). *The Guide to Simulations/Games for Education and Training*, 4th ed. Beverly Hills, California: Sage. Lists 1100 games.

Rohn, W. E. (1984). *Deutsche Planspiel-uebersicht*, 2 Auflage. Wuppertal, Federal Republic of Germany: Deutsche Planspiel-Zentrale. Lists 200 German games.

Stadsklev, R. (1975). *Handbook of Simulation Gaming in Social Education—Part 2: Directory*. Tuscaloosa, Alabama: Institute of Higher Education Research and Services, University of Alabama.

Gaming Bibliographies

Cruickshank, D. R., and R. A. Telfer (1979). *Simulation and Games: An ERIC Bibliography*. Washington, D. C.: ERIC Clearinghouse.

Duke, R. D. (1969). *Operational Gaming and Simulation in Urban Research: An Annotated Bibliography*. Ann Arbor, Michigan: Environmental Simulation Laboratory.

Graham, R. G., and C. F. Gray (1969). *Business Games Handbook*. New York: American Management Association.

Greenblat, C. S. (1972). Gaming and simulation in the social sciences: A guide to the literature. *Simulation and Games* 3, 477–491.

————, and R. D. Duke (1981). *Principles and Practices of Gaming Simulation*. Beverly Hills, California: Sage. Bibliography, pp. 255–280.

Guyer, M., and B. Perkel (1972). *Experimental Games: A Bibliography (1945–1971)*. Ann Arbor, Michigan: Mental Health Research Institute.

Kidder, S. J. (1971). *Simulation Games: Practical References, Potential Use. A Selected Bibliography*, Report No. 112. Baltimore, Maryland: Center for Social Organization of Schools, The Johns Hopkins University.

Publications Department, The Rand Corporation (current). *Special Bibliography on Gaming*. Santa Monica, California: The Rand Corporation.

Rohn, W. E. (1984). *Literaturliste: Planspiel und Simulation*, 3 Auflage. Wuppertal, Federal Republic of Germany: Deutsche Planspiel-Zentrale.

Shubik, M. (1975). *The Uses and Methods of Gaming*. New York: Elsevier. See Chapter 7: A guide to information sources on gaming and related topics.

————, and G. D. Brewer (1972). *Reviews of Selected Books of Gaming and Simulation*, R-732-ARPA. Santa Monica, California: The Rand Corporation.

————, ————, and E. Savage (1972). *The Literature of Gaming, Simulation, and Model Building: Index and Critical Abstracts*, R-620-ARPA. Santa Monica, California: The Rand Corporation.

Smith, P. (1975). *Bibliographie rond Operationele Spielen*. Utrecht: University of Utrecht.

Chapter 5
Social Experimentation
Some Whys and Hows

Rae W. Archibald and Joseph P. Newhouse

5.1. Introduction

Social experimentation should occupy an important place in the system analyst's toolkit. Riecken and Boruch (1974) offer the standard definition of a social experiment:

> An experiment is one or more treatments (programs), representing intervention into normal social processes, that are administered to some set of persons (or other units) drawn at random from a specified population; observations or measurements are made to learn how (or how much) some relevant aspects of their behavior differ from those of a group receiving either another treatment or no treatment, also drawn from the same population.

Social experimentation is sometimes likened to the laboratory experiment in the natural sciences. But because the social experimenter has somewhat less control than the laboratory scientist normally has, a better analogy is with the clinical trial in medicine (Cochrane, 1972).

In comparison with other methods, the great advantage of an experiment is that from it, when it is properly designed and executed, the analyst can infer confidently that a given intervention (program) actually caused a given result (Riecken and Boruch, 1974; Gilbert, Light, and Mosteller, 1975).

In many situations, however, the analyst is called in after the fact, to work with data already collected by some experimental program or treatment in which the analyst may have had little or no say in the selection of sample groups. For example, the program may have been administered to those who applied for it first, or who were judged to be in greatest need; the analyst's task may be to infer what effect the program might have if extended to the general population. The analyst may not even have had much say over what data were collected.

Such situations are termed quasi-experimental or demonstration projects; appropriate research protocols for these situations are described by Campbell and Stanley (1966). We do not discuss such protocols here, although we do describe why a true experiment will usually be more informative.

This chapter consists of three parts: Sections 5.2, 5.3, and 5.4 discuss when to conduct a social experiment and when not to; Sections 5.5 through 5.11 explain how to manage a social experiment; and Section 5.12 gives some practical advice ("tips") to experimenters. Section 5.13 offers a brief summary of these three parts. This chapter's intended audience consists of people who face decisions on undertaking social experiments and persons who will either monitor their progress or actually conduct them.

This chapter reflects the authors' interpretation of views dominant in the literature, but readers should be aware that alternative views exist.

5.2. An Example of an Experiment's Potential

A nonexperimental study done by Kessner et al. (1973) illustrates the value of an experiment in making causal inferences. These researchers studied infant mortality among mothers who did and did not receive adequate prenatal care, according to the researchers' criteria. Controlling for the mothers' socioeconomic status, the mortality rate among children whose mothers received adequate care was found to be 16 percent below the rate among children whose mothers did not.

On its face, this result suggests that a large-scale prenatal care program could substantially lower the infant mortality rate. Yet one would be rash to draw this inference, because mothers who received adequate care probably differed systematically from those who did not—even controlling for socioeconomic status. For example, one criterion for adequate prenatal care was that the mother voluntarily sought care in the first trimester. It is likely that mothers who did so were also more motivated to exercise and keep their weight down than were mothers who did not. In such a case, a difference in infant mortality rates might well have been observed between the two groups of mothers even if they had all received adequate care. In other words, some of the difference may have been due not to prenatal care but to characteristics of the mothers.

How much a prenatal care program for all mothers might reduce infant mortality rates cannot be ascertained using Kessner's methods, but a properly conducted social experiment could yield the answer. What would the experiment look like? The new program could be instituted with a randomly chosen group of mothers. If, after a period of time, the infant mortality rate differed between that group and a similar group who did not receive the program (but could, of course, seek care on their own), the program could be instituted generally. If differences did not appear (and the analysis were precise enough to rule out all but negligible differences), a more effective program design could be attempted.

If one chose not to conduct an experiment, the alternatives would then be: (1) not to institute the program at all; (2) to institute it universally; (3) to institute it partially, but not on a randomized basis. Let us assume that enough is known to make the potential program appear promising; in particular, the odds that the program will successfully reduce the infant mortality rate justify the resources that an experiment would require. Otherwise, the program should probably not be instituted at all.[1]

To institute the program universally, one must be confident that the program will "work"; but it may be infeasible to do so for administrative or budgetary reasons. If so, an experiment is almost certainly preferable to nonrandom partial introduction, because it will provide evidence on whether the program should be continued.[2] Even if one could implement the program universally, it is often preferable to evaluate it experimentally, and this requires nonuniversal implementation.

This chapter considers the decision to undertake an experiment, its organization, and some possible pitfalls in designing it. The discussion is based on the authors' experience at Rand with designing and managing a social experiment, the Health Insurance Study (HIS), which began in 1971 and the field work for which ended in 1982. (For a description of the study's experimental design, see Newhouse, 1974.) We do not describe basic experimental design, because many texts and monographs are available.[3]

---

[1] Justification for introducing the program in spite of its high cost would have to rest on assumed improvement in nonmeasurable outcomes, for example, social cohesion. If this were the case, an experiment of the type described would not be useful, as pointed out below.

[2] If it is argued that some mothers are clearly more in need, it may be possible not to randomize the neediest of mothers and still conduct an experiment.

[3] Aigner, 1979; Campbell and Stanley, 1966; Cochran and Cox, 1957; Cook and Campbell, 1976; Cox, 1958; Federov, 1972; John, 1971; John and Quenouille, 1978; Kempthorne, 1952; Kendall and Stuart, 1968; Scheffé, 1959; Srivastava, 1975. Discussions of social experimentation, including many examples of experiments not discussed in this paper, may be found in Boruch and Riecken, 1975; Boruch et al., 1978; Ferber and Hirsch, 1978; Orr, 1974; Orr, Hollister, and Lèfkowitz, 1971; Plott, 1979; Riecken, 1977; Riecken and Boruch, 1978; Rivlin, 1974; and Wilson, 1974, in addition to the other references herein. Boruch (1974) presents a list of illustrative experiments. Greenberg and Robins (1986) give an overview of 38 social experiments conducted in the United States since 1968; tables summarize the key facts about each.

5.3. The Advantages of Social Experimentation

Determining Causality

The most important advantage of social experimentation has already been identified: A properly designed and executed experiment can provide strong evidence that certain programs or policy actions actually cause or, if implemented, would cause certain outcomes. Because the experimenter controls the program policy or treatment and who receives it, one can be more confident than with other methods that an association between program and outcome was not spurious. (In statistical terms, the program or policy is definitely exogenous.) Several social experiments have been undertaken because existing data did not permit analysts to determine causal relationships with sufficient confidence. Recent examples include studies of income maintenance and Rand's HIS study.

Income maintenance. A variety of experiments in income maintenance have been conducted in the United States.[4] In these experiments, families were guaranteed varying levels of income, even if they earned nothing. Payments to the family were reduced ("taxed") as family income rose. After a certain income level was reached, payments ceased altogether. At issue was the degree to which families might reduce their hours of work as support levels (guarantees) and alternative tax rates varied. Would workers, for example, quit their jobs and simply live off the guaranteed income?

The economics literature contains numerous nonexperimental studies of this matter (for example, Cain and Watts, 1973). These studies infer how families might respond to proposed income maintenance programs by analyzing how much work is performed by people with differing wage and property incomes. Such an analysis might show, for example, that workers earning $6 an hour worked 5 percent less than workers earning $8 an hour. One might then hypothesize that if the tax rate on the $8-an-hour worker were raised enough to reduce after-tax wages to $6 an hour, he or she would work 5 percent less. Unfortunately, one requires a strong assumption to reach this conclusion. People who prize material goods may both invest in training that yields a high wage rate and then willingly work long hours. People who believe that the best things in life are free may prize their leisure time more than the rewards of work. It would be rash to assume, then, that high-wage workers would slacken their efforts if their wage were lower, or that higher wages would inspire the others

---

[4] Bawden and Harrar, 1976; Keeley et al., 1978a, 1978b; Kehrer, 1977; Kershaw, 1972; Kershaw and Fair, 1976; Palmer and Pechman, 1978; Pechman and Timpane, 1975; Rossi and Lyall, 1976; Spiegelman et al., 1983; Watts and Rees, 1977a and 1977b.

to work more. Nonetheless, nonexperimental studies had to make this assumption, whereas the income maintenance experiments induced variation in the actual tax rate and studied the number of hours that people worked.[5]

The health insurance study. In the Health Insurance Study (HIS), one objective was to measure how the use of medical care varied with the completeness of insurance coverage (for example, how much use would increase if medical services were free). A nonexperimental method for achieving this objective would be to compare, say, the numbers of visits to physicians made by people who do and who do not have complete health insurance. One trouble with this measure, however, is that people in the United States who have more complete insurance rate themselves less healthy than do those with less complete insurance (Phelps, 1976). Because the two groups therefore differ in more ways than their insurance coverage, a comparison of the two could overstate the responsiveness of use to insurance. Statistical methods have been designed to correct for such bias (that is, simultaneous equation methods), but in this instance they yielded highly imprecise estimates of the effect of insurance (Newhouse and Phelps, 1976). In the experiment, however, similar groups could be assigned to each insurance plan; consequently, one could be reasonably confident that any difference in the use of medical services across plans was attributable to the insurance.[6]

Social experimentation offers a number of other advantages over analysis of nonexperimental data; these include: expanding the range of variation in existing data; making analysis of some problems more tractable; ensuring data availability; and providing evidence on the administrability of a proposed program.

Expanding the Range of Variation in Existing Data

Existing data are often insufficient for evaluating a proposed program because they do not apply to the relevant range.

In the income maintenance experiments, a key issue was how much low-income households would reduce their work effort as the size of the guaranteed income rose.[7] An estimate can be derived from nonexperi-

---

[5] Technically, the taste for leisure was kept independent of variation in after-tax wage rates by varying the rate at which the guarantee was taxed away and by randomly assigning individuals to different tax rates. There was exogenous variation in "after-tax" wage rates.

[6] Technically, the insurance plan is exogenous and, to the degree possible with a finite sample, orthogonal to other covariates.

[7] Technically, what was desired was an estimate of the income effect on labor supply among low-income households.

mental data by observing how hours of work vary among households with differing amounts of property income. (Property income is economically analogous to a guarantee.) But low-income households rarely have much property income, and almost none had as much as the income maintenance program proposed to give them as a guarantee. Hence, estimates using nonexperimental data required extrapolation of an estimated relationship well outside its observed range.

In the peak-load electricity pricing experiment reported by Manning, Mitchell, and Acton (1976), an estimate was desired of how much households would shift their electricity consumption from peak hours to off-peak hours if electricity cost more during peak hours. If some consumption shifted to off-peak hours (for example, if households used washing machines only at night), utility capacity costs could be reduced. No American data addressed this issue, because peak-load pricing had never been tried in the United States; that is, a kilowatt-hour cost the same around the clock. Some European data did exist, but because of different conditions (for example, differences in climate and in stocks of household appliances), it was decided to undertake an experiment in the United States.

Of course, the necessary data for an experiment may not exist because the proposed experimental treatment is infeasible or undesirable. In such cases the experimental information can have little value. Before mounting an experiment, then, the experimenter needs to ask why data do not exist—that is, why existing institutions do not provide the program or treatment.

But a treatment might not be feasible at a universal level and still yield useful information, because it may show whether moving in a certain direction has any desirable effect at all. For example, it may be impractical to triple the number of foot patrolmen in a police system, but one may conduct a limited experiment to see if tripling has any measurable effect on crime at all. If it does not, one can dismiss the idea of increasing the number of foot patrolmen by a lesser amount.

Improving Analytical Tractability

Sometimes data exist in a form difficult to analyze, or the particular data necessary have not been collected. In such a case, the assumptions needed to estimate the effects of a program or policy can raise suspicions about the conclusions. Starting from scratch is therefore a great advantage: One can design the experiment and specify the data to be collected in ways that render the experimental results relatively easy to analyze. An example is provided by the Health Insurance Study.

In the United States, many health insurance policies vary the fraction of the bill reimbursed according to how many dollars have already been spent. One example is an initial deductible; the insured pays for the cost of medical care up to a fixed amount, beyond which insurance benefits begin. Policies that vary the fraction of expenditures that are reimbursed are difficult to analyze (Keeler, Newhouse, and Phelps, 1977). The experimental insurance policy was designed to avoid these analytical problems as much as possible by keeping the fraction of expenditures reimbursed constant except for families with infrequent, very large expenditures.

It might be argued that designing the experiment to allow simple analysis limits generalizability—for example, that experimental findings cannot be used to predict behavior under real-world insurance policies. (This generic argument has been made by Roos, 1975.) Although in a strict sense this objection is correct, it ignores the principle of divide and conquer. Analysis in this particular experiment examines (among other things) how the use of medical care services responds to rather large variations in how much the patient must pay (net of reimbursement) for medical services. If analysis of the experimental data were to show that use responded very little to changes in the extent of insurance (that is, that people tend to consume about the same amount of medical care regardless of their insurance coverage), one could generalize to real-world policies with some confidence. Because the variation in the real-world policies is almost certainly smaller than variation in the experimental plans, it can be inferred that the variation in real-world policies has little effect upon demand.

In fact, the experimental results may show that insurance coverage makes a great deal of difference in the amount of medical care a policyholder consumes (Leibowitz et al., 1985; Manning et al., 1984, 1985; Newhouse et al., 1981). In this case, the analyst who wishes to estimate use for a real-world policy must decide which experimental policy best approximates the real-world policy.[8] Although this endeavor must necessarily be somewhat imprecise, the experimental results have substantially reduced uncertainty about the effects of different insurance plans.

Availability of Data

The analyst may not have the data necessary to make sound predictions of a proposed program's effects; new data therefore must be sought to enable an empirical analysis prior to implementing a program. For example, the Health Insurance Study sought to relate people's health status to their insurance coverage. Because no existing data set permitted a

---

[8] Technically, the analyst must translate the real-world policy into the price space spanned by the experiment.

comprehensive analysis of this issue, the choice was between running an experiment or collecting data by observing households with whatever insurance policies they held. In either event, new data had to be collected; in this case, the additional cost of an experiment may not be very large, and the experiment will offer all the other advantages discussed here. (Results of this analysis are described in Bailit et al., 1985; Brook et al., 1983; Valdez et al., 1985.)

Implementation Issues

When a new program is being contemplated, an experiment can sometimes reveal unsuspected implementation issues. Again, the income maintenance experiments serve as an example. Although a number of proposals had been made for a guaranteed income in the United States (for example, Friedman, 1962; National Commission on Technology, Automation, and Economic Progress, 1966; The President's Commission on Income Maintenance Programs, 1969), little or no thought had been given to the demarcation of income-accounting periods at the time the first income maintenance experiment began (1968).

When the experiment was on the drawing boards, an issue arose concerning the time period over which income should be measured (accounting period). On the one hand, if the payments to the family were to be a function of the previous year's income, the family could be left in considerable need if, near the beginning of the present year, its income fell suddenly but it qualified only for low payments because of its previously high income. On the other hand, if payments were to be a function of income over a shorter period of time, say a month or a quarter, normal seasonal fluctuations in the family's income (for which the family could plan) could inflate the family's payments. (An example of seasonal fluctuation is the situation of a farm family that receives its income when it sells its crop.) Not until serious experimental design began was it realized that alternative rules for income-accounting periods could have important consequences for cost and behavior.

An alternative to an experiment that will yield information about feasibility and administrability is a demonstration project. In such a project, one simply launches a program without attempting to obtain data about a comparable group of people who do not receive the program. But a demonstration project does not offer the other advantages of an experiment. In particular, it cannot answer the question of what difference the intervention made, and should probably not be used unless the sole issue is feasibility. Even then, an experiment might cost little more than did the demonstration project.

5.4. When Not to Experiment

Problems that could preclude an experiment include: adequacy of existing data, inability to obtain sufficient observations, inability to define or control a relevant treatment variable, unresolvable ethical issues, and important outcomes that are not measurable.

Adequacy of Existing Data

If existing data are adequate, an experiment should not be undertaken. The experiment will be much more expensive and time-consuming than analysis of existing data. It should therefore be undertaken only when existing data do not permit satisfactory answers and the importance of the issue justifies the resources that an experiment will require.

Insufficient Observations

The simplest experiments to implement are those that vary the program or treatment across individual families. The experimenter probably will have good control over the treatment, and the number of observations will usually suffice to provide the desired information. The experiments described above are of this type. By contrast, if the unit of observation is an entire geographic area, cost considerations may permit only a few observations and interfere with reliable estimation.

For example, several proposals have recently been made in the United States to restructure health insurance, with the intent of strengthening price competition in the medical care market (Ellwood, 1978; Enthoven, 1978; Newhouse and Taylor, 1971). These proposals share the feature that the insured's premium goes up if he or she chooses doctors or hospitals that charge a high price for their services or deliver a great many services. (In the United States, as in most countries, the insured patient typically bears little, if any, of the cost differences among doctors or hospitals that the patient might use.) The advocates of these proposals hope that consumers will take costs into account when choosing a doctor, thereby creating incentives for efficient production. They also hope the schemes will not cause access problems for people in chronic ill health, although they potentially could if doctors are rated by the billings they make to the insurance company.

Consider an experiment with a plan that varied insurance premiums with the doctor used. Most people in a given local medical market would have to be covered by the plan; otherwise, medical providers would have no incentive to give much attention to the prices they charged or quantity of services delivered. Thus, an experiment would require that the plan be instituted in some markets but not in other, comparable ones. But it may not be feasible or practical to introduce an experiment into many local markets; millions of people might have to be enrolled. Alternatively, one could perhaps implement the program in two or three localities. Although such a small sample probably would not yield very precise estimates of the effects of a national program, one might well garner valuable information (especially about administrative problems), thereby permitting a more informed choice about the desirability of these proposals. In that case, however, the project would more resemble a demonstration than an experiment.

Inability to Define or Control the Treatment Variables

In some cases, the experimenter can only specify the treatment or program partially; those administering the treatment to participants will also shape it importantly. Sometimes, for example, it may be advisable to allow considerable discretion to local program managers in adapting the program to their locality. This situation can pose a dilemma for the experimenter.

On the one hand, the experimenter may want full control, and could specify the treatment in such detail as to deny discretion to the local manager. On the other hand, if the experimenter wants to know what will happen if local managers have discretion, he or she may have to surrender a considerable degree of control. (Rigid control is not always possible, anyway.) The delegation of control also has its pitfalls. If the experimenter leaves the treatment poorly or incompletely specified to those implementing it, the results of the "experiment" may be uninterpretable and unreplicable, because no one knows what actually was implemented.

This issue did not arise to any appreciable degree in the income maintenance, health insurance, or peak-load pricing experiments described above. The treatment was well defined, and reasonably similar application in a national program could be assumed. In the income maintenance experiments, for example, the issue was whether the guarantees and tax rates were as the experimenter claimed (that is, whether the experimenter maintained control of the treatment).

But when the ultimate program does not "reasonably" resemble the experimental treatment, the experiment will lose its value (Guttentag, 1973, 1977). Sometimes this is due to poor experimental management; we take up management issues in Sections 5.5 through 5.11. Sometimes the problem is discretion in implementation at the local level. If the proposed program includes an element of local discretion, the experimenter could implement the program in a group of randomly chosen localities or organizations (allowing local administrators leeway) and compare those localities or organizations with a control group. Such a design can be difficult to carry out; there may be too few areas for a reliable comparison, or residents of areas without the program may obtain its benefits by moving.

Problems of this sort would occur if one were to experiment with unrestricted revenue transfers from one level of government to another—for example, from the federal government to states. If the experimenter chose recipient states at random, probably little could be gained from comparing them with states that did not receive such revenues. The recipient states may have done very different things with the funds (that is, implemented quite different programs or treatments), and any particular state program could have had numerous and different effects. The path between a decision to make unrestricted transfers to certain states and their ultimate effects is too long to make experimentation a very useful tool for evaluating such transfers. Nonexperimental efforts to understand how the lower levels of government spent any additional resources afforded them seem more appropriate.

Inability to Resolve Ethical Issues

Ethics should seldom bar social experimentation (see Rivlin and Timpane, 1975a and 1975b). In general, we are speaking about experimenting with programs that are being seriously proposed as public policy. If such a program is unethical, then it should not be enacted into law, much less undertaken as an experiment. In the case of clinical trials in medicine, the experimenter must determine whether the known benefits of a treatment are so clear as to make it unethical to *withhold* the treatment. Rarely will enough be known about a proposed social program to make it unethical to withhold the treatment; even if it is thought that enough is known, the program possibly cannot be implemented immediately for the entire population anyway.

Nonexperimental studies also encounter ethical problems, although they are often largely unrecognized. Gilbert, Light, and Mosteller (1975) point out that experiments frequently show that a treatment had no effect (or perhaps even a deleterious effect) whereas nonexperimental evidence had indicated that the program would be beneficial. Errors in the other direction may also occur. If the nonexperimental studies are numerous enough, it may be difficult on ethical grounds to justify a controlled experiment. For example, if a nonexperimental study has concluded that a certain program is harmful, it is possible to argue that a controlled experiment would be unethical. Rightly or wrongly, then, part of the price of a nonexperimental study is the decreased likelihood of later experimentation.

Although ethics will seldom preclude an experiment, they should shape the experimental design. Ethics dictate that the design minimize risk to participants; reduction of risk will also lower refusal and attrition, thereby increasing the value of the data collected. Some techniques to reduce risk are described in more detail in Section 5.12; they include unconditional payments to families to protect them against any financial loss from participating in the experiment, insurance against untoward events for participants, and clear wording of written materials and oral presentations in explaining the benefits and obligations of participation.

The doctrine of informed consent has traditionally been the ethical guideline; in the United States, however, institutional review boards (composed of scientific peers) supplement the experimenter's obligation to obtain informed consent. The review boards (now called Human Subjects Protection Committees) assess the risks to the participants and the benefits to society from the proposed experiment (or other data-collection project). In many instances, governments will not grant financial support to a project if a Human Subjects Protection Committee has given it a negative assessment.

Nonmeasurable Outcomes

Finally, an experiment will not be appropriate if important outcomes will not be measurable. Government agencies, especially, sometimes launch programs for symbolic purposes, or as proof of the government's concern about some issue or group of people (Glennan et al., 1978). Although these actions are sometimes termed "experiments," they are often hastily put together, whereas the lead time required to field a scientifically sound experiment makes it unlikely that such actions will produce useful knowledge. For this reason, the principles expounded in this report are not meant to apply when the primary intent of the policymaker is symbolic.

5.5. Organizing and Managing an Experiment

Writing about the management of a social experiment is much like writing about management in general. Although specific, substantive knowledge is useful, classical management skills are nonetheless required to operate a social experiment successfully. The experiment manager can be likened to the executive vice president of a private enterprise, who must possess a full range of planning, operations, financial management, personnel management, public relations, and communication skills.

Others who have written about social experimentation have stressed the importance of management to an experiment's success (Riecken and Boruch, 1974). We heartily support this point. However, we look dubiously on the mystique that seems to have grown up around the management role in the literature. The literature tends to dramatize the innovative nature of social experimentation at the expense of the many mundane but vital tasks of operating an experiment. In our judgment, the management skills required for social experiments are not unique. Although special expertise is helpful, application of traditional organizational and managerial principles is essential.

The danger in our starting out with such an assertion is that it may mislead the reader into the belief that we are embarking on an "inspirational" treatise, in which we try to put new life in hoary management maxims, such as "plan ahead." Given our perspective, this is to a certain extent true—and inevitable—but there is much more to it than that.

Sections 5.6 through 5.11 elaborate on aspects of the organization and management of social experiments that we have found important.

5.6. Organizational Design

Because social experiments typically involve several institutions and extend over several years, the experimenter must explicitly consider organizational design at the very outset. The apportionment of responsibility and authority makes a difference; it is one of the experimenter's most important jobs, and it continues as the experiment evolves.

Although many organizational arrangements can be suitable for a social experiment, an important tenet is that a single organization should be vested with total responsibility and authority for completing the work—from experimental design through program operation and research. The generally large scale of social experiments, and the broad range of skills required, frequently lead to the formation of consortia and partnerships. These can be viable organizational forms, but the high degree of coordination that an experiment calls for suggests that ultimate authority be as unambiguous as possible. From a management perspective, none of the difficulties in assembling the requisite mixture of talent and job assignments should be allowed to weaken the concept of a sole source of accountability.

Even though one organization should have ultimate responsibility, a team approach to implementing an experiment is an absolute necessity. Riecken and Boruch (1974) have defined six roles that appear in the social experimentation process: initiator, sponsor, designer-researcher, treatment administrator, program developer, and audience-user. Field and Orr (in Boruch and Riecken, 1975) have pointed out that the initiator and sponsor are often one and the same, and that the roles of designer-researcher, treatment administrator, and program developer may be executed by people from the same organization. The primary reason for defining specific roles, however, is to stress the variation in perspective, motivation, and skills required in an experiment. It is unlikely that the staff of any institution will have all the necessary skills. Even if it does, the task of coordinating these skills is far from trivial.

5.7. The Team Approach

The team approach is designed to overcome the most frequent source of dissonance cited by those who have analyzed social experiments: a conflict between the *Weltanschauung* of the designer-researchers and of those who must carry out the day-to-day activities. This has been characterized as the conflict between the "research" team and the "action" team (Riecken and Boruch, 1974).

The team approach is easy enough to describe, but hard to realize in practice. First, representatives of each component of activity must participate fully and as early as possible in the experimentation process. In particular, the "action" perspective should be represented in the design phase because the choice of experimental treatments must include assessments of their operational feasibility. (The data-processing group should also be included in this phase.) Treatments not considered by the action team have a lower probability of success, not only because of objective difficulties, but also because the action team has less stake in their success.

For example, a treatment that routinely requires substantial judgment by an action-team member may be so difficult to administer consistently that the team member loses faith in the experiment. This may happen because the team member is aware of the inconsistency and cannot conceive how the information gathered can be useful to the researcher. A probable consequence of such a loss in faith is even less consistent administration than might otherwise have been the case. Eventually it may result in an inferior treatment for the researcher to analyze. In this case, participation in the design by the action team may lead to a redesigned treatment with less action-team judgment required. Or, equally important, the action team may be informed clearly that the treatment is supposed to include administrative judgment and that whatever inconsistency is incurred by well-intentioned administration will be measured and analyzed. In either case, the participation in the design process by the action team should improve the experiment.

Second, the experiment's manager(s) must cultivate and perpetuate mutual respect for each perspective within and across participating organizations. Unproductive conflict between the two perspectives is a constant hazard. As Riecken and Boruch (1974) have stated, "Nothing is more crucial for the successful management and execution of the social experiment than good cooperation between action and research teams." Strong disagreements are bound to arise, however, and little can be done to suppress them—and rarely, in our judgment, should they be. Rather, they should be turned into productive exercises that in the end contribute to the success of the experiment. Close, daily communication between the head of the research team and the head of the action team is essential.

Third, routine—as opposed to *ad hoc*—mechanisms must be designed for resolving differences. Such mechanisms should enable ready contact between action- and research-team members; this can promote a useful broadening of the perspectives of both parties that will be especially helpful when difficult differences in approach must be resolved. Routine interaction also makes for consistent decisions and actions. Reliance on *ad hoc* resolutions risks too much "hardening" of positions because of one group's unfamiliarity with the objectives and constraints facing the other group. Conflict and tensions are inevitable, and a strategy that waits for them to become intolerable to either side risks more than does a strategy that tries to resolve them as early as possible.

For example, research designs frequently call for repetitive measurement. At times these checks may seem demeaning or burdensome to the action-team member. Having to call back people recently interviewed to ask their birthdate when the action-team member is positive the initial interviews were correct can cause resentment of the team member and the people interviewed—especially if the reasons for the procedure are not clearly understood. Similarly, requests to the research team from the action team for seemingly simple advice on how to respond to irate experiment participants can cause the research team to lose faith in the ability of the action team and to make unnecessary or unproductive changes in design or analysis to attempt to compensate for perceived deficiencies in conducting field operations. In either case, routine mechanisms for interaction and conflict resolution can forestall resentment or loss of faith by either team that might jeopardize experimental goals.

Fourth, although sustained interaction among groups is essential, team members and groups must be given as much latitude as possible to carry out their designated tasks without interference. Considerable division of labor is necessary in a social experiment. Some of this division can and will take place naturally, but some must be managed—even forced. Members or groups within teams ultimately derive their value from the special talents they contribute to team performances. Too much interference by one team in the activities of the other is likely, in the long run, to be detrimental to the exercise of that talent—and thus to the experiment.

For example, research team instructions to the action team on how to order postage stamps, install telephones, or purchase desks do little to generate the respect of the action team. The action team is bound to consider itself capable of handling such routine administrative tasks. Similarly, efforts by the action team to alter data-collection activities because it thinks the data so collected will be more suitable for analysis frustrates the research team and possibly the goals of the experiment. These examples may seem trivial, but our experience suggests that such actions are not atypical in social experiments.

The research team will not be as successful as it might if research possibilities seem constantly to be constrained by dictates of the action team. Similarly, the action team cannot carry out its tasks if guidance and direction from the research team appear unduly constraining. An exhortation to both sides to exercise tolerance may be all that is needed to smooth over difficulties. When push comes to shove, however, this advice breaks down. The research team has ultimate responsibility for the experiment, and cannot relinquish that responsibility.

5.8. The Key Focus: Preserving the Integrity of Experimental Treatments

Of all the managerial challenges, perhaps the most compelling is the need to preserve the integrity of the experimental treatments in the face of both operational and analytic difficulties and opportunities. We have tried to apply a fundamental test whenever that challenge arises: To what extent do alternative courses of action strengthen or threaten the integrity of the experimental treatments? This question may seem mundane out of context; perhaps only those who have experienced the challenge of managing a social experiment will sympathize with the need to make this point explicitly.

In the absence of other guidelines, many operations decisions will be made to suit administrative convenience. However, social experimentation also requires considering the decision's effect on participant behavior and the resulting ability of the analyst to make inferences. Thus, the action team must think about research implications as well as the convenient or expedient method to accomplish a task. We do not deny the utility of administrative simplicity, but research implications are also important. Taking account of research implications is not easy, and not always successful; but the rewards for doing so are high. (Perhaps more accurately, the penalties for not doing so are high.) Training, well-thought-out procedures, and an organizational structure that facilitates sustained interaction between the action and research groups throughout the operation of the experiment will foster the goal of taking account of research considerations. Such a simple matter as placing the research and action teams in close physical proximity has proved very helpful. At a more general level, frequent use of the question "How will my actions affect participant behavior?" has been a useful guideline for action-team decisionmaking.

Such a guideline is helpful; unrealistic expectations are not, such as expecting the action team to be trained in statistical inference or to have more than a moderate appreciation of the technical research skills that will be used to analyze the experimental data.

One great danger is that the research team may never know what field decisions were made without considering the possible impact on analysis. Most social experiments involve field operations so extensive that written procedures cannot cover every situation; virtually all experiments have crises that appear to require immediate action. Thus, it is unrealistic to expect that the research team can participate in all the day-to-day decisionmaking of the field staff. Consequently, mechanisms must be designed to trigger communication between the teams when important issues arise.

How can the action team distinguish innocuous situations from those it should take to the research team? The action team should ask whether its actions may affect participant behavior in a fashion not already known to be accounted for. If the question is raised, the proper disposition will usually be clear. The moderately simple instruction to raise questions based on this guideline should help assure that interteam conflict takes place only on significant issues and that the research team does not squander its time on routine matters.

Inevitably, the need for some crisis management and quick judgment will arise, and the actions taken are highly likely to affect other matters or actions taken later on. When time and resources are available to collect information for decisionmaking, these effects may well be foreseen; often, however, neither sufficient time nor resources are available for as full an investigation of a matter as one might desire. Even then, the effects may remain highly uncertain. Furthermore, experiment participants frequently have little tolerance for delays in decisions. Consequently, both the action and research teams can feel pressured to act quickly.

That happened to us, for example, during the Health Insurance Study when we got a telephone call in the late afternoon informing us that a participant, from whom we had collected 75 percent of the expected experimental data at a cost of more than $10,000, planned on withdrawing "tomorrow" unless we resolved his immediate family budget crisis. And, as might be expected, his case was a complex and "different" one that our standard guidelines did not cover. When faced with such decisions, our course of action is to gather as much information as possible in the limited time available and then speculate about the effect of alternative actions on treatment integrity. This approach can sometimes lead to unorthodox (even sloppy) administrative practice, and frequently complicates analysis (for example, preserving the integrity of the treatment may require gathering data outside of the established system and identifying the participant for special treatment by the analyst). Nevertheless, failure to remember the fundamental purpose of the experiment in the "heat of battle" can be extremely costly.

5.9. Adaptability: A Necessary Ingredient

The transformation of any experimental design into a field operation is likely to generate unanticipated consequences. Both the research design and operations plan must be adaptable. Protocols that are too strict are doomed to failure as the action team struggles to force human behavior into the prescribed treatment. Operations plans carried out only according to administratively precise procedures can irreversibly compromise the possibility of valid treatment comparisons before the first set of data reaches the research team, if those procedures later prove inappropriate. We reemphasize that research and action teams operate in different worlds. We have already stressed the importance of bridging the gap between these worlds. Adaptable experimental designs and field operations make this easier.

Expecting the unexpected, the experimenter must know when to modify design or operating procedures. Thus, defining good performance measures is important. How does the action team know when changes are necessary? Standard administrative performance measures are of little help. The experiment manager seeks consistent treatment of participants, uniform application of data-collection procedures, careful organization and delivery of services according to specifications, and excellent early warning systems of potential flaws in the design. Unfortunately, conventional efficiency measures, such as cost per unit output, number of units processed per month, or number of complaints per unit of output do little for the experiment manager.

In lieu of such indicators, the experimenter must design measures to judge the effect of operations on participant behavior and the integrity of experimental treatments; the manager will then have the opportunity to make "mid-course" corrections. If the performance measures are ill defined or badly reported, the manager simply cannot know when to intervene. Performance measures are also needed to monitor the effect of any intervention. The manager must know if adjusting one part of the system is wreaking havoc with other parts. Such knowledge requires performance measures that allow sufficient oversight and control to confirm that global experimental objectives are being met.

Creating timely feedback and establishing a basis for understanding participant behavior can help in the design of such measures. Most data-collection systems used in social experimentation result in machine-readable data that are batch-processed after considerable preparation and editing. This inevitably takes time, often creating delays of many months between initial data collection and subsequent analysis by the research team. In the intervening period, the manager needs mechanisms to get a quick reading on the success of data-collection activities. Elaborate measures that reveal only after the fact how well the experiment was managed can aid the analysis, but they do not help the manager to intervene at the right times. Section 5.12 discusses several analytical methods used in the Health Insurance Study to test our success in being able to make valid inferences from the experimental data. However, we also have used devices such as weekly progress reports, periodic system audits, site observation, and analysis of hand-tallied data to provide more immediate feedback about the progress of experimental operations.

We also have tried, when possible, to anticipate participant behavior and to be in a position to react when participant behavior appeared to deviate from our expectations. A simple case in point, when we failed initially but took corrective action, occurred when we mailed two complex self-administered questionnaires at the same time to one of our sites, but mailed them at different times to the other sites. Response rates were lower and data-retrieval costs higher at the site to which we mailed the questionnaires together. We corrected the situation by separating the mailing periods for subsequent questionnaires. Since then, we have considered much more carefully how the burden on the respondent will affect the success of our data-gathering activities—even though we were aware of that factor from the beginning of our project. The point is that such signals should not be missed; continuous monitoring of the effect of operations on respondent behavior helps ensure that they will not be.

This example also illustrates the importance of an adaptable design and field operation. The questionnaires were mailed together in the first place because the design called for administration of one (an income report) in the spring, when income-tax forms are due in the United States, and the other (a health questionnaire) at a yearly interval after enrollment—which also happened to be in the spring at this site. Being able to adjust the interval because of the robustness of the design, and having an action team that could rearrange its activities without trauma, made this (rather minor) mid-course correction possible. The alternative without this flexibility may well have been to accept a lower quantity and quality of data.

Another specialized technique for creating useful intelligence is the use of extensive quality-control systems. Because quality-control approaches are common to survey research and data processing, two key activities in any social experiment, we do not review their development and use here. Rather, we highlight quality control as a special case of designing performance measures to focus attention on data quality as well as administration.

Quality control plays two important roles in a social experiment. Because of the long period between data collection and analysis, field quality-control operations must carry the principal burden of maintaining data quality. Most decisions regarding data quality will be effectively made at the time of collection—with little prospect for reversibility. Therefore, the research team should participate fully in the design of data quality-control checks and monitoring of results.

Establishing a quality-control system offers an opportunity for productive cooperation between the action and research teams. It is an especially difficult challenge for the researcher, who will be required to make judgments without having the opportunity to analyze data in any sophisticated manner, and who may be forced to make decisions that run some risk of compromising analytic possibilities. But intensive participation by the researcher at the data-collection stage of an experiment can preserve analytic options that may otherwise be unintentionally foreclosed. In addition, the greater the participation of the researcher, the better the chances for adapting appropriately to changing or unforeseen circumstances.

Quality-control procedures also act as important communication devices between the research and action teams. They permit the researcher to communicate intended outcomes to the action team in a concrete, useful manner. Quality-control instructions signal to the action team what is important in the field operation, and they help the action team react to field circumstances.

Certain measurement problems are common to many social experiments. Many economic experiments, for example, must deal with the problem of how to measure income. Considerable folklore and some research suggest ways to measure income, but not all the difficulties have been resolved. Lectures and elaborate statements of principle are rarely as successful as detailed procedures in communicating to the action team how to define income and when to seek policy advice from the research team.

5.10. Planning and Managing the Information Flow

Experimental operations require day-to-day management of a large amount of information, something to which many researchers may be unaccustomed. Much of the information requires action. From a management perspective, this fact of life is not particularly troublesome; well-trained managers are expected to design and operate information systems without difficulty. However, the information flow does need to be planned and managed; if it is ignored for very long, it can get out of control and chaos may result.

The information flow, when properly structured, can serve to create a written record of what has already transpired in the experiment. Especially if administration of the treatments is complex and the experiment runs for several years, it is difficult, if not impossible, to retain needed information for decisionmaking in personal memories. The written record can help counter both threats to consistent application of the experimental program—which are legion—and changing perspectives that accompany the maturation of the experiment. We caution, however, that seemingly simple tasks, such as careful file organization, rigorous documentation, and inclusion of all relevant parties in decisionmaking, are difficult to maintain in practice.

A well-structured, written record helps in another way. Many decisions cannot be reduced to small, discrete issues susceptible to unilateral consideration. As a result, several parties will participate, and so more thorough and detailed information will be required to support the decisionmaking process than would be the case if decisions were made unilaterally.

The management information system not only has to provide timely, relevant information (as usual), but it also should simplify accounting for the effects of change in one part of the system on other parts. In communicating the consequences of changes in design or procedures to various teams, large-scale experiments will be forced to adopt some bureaucratic characteristics. Although we would not wish the well-known bureaucratic pathologies on an experiment, neither do we believe that most experiments can be run in the collegial style more familiar to academic researchers. The need to structure information flow, accept considerable division of labor, and purposefully design communication and decisionmaking procedures to account for the variety of activities and actors involved must be recognized at the outset.

5.11. Skill Requirements

We noted in Section 5.5 the wide set of skills that will be drawn upon in the course of a social experiment. Although these skills are not mysterious, we discuss them briefly here as a reminder that they should be required for experimental designers and managers.

Financial Management Skills

Financial skills include budgeting, cost-estimation, accounting, cash-flow management, and financial auditing. Naturally, a social experiment should have a budget. However, especially if the research or the treatments involve considerable uncertainty, "the" budget is likely to consist of many budgets, adjusted over time to reflect changing costs, the operating environment, and research progress.

One key difficulty facing the experiment manager is how to trade off research possibilities in the face of the uncertainty of the cost of field operations. Pessimistic cost estimates may suggest immediate budget constraints that do not materialize over the long run, leading the manager to cut back activities prematurely; or actions based on optimistic cost estimates may force later compromises to the research design that a more conservative approach could have avoided. Two implications follow: (1) Without careful management, cost estimates and budget realities can end up dictating the course of the experiment to a much greater degree than might be imagined and (2) because the manager's tradeoffs between cost and research possibilities may have more irreversible consequences than do many other decisions, a high premium attaches to sound cost information. The timing of adjustments to research plans, and consequently to field operations, can profoundly affect the richness of experimental outcomes.

The most important building blocks to a budget are good cost-estimating procedures. Cost uncertainty is likely to afflict both the administration of the experimental treatments and whatever survey activities are undertaken—to say nothing of the cost of the research. Also, in lengthy longitudinal studies, data processing accounts for a large portion of total expenditures. Managing an experiment therefore involves scores of cost estimates for various components of the experiment, and periodic projections of the total cost of the experiment. Revised budgets will have to be issued frequently to reflect the results of these analyses—perhaps as often as quarterly instead of the more common annual budget exercise.

Accounting and auditing skills are also important. Because experiments often are funded with public monies, there is an obligation to design financial systems that are easily audited and that organize expenditures into standard systems of accounts. The peculiarities of government contracting and accounting in the United States mean that grants or contracts may not be audited and "closed out" until several years after the expenditures have been made. Actions taken in the "heat of the battle" by the experiment manager should be supported routinely by sound accounting practices so that issues of liability never have to cause concern later.

One perhaps unexpected skill that some social experimenters have found necessary is cash-flow management. When large sums of money are involved, the timing of transfer of funds from one agency or unit to another can be important to the experiment manager. Managing that flow wisely can effectively increase the budget of the experiment.

Since these financial management techniques are commonplace to executives in their everyday dealings, there is no reason for a mystique to surround their use in social experiments. But familiar as they are, they represent a set of skills that must be incorporated into the team conducting a social experiment.

Coordination Skills

Most social experiments call for skills in planning, scheduling, and project management. In a general sense, these skills amount to providing leadership and coordination. In a more specific sense, they are central to the day-to-day operations of an experiment, and require more than exhortation. Not only do schedules and project plans affect costs, they inevitably set the pace for the analytic effort.

Long-range planning should be a natural part of the experimental design process. But just as a social experiment is likely to have many budgets, so is it likely to have many plans. The primary skill required is to integrate each component of experiment operations into the larger design of the experiment. This calls for detailed projections of resource needs, environmental constraints, task definitions, task completion times, and interrelationships. Although it is not likely that computerized networks and critical-path algorithms such as those provided by PERT and CPM will prove cost-effective in most social experiments (the cost of input and update may well exceed the cost of more conventional methods), the basic principles underlying the use of these techniques will have to be applied in one form or another.

Scheduling activities will prove to be a nontrivial task, and at the very least Gantt charts and similar scheduling tools will be necessary. Many organizational units may require detailed schedules of their own, with relevant unit outputs or needed inputs forming the basis for a more global project schedule. The advent of scheduling software packages for microcomputers should ease future scheduling burdens.

The particular technique used is less important than careful planning and scheduling; otherwise substantial delays and inefficiencies are likely. And, of course, any plans and schedules require monitoring and updating. Obviously, failure to do so penalizes the experimenter most severely when operations have uncertain outcomes or when the design of future activities depends on results from preceding activities.

Other Executive Skills

Brief mention should be made of a series of other skills that may be needed at one time or another during a social experiment. The need is likely to arise for public-relations skills, personnel-management skills, legal and contract-negotiation skills, data-processing skills, and, as usual, good written and oral communication skills.

People and their behavior are the subject of any social experiment. In reporting its results to a wide audience, the experimenter will want to explain clearly the purpose of the experiment, the details of treatments, the obligations of participants and cooperators, and a myriad of other details. Some programs will use multimedia campaigns and various communication techniques to recruit participants. (For example, the Housing Assistance Supply Experiment [Lowry, 1983] conducted extensive "outreach" campaigns for that purpose.) In other cases, cooperation of various actors in the environment may be essential. (The Health Insurance Study could succeed only if physicians and other health care providers agreed to treat its participants.) In any case, experimenters have found it necessary to design, take part in, or at the minimum approve campaigns to "sell" their program. Researchers do not necessarily have the skill or the time to mount successful public relations campaigns.

Relations with the sponsor are also important for the experimenter, who will probably have frequent interactions with him. Most experiments deal with questions of public policy, and the researcher may find it hard to maintain the longer-term, more reflective view of the scientist if the sponsor is pressing for shorter-term results that are directly relevant to policy. During the course of the study, the experimenter is likely to need communication skills other than those that suffice for writing journal articles or final reports.

The number and diversity of personnel in some experiments will require more time and attention to personnel management than the researcher may be accustomed to. Also, a good deal of personnel turnover may occur in both the action team and the research team over the life span of prolonged experiments. Although much of the personnel function may be delegated to others, the experiment manager is likely to spend considerable time seeing that staff members have comfortable working conditions, appropriate compensation, and challenging responsibilities. Because many people will outgrow the jobs they held at the beginning of a long experiment, the manager will find it necessary to provide for their growth and development and find replacements for them when they move up or leave. Again, these functions are commonplace, but they may take much more time and cause more worry than a researcher may expect.

Legal skills will be used throughout the course of most social experiments. An analogy is the business enterprise that has counsel on a retainer and uses services when necessary. Since legal advice usually is more valuable before the trouble or problem develops, rather than after, the experimenter should be alert to situations in which legal advice will be helpful and learn how to use it.

Many field operations involve routine legal processes such as obtaining business licenses, filing reports with regulatory or other administrative agencies, and checking for compliance with local laws. More substantial legal help may be needed in negotiating contracts or in establishing and maintaining data-safeguarding procedures.

It is not uncommon to subcontract work in a social experiment. Legal aid may be needed for writing and negotiating contracts, assuring conformance with government or other sponsor procurement regulations, and resolving disputes among parties. At least in the United States, an attorney is likely to be a standard member of the experiment team.

Perhaps most important, legal help may be needed in establishing procedures to protect the confidentiality of data provided to the experiment. Social science research in the United States has not reached the point where statutes, regulations, case law, or common practice provide an unambiguous guide for action. It is the goal of many experimenters to protect from third parties both the identity of participants and any personally identifiable data that they provide. In addition to being a matter of ethics, such protection may reduce refusal rates and improve the quality of data that the participants provide. Except in certain narrowly defined circumstances, however, full protection is not available in the United States. Information provided to the experiment can be subpoenaed, and may have to be made available to others. Disclosure can be minimized, but establishment of proper procedures, including the wording of any promises of confidentiality, will require legal help.

Data-Processing Capability

Data-processing capability is crucial to the success of a social experiment. Data-processing design and management require: (1) organizing a data base in which all data collected by the experiment are stored; (2) creating a sample-maintenance system to keep track of the status of individuals or institutions participating in or cooperating with the experiment; (3) designing a system to match the status (e.g., married, divorced) of persons, families, or institutions to the data collected from them; (4) designing an efficient method of extracting data and status information for many different combinations of variables of interest to the researcher; and (5) providing for disseminating the data base to other researchers.

Typically, two generic types of data are collected in an experiment: survey data and program (treatment administration) data. Survey data include data from self-administered questionnaires or personal interviews conducted at one point in time, or at intervals, to gather sociodemographic or attitude information. Program data include information collected in the process of administering the experimental treatment (medical-care-expenditure information, income-support payments, and the like). Program data may be reported sporadically at the participant's initiative (for example, when a participant files a medical insurance claim), or may be collected routinely for program monitoring (for example, monthly income reporting as part of the calculation of monthly income maintenance benefits). Survey data are usually collected at the initiative of the experimenter. Although the data-collection forms and procedures may be quite varied, at some point the data will need to be archived in a consistent, easy-to-access, machine-readable form. This is a significant challenge that requires skills in data-base design and management.

But the task does not end with the organization of a data base. Perhaps the greatest challenge for the data-processing system is to design a sample-maintenance system for keeping track of the status of each participant. At any one point in time a number of variables will define a person's status for the analyst (age, sex, employment status, marital status, location, and the like). During the course of an experiment, people's status will change—rather more frequently than might be expected. For example, they may change their address or employment, become separated or divorced, acquire children, leave the experiment voluntarily, or die.

The design of this system is enormously important to the experiment. The following example illustrates but one aspect of the problem. Suppose a three-person family is receiving health insurance benefits as an experimental treatment. The benefits are related to family income, and are paid when a member of the family files a health insurance claim. During the course of a year, the family splits up and forms two new families, one with two members and one with one member. However, neither family informs the experiment of the split until sometime after it has taken place. In the meantime, the experiment has paid some health insurance claims as if the original family were intact. Once the experiment learns of the change in family composition, how does it identify data collected between the time of the change and when it learned of the change? Alternative associations of family status with the program data will present different pictures to the researcher. The ideal system will provide information so the researcher can define family status in the manner most appropriate to the specific analysis at hand. However, the number of changes that can occur in the field make this ideal sample-maintenance system quite complicated. Providing for a wide range of options is easier said than done and will impose additional costs.

Given a good data-base design and sample-maintenance system, the problem remains of matching the program and survey data to relevant participant status variables. Generally, the program data-collection system cannot gather a full range of sociodemographic information (status variables) every time some program information is collected (for example, is the participant employed on a certain date?). Some of this information will have been gathered from survey data or from periodic checks of program data. Typically, except when doing cross-sectional analysis, the analyst will always face uncertainty in associating data gathered from one form at one point in time with data gathered from another form at another point in time. The better the system for matching data to status variables (that is, for tracking status across time), the higher the quality of analysis.

Most experiments have analytical agendas requiring different combinations of data for different analytical tasks. Even given consistent, efficient storage of data and ready matching to status variables, researchers must be able to extract different combinations of data for different tasks easily and cheaply.

In the United States, one of the selling points of experimentation has been the creation of large, longitudinal data bases that other researchers can eventually use to validate or extend the experiment's findings or to conduct research on new topics. This value of experimentation is lost if later researchers find it prohibitively expensive or complex to extract data from the data base for their own uses.

5.12. Tips for the Experimenter

Create a pilot sample. Typically feeling pressed, at the beginning, to produce results as soon as possible, the experimenter may be tempted to proceed rapidly with implementation. Despite the natural urge to get on with the work, the experimenter should establish a pilot sample but assume that the data from that sample probably will not be useful for formal analysis. The pilot sample should precede the regular sample by several months.

When purchasing a fleet of new aircraft, it is wise to "fly before you buy"—to develop a prototype to determine cost and feasibility before committing to a full-scale production run. Similar considerations apply to social experimentation. In particular, a pilot sample can:

1. Establish the feasibility of enrolling participants. It even may be possible to estimate two or three points on a "supply curve" of participants; that is, discover how the refusal rate varies as a function of payments or obligations required of participants. Such knowledge may prevent either excessive refusal rates or monies paid to participants in excess of what the participant would require to enroll. Techniques for enrolling participants may also be tested.

2. Establish the ability to resolve procedural and definitional difficulties with the treatment. In some experiments, for example, there may be apparent legal problems.[9] A pilot sample should clarify the situation. More generally, the group responsible for conducting the experiment will gain experience they have probably not had in administering the treatment (for example, in the income maintenance experiment, the experimenters had no prior experience in determining the amount of payment due to a participating family). By first

---

[9] For example, when we began the Health Insurance Study we did not know whether the experiment would be subject to state insurance regulation, and if so, what the consequences would be.

administering the treatment on a small scale, lessons may be learned that will permit more efficient operation when scale increases. And it may turn out that the experimenters are incapable of operating the experiment, in which case a new group must be found or the experiment abandoned.

3. Provide a pretest group for testing interviews and other data-collection instruments. Rarely can one design a flawless data-collection instrument on the first try. It is axiomatic in survey design to pretest virtually all instruments. A pilot sample provides a convenient group on which to pretest such instruments.

A pilot sample, then, should show if an experiment is feasible: It can identify operational difficulties and problems with data operations and data-collection instruments, and it will prevent the waste of resources that would occur if the first several months of data from an experiment without a pilot sample proved impossible to analyze because of continuing adjustments to unforeseen problems.

The experimenter should not regard the pilot sample as a means to demonstrate the analytical worth of the experiment or even, necessarily, to provide estimates of variances that will be useful in experimental design. Rather, the pilot sample should be used to test operational feasibility. Although the principle of using the pilot sample as the first step in a sequential experimental design appears attractive, it will probably not succeed in practice. There is a good chance that the experimenter will wish to change the protocol (and maybe even the treatment) after beginning the pilot sample. For example, results from pretesting interviews on the pilot sample may dictate the revision of interview instruments. It is unlikely that one can keep the data from the pilot sample comparable to data that will be generated later. Thus, although the statistical design of the experiment may benefit from the pilot sample, most of the benefit should accrue to operational aspects.

Build into the design an ability to measure effects that are an artifact of the experiment (methods or Hawthorne effects). At the conclusion of an experiment, those conducting it may have to deal with the common criticism that the participants' behavior might have been different in a real program because experiments and programs differ, if only in the amount of data collection. Anticipating that criticism, the experimenter can often build into the experimental design, at relatively little expense, an ability to detect and measure effects that may be peculiar to the experiment (methods effects). The standard technique for doing so is to collect data on two comparable groups, one that receives the experimental treatment and one that does not. Sometimes it will be impossible to make a comprehensive assessment of the methods effects in the experimental design, but this does not absolve the experimenter from trying to quantify the effects that can be measured.

Three examples arose during the Health Insurance Study that illustrate the possibility of measuring experimental or methods effects; Newhouse et al. (1979) discuss them—and others—at greater length.

1. At first, the Health Insurance Study (HIS) proposed to measure the health status of all participants by means of a medical screening examination at the beginning and the end of the experiment. This would have yielded much more precise information than would a single examination at the end. But one objective of the experiment is to measure how medical care use varies as a function of the health insurance plan. It is easy to see that a screening examination could affect that use. For ethical and legal reasons, for example, HIS staff are obliged to report any medical problems or abnormal symptoms uncovered by the examination to the physician the participant designates. The physician may well wish to follow up such results. Any resulting treatment would not have occurred without the screening examination; moreover, follow-up may be more aggressive if the participant is in a treatment plan that requires less out-of-pocket payment for physician visits.

To measure the amount of follow-up, the HIS sample was split into two groups at the time of enrollment: those who were to take an examination and those who were not. (Reflecting the importance of the data to subsequent analysis of health status changes, about 60 percent of the participants were given an examination.) Everyone was asked to take an examination at the end of the study, when follow-up would not contaminate the data.

2. Because an experiment usually lasts for only a limited period of time, families may not behave as they would if the experiment were longer (Metcalf, 1973; Burtless and Greenberg, 1978). The experimenter presumably wishes to generalize to steady-state behavior and therefore must estimate the effect of experimental duration on behavior. Two techniques are useful for the purpose: (a) The length of participation can be varied within the experiment; for example, in both the HIS and the Seattle-Denver Income Maintenance Experiment, some families participate for three years, others for five years. (b) Data may be collected after the experiment is over; comparison of these data with data from the experimental period permits estimates of transitory behavior. Such behavior may include "crowding in" during the experiment (for example, purchasing medical care while it is cheap) or postponing certain actions until after the experiment is over. Arrow (1975) provides a rigorous discussion of the use of postexperimental data to estimate transitory behavior.

3. The HIS pays families for participating. The issue arose whether such money payments would cause families to act differently from how they would in a national insurance plan that did not pay such monies.

Again, an "experiment-within-the-experiment" was set up; the amount of money paid to families was deliberately varied to measure the effect of the payments upon behavior.

Each of the preceding three examples illustrates an aspect of the experimental design that could alter behavior in a manner that would not be replicated in a general program. The most favorable outcome one could hope for from these experiments-within-experiments would be to demonstrate that certain measurement devices changed behavior negligibly, at most. Even if such an outcome was obtained (and certainly if it did not), one would like a test of how much, if at all, participants altered their behavior because of their awareness of being observed or studied.

Such a test can be hard to perform, for it implies obtaining information on the behavior of a group of people (a "control group") without enrolling them in the experiment. Sometimes this will be impossible to do, because one simply cannot collect data without enrolling individuals, thereby making them aware that their behavior is being studied. On other occasions it will be possible to create a "control-on-control" group that is similar to a control group except it is not enrolled and does not receive the data-collection instruments that a control group would.

Sometimes a one-time interview with a control-on-control group will serve this purpose. In the case of the HIS, for example, one may wish to know if asking a person about his or her health status annually for five years makes the person more health-conscious and therefore more likely to visit a physician. To test this hypothesis, one could in principle survey people not enrolled (or rely on existing surveys of similar populations) to ascertain their physician-visit rates. For such a method to succeed, data on people in the experiment should be collected by a similar method (that is, a similar survey document); otherwise, differences between the experimental and nonexperimental samples could be attributable to a difference in data-collection methods.

Sometimes it may be possible to obtain information on a control-on-control group from records. Again, an example comes from the HIS. Part of the enrolled sample is a random sample of a prepaid group practice's members;[10] these people are formally enrolled in the experiment and are given all interviews. Their use of services at the prepaid group practice can be compared with that of other practice enrollees who are not enrolled in the experiment, using the medical records of the prepaid group practice. (To date no measurable effects have been found.) A second example is provided by the electricity rate experiment. This experiment enrolled a few hundred families in a control group and compared their consumption

---

[10] A prepaid group practice, sometimes called a Health Maintenance Organization, agrees to provide its members all necessary medical care for a fixed monthly price.

with that of a comparable group who were not enrolled, using billing records from the electric utility. Any systematic differences between these control groups should be attributable to the experiment.

Structure the experiment to keep refusal and attrition at low levels, especially refusal and attrition correlated with treatment. The desirability of keeping refusal and attrition at low levels is obvious, the means to do so less obvious. One way is to design the experiment so that no one is financially worse off from participating. In addition, payment for interviews can compensate the participant for the time and trouble taken to provide data.

Any payments made to minimize refusal and attrition rates cannot be conditional upon behavior the experimenter seeks to measure. The payments must be made solely for participating in the experiment, taking interviews when requested, mailing back self-administered questionnaires, and so forth. For example, the HIS families are paid a participation incentive, a sufficient amount of money to ensure that the families cannot be worse off financially from participating. Such a payment is necessary because the family's own health insurance plan may in some cases be better than the experimental plan to which it is randomly assigned. Without such a payment, then, the family would be worse off in some situations and it would be in its interests to withdraw.

Payments to families can be structured so as to give them an incentive to complete the experiment. In the HIS, part of the participation incentive is withheld until the end of the experiment, whereupon cooperating families also receive an additional amount, the completion bonus. Otherwise, situations could arise in which it is in the family's best interest to withdraw. One cannot completely eliminate attrition, of course. Random attrition at a low level should not be worrisome; it will cause only a small loss in efficiency of estimation. Nonrandom attrition, however, even at a seemingly low level (for example, 5 percent of the sample), can lead to nontrivial bias. Methods for analyzing censored samples (Heckman, 1979; Hausman and Wise, 1976, 1977; Nerlove, 1978; Griliches, Hall, and Hausman, 1978) can be used to test for the possibility of nonrandom attrition and refusal. Unless one can specify a variable that will affect attrition but not affect the behavior one is attempting to measure (for example, health services utilization), such methods require the analyst to assume what the statistical distribution of observations would have been without the attrition, and the analyst usually has little or no basis for choosing among alternative assumptions. One can minimize dependence on such methods by designing the experiment to keep attrition at minimal levels.

Do not attempt to eliminate bias. Although prudence dictates an experimental design that will keep refusal and attrition—and thus possible bias—at a low level, it is almost certainly not desirable to attempt to eliminate bias; the expense is likely to be high, and the money could probably be used to greater advantage elsewhere.

Bias can arise in an attempt to balance families across treatments (as an efficient design requires), that is, to ensure that the distribution of families within treatments is as similar as possible for each treatment (see below). To achieve this balance, it is necessary to collect and process information about candidate families, and return to them later with an offer to participate on a specified treatment. Unresolved issues of sampling are raised if the family's composition changes (someone moves in with the family, the heads separate, the family moves out of the area). Morris, Newhouse, and Archibald (1979) discuss these problems; it suffices here to note that it is virtually impossible in practice to maintain an unbiased sample frame. To do so, for example, would require following families who move from the area. Although that may be practical once the families have been enrolled in the experiment (using self-administered interview instruments), it is expensive to do so prior to enrollment. The amount of bias, if any, introduced into the sample by not following movers out of the area prior to enrollment (and, say, substituting into the sample the family that has moved into the dwelling) is almost certain to be small in most applications; it therefore will not be a wise use of funds to eliminate it.

Usually balance the sample across the treatments. Balancing the sample means ensuring that the characteristics of participants assigned to each treatment are as similar as possible to those of participants assigned to other treatments. An algorithm for balancing the sample, the Finite Selection Model, is now available and can be used for this purpose (Morris, 1979). Its use can achieve substantial gains in precision over simple random allocation of subjects to treatments, and gains over simple blocked allocation schemes may also be possible.

The income maintenance experiments in the United States used the Conlisk-Watts model to allocate families to treatments (Conlisk and Watts, 1969; Conlisk, 1973). This model can be used to determine the optimal number of families to receive each treatment, and we recommend its use in this manner. In the income maintenance experiments, however, the model was used in another way that had the effect of unbalancing the allocation of the sample to treatments. The experiments' designers used the Conlisk-Watts model to exploit the dependence between costs of different types of families and the treatments to which they are assigned. High-income families are relatively less expensive if placed on generous income maintenance plans because they receive fewer benefits than low-income families from a generous plan, whereas neither type of family receives many benefits from a stingy plan. (A generous plan has a high guarantee and low tax rate.) As a result, high-income families were disproportionately allocated to generous plans. If the effect of the plan is independent of income (that is, there is no interaction between income and treatment), this disproportionate allocation will improve precision. But the assumption of no interaction may be incorrect, and the unbalanced allocation makes it difficult to test the assumption (Keeley and Robins, 1978; Cogan, 1978). This problem can be avoided by balancing the sample.

Although we recommend use of the Finite Selection Model to assign families to treatments, the important point is to ensure a robust design. A balanced design will be robust; that is, it should prove suitable not only for analyses where the designer is fairly confident that there is no dependence between treatment and family characteristics, but also for unforeseen analyses where such an assumption may be less appropriate or may be questioned by others.

Do not strongly oversample a group whose membership is not well defined. The experimenter is often more interested in a program's effect on one population subgroup than on another. It is appropriate to oversample the subgroup of interest (the favored group) if it can be defined with little error (for example, people more than 65 years old). But if a favored group can be defined *a priori* only with substantial error (for example, those who will consistently have incomes less than $5000 per year), disproportionate sampling can reduce the experimenter's precision of estimation, even for the favored group (Morris, Newhouse, and Archibald, 1979). We recommend that the experimenter ascertain the reliability of any measures used for disproportionate sampling before setting sampling fractions, and that sampling fractions not depart far from proportionality if the classification contains measurement error.

Go to some lengths to inform participants about the treatment. Informed consent should be an ethical precept for an experimenter. Apart from the ethical argument, however, the experimenter's self-interest generally dictates informing the participants fully about the details of an experimental treatment. First, full information at the time of enrollment will help minimize attrition and, in a well-designed experiment, refusal rates as well. Attrition should be low because participants will receive no unpleasant surprises after enrollment; refusals should also be few because proper experimental design and full information should make it clearly in the participant's interest to participate. Second, the experiment will almost certainly last for only a limited time (rarely more than a few years). If behavior under the experiment is to be generalized to behavior under some kind of permanent or open-ended program, it is important that the participants understand the experiment as fully as it will the later programs. To convey such understanding, the treatment should be explained to the participant as clearly and simply as possible.

Choose the number of sites and length of enrollment to minimize variance, given a fixed project budget. If there are no fixed costs to operating a site, the optimal sample will be completely dispersed, that is, not concentrated in a small number of geographic areas. Usually, however, there will be fixed costs for each site. For example, a sampling plan for the site may need to be created, and a field office may be necessary during experimental operations. If there are fixed costs for a site, there is a tradeoff between the number of sites (locations of participants) and the total number of participants. This tradeoff can be made to minimize the variance of variables to be estimated.

In unpublished work, Morris has shown that the optimal number of sites K equals $BV^{1/2}/F(1 + V^{1/2})$, where B is the budget available, F is fixed costs per site (assumed invariant to the number of sites), and $V = (F/M)[t^2(1 - r + rL)/s^2d]$, where M is the marginal cost per family (assumed constant), t^2 and s^2 are between-site and within-site variance, respectively, r is the correlation coefficient among treatment means within a site, L is the proportion of "level" estimates desired as opposed to "treatment contrasts" (a level estimate is the mean for each treatment; a contrast is the difference between treatments), and d is $\sum p_i c_i/\sum p_i^2 c_i$, where p_i is the sample proportion on the ith treatment and c_i is the relative cost of the ith treatment. The parameter d is the effective number of treatments of equal interest (d equals the number of treatments if sample size and cost per treatment are equal for each treatment; otherwise it is less than the number of treatments).

The optimal number of sites K rises with the total budget available and falls as fixed costs per site F rise; K rises as between-site variance rises relative to within-site variance. A larger value of r means that treatment differences in one site can be used to predict differences in other sites, and therefore fewer sites are needed if one is interested in differences, but this correlation is irrelevant if one is interested only in levels ($L = 1$).

Length of enrollment is typically a difficult variable to optimize *a priori*. In general, the greater the intertemporal correlation in the variable of interest, the greater the gain from a larger sample participating in the experiment for a shorter period of time. If there is zero intertemporal correlation (unlikely in practical cases), a small sample participating for a long period of time will ultimately provide the same information as a large sample participating for a short period of time. Because the large sample participating for a short period will provide data earlier, however, it should be preferred. But three potential problems must be noted. First, if one wishes a measure of the long-term effect of the treatment (as one usually will), participants must be given enough time to adjust their behavior in response to the experimental treatment. This may require operating the experiment for several years. Second, as explained above, it is frequently desirable to vary the period of participation in order to measure the effects on behavior of a limited time of enrollment. The proportion participating for a longer period should reflect the weight accorded to determining the experimental effect of length of participation, relative to the weight given to having results early and having results with less variance when there is intertemporal correlation. Third, for an equal number of person-years, costs will probably be lower if a larger number of families is enrolled for a shorter period of time. Although there are one-time enrollment costs for each family, management costs are incurred for lengthening the duration of the experiment. In general, the additional management costs of a longer experiment dominate the additional enrollment costs.

Do not attempt too much. Most experiments have multiple objectives and use interdisciplinary research teams. Data gathered to serve one discipline or type of analysis may not be suitable for another. Assuming that resources are scarce, tradeoffs are inevitable. Naturally, the needs of the sponsor, the skills and preferences of the analyst, and the feasibility of alternative courses of action affect the manager's resource allocation decisions. But one should guard against attempting to do too much with one experiment, thereby risking doing too little. We advise the experiment manager to concentrate principally upon data-collection activities that support sound analysis of treatment comparisons. Only after that is assured should one invest in potentially interesting methodological studies or auxiliary use of the data.

Although such a caveat may seem banal, our experience suggests that the unique setting of most social experiments can induce those responsible for managing it to overextend themselves. And once activated, the cycle of mistakes can be unforgiving. An increased research agenda usually demands increased field operations. All of the social experiments we are familiar with have had operation staffs functioning under extreme time pressure and severe budget constraints. When care is not exercised with the research agenda, performance can deteriorate severely in field operations. Indeed, the field operations may not be able to recover from an overambitious research design, thus risking contamination of the primary treatment comparisons.

Experiment managers should be conscious of their risk-taking behavior. They are likely to feel a pull between using accepted, "safe" analytical approaches and more ambitious approaches that push the state of the art. Alas, analysis plans that truly push the state of the art will fail some of the time. To guard against overextension, but still engage in some research that pushes the state of the art, we have tried to do two things. First, whenever possible we avoid research activities that threaten our ability to complete the primary tasks of treatment comparisons. Second, we have tried to apply the concept of "nested" research, whereby results from small pieces of work feed into larger and larger tasks in such a way that failure in one task need not threaten the ultimate success of the important, broader tasks.

A key aspect of this approach is planning "fallback positions." In the HIS, for example, we have collected data to calculate a price index for each of our sites, even though existing regional data on prices could be used as an approximation if such an approach turned out to be infeasible. Our analysis will be better with the local index, but the viability of the experiment is not threatened without it.

Do not be easily discouraged. We believe social experimentation, if properly used, can be an extraordinarily valuable tool. To be sure, it is time-consuming, sometimes frustrating, expensive by the usual standards of social science research, and risky—a mistake in the design or its application may vitiate the entire endeavor. But new knowledge is seldom easily achieved.

5.13. Summary

Social experimentation compares people who receive a certain program (experimental treatment) with similar people who do not or who receive some other treatment. Some form of randomization is typically used to maximize the likelihood that the various groups are in fact similar. This chapter had three parts: Sections 5.2, 5.3, and 5.4 discussed when to conduct a social experiment and when not to; Sections 5.5 through 5.11 explained how to manage a social experiment; and Section 5.12 offered some practical advice for experimenters. The principal points made in these three parts are summarized below.

When to Experiment

The experimental method is the best available for establishing that a certain program or intervention actually caused a given result or set of results. Other methods often encounter difficult problems. Typically, they compare groups that are (or may be) different and attempt to adjust for differences; if the adjustment is not fully satisfactory, it may leave the cause-and-effect relationship ambiguous. Furthermore, other methods often have to work with data already in existence, which may be difficult to analyze and which may not bear on the proposed program or treatment, or can be brought to bear only by making unverifiable assumptions. When such problems prevail, inferences about program effects may be questionable.

Obstacles may also impede or even preclude social experimentation. Perhaps most important, it is usually more expensive and time-consuming to collect and analyze experimental data than it is to use existing data; consequently, the experimenter should be sure that existing data will not suffice for the purpose before deciding to go ahead with an experiment—and should also be sure that the problem to be studied is important enough to warrant the expenditure of resources. Problems that could vitiate an experiment include: insufficient observations (usually a subcase of the experiment's being too expensive for the problem), inability to define or control the treatment variable, and important outcomes that are not measurable.

How to Manage an Experiment

Because of its unique aspects, the considerable attention given to social experimentation in the literature has surrounded the manager of an experiment with an exaggerated mystique—exaggerated because the manager must possess, above all, the traditional managerial skills needed for planning, operations, financial management, personnel management, public relations, and communications.

Experiment personnel are likely to be organized into a research team or teams and an action or operations team or teams (including a data-processing group). Because it will be responsible for analyzing the data, the research team must have ultimate authority over the experiment, but early and frequent interaction between the teams is necessary for success. The teams must respect each other and understand the constraints imposed on each other. One of the manager's crucial duties is to preserve the integrity of the treatments, a task for which it is essential to create information systems that permit monitoring the field operations.

Planning and cost estimation are both important skills; deficiencies in them can undermine the best-designed experiments. Most experiments will have highly interrelated field operations; failure to consider dependencies among them is virtually certain to lead to poor-quality data. Accurate cost estimation is essential if the experiment has a fixed budget. If costs prove higher than estimated, necessary *ad hoc* changes in the design of the experiment may cause much more damage than would have occurred if the higher costs had been foreseen. If costs prove lower than estimated, opportunities to collect useful information may have been missed.

The data-processing group should participate in the early stages of design if its operations are to be efficient and effective. Most social experiments generate a large data base, and unless the data are gathered with an eye toward the methods by which they will ultimately be processed, the research team may well find itself able to use only a portion of the data.

Practical Advice to the Experimenter

Several "tips" can be given to the prospective experimenter: Create a pilot sample of people with whom to pretest the operational feasibility of the experiment. Build into the design an ability to measure effects that are an artifact of the experiment ("methods" effects or Hawthorne effects). Construct the experiment to keep refusal and attrition at low levels, especially refusal and attrition correlated with treatments. Usually, balance the sample across treatments. Do not strongly oversample a group whose membership is not well defined. Go to some lengths to inform participants about the treatment and their obligations. Choose the number of sites and length of enrollment to minimize variance, given a fixed project budget. Do not attempt too much, and do not be easily discouraged.

This chapter was prepared as one product of the Health Insurance Study at The Rand Corporation, Santa Monica, California, and it was supported by a grant from the U.S. Department of Health, Education, and Welfare. The authors acknowledge helpful comments from Henry Aaron, Gene Fisher, James Gaither, Thomas Glennan, David Novick, Edward Quade, Henry Riecken, Noralou Roos, and Peter Szanton. The views expressed are those of the authors and do not necessarily represent those of the sponsors of the work.

References

Aigner, Dennis J. (1979). A brief introduction to the methodology of optimal experimental design. *Journal of Econometrics* 11, 7–26.

Arrow, Kenneth J. (1975). *Two Notes on Inferring Long Run Behavior from Social Experiments*, P-5546. Santa Monica, California: The Rand Corporation.

Bailit, Howard L., et al. (1985). Does more generous dental insurance coverage improve oral health? *Journal of the American Dental Association* 110, 701–707.

Bawden, D. Lee, and William S. Harrar (1976). *Rural Income Maintenance Experiment: Final Report*. Madison, Wisconsin: Institute for Research on Poverty.

Boruch, Robert F. (1974). Bibliography: Illustrative randomized field experiments for program planning and evaluation. *Evaluation* 2, 83–87.

———, and Henry W. Riecken, eds. (1975). *Experimental Testing of Public Policy: The Proceedings of the 1974 Social Science Research Council Conference on Social Experiments*. Boulder, Colorado: Westview Press.

———, et al. (1978). Randomized experiments for evaluating and planning local programs: A summary on appropriateness and feasibility. In *Policy Studies Review Annual*, vol. 2 (Howard E. Freeman, ed.). Beverly Hills, California: Sage.

Brook, Robert H., et al. (1983). Does free care improve adults' health? Results from a randomized controlled trial. *New England Journal of Medicine* 309, 1426–1434.

Burtless, Gary, and David Greenberg (1978). *The Limited Duration of Income Maintenance Experiments and Its Implications for Estimating Labor Supply Effects of Transfer Programs*, Technical Analysis Paper No. 15. Washington, D.C.: Office of Income Security Policy, OASPE, Department of Health, Education and Welfare.

Cain, Glen, and Harold Watts (1973). *Income Maintenance and Labor Supply*. Chicago, Illinois: Markham.

Campbell, D. T., and J. C. Stanley (1966). *Experimental and Quasi-experimental Design for Research*. Chicago, Illinois: Rand McNally.

Cochran, W. G., and Gertrude M. Cox (1957). *Experimental Designs*, 2nd ed. New York: Wiley.

Cochrane, A. L. (1972). *Effectiveness and Efficiency: Random Reflections on Health Services*. London: Nuffield Provincial Hospitals Trust.

Cogan, John F. (1978). *Negative Income Taxation and Labor Supply: New Evidence from the New Jersey-Pennsylvania Experiment*, R-2155-HEW. Santa Monica, California: The Rand Corporation.

Conlisk, John (1973). Choice of response functional form in designing subsidy experiments. *Econometrica* 41(4), 643–656.

———, and Harold Watts (1969). A model for optimizing experimental designs for estimating response surfaces. *Proceedings of the Social Statistics Section* (American Statistical Association), 150–156.

Cook, Thomas D., and Donald T. Campbell (1976). The design and conduct of quasi-experiments and true experiments in field settings. In *The Handbook of Industrial and Organizational Psychology* (M. D. Dunnette, ed.). Chicago, Illinois: Rand McNally.

Cox, D. R. (1958). *Planning of Experiments*. New York: Wiley.

Ellwood, Paul M., Jr. (1978). The health care alliance. In *Report of the National Commission on the Cost of Medical Care, 1976–1977*, vol. 2. Chicago, Illinois: American Medical Association.

Enthoven, Alain (1978). Consumer choice health plan. *New England Journal of Medicine* 298(12 and 13), 650–658 and 705–720.

Federov, V. V. (1972). *Theory of Optimal Experiments*. New York: Academic Press.

Ferber, Robert, and Werner Z. Hirsch (1978). Social experimentation and economic policy: A survey. *Journal of Economic Literature* 16(4), 1379–1414.

Field, Charles G., and Larry L. Orr (1975). Organization for social experimentation. In *Experimental Testing of Public Policy: The Proceedings of the 1974 Social Science Research Council Conference on Social Experiments* (Robert F. Boruch and Henry W. Riecken, eds.). Boulder, Colorado: Westview Press, Ch. 4.

Friedman, Milton (1962). *Capitalism and Freedom*. Chicago, Illinois: University of Chicago Press.

Gilbert, John P., Richard J. Light, and Frederick Mosteller (1975). Assessing social innovations: An empirical base for policy. In *Evaluation and Experiment: Some Critical Issues in Assessing Social Programs* (Carl A. Bennett and Arthur A. Lumsdaine, eds.). New York: Academic Press.

Glennan, T. K., Jr., et al. (1978). *The Role of Demonstrations in Federal R&D Policy*, R-2288-OTA. Santa Monica, California: The Rand Corporation.

Greenberg, D. H., and P. K. Robins (1986). The changing role of social experiments in policy analysis. *Journal of Policy Analysis and Management* 5, 340–364.

Griliches, Zvi, B. H. Hall, and Jerry A. Hausman (1978). Missing data and self-selection in large panels. In *The Econometrics of Panel Data* (Marc Nerlove, ed.), *Annales de l'Insee*, April–September. Paris.

Guttentag, Marcia (1973). Evaluation of social intervention programs. *Annals of the New York Academy of Sciences*, 3–13.

———— (1977). Evaluation and society. *Personality and Social Psychology Bulletin* 3, 31–40.

Hausman, Jerry A., and David A. Wise (1976). The evaluation of results from truncated samples: The New Jersey income maintenance experiment. *Annals of Economic and Social Measurement* 5(4), 421–445.

————, and ———— (1977). Social experimentation, truncated distributions, and efficient estimation. *Econometrica* 45(4), 919–938.

Heckman, James F. (1979). Sample selection bias as a specification error. *Econometrica* 47(1), 153–161.

John, J. A., and M. H. Quenouille (1978). *Experiments: Design and Analysis*, 2nd ed. London: Charles Griffen.

John, Peter W. M. (1971). *Statistical Design and Analysis of Experiments*. New York: Macmillan.

Keeler, Emmett, Joseph P. Newhouse, and Charles E. Phelps (1977). Deductibles and the demand for medical care services: A theory of a consumer facing a variable price schedule under uncertainty. *Econometrica* 45(3), 641–655.

Keeley, Michael C., et al. (1978a). The estimation of labor supply models using experimental data. *American Economic Review* 68(5), 873–887.

————, et al. (1978b). The labor supply effects and costs of alternative negative income tax programs. *Journal of Human Resources* 13(1), 3–36.

————, and Philip K. Robins (1978). *The Design of Social Experiments: A Critique of the Conlisk-Watts Assignment Model*, Research Memorandum 57. Menlo Park, California: SRI International Center for the Study of Welfare Policy.

Kehrer, Kenneth C. (1977). *The Gary Income Maintenance Experiment—Summary of Initial Findings*. Gary, Indiana: University of Indiana (Northwest).

Kempthorne, Oscar (1952). *The Design and Analysis of Experiments*. New York: Wiley.

Kendall, Maurice G., and Alan Stuart (1968). *The Advanced Theory of Statistics*, vol. 3. New York: Hafner.

Kershaw, David N. (1972). A negative income tax experiment. *Scientific American* 227(4), 19–25.

————, and Jerilyn Fair (1976). *The New Jersey Income Maintenance Experiment: Vol. 1. Operations, Surveys, and Administration*. New York: Academic Press.

Kessner, D. M., et al. (1973). *Infant Death: An Analysis by Maternal Risk and Health Care*. Washington, D.C.: Institute of Medicine, National Academy of Sciences.

Leibowitz, Arleen, et al. (1985). The effect of cost sharing on the use of medical services by children: Interim results from a randomized controlled trial. *Pediatrics* 75, 942–951.

Lowry, Ira S., ed. (1983). *Experimenting with Housing Allowances: The Final Report of the Housing Assistance Supply Experiment*. Cambridge, Massachusetts: Oelgeschlager, Gunn, and Hain.

Manning, Willard G., Bridger M. Mitchell, and Jan P. Acton (1976). *Design of the Los Angeles Peak Load Pricing Experiment for Electricity*, R-1955-DWP. Santa Monica, California: The Rand Corporation.

———, et al. (1984). Cost sharing and the use of ambulatory mental health services. *The American Psychologist* 39, 1090–1100.

———, et al. (1985). The demand for dental care: Evidence from a randomized trial in health insurance. *Journal of the American Dental Association* 110, 895–902.

Metcalf, Charles E. (1973). Making inferences from controlled income maintenance experiments. *American Economic Review* 63(3), 478–483.

Morris, Carl (1979). A finite selection model for experimental design of the health insurance study. *Journal of Econometrics* 11(1), 43–61.

———, Joseph P. Newhouse, and Rae W. Archibald (1979). On the theory and practice of obtaining unbiased and efficient samples in social surveys and experiments. In *Experimental Economics* (Vernon Smith, ed.). Westport, Connecticut: JPI Press.

National Commission on Technology, Automation, and Economic Progress (1966). *Technology and the American People*. Washington, D.C.: Government Printing Office.

Nerlove, Marc, ed. (1978). *The Econometrics of Panel Data. Annales de l'Insee* 30–31, April–September, Paris.

Newhouse, Joseph P. (1974). A design for a health insurance experiment. *Inquiry* 11(1), 5–27.

———, et al. (1979). Measurement issues in the second generation of social experiments: The Health Insurance Study. *Journal of Econometrics* 11, 117–129.

———, et al. (1981). Some interim results from a controlled trial of cost sharing in health insurance. *New England Journal of Medicine* 305, 1501–1507.

———, and Charles E. Phelps (1976). New estimates of price and income elasticities. In *The Role of Health Insurance in the Health Services Sector* (Richard N. Rosett, ed.), Universities-National Bureau Conference Series No. 27. New York: National Bureau of Economic Research.

———, and Vincent Taylor (1971). How shall we pay for hospital care? *The Public Interest* 23, 78–92.

Orr, Larry L. (1974). The Health Insurance Study: Experimentation and health financing policy. *Inquiry* 11(1), 28–39.

———, Robinson G. Hollister, and Myron J. Lefkowitz, eds. (1971). *Income Maintenance: Interdisciplinary Approaches to Research*. Chicago, Illinois: Markham. (See especially the papers by Orr and Kershaw.)

Palmer, John L., and Joseph A. Pechman, eds. (1978). *Welfare in Rural Areas: The North Carolina-Iowa Income Maintenance Experiment*. Washington, D.C.: The Brookings Institution.

Pechman, Joseph A., and P. Michael Timpane, eds. (1975). *Work Incentives and Income Guarantees: The New Jersey Negative Income Tax Experiment*. Washington, D.C.: The Brookings Institution.

Phelps, Charles E. (1976). Demand for reimbursement insurance. In *The Role of Health Insurance in the Health Services Sector* (Richard N. Rosett, ed.), Universities-National Bureau Conference Series No. 27. New York: National Bureau of Economic Research.

Plott, Charles R. (1979). Experimental methods in political economy: A tool for regulatory research. Unpublished.

President's Commission on Income Maintenance Programs, The (1969). *Poverty Amid Plenty: The American Paradox*. Washington, D.C.: Government Printing Office.

Riecken, Henry W. (1977). Principal components of the evaluation process. *Professional Psychology*, November, 392–410.

———, and R.F. Boruch (1974). *Social Experimentation: A Method for Planning and Evaluating Social Intervention*. New York: Academic Press.

———, and ——— (1978). Social experiments. *Annual Review of Sociology* 4, 511–532.

Rivlin, Alice M. (1974). How can experiments be more useful? *American Economic Review* 64(2), 346–354.

———, and P. Michael Timpane, eds. (1975a). *Ethical and Legal Issues of Social Experimentation*. Washington, D.C.: The Brookings Institution.

———, and ——— (1975b). *Planned Variation in Education: Should We Give Up or Try Harder?* Washington, D.C.: The Brookings Institution.

Roos, Noralou P. (1975). Contrasting social experimentation with retrospective evaluation: A health care perspective. *Public Policy* 23(2), 241–257.

Rossi, Peter H., and Katherine C. Lyall (1976). *Reforming Public Welfare: A Critique of the Negative Income Tax Experiment*. New York: Russell Sage Foundation.

Scheffé, Henry (1959). *The Analysis of Variance*. New York: Wiley.

Spiegelman, Robert G., et al. (1983). *Final Report of the Seattle-Denver Income Maintenance Experiment*, Publication No. 0-389-134. Washington, D.C.: Government Printing Office.

Srivastava, Jagdish N., ed. (1975). *A Survey of Statistical Design and Linear Models*. Amsterdam: North-Holland.

Valdez, R. Burciaga, et al. (1985). The consequences of cost sharing for children's health. *Pediatrics* 75, 952–961.

Watts, Harold W., and Albert Rees (1977a). *The New Jersey Income Maintenance Experiment: Vol. II. Labor Supply Responses*. New York: Academic Press.

———, and ——— (1977b). *The New Jersey Income Maintenance Experiment: Vol. III. Expenditures, Health, and Social Behavior, and the Quality of the Evidence*. New York: Academic Press.

Wilson, John O. (1974). Social experimentation and public-policy analysis. *Public Policy* 22(Winter), 15–37.

PART III
SETTING BOUNDARIES
FOR THE ANALYSIS

Chapter 6
Generating and Screening Alternatives

Warren E. Walker

6.1. Introduction

Where do the alternatives to be examined in a systems-analysis study come from? And, if there are too many alternatives to enable each to be examined in detail, how can the least promising be eliminated quickly and easily? The purpose of this chapter is to examine these two questions. To do so, it begins with some definitions and a framework for the discussion.

In systems and policy analysis, actual examples of analyses addressed to differing problems appear to show little similarity on the surface, but they almost always involve performing logical steps from a typical generic set (see the Appendix to Chapter 1 and the *Overview*, Chapters 1 and 4). In other sources, the names and schematic arrangements of the steps sometimes differ, and in actual analyses some steps may be included implicitly rather than explicitly—and some parts of the schema may be expanded and others contracted, depending on the purpose of the discussion it supports. Thus, for the purposes of this chapter, Figure 6.1 rearranges what has been presented elsewhere to emphasize the portion of the outline devoted to generating and screening alternatives; it condenses the other parts of the analysis (such as discussed elsewhere in this volume and the *Overview*). The focus of this chapter is on the steps shown inside the dashed-line box.

Figure 6.1 and the discussion of this chapter depend on definitions of tactics, strategies, and policies, and the distinctions among these concepts.

Tactics. A tactic is a single thing to do to meet an objective. For example, in developing a water-management policy, the tactics might include

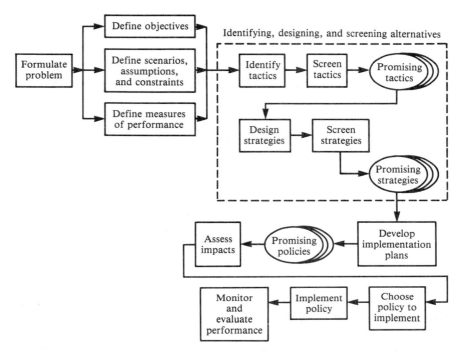

Figure 6.1. Steps in the systems or policy analysis process. This schema condenses and rearranges the one shown in Figure 1.1 while expanding the portion devoted to identifying, designing, and screening alternatives. The step labeled "develop implementation plans" is added after the dashed-line box because it is sometimes an important issue in assessing the promise of a policy. The term policy used here also includes the concept of program.

building particular waterways, imposing charges on particular discharges, and issuing permits for particular uses. In an urban transportation study, tactics might include specifying one-way streets and parking regulations. Tactics are the building blocks for strategies.

Strategies. A strategy is a combination of tactics; it describes one possible set of actions to meet an objective. For example, a transportation strategy for a city might comprise the set of changes it plans to make in its entire transportation system over the next ten years (bus routes, new highways, parking structures, new rail lines, and the like). A strategy is a description of *what* is to be done, not how to get it done.

Implementation Plans. As Goeller defines it, an implementation plan deals with *how* to do something (see House and McLeod, 1977, and a slightly revised form in the *Overview*, pp. 214–216). Such a plan specifies what actions by what persons and institutions will bring a particular preferred strategy into being. For example, if a strategy involves building a subway system through the center of the city, the implementation plan describes where the stations will be, and how the necessary funds will be raised.

Policies. A policy is a strategy plus an implementation plan. Thus, this term also comprehends what is elsewhere called a program.

The framework described above and shown in Figure 6.1 provides a hierarchical structure to the formulation of policies, and an efficient structure for generating and screening alternatives. It delays considering detailed implementation issues until the impact assessment phase of the work, an arrangement that has several advantages (Goeller and Walker, 1988):

> Generating tactics and strategies can be done more creatively, since this process does not have to be distracted by considering details of implementation (although it might well have to include basic questions of implementation feasibility).

> It will probably be easier to get decision- and policymakers to agree on promising tactics and strategies, since some of their more controversial aspects are being deferred to the detailed implementation-planning stage. For example, a policymaker might favor a strategy to reduce smokestack emissions by 25 percent, but might not if it were proposed in a form that implied a tax on polluters.

> Deferring implementation planning makes the entire analysis process more economical, because implementation planning, which is resource-intensive, is performed for only a small set of strategies and tactics, rather than for all those originally considered.

6.2. Generating Alternatives

How do the tactics and strategies to be evaluated get identified and designed? This question poses perhaps the most important issue faced by systems and policy analysts, and it is the least discussed. As Alexander (1979) points out:

> Alternatives design is a stage in the decision process whose neglect is unjustified in terms of its possible effect on decision outcomes. . . . If the choices which determine outcomes . . . are made informally and intuitively before the evaluation phase begins, then attempts at formal-

ized and rationalized evaluation, however praiseworthy, are made in vain.

Enthoven (1975, p. 463) makes a similar point: "A design of a good new alternative is likely to be worth a lot more than a thorough evaluation of some unsatisfactory old alternatives."

It should be noted that the design of alternatives does not necessarily take place at one point in a study; the analysis itself will inevitably generate ideas for alternatives that may be worth exploring.

The literature on generating alternatives is sparse, but some sources are worth noting. Quade (1982, Chapter 7), Mood (1983, Chapter 3), Alexander (1979, 1982), Hogwood and Gunn (1984, Section 10.2), and the *Overview* (Sections 4.4 and 6.3) give brief general discussions; Brewer and deLeon (1983, Chapter 3) and Friend and Hickling (1987, Chapter 6) deal with it in somewhat greater depth. This section synthesizes the literature and offers some specific suggestions.

Rules of Thumb

Generating alternatives requires a thorough understanding of the problem and its surrounding situation as well as a good deal of creativity and imagination. At this stage of the analysis process, it is important to specify a wide range of alternatives for consideration and to include as many as stand any chance at all of being worthwhile. If a tactic is not included in this step, it will probably never be examined, nor become part of a later strategy. If it is not examined, there is no way of knowing just how good— or bad—it may be. The best strategy may not be chosen by the analysis because it was never considered as an alternative, perhaps because the tactics to make it up were not included in the analysis.

Alternatives should not be excluded merely because they seem impractical or run contrary to past practice. Personal judgments on such issues should be withheld. The analysis will show whether the benefits to be derived outweigh the costs of making such radical changes. If some alternative proves to be very valuable, it may be worthwhile to modify the environment in order to get it implemented and to achieve its benefits.

The importance of the need to withhold personal judgments and withstand political pressures (and the difficulty of doing so) cannot be overemphasized. The *Overview* (p. 184) explains the dilemma:

> There are many forces that tend to restrict the range of alternatives likely to be examined. Some of the strongest are biases of various sorts due to the unconscious adherence to an organization's "party line" or cherished beliefs or even mere loyalty. . . .
> It can also happen that the analyst, in talking with the decisionmaker or his staff, becomes aware that the decisionmaker (or his superior)

doesn't like certain kinds of alternatives. He may sense that it is both useless and hazardous to even give the impression that he might advocate these alternatives as possible solutions. As a result, the development of such alternatives is likely to be neglected or forgotten, thus leading to inferior results.

Alexander (1979) shows how the development and review of options for the United States' Vietnam policy between 1961 and 1968 was restricted by exactly such limitations. "It is striking how few, and how similar, the options were which emerged from the review process and its various iterations. . . . The policy establishment repeatedly closed ranks against ideas suggested by what they perceived as a hostile environment, or proposed by establishment mavericks" (loc. cit., p. 391).

It is important to include the current strategy in the set of alternatives as the "base case." By comparing the effects of other strategies to the base case it is possible to determine if a proposed strategy will be better, and how much of an improvement can be expected. Sawicki (1983) discusses the importance (and historical tradition) of including the "do-nothing" strategy in the analysis. He also notes that there has recently been more receptivity for considering the do-nothing alternative (for example, the U.S. National Environmental Policy Act requires that a no-action alternative be generated as part of each environmental impact statement), but that in most policy studies the no-action alternative is poorly articulated and not seriously analyzed. Brewer and deLeon (1983, p. 64) point out: "There are times when inaction is the proper course, permitting exogenous forces to produce the desired effect; not only is this a low-cost policy, but it also shields the organization from any adverse publicity that might follow."

On the other hand, Fisher (1981) sees alternative generation as a proactive approach. The alternatives need not only react to the perceived environment (disaster avoidance) but may actually be able to help shape a desirable future (opportunity exploitation), especially if the lead time to implementation of the strategy is long. If it is, the implementation plan for the strategy should include ways to change the environment (such as a research and development program) to produce the environment needed for the strategy to succeed.

The *Overview* (Section 6.3) discusses characteristics of alternatives that, although not explicitly suggested by the decisionmaker (such as cost), are important and are likely to be considered later in the evaluation:

Insensitivity—the degree to which attainment of the objectives will be sustained despite disturbances encountered in normal operation (such as varying loads, changing weather conditions, changes in the environment, and so on).

Reliability—the probability that the system is operating at any given time (as opposed to being out of order).

Invulnerability—the degree to which the performance of the system remains unaffected by damage or failure of one of its parts.

Flexibility—the degree to which an alternative designed to do a certain job can be used with reasonable success for a modified, or even an entirely different, purpose.

Archibald (1979) adds these characteristics to the list:

Risk of failure—the riskier the alternative, the greater the chance for failure.

Communicability—the ease with which the alternative can be explained to persons not involved in the analysis.

Merit—the intrinsic value or face validity that the alternative is perceived to have. An analyst proposing an alternative that goes against the common wisdom must expect tremendous resistance. Such a strategy will face more intensive scrutiny throughout the analysis and communication processes than one easily (although not necessarily correctly) seen as having merit.

Complexity—the simpler the alternative, the easier it is likely to be to gain acceptance and implementation for it.

Compatibility with existing norms and procedures.

Reversibility—the degree to which it is possible to return to the conditions that existed before a change if the change does not work out to be effective.

To this set many have added (see, for example, Arguden, 1982):

Robustness—the degree to which the strategy can succeed in a wide range of future environments. The future is uncertain and usually changing. Thus, if an alternative will succeed well only under a certain specific scenario, it is likely never to succeed at all. Fisher (1981) places strong emphasis on this characteristic, which he calls "option-preserving."

Identifying Tactics

Where do the alternatives to be examined come from? Alexander (1979, 1982) suggests two different, but consistent, generation processes: search and creativity.

Search. This approach to alternative generation assumes that the solution to the problem is available and needs only to be identified by discovering and perhaps recombining its preexisting constituent parts. For example, the informed judgments of the decisionmaker and his staff are one of the best sources of alternatives. These people usually have a wealth of ideas for policy changes, but lack hard evidence to warrant choosing the best one and justifying its implementation. Some of the changes they may have in mind might be beneficial, others might not. The availability of methods for assessing the impacts of the various alternatives permits these ideas to be tested and evaluated.

Mood (1983, Chapter 3) suggests generating alternatives by listing all of the different groups that would be affected by a change in policy and trying to devise a specific solution that would be viewed as desirable for each of the groups.

In a water-management policy study for the government of the Netherlands, Walker and Veen (1981) describe how technical tactics were generated that were aimed at resolving the country's major water-shortage and water-salinity problems. They used four approaches:

A review of technical reports and documents dealing with tactics that were already being considered.

Designing new tactics when the search for tactics to solve a specific problem did not reveal any candidates that had been previously proposed.

Discussions with Dutch water-management experts.

Suggestions and guidance from the management of the client organization, from other Dutch ministries, and from provincial and local governments and various other organizations and interest groups that were likely to be affected by changes in the country's water-management policy.

A common problem in systems and policy analysis is that there may be many excellent alternatives whose performances with respect to modeled objectives are similar, but that are considerably different from each other in the decision space, thus presenting the decisionmaker with quite different decision problems. In such a case, the decisionmaker would like to see as many of the attractive possibilities as he can, so that he can choose among them based in part on attributes that may not be in the models used for evaluation (because, for example, the criteria are not quantifiable or the decisionmaker is reluctant to make the attributes explicit). Too, displaying several different "good" alternatives may serve as a catalyst for generating more creative solutions. The *Overview* (pp. 183–184) describes how two desirable alternatives in an energy study looking to the long-range future were seen to offer a combined option more desirable than either: A nuclear plant was used to supply the large amount of heat needed for a coal-liquefaction process, thus greatly reducing the amount of coal needed for an all-coal process and the volume of objectionable gases it would give off.

Brill et al. (1982) propose a mechanical procedure (using mathematical programming) to produce alternatives sequentially that are very different from previously generated ones. Each alternative generated is good in the sense that it meets targets specified for the modeled objectives. The procedure, which they call the Hop, Skip, and Jump (HSJ) method, maximizes the difference (in decision space) between successive solutions subject to constraints on feasibility and either constraints or tradeoffs with model objectives.

Creativity. As the *Overview* (p. 132) points out: "It can hardly be overstated that generating alternatives is, in systems analysis, an exercise of creativity and imagination, appropriately tempered by a thorough and broad knowledge of the issues." Brewer and deLeon (1983, p. 62) define creativity in the policy context as "the capability to view a problem and propose alternatives in perspectives and methods that are unique and unlike prior attempts to solve similar problems." Numerous techniques for exploiting human creativity have been developed, and some have been applied to generating policy and program alternatives. Summers and White (1976) review some of them.

Brainstorming is probably the best known of these techniques. Osborn (1963) developed it systematically to help solve advertising problems. A typical brainstorming group consists of several persons who get together to identify solutions to a problem. Osborn suggests four rules, all of which have already been listed here as good rules of thumb:

withhold judgment until the evaluation stage,

encourage wild ideas,

encourage numerous ideas, and

utilize the ideas of others to develop additional ideas.

The concept is to assemble persons with a diverse set of perspectives on the problem and to place few constraints on the characteristics of the solutions.

Luckman (1967) points out that, although the literature of operations research and policy analysis deals very little with generating alternatives, ideas about how to design good alternatives can be obtained by studying the processes of engineering and architectural design. Alexander (1982, p. 288) is even more positive in recommending this approach: "Systematically applied design methods do seem to offer the same prospect of raising the level of policy analysis and policy debate that the introduction of more sophisticated evaluative techniques provided thirty years ago."

The Design of Strategies

In many systems and policy studies, tactics and strategies are equivalent, so that there is no hierarchical building of strategies out of tactics. In fact, in many cases, the flow shown in Figure 6.1 is short-circuited, so that there is no identification of tactics, design of strategies, screening, or implementation planning. More complex systems and policy studies, however, require the full structure. In fact, some need an even richer structure involving pure and mixed strategies, as shown by the studies of oxidant control in Los Angeles and of water management in the Netherlands, both of which are discussed below. In these cases, the flowchart in Figure 6.1 must be changed to add a third screening step. Promising tactics of each of several pure types (such as pricing tactics) would feed into a process that would design and screen pure strategies (such as pricing strategies) and produce promising pure strategies. These would then feed into a process that would design and screen mixed strategies in order to produce promising mixed strategies.

There is no set of rules or procedures that can be specified in advance for generating good strategies. Strategy design, like systems and policy analysis, is more art than science; therefore, perhaps the best way to suggest how it might be developed is to describe some examples of approaches used in actual studies. Section 10.4 gives an extended discussion of a study aimed at ways of achieving cleaner air in San Diego, California; this description shows how the analysts combined tactics into strategies that analysis showed to be attractive (see also Goeller et al., 1973a). Three additional examples are described below. Finally, the next subsection discusses a general approach to strategy design called the analysis of interconnected decision areas (AIDA).

Oxidant control in Los Angeles. Goeller et al. (1973b) describe a study to generate and evaluate alternative strategies for meeting the U.S. national primary air quality standard for oxidant in the Los Angeles Air Quality Control Region. In this work, the analysts made up an overall strategy for the region from a large number of tactics, each of which was a single action for improving air quality.

An overall strategy (termed a *mixed strategy*) was viewed as being made up of three components (termed *pure strategies*), each of which was a set of tactics chosen from one of three classes of controls: (1) fixed-source controls, (2) retrofit control devices and inspection-maintenance tactics, and (3) transportation management. Promising mixed strategies were designed for detailed impact assessment by generating a small number of promising pure strategies for each of the three strategy components and finding the best combination using a tradeoff model. The analysis used a sophisticated procedure to screen the tactics within each component to reduce the number of pure strategies that had to be considered in impact assessment.

Los Angeles 2000. Goeller and Walker (1988) propose a conceptual approach to help the Los Angeles 2000 Committee develop a strategic plan for the Los Angeles area for the year 2000. The structure revolves around goals, functional areas, tactics, and strategies. They define five general thematic goals for the city: enriched diversity, livable communities, crossroads city, ecosystems maintenance, and individual fulfillment. To achieve these goals, although they are not mutually exclusive, would require different, often conflicting, strategies. An overall strategy is made up of seven components (that is, functional area strategies), one for each of seven functional areas: economic development, international flows, education, culture and arts, transportation, housing, and water. A functional-area strategy for one of the goals identifies the tactics that should be adopted for a functional area to reach the specified goal.

The members of the Los Angeles 2000 Committee (volunteers from the business, academic, arts, and other communities) have been divided into five goal committees, each of which is charged with designing a preferred overall strategy (plus an implementation plan) for achieving its goal. An executive committee will then compare the resulting single-goal strategies, design a number of hybrid strategies by combining pieces of the single-goal strategies, and then select a preferred hybrid strategy (plus its implementation plan) to recommend.

Water management in the Netherlands. Goeller et al. (1983) describe a study to examine the consequences of alternative strategies for managing the water resources of the Netherlands. A water-management strategy involves a mix of tactics, each of which is a single action to affect water management (such as building a particular canal). Four kinds of tactics were considered: technical, managerial, pricing, and regulation. A pure strategy was a combination of tactics of the same kind. A mixed strategy was a combination of all four kinds of strategies, and could contain any number of tactics.

Each kind of water-management tactic offers many alternatives (for example, technical tactics could modify the water distribution system in hundreds of different ways). It was, therefore, impractical to evaluate the detailed consequences of all of them; this problem was dealt with by performing the analysis in a series of stages. The next section will discuss how the analysts screened the technical and managerial tactics.

Bigelow (1982) describes the Managerial Strategy Design Model (MSDM), a sophisticated decomposable nonlinear programming model that the analysts used to design managerial strategies. Given a fixed set of facilities (current infrastructure plus technical tactics), the MSDM finds the day-to-day water-management actions (such as lake-level management, amount of water to use to cool power plants, and so on) that result in the greatest possible net benefit to water users and the nation.

In the MSDM, the variables are the flows in rivers and canals, the water levels in lakes and reservoirs, and the concentrations of pollutants at numerous locations. The constraints relate primarily to maximum and minimum flows and water levels and to pollution standards. The objective function to be minimized is the sum of all direct monetizable economic losses (disbenefits) to water users, plus the direct costs of the managerial tactics (such as, for example, pumping costs). Goeller and the PAWN Team (1985) give a useful overview of this project aimed at improving water management in the Netherlands.

The Analysis of Interconnected Decision Areas (AIDA)

Luckman (1967) proposes a structured approach to strategy design using network models. He calls this technique the analysis of interconnected decision areas (AIDA) and shows how it can be used in designing a house. He begins by defining the process of design as "the translation of information in the form of requirements, constraints, and experience into potential solutions which are considered by the designer to meet required performance characteristics." The process must include some creativity or originality: "If the alternative solutions can be written down by strict calculation, then the process that has taken place is not design."

To apply AIDA, one assumes that there are n "decision areas" (such as the type of external walls or type of roof). A design is a choice of a single tactic from each decision area. Decisions from many decision areas are interdependent; for example, the decision about the roof affects the decision about the external walls. AIDA offers a structured approach to generating feasible combinations of tactics. Friend and Hickling (1987) provide an up-to-date picture of AIDA and illustrate its use in the process of strategic choice.

6.3. Screening Alternatives[1]

Any large systems- or policy-analysis study is likely to have to consider numerous possible policies and a considerable number of measures of performance, so many in fact, that enumerating and evaluating all of the alternatives will be impossible to carry out within reasonable time and resource constraints. In addition, as de Neufville and Stafford (1971, p. 9) point out: "Even if it were feasible to think of all variations, common sense suggests that the investigation of all alternatives is not worthwhile: some are not sufficiently different to warrant separate treatment for each and some are clearly dominated by others."

The situation poses a problem for systems and policy analysts. How can they carry out a study that considers the full range of alternatives within their time and resource constraints? In most cases, they solve this

---

[1] This section is based in large part on Walker (1986).

problem by performing an incomplete analysis directed toward what they judge to be the most important issues and variables in the problem situation; they examine a small subset of the alternatives or ignore some impacts, or both. In a public policy study, it is common for many—if not most—of the impacts of a policy to be hard to quantify (such as changes in the quality of life) or hard to represent in common units such as dollars (for example, the loss of a game preserve or an increase in noise pollution). The usual—and often doubtfully adequate—response to this difficulty is to ignore all but a few of the impacts in performing the analysis. This approach is generally taken when cost-benefit analysis or optimization models are used to find the "best" policy.

A more acceptable solution to the difficulty—and one that makes it feasible to consider the full range of alternatives and impacts—is to perform the analysis in the hierarchical way shown in Figure 6.1. This approach develops promising policies through a process of generation and screening.

The first step in this approach, as already discussed, is to generate large numbers of tactics. These are then screened to eliminate the least promising (or to select the most valuable) alternatives. The next step is to design strategies composed of promising tactics. If this step yields a large number of strategies, a screening process can reduce their number. The goal is to achieve a small subset of strategies, each of which appears to be promising, that is, sufficiently sensible and beneficial to merit thorough evaluation. The analysis team can then carry out detailed examinations of these remaining alternatives, a process often called *impact assessment*. As Goeller et al. (1983) note, this hierarchical approach leads to the design of increasingly promising alternatives as the analysis progresses. The process may therefore be called the explicit hierarchical design and evaluation of alternatives.

Although screening is implicit in most large systems- and policy-analysis studies, it is usually not performed in a rigorous or systematic way. There are good reasons why it is not usually possible to seek a policy that can be regarded as optimal in some analytic sense (see also Section 2.6). First, the analysis team has no assurance that the alternative that would achieve the optimum has been conceived and included in the set that was considered, or that, if so, it passed through the somewhat partial screening process to be considered in the detailed impact-assessment process. Second, since it is invariably the case in major studies that many interests will be affected by what course of action is adopted, Arrow's theorem says that there is no completely satisfactory way to combine individual ranking functions into a ranking function covering all of the interests affected (Arrow, 1951; Mood, 1983, pp. 12–13). Nevertheless, experience shows that systems and policy analysis can usually identify courses of action that show promise of improving matters, even if an optimum cannot be claimed—and executives have been taught by experience to content themselves with demonstrable improvements in the operations over which they have control.

Although it is not normally feasible to design a systems- or policy-analysis study with a view toward achieving a policy that can be claimed to be optimum in some analytic sense, it is nevertheless important to carry out as rational a process as possible for identifying good policies. As Manheim (1964, p. 182) puts it: "In such a process, we try to find as good a solution as possible, balancing the prospects of improving the solution against the expenditures of . . . resources required to do so."

The literature on screening is sparse. Mood (1983) and the *Overview* make passing references to the concept. One approach to screening is known as *satisficing* (see Section 2.6 and the *Overview*, pp. 222–223). Dorfman (1965) and Jacoby and Loucks (1974) describe how screening was used in designing water-resource systems. Martland et al. (1977) describe the use of screening in railroad network planning, and Rogers and Berry (1982) use it in designing urban telephone routes. Most of the examples of the use of screening in public policy studies appear in reports describing studies led by Bruce Goeller of The Rand Corporation (1973a, 1973b, 1977, 1983, 1985). With a few exceptions, the descriptions of the screening process in this literature tend to be specific to the application being described and not easily generalizable.

The remainder of this chapter takes a step toward specifying a general approach to screening. It first identifies two screening strategies and then gives an example showing how they were used in a major policy study.

Strategies for Screening Alternatives

Since the purpose of screening is to narrow the range of alternatives that will need to be examined in detail, its value increases as it becomes easier, cheaper, or faster to eliminate alternatives, and as the basis for eliminating them becomes more discriminating and reliable. Conversely, its value decreases as it becomes more costly, less discriminating, and less reliable.

Any worthwhile screening procedure should possess these two properties: (1) no very good alternative will be missed and (2) the number of alternatives to be evaluated in the impact assessment will be relatively small. Mood (1983, p. 33) and Goeller et al. (1983, p. 240) mention several other generally accepted principles to be used in the screening process, including:

Infeasibility. If some technical, economic, or administrative difficulty or some organizational constraint is likely to prevent implementing an alternative, it is probably not worthwhile to pursue it. (It might, however, be useful to consider what steps might be taken to obviate such difficulties and constraints in the longer term.)

Political Unacceptability. If an alternative appears to be unacceptable to large segments of the public or to the decisionmakers who must ultimately decide on its adoption, it might as well be left out of consideration. (It must be admitted that there may be cases where the potential benefits from an alternative may be so large that opposition will dissipate when the analysis makes them clear—but such a situation is rare.)

Dominance. If two alternatives *A* and *B* have the same purpose and the benefits from alternative *A* are at least as good as those from *B* for all important measures of performance, then alternative *B* can be dropped from further consideration. Dominance, however, is usually not easy to find or to show.

Practically all structured screening procedures employ one or both of these two "pure" strategies:[2]

1. *Bound the space of promising alternatives.* This strategy eliminates sets of potential alternatives by placing constraints on the characteristics of promising alternatives. It takes place before any extended impact assessments have been carried out. Its major benefit is to focus the search for the alternatives that will be chosen for detailed evaluation.
2. *Use a simplified assessment model.* This strategy eliminates unpromising alternatives (often one at a time) by using a simple, but efficient, model. One common simplification is to evaluate an alternative's performance on only a few key measures. The major benefit is to speed up the assessment process.

Most studies employ a mixed strategy. The relation between the two pure strategies is similar to the relation between primal and dual linear programs: the primal evaluates a sequence of feasible programs; the dual drives through nonfeasible solutions toward the feasible space. Strategy 2 evaluates many alternatives within the decision space, whereas Strategy 1 constrains the space in which "good" alternatives are to be found.

Strategy 1. Bound the space of promising alternatives. The essence of

---

[2] Tuggle (1986) notes that these two strategies correspond closely to approaches taken in the field of artificial intelligence to prune large decision trees, where evaluation functions are used to direct attention in searching and abstracting from a problem space to a planning space.

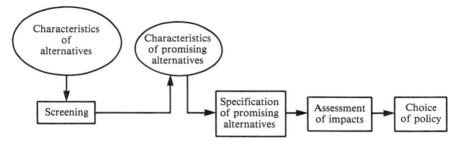

Figure 6.2. A schematic representation of the strategy of screening by bounding the space of promising alternatives.

this strategy is to place restraints on the characteristics that promising alternatives must possess. If this screening strategy is used uniquely, it should be designed so as to yield only a few alternatives to be subjected to detailed assessment procedures. It may, indeed, identify a single promising alternative, in which case the process yields its result by a form of satisficing (see, for example, Section 2.6). If more than one alternative survives the screening process, then the remaining alternatives must be tested further for feasibility, with the feasible ones then given detailed assessment. Figure 6.2 shows a schematic summary of the process.

The art in this strategy lies in specifying the constraints on the alternatives; many, indeed, may be quite arbitrary. However, if the constraints are well chosen, this strategy can be very efficient: Each constraint can eliminate entire sets of alternatives. For example, if all promising alternative highway routes between city A and city B have to pass through city C, large numbers of possible routes between A and B are eliminated immediately.

Manheim (1964) provides a highly structured example exhibiting this strategy applied to highway route location. He presents an optimal search branch-and-bound algorithm that concludes with a restricted decision space containing a single alternative. In his example, a full specification of a single alternative may require assigning values to over 30,000 variables. In his approach, however, the location process passes through several stages; in each, the degree of detail with which the alternatives are specified is increased, so that only in the final phase (called *final design*) are specific values assigned to all variables.

This strategy for screening can be compared to the approach commonly used in designing a building. The process starts with specifying the building's functional requirements, a step that implicitly rules out the many possible designs that would not satisfy these requirements. The next steps are specifying the building's physical requirements, sketching the general plans, and drawing the detailed blueprints. At each step, large numbers of possible blueprints are eliminated without their having to be drawn.

Variations of this strategy have been used in several policy studies besides Manheim's. Walker and Veen (1981, 1987) applied it as a first step in their screening of technical and managerial tactics for water management in the Netherlands. They calculated an upper bound on the expected benefits to agricultural production that tactics could produce in each of the country's eight regions. This upper bound was so small in two of the regions that no tactics pertaining to these regions survived to be evaluated. (The next subsection describes this screening work.)

Martland et al. (1977, p. 14) report that the U.S. Department of Transportation (DOT) used this strategy to determine which parts of the railroad system in the midwest and northeast regions should be subjected to detailed analysis to determine if there was excess capacity. Using a statistical study of branchline operations, DOT was able to determine the relations between a branch's profitability and the independent variables of line length and traffic volume. As a result, it could identify, for branch lines of any length, the range of traffic volumes that would be potentially unprofitable. After analyzing every branch line in the regions using this criterion, DOT found that 25 percent of the mileage was "potentially excess." It then concluded: "These are the lines which should receive the closest analysis in the process of restructuring the local rail service network."

Rogers and Berry (1982) used a similar idea to find urban telephone feeder routes whose lengths and population distributions make installing pair gain systems (PGS) economically feasible. (In a pair gain system the signals from many copper-wire pairs, that is, individual subscribers' lines, are concentrated and multiplexed onto many fewer pairs.) For a route of a given length and population distribution, their screening model defines the set of locations at which installing PGS is cheaper than using only additional copper-wire pairs to meet subscriber growth.

Strategy 2. Use a simplified assessment model. The essence of this approach to screening is (1) to construct an extensive and diverse set of alternatives (actual or implicit) that may or may not turn out to be promising and then (2) to use a limited, broad-brush form of assessment based on a small number of impact measures to produce a short list of promising tactics. Manheim (1979, p. 585) calls this form of screening *sketch planning*. Figure 6.3 offers a schematic depiction of this approach.

To illustrate, the process can be compared to the steps people take when searching for a house to buy. A broker generally has more houses listed than a potential buyer could possibly visit. However, a few basic criteria such as neighborhood, purchase price, and number of rooms will significantly reduce the number of alternatives. More detailed criteria that

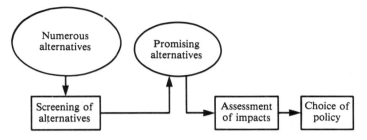

Figure 6.3. A schematic representation of the strategy of screening by using a simplified assessment model.

assess the impacts on family living or future expenditures, such as the room layout, condition of wiring and plumbing, and so on, can be applied to the remaining houses. This strategy may rule out one's dream house at its first stage (perhaps because the purchase price is slightly higher than the limit). However, the result is generally satisfactory. Unsatisfactory results can usually be avoided by careful choice of limits of acceptability.

Walker and Veen (1981) applied this strategy to screening policy options (which they called "technical and mangerial tactics") that would change the movement and storage of water in the Netherlands' rivers, canals, and lakes. The tactics were designed to alleviate problems caused by shortages of surface water and/or by salinity. Most of the screening used only two measures: the tactic's construction cost and the resulting reduction in the country's agricultural losses. The next subsection presents a more complete description of this screening analysis.

Three Rand Corporation studies used similar approaches. Two of them (Goeller et al., 1973a and 1973b) concerned ways to reduce air pollution. They analyzed air-pollution-control alternatives in terms of a wide variety of environmental, transportational, economic, and distributional impacts. In each study, the number of feasible alternatives was so vast that it was impractical to evaluate all of them in terms of their detailed impacts; therefore, the alternatives were screened after calculating only their costs and their effectiveness in reducing emissions. (Section 10.4 gives an overview of one of these studies.)

The third Rand study (Goeller et al., 1977) examined the possible consequences of three alternative strategies for protecting the Oosterschelde, an estuary in the southwestern Netherlands, against North Sea flooding. The impacts considered in the analysis included the security of people and property from flooding; the financial costs of constructing and operating protective facilities; the changes in the ecology of the region; the economic effects; the water-management impacts; and the various social effects, including the displacement of households. The screening analysis identified a promising design for each strategy, concentrating on a few impacts related to security, ecology, and construction costs. For example, any design that would provide the desired level of security with a higher cost and a less desirable ecology than some other design was ruled out of consideration. (Section 2.2 gives a brief summary of this study, but does not describe the screening portion of it.)

Jacoby and Loucks (1974) applied this screening approach to evaluating alternative river-basin designs, where a design included the size, location, and operating rules for dams, canals, irrigation systems, municipal and industrial water supplies, power stations, and pollution-control works. They made simplifying assumptions that allowed a relatively simple optimization model to be used to screen out the less promising alternatives, leaving only a few to be examined by means of a detailed simulation model.

A Case Study of Screening

A severe drought in 1976 had important negative impacts in the Netherlands: the agricultural losses amounted to more than $2000 million, low river levels caused serious shipping delays and substantial economic losses because ships could not navigate the waterways with full loads, and the water shortage worsened water-quality problems. In 1977 the Dutch government commissioned a broad study of the country's water-management system. The resulting project, called Policy Analysis of Water Management for the Netherlands (PAWN), was a three-year, 125-man-year effort conducted jointly by The Rand Corporation, the Rijkswaterstaat (the government agency responsible for water control and public works), and the Delft Hydraulics Laboratory (a Dutch research organization).[3]

PAWN had two objectives:

To develop a methodology for assessing the multiple consequences of Dutch water-management policies.

To apply this methodology to evaluate a number of alternative water-management policies.

The study considered a wide range of consequences of the water-management policies, these consequences being called *impacts*. The work

---

[3] The results of the study are documented in a series of 21 reports. Goeller et al. (1983) and Goeller and the PAWN Team (1985) provide overviews of the methodology and summaries of the results. Walker and Veen (1981, 1987) give details of the screening of technical and managerial tactics.

developed impact measures to cover the objectives of all the affected groups (farmers, shippers, industrial firms, the country's ecology, and public health), and to reflect both equity and efficiency. Table 6.1 lists a subset of the measures that PAWN used in assessing the impacts of alternative water-management policies.

As mentioned previously, a water-management policy involves a mix of four kinds of tactics:

Technical tactics, which add to or modify the current water-management infrastructure (such as dig a new canal, enlarge a pumping station, or build a dike).

Managerial tactics, which change the rules by which a particular part of the infrastructure is operated (such as change the operation of weirs, pumps, and sluices so as to affect the water levels in lakes or the flows in rivers and canals).

Pricing tactics, which impose charges on water use or discharges.

Regulation tactics, which use legal and administrative measures to control water uses or discharges.

The first two kinds of tactics have direct effects on the supplies of water to users; the last two have direct effects on users' demands for water. One screening approach was used for technical and managerial tactics, another for pricing and regulation tactics. The remainder of this discussion deals with the former screening approach; the latter approach is described in Goeller et al. (1983, Chapter 23).

Estimating costs and benefits. A comparison of costs and benefits determined whether tactics were promising or unpromising. The cost assigned to a tactic was its annualized fixed cost (AFC), which was obtained by applying a capital recovery factor of 0.10 to the investment cost and adding the fixed annual operating cost. A capital recovery factor of 0.10 reflects a useful life of approximately 50 years and a discount rate of 10 percent. Operating costs were ignored, because they were very small relative to the fixed costs. (For a discussion of cost concepts, see Chapter 10.)

The benefits and costs of alternative water-management policies accrue primarily to its direct users: farmers, shippers, industries, and drinking-water companies. The policies also have more global impacts, for example, on consumers, the environment, and other countries. The screening process focused attention on agriculture, since it is the largest consumer, and it suffers most from shortages and salinity. Agriculture's water use is about 9 times that of the drinking-water companies and 50 times that of industries. It is estimated that in 1976 agriculture suffered losses of over 6000 million guilders owing to water shortages and about

Table 6.1. Some of the Measures Used by PAWN in Assessing the Impacts of Water-Management Policies in the Netherlands

Impacts on water-management system
  Costs of tactic
  Risk of flooding

Direct impacts on users
  1. Agriculture
    Gross benefits
      By crop (13 crops)
      By region (8 regions)
    Sprinkler costs (fixed and variable)
    Net benefits (by who receives benefits)
      Dutch (producers, consumers, government)
      Foreign (producers, consumers, government)
  2. Shipping
    Low-water losses (Dutch and foreign)
    Lock-delay losses (Dutch and foreign)
    Annualized cost of changing fleet (Dutch and foreign)
  3. Power plants
  4. Industrial firms and drinking water companies
    Revenues of drinking water companies
      From commercial firms
      From industrial firms
      From households
    Cost of supplying drinking water
    The percentage of groundwater in drinking water
    Total industrial water consumption (by source)
    Industrial cost increases (taxes, drinking water costs, etc.)
  5. Other users
    Lock delays for recreational boats
    Miles of fresh water beaches

Impacts on the environment
  Violations of water quality standards (salinity, chromium, biochemical oxygen demand, phosphates, heat)
  Changes in groundwater levels
  Algae growth
  Damage to nature from constructing facilities

Impacts on entire nation
  Jobs
  Sales
  Trade
  Public health
  Government revenues

500 million guilders owing to salinity of the water used to irrigate crops.[4] The impacts of the tactics on shipping costs (which are increased by low water levels) were also considered.

The benefits from a tactic are defined as the difference between the losses to agriculture and shipping if the tactic is implemented and the losses that would occur without the tactic being applied (it being presumed that the tactic would only ameliorate losses, not eliminate them entirely). The analysts considered three major categories of losses in the screening analysis:

Agriculture shortage losses. If a tactic increases the supply of water to an area, then, in years with low rainfall, a smaller proportion of the crops will die owing to an insufficient amount of water. The benefits from an increased supply are therefore measured in terms of reduced agricultural water-shortage losses.

Agriculture salinity losses. If a tactic is able to reduce the salinity of water supplied to agriculture, then a smaller proportion of the crops will be damaged by salt. The benefits from reduced salinity are therefore measured by the reduced crop losses due to salinity damage.

Low-water shipping losses. If the depths of the waterways are not sufficient, ships cannot carry their maximum loads, but must travel with less cargo to reduce their drafts. This means more trips and higher operating costs for carrying a given amount of goods. Tactics that change flows and water levels on certain waterways will affect shipping depths. The benefits from these tactics are measured by the reductions in shipping losses.

The benefits from a tactic vary from year to year, depending on the rainfall levels and the river flows. For example, tactics designed to reduce shortage losses will produce higher benefits in dry years than in wet years. Thus, since rainfall levels and river flows vary from year to year, the benefits of a tactic for any given year are unpredictable. To compare a tactic's benefits to its cost, one needs an estimate of its expected benefits over many years. The PAWN team used an approach (described below) that produces upper and lower bounds on the expected annual benefits by taking weighted averages of the benefits in four years that had very different rainfall and river-flow patterns: 1943, 1959, 1967, and 1976.

Based on the losses to agriculture for the years 1933 through 1976, it was estimated that both the shortage and salinity losses associated with 1976 would be exceeded only one year in 44, or in approximately 2 percent of all years. Table 6.2 shows the probability of exceeding the losses for all four of the chosen years (called the *external supply scenario years*).

---

[4] In 1976 a guilder was worth approximately $0.40.

Table 6.2. The Probabilities of Annual Losses
Exceeding Those of the Four External
Supply Scenario Years

	1976	1959	1943	1967
Shortage losses	0.02	0.07	0.21	0.63
Salinity losses	0.02	0.09	0.13	0.57

To obtain the upper bound for expected annual benefits, it was assumed that the benefits in 1976 would be obtained in all years drier than 1959, the benefits in 1959 would be obtained in all years between 1959 and 1943 in dryness, and so on. For the lower bound it was assumed that the 1976 benefits would be obtained only in years drier than 1976, 1959 benefits would be obtained in years between 1959 and 1976 in dryness, and so on. It was also assumed that there would be no benefits from the tactics in any year wetter than 1967. Treating a tactic as promising if its annualized fixed cost is less than the upper bound on its expected annual benefits is very conservative. A year as bad as 1976 is treated as if it is expected to occur once every 11 to 14 years instead of once every 50 years. Therefore, only very poor tactics would be screened out by this criterion.

If $B(y)$ is the benefit obtained in year y from implementing a tactic, then Table 6.3 gives the formulas for the upper and lower bounds on the expected annual benefits (EAB).

The analysts estimated the benefits for each of the four external-supply-scenario years by using a model called the *distribution model*, which simulates the major components of the surface water system on the Netherlands (Wegner, 1981). Given information over time on how much water enters the country (from rivers, rainfall, and groundwater flows), and how much water is extracted by the various user groups, the model calculates the water flows in the major rivers and canals, the levels of the lakes,

Table 6.3. The Upper and Lower Bounds on the Expected Annual Benefits
(EAB) for the Benefits $B(y)$ Obtained in Year y from Implementing
a Tactic

For shortage losses:

$0.02B(1976) + 0.05B(1959) + 0.14B(1943) + 0.42B(1967) <$ EAB

$$< 0.07B(1976) + 0.14B(1959) + 0.42B(1943) + 0.37B(1967)$$

For salinity and shipping losses:

$0.02B(1976) + 0.07B(1959) + 0.04B(1943) + 0.44B(1967) <$ EAB

$$< 0.09B(1976) + 0.04B(1959) + 0.44B(1943) + 0.43B(1967)$$

Figure 6.4. The eight Netherlands regions considered in the PAWN analysis.

and the concentrations of salt in these waters. Using this information, it calculates the agricultural shortage and salinity losses, and the losses to shipping from low water flows and lock delays.

Screen 1: Bounding the space of promising alternatives. For the purposes of the analysis, the Netherlands was divided into eight regions, as shown in Figure 6.4. The general approach to screening was to prepare a list of potential tactics for each region and to evaluate each one using a small subset of the impact measures (as described under the heading Screen 2). Before preparing these lists, however, the analysts conducted

Table 6.4. The Upper Bounds on Expected
Annual Preventable Losses for Low-
Demand and High-Demand Scenarios

Region	Losses in millions of guilders	
	Low demand	High demand
1	0.0	7.5
2	0.6	19.9
3	0.0	0.2
4	0.0	0.7
5	0.8	4.2
6	0.0	0.3
7	0.0	0.5
8	3.7	28.1
All	5.1	61.4

an analysis (Screen 1) to identify the regions and scenario assumptions for which the costs of implementing tactics were unlikely to be offset by the expected annual benefits in terms of reduced crop losses. These regions could then be ignored in the subsequent analysis, unless they had serious problems with respect to other impact measures. (A scenario consisted of factors that are uncertain and whose value will be determined by outside processes, political or natural, beyond the control of water-management decisionmakers. The scenario variables included the amount of sprinkler equipment that would be installed by farmers, and the amount of salt that would be dumped into the Rhine River outside the Netherlands.)

For each region and several demand and supply scenarios, estimates were made of the agricultural shortage losses that would occur (1) with the existing water-management infrastructure and (2) if the infrastructure were expanded so that there would be no shortage losses in areas with access to surface water (that is, so that the demands for irrigation water would be met fully).

The difference between these two estimates of shortage losses, which was called the preventable losses, is the maximum benefits that could be expected from any tactics designed to reduce water shortages in the region. Using the formula for shortage losses shown in Table 6.3, one can obtain an upper bound on the preventable losses; if $L(y)$ is the preventable losses in year y, an upper bound on the expected annual preventable losses (UB) is given by

$$UB = 0.07L(1976) + 0.14L(1959) + 0.42L(1943) + 0.37L(1967).$$

Table 6.4 presents the maximum benefits by region for two demand scenarios: a low-demand scenario (corresponding to the current use of water for irrigation) and a high-demand scenario (representing projected future demand). The results show that, unless the demand for irrigation water increases substantially, preventable shortage losses are expected to be small, so few if any tactics are needed. Thus, the next stage of the screening analysis (Screen 2) was limited to the high-demand scenario, except for tactics involving Region 8. In addition, since the maximum benefits for Regions 3 and 6 were very low in both scenarios, no technical tactics for reducing agricultural shortage losses in these two regions were analyzed.

Screen 2: Using few impact measures. The second phase of the screening analysis began with compiling a list of technical and managerial tactics designed to solve shortage and salinity problems at the national level and in the regions not eliminated from consideration in Screen 1. The major sources of tactics were reports prepared by the Rijkswaterstaat and discussions with Dutch water-management experts. Screen 2 analyzed a total of 57 tactics.

The distribution model was used to estimate the expected annual benefits from each of the tactics. It was run four times for each of the tactics, once for each supply scenario (1976, 1959, 1947, and 1967). For the situation in which no new tactics were implemented, four base-case runs were made. The predicted loss reduction from a tactic was obtained by subtracting the losses in the run with the tactic from the losses in the corresponding base-case run. Then the formulas given in Table 6.3 were used to obtain upper and lower bounds on the expected annual benefits, that is, the sum of the expected reductions in annual shortage losses, salinity losses, and shipping losses.

The expected annual benefits were then compared with the tactic's annualized fixed cost. If the annualized fixed cost exceeded the upper bound on the expected annual benefits, the tactic was clearly not promising and was screened out. If the annualized fixed cost was less than the lower bound on the expected annual benefits, the tactic was clearly worth further consideration and was passed through the screen. These steps left for consideration all of the tactics whose annualized fixed costs were between the upper and lower bounds for the expected annual benefits; it was decided to pass them all along to the impact-assessment phase, recognizing that their expected annual benefits might in fact turn out to be less than their annualized fixed costs in some cases.

A summary of how one of the tactics was analyzed will illustrate how Screen 2 was applied. This tactic involved building a canal to reduce agricultural salinity losses in Region 5. Some of the country's most valuable crops grow in this region: the world-renowned Dutch bulbs (tulips, hyacinths, and the like), greenhouse vegetables (tomatoes, cucumbers, and lettuce), and greenhouse flowers (roses, carnations, and chrysanthe-

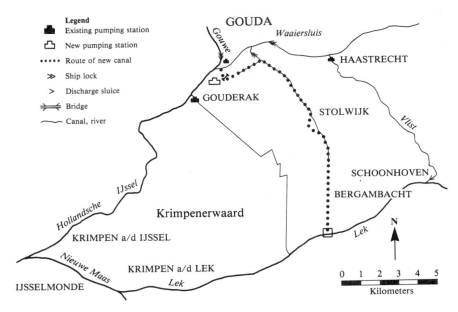

Legend
- 🏭 Existing pumping station
- 🏠 New pumping station
- ••••• Route of new canal
- ⟫ Ship lock
- ⟩ Discharge sluice
- ⟩⟨ Bridge
- ⌒ Canal, river

Figure 6.5. The route of the Krimpenerwaard canal proposed as a tactic in the PAWN study.

mums). All are quite sensitive to salt. The principal inlet point for the region's surface water is located along the Hollandsche IJssel at Gouda (see Figure 6.5).

The proposed solution to the salinity problem in this region was to extract water for the region farther upstream, where there was no danger of contamination from North Sea salt water. The water would be transported to Gouda by a new canal to be built through the portion of the country known as the Krimpenerwaard, as Figure 6.5 shows.

The benefits from the canal were measured by the reduction in agricultural salinity losses in Region 5. These benefits were partially offset by increases in shipping losses caused by reduced flows in one of the country's major rivers because it would supply the water for the canal. Distribution model runs were made with and without the Krimpenerwaard Canal using the high-demand scenario and the 1976, 1959, and 1943 external supply scenarios. (No year wetter than 1943 was considered because it was clear that the canal would not be needed in such a year.) The differences between the salinity and shipping losses in the two sets of runs (with and without the canal) were used as the benefits to be derived from building the canal. Table 6.5 shows the results derived from these runs, as well as the upper and lower bounds calculated from them.

Table 6.5. Comparison of the Costs and Benefits of Building the
          Krimpenerwaard Canal

Type of benefit	Annual benefits				Expected annual benefits		Annualized fixed cost
	1976	1959	1943	1967	Upper bound	Lower bound	
Salinity	41.5	0.2	0.0	0.0			
Shipping	(6.9)	(2.4)	(0.3)	0.0			
Net	34.7	(2.2)	(0.3)	0.0	2.9	0.5	5.7

Note: The benefits shown are in millions of guilders. Figures in parentheses are disbenefits. The annual benefits shown are for the four external supply scenarios for 1976, 1959, 1943, and 1967.

Table 6.5 offers several important insights. First, North Sea salt is a serious problem only in an exceedingly dry year (such as 1976). Second, the increased losses to shipping from this tactic are significant; in fact, in 1959 and 1943 they outweigh the reductions in salinity losses that the tactic achieves. Finally, the upper bound on the tactic's expected annual benefits is less than its annualized fixed cost, which means that it is not a promising tactic. It was therefore screened out and not given further consideration in PAWN's later stages.

The screening results. In the PAWN project screening was a simple, inexpensive but powerful way to reduce the effort that otherwise would have been required to evaluate many potential alternatives against scores of impact measures. The purpose of screening was to select from the large number of potential tactics a reasonably small number that merited a more detailed evaluation—and it achieved this objective. But screening accomplished more than just producing intermediate results to be used as inputs to the next phase of the project: it produced insights that helped guide later water-resources planning in the Netherlands.

Screen 1 showed that relatively few tactics for reducing agriculture shortage losses were worth considering further, unless the demand for surface water were to increase significantly above current levels. It also showed that it would not be worth the effort to generate or evaluate any alternatives for reducing shortage losses in two of the eight regions.

The somewhat surprising insights from Screen 2 were even more valuable. Prior to the PAWN study the Dutch had been considering several large and expensive construction projects. Almost all of them were screened out because the costs of construction were greater than the most optimistic assessment of the potential benefits. The savings to the Dutch government from not proceeding with these projects were estimated to total hundreds of millions of dollars.

Except for a few very inexpensive national tactics, most of the promising tactics affected primarily only a single region, a fact suggesting that the national water-management infrastructure is functioning rather well and that future water-management analysis and policymaking should focus more on regional problems than on national problems.

For the high-demand scenario, nine inexpensive regional construction projects and five national tactics were shown to be highly cost-beneficial, and they were passed on to the impact-assessment phase. Six of these fourteen tactics were also promising for the low-demand scenario; they included two aimed at reducing agriculture salinity losses, one that will help reduce agriculture shortage losses, two that will reduce low-water shipping losses, and one designed to reduce the chance of flooding. The combined annualized fixed cost for all six tactics was less than $3 million (in 1979 dollars at the 1979 exchange rate). For the high-demand scenario, in addition to the fourteen tactics mentioned above, eleven others were found to be promising (that is, cost-beneficial), but they were screened out because they were each dominated by one of the other fourteen (that is, they attacked the same problem, but had higher costs and/or lower benefits).

Conclusions on Screening

This section has demonstrated the usefulness of screening and presented systematic strategies for carrying it out. Screening significantly reduces the cost and time needed for carrying out systems and policy analyses involving large numbers of alternatives and impacts. In fact, screening makes it feasible to carry out a study that might otherwise appear, owing to the "curse of dimensionality," to be intractable.

The PAWN study illustrates the power of screening. Screen 1 eliminated two of the country's eight regions from detailed analysis before any tactics were evaluated by showing that no tactics in them could be cost effective. Screen 2 achieved a broad-brush, fast, and simple analysis of about 100 tactics, and eliminated 85 percent of them from further analysis. Although many of them had been under serious consideration for implementation, the analysis was convincing enough for their original sponsors to accept the results.

References

Alexander, E. R. (1979). The design of alternatives in organizational contexts: A pilot study. *Administrative Science Quarterly* 24, 382–404.

——— (1982). Design in the decision-making process. *Policy Sciences* 14, 279–292.

Archibald, R. W. (1979). Managing change in the fire department. In Walker et al. (1979), Chapter 5.

Arguden, R. Y. (1982). *Are Robustness Measures Robust?* P-6734. Santa Monica, California: The Rand Corporation.

Arrow, K. J. (1951). *Social Change and Individual Values.* New York: Wiley.

Bigelow, J. H. (1982). *Policy Analysis of Water Management for the Netherlands: Vol. V. Design of Managerial Strategies*, N-1500-5/NETH. Santa Monica, California: The Rand Corporation.

Brewer, G. D., and P. deLeon (1983). *Foundations of Policy Analysis.* Chicago, Illinois: The Dorsey Press.

Brill, E. D., Jr., S.-Y. Chang, and L. D. Hopkins (1982). Modeling to generate alternatives: The HSJ approach and an illustration using a problem in land use planning. *Management Science* 28, 221–235.

de Neufville, R., and J. H. Stafford (1971). *Systems Analysis for Engineers and Managers.* New York: McGraw-Hill.

Dorfman, R. (1965). Formal models in the design of water resource systems. *Water Resources Research* 1, 329–336.

Enthoven, A. C. (1975). Ten practical principles for policy and program analysis. In *Benefit Cost and Policy Analysis 1974* (Richard Zeckhauser et al., eds.). Chicago: Aldine, pp. 456–465.

Fisher, G. H. (1981). Comments on an outline for Project Air Force 2000, attachment to a letter to the Air Force dated November 18, 1981. Santa Monica, California: The Rand Corporation.

Friend, J., and A. Hickling (1987). *Planning under Pressure: The Strategic Choice Approach.* Oxford, England: Pergamon.

Goeller, B. F., et al. (1973a). *San Diego Clean Air Project: Summary Report*, R-1362-SD. Santa Monica, California: The Rand Corporation.

———, et al. (1973b). *Strategy Alternatives for Oxidant Control in the Los Angeles Region*, R-1368-EPA. Santa Monica, California: The Rand Corporation.

———, et al. (1977). *Protecting an Estuary from Floods—A Policy Analysis of the Oosterschelde: Vol. I. Summary Report*, R-2121/1-NETH. Santa Monica, California: The Rand Corporation.

———, et al. (1983). *Policy Analysis of Water Management for the Netherlands: Vol. I. Summary Report*, R-2500/1-NETH. Santa Monica, California: The Rand Corporation.

———, and the PAWN Team (1985). Planning the Netherlands' water resources. *Interfaces* 15 (January–February), 3–33.

———, and W. E. Walker (1988). *Strategic Planning for a City: A Conceptual Approach for LA 2000*, P-7355. Santa Monica, California: The Rand Corporation.

Hogwood, Brian, and Lewis Gunn (1984). *Policy Analysis for the Real World.* Oxford, England: Oxford University Press.

House, P. W., and J. McLeod (1977). *Large-Scale Models for Policy Evaluation.* New York: Wiley.

Jacoby, H. D., and D. P. Loucks (1974). The combined use of optimization and simulation models in river basin planning. In *Systems Planning and Design* (Richard de Neufville and D. H. Marks, eds.). Englewood Cliffs, New Jersey: Prentice-Hall, Chapter 17.

Luckman, John (1967). An approach to the management of design. *Operational Research Quarterly* 18, 345–358.

Manheim, M. L. (1964). *Highway Route Location as a Hierarchically-Structured Sequential Decision Process* (Ph. D. dissertation). Cambridge, Massachusetts: Department of Civil Engineering, Massachusetts Institute of Technology.

——— (1979). *Fundamentals of Transportation Systems Analysis: Vol. 1. Basic Concepts*. Cambridge, Massachusetts: The MIT Press.

Martland, C. D., Richard Assarabowski, and J. R. McCarren (1977). *The Role of Screening Models in Evaluating Railroad Rationalization Proposals*, Studies in Railroad Operations and Economics Series, Vol. 21. Cambridge, Massachusetts: Department of Civil Engineering, Massachusetts Institute of Technology.

Miser, H. J., and E. S. Quade, eds. (1985). *Handbook of Systems Analysis: Overview of Uses, Procedures, Applications, and Practice*. New York: North-Holland. Cited in the text as the *Overview*.

Mood, A. M. (1983). *Introduction to Policy Analysis*. New York: North-Holland.

Osborn, A. F. (1963). *Applied Imagination*. New York: Charles Scribner's Sons.

Overview. See Miser and Quade (1985).

Quade, E. S. (1982). *Analysis for Public Decisions*, 2nd ed. New York: North-Holland.

Rogers, J. S., and D. G. Berry (1982). *Planning Pair Gain System Installation Capacity on an Urban Telephone Route: An Analytic Model*, Working Paper 82-18. Toronto, Canada: Department of Industrial Engineering, University of Toronto.

Sawicki, D. S. (1983). On the virtues of the policy of doing nothing. *Journal of Policy Analysis and Management* 2, 454–457.

Summers, Irvin, and Major D. E. White (1976). Creativity techniques: Toward improvement of the decision process. *The Academy of Management Review* 1, 99–107.

Tuggle, F. D. (1986). Private communication.

Walker, W. E. (1986). The use of screening in policy analysis. *Management Science* 32, 389–402.

———, J. M. Chaiken, and E. J. Ignall, eds. (1979). *Fire Department Deployment Analysis: A Public Policy Analysis Case Study*. New York: North-Holland.

———, and M. A. Veen (1981). *Policy Analysis of Water Management for the Netherlands: Vol. II. Screening of Technical and Managerial Tactics*, N-1500/2-NETH. Santa Monica, California: The Rand Corporation.

———, and ——— (1987). Screening tactics in a water-management policy analysis for the Netherlands. *Water Resources Research* 23(7), 1145–1151.

Wegner, L. H. (1981). *Policy Analysis of Water Management for the Netherlands: Vol. XI. Water Distribution Model*, N-1500/11-NETH. Santa Monica, California: The Rand Corporation.

Chapter 7
Uncertainty
Coping with It and with Political Feasibility

Yehezkel Dror

7.1. Introduction

This chapter is divided into two parts. The first considers uncertainty in general and offers a broad view of how it can be responded to in professional practice; the second considers political feasibility as an important special case.

PART 1. SYSTEMS ANALYSIS AS A DECISION-GAMBLING AID

7.2. The Uncertain Context for Decisionmaking

Uncertainty saturates reality and dominates many decision issues. Spoiling decisionmaking with a multitude of errors deeply grounded in psychology, group dynamics, organizational behavior, politics, societal propensities, and culture and producing vast amounts of unanticipated and undesired decision consequences (Boudon, 1982; Sieber, 1981), uncertainty can rightly be regarded as the main affliction of choice. Therefore, systems analysis (a term that I use in its broader meaning, including policy analysis, the less technical parts of operations research and management sciences, applied decision theory, and the like) must largely focus on uncertainty and supply modes for coping with it effectively. But, as a matter of fact, uncertainty is usually undertreated, and often mistreated, in systems-analysis theory and practice (for example, by inappropriate reliance on subjective probabilities). As a result, the domain of applicability of systems analysis is sharply reduced and its actual utilization as a main aid to decisionmaking is often limited to relatively simple issues.[1]

It follows that improved handling of uncertainty constitutes a main imperative for upgrading systems analysis as a professional activity. But, in order to move ahead with an appropriate systems-analysis praxis, it is necessary to develop a suitable theory on which professional "knowledge-in-action" (Schön, 1983) can be based. Accordingly, this chapter first explores some main notions of uncertainty and provides a concept package for mapping it. Examination of reality leads to the crucial conclusion that nearly all decisions are inherently *fuzzy gambles*, and then to the reformulation of much of applied systems analysis as a *decision-gambling aid*. On these bases, the chapter proceeds to a number of operational proposals for improving how systems analysis handles uncertainty.

Since this chapter is oriented toward professional practice, it hints at the underlying deep theories but leaves them for full exploration elsewhere (Dror, in preparation), but it derives recommendations from a theoretical analysis and presents them on several levels of applicability. The second part of the chapter provides some applications dealing with political feasibility. To permit a compact treatment of a complex subject, the main ideas are presented concisely, with references to the relevant literature.

7.3. A Concept Package for Mapping Uncertainties

As distinguished, for instance, from the use of probabilistic terminology for evaluating evidence and proof (Eggleston, 1983) and from such special forms of predictive uncertainty as reactive risk arising from the reactions of others to our actions (Heimer, 1985), conventional terminology for handling uncertainties in the predictive context of decisionmaking is well known to systems analysts. For example, a scale of 0.0 to 1.0 denotes certainty at the extremes (0.0 = certainly not; 1.0 = certainly yes), whereas in-between notations, which can be simplified, express various degrees of what technically is called *risk*, that is, a definite measurement of the probability that the considered phenomena will happen, given explicit or implicit assumptions. Despite the glaring inadequacies of this one-dimensional concept package, it underlies most of systems analysis, with much reliance on subjective probabilities to supply risk assessments for all uncertainties (for example, Raiffa, 1968; Behn and Vaupel, 1982).

---

[1] In a field study of 34 offices of heads of governments in the Organization for Economic Cooperation and Development (OECD) and Third World countries, I failed to find even one case of institutionalized use of systems analysis as a decisionmaking aid. Although this finding has a number of converging explanations, a main one is the inadequacy of most of present systems analysis for helping with high-level decisions (Dror, 1984), *inter alia* because of inadequate methods for handling uncertainty.

Although systems-analysis practice is sometimes more advanced than the literature, the inadequacies of available concept packages and theoretic scales constitute a rigid barrier to its advancement. At the very least, a minimum concept package for handling and mapping uncertainties for the purposes of systems analysis must add to the simple certainty-risk-uncertainty scale the following main distinctions (for a parallel treatment, see Dror, 1986b):

Quantitative and qualitative uncertainty must be differentiated, "quantitative uncertainty" referring to situations where the alternative possible futures are known, but their probability distribution is unknown; and "qualitative uncertainty" referring to situations where the very shape of possible futures is not known. The term *quantitative uncertainty* is identical with most of the conventional uses of the term *uncertain* at present, whereas qualitative uncertainty leads into a dimension not recognized in accepted terminology, although proposed under different terms by a number of authors (noteworthy is Shackle, 1972).

A distinction is needed between situations where we can know at least the shape and sometimes the direction of alternative trends, with various degrees of certainty and uncertainty, and situations where the very shape of alternative curves and trends is in qualitative uncertainty. An illustration of the latter are what can be called *explosive situations*, where sudden and unpredictable rapid and radical shifts may happen. Smooth curves, still ornamenting texts in systems analysis and related disciplines and, even more so, in economic theory conditioning the frames of appreciation of systems analysis, are very misleading when nonsmooth dynamics prevail, as is often the case. Instead, catastrophe topology and chaos mathematics (Woodcock and Davis, 1979; Thom, 1983) should serve much of systems analysis.

The idea of declining curves is important when trends move in an undesired direction (an interesting historic illustration is supplied by Young, 1977). In such cases, incremental improvements are of little use and systems analysis must search for breakthroughs that mutate historic processes into desired directions (Dror, in publication).[2]

---

[2] An additional, even more important perspective is to view decisions as attempts to intervene with historic processes, whether on small or grand scales. This implies that all but banal systems analysis must be based on study and maximum achievable understanding of relevant historic processes, not only on the level of events, but also on the level of societies and institutions and even on the underlying level called by the French *Annals* school *longue durée* (Braudel, 1980, especially Part 2). Consequently, systems analysis urgently needs methods for integrating historic perspectives and better understanding of deep societal dynamics, as a practical necessity for real decision issues. (An important first step in the needed direction is made by Neustadt and May, 1986.) Better integration of historic perspectives into systems analysis in turn requires and depends on more sophisticated handling of uncertainty.

In paradoxical language, some situations are characterized by a high probability of low-probability "surprise events," each one of which, insofar as it can be imagined in advance, has a very low chance of occurring. Such situations can conveniently be denoted as "surprise-prone."

It is crucial to note the differentiation between situations in which uncertainties are "hard" (that is, structurally inbuilt into the dynamics of the phenomena, which behave, at least partly, in a chaotic, indeterminate, and random mode from the perspective of present human thinking (Suppes, 1984), and situations where uncertainties are "soft." When the dynamics of phenomena follow some orderly fashion, even if complex ones such as stochastic chains or long-range swings, uncertainty results from a scarcity of good prediction methods and our lack of adequate understanding rather than from structural features of the phenomena. This is a very important distinction, because hard uncertainty imposes an absolute limit on predictability and on the outputs of even ideal prediction methods. In other words, the prevalence of situations in which even a "perfect witness" (that is, an entity that knows all facts and their dynamics and will say "I know that one cannot know") must be recognized. In such situations, any reliance on subjective probabilities is a dangerous illusion and even a delusion.

Finally, a distinction is needed between phenomena in respect to which the above mapping categories can be applied, with various levels of reliability, and phenomena concerning which predictability or nonpredictability itself is unknown, and perhaps unknowable—to which the term *ignorance* may be usefully applied (although used in different meanings in some of the literature that tries to move beyond conventional uncertainty scaling).

These (and other) dimensions of uncertainty conceptualization, guesstimation, and mapping provide a multidimensional matrix that goes beyond the equipment of systems analysis in its present form. Some of the cells of the matrix are logically impossible, but most of them provide categories essential for making systems analysis relevant for more than micro-issues.

7.4. The States of Uncertainty of Main Systems-Analysis Variables

When uncertainty is limited to variations within rather narrow ranges and is quantitative in nature, the difficulty is not too great and the necessary "gambles" are well structured and can be handled with available systems-analysis methods, such as in risk analysis (for example, Fischhoff, 1981; Kobrin, 1982; Merckhofer, 1987), subject to some upgrading. The situation is quite different when a number of variables of critical importance to a decision are "jumpy" and pose hard qualitative uncertainty. This is the case in the foreseeable future, which is characterized in many domains by much turbulence (Dror, 1986a, especially Chapter 2) with many features of politics, economy, technology, social values, external relations, and more, showing (in terms of present human perspectives)[3] much inbuilt uncertainty. Acquired immune deficiency syndrome (AIDS) and the suddenly discovered possibility of relatively cheap superconductors illustrate surprise events of far-reaching implications, even in relatively stable and usually predictable domains.

The amounts and degrees of uncertainty in respect to different variables vary between domains, periods, and countries and depend on the time spans under consideration. Exaggeration of unpredictability should be avoided, as many variables are relatively stable for time streams relevant to many decisions. Still, once one moves beyond simple physical projects and short-range micro-issues, uncertainty is pervasive and much of it takes the form of hard and qualitative uncertainty, often coupled with ignorance. Recent thinking on the nature of social change (Boudon, 1986) reinforces this conclusion, tending to the opinion that general laws are only of very limited validity for social processes, that social dynamics depend a lot on specific historic and contextual settings, and that chance seems to play an important role, at least on the level of events and contemporary history important for most of decisionmaking—and perhaps also on deeper societal and institutional levels (on the distinction, see Braudel, 1980).

7.5. A Fuzzy Gambling Model of Decisionmaking

Every decision under conditions of uncertainty constitutes, in essence, a gamble. Improving simple decision gambles involves estimating the payoffs of the different options with their risk and time distributions on one

---

[3] Modern quantum physics and evolutionary theory seem to support a *Weltanschauung* in which important phenomena have inbuilt dynamics that include chance jumps, leading to a *probabilistic and chance metaphysics* (going beyond Suppes, 1984) in which reality is between "chance and necessity" (Monod, 1971). But my treatment of hard uncertainty and its implications for the purposes of systems analysis does not depend on absolute validity of one or another metaphysical picture of reality, which in any case cannot be determined with present human capacities. It is enough for our purposes that significant parts of decision-relevant reality behave as if in hard uncertainty. True, this makes the distinction between hard and soft uncertainty pragmatic and elastic. But, for the theory and praxis of systems analysis, this makes no difference other than emphasizing what in any case is true, namely, that no hard uncertainty should be admitted until all possibilities for reducing uncertainty are exhausted and positive support is provided for mapping specific domains as in hard uncertainty, for instance, by examining underlying change dynamics.

hand and, on the other, a value choice between the various predicted payoffs, including "lottery-value" preferences between various bundles of certainty, risk, and uncertainty (Raiffa, 1968) as well as different time preferences not considered in this chapter. Much of systems analysis, including some of the more advanced notions (for example, most of Miller, Kleindorfer, and Munn, 1986), clings to such a relatively simple image of decision gambling, and relies on methods limited in their applicability to situations of risk and, in part, quantitative uncertainty.

But, when decisions face hard qualitative uncertainty, surprise propensities, and ignorance, then the decision involves choice between in part unknowable and not predetermined—but often very significant—consequences. This widespread situation I propose to call a *fuzzy decision gamble*.[4] Most available tools of applied systems analysis are quite useless in the face of fuzzy decision gambles, resulting either in spoiling decisions rather than improving them,[5] or in limiting the domain of utilization of systems analysis.

The postulated conditions of fuzzy decision gambling prevail, more or less, in the vast majority of decisions other than the most simple ones. Therefore, *much of decisionmaking should be viewed as constituting, or at least including important elements of, fuzzy gambling* (that is, the main perspective of Dror, 1986a). This conclusion has far-reaching implications for systems analysis, including, on the applied level, the need for innovative protocols for handling uncertainty.

7.6. Protocols for Handling Uncertainty

Once decisions are correctly recognized as, at least in part, fuzzy gambles, the basic mission of systems analysis as an approach to improving choice must be reformulated: *One of the main tasks of systems analysis is to help in fuzzy decision gambling.* To be emphasized is the impossibility of "ungambling" most decisions. Rather, systems analysis must develop

---

[4] The proposed term *fuzzy gamble* is not to be mixed up with fuzzy set theory, despite some affinities. At this stage of my thinking on the subject, I leave open the question of whether some ideas in fuzzy set theory may perhaps be developed so as to help with the analysis of fuzzy decision gambling.

[5] Nietzsche correctly pointed out, when discussing Hamlet, that a veil of illusion is essential for human action. Decisionmakers may therefore welcome transformation of hard qualitative uncertainty into risk maps by systems analysis relying on one or another method producing subjective probabilities, such as Delphi (Linstone and Turoff, 1975), in spite of their frequent lack of an adequate epistemological basis. This interest by clients is shared by practitioners of systems analysis, most of whom quite naturally prefer scientific-looking and easily usable and teachable techniques to complex qualitative protocols and approaches. The common interest of producers and clients in inadequate and misleading treatments of uncertainty helps to explain the persistence of obviously wrong methods.

approaches, methods, and methodologies fitting the inherent fuzzy-gambling features of most decisions and help to improve decision gambling in ways fitting its true nature.

This view has multiple implications, ranging from the need to revise the very notions of "rationality" underlying much of systems analysis to changing the professional training of systems analysts so as to prepare them better for serving as improvers of fuzzy decision gambling. Indeed, all of the work program of systems analysis as an interdiscipline and a profession needs revision so as to focus in part on advancement of knowledge fitting the needs of fuzzy decision-gambling improvement. Leaving such tasks to some other occasion, within the praxis orientation of this handbook some main operational implications of the perspective developed in this chapter can best be presented in the form of seven main, interrelated systems analysis protocols (in the sense of IIASA, 1985) recommended for handling uncertainty:

1. Preparing a set of broad, coherent, long-range alternative contextual futures.
2. Mapping and reducing uncertainties.
3. Including uncertainties adequately in models.
4. Handling low-probability, high-impact contingencies in a special way.
5. Developing robust, resilient, and uncertainty-absorbing options and strategies.
6. Debugging the handling of uncertainty in decisionmaking.
7. Interfacing with decisionmakers.

Each of these protocols is discussed separately in what follows.

Preparing a Set of Broad, Coherent, Long-Range Alternative Contextual Futures

Much of real uncertainty is repressed in most of current practice with the help of the assumption of all other things being equal. This is a serious error when consequences of different alternatives will, in fact, be shaped by changes that are sure to occur in a large range of variables exogenous to the decision[6] issue in the narrow sense of that term but forming its embracing context. To overcome this widespread error, it is necessary to place analysis of any decision issue within a broad setting of alternative

---

[6] The term *decision* is used in this chapter in a broad sense, including single discrete choices, integrated sets of decisions as in the form of plans, programs of action including contingent ones, and more.

futures of main relevant variables and environments. These alternative futures must cover a broad range of possibilities and deal with a time stream relevant to the decision under consideration; and each alternative future must be coherent in its main features. Only when such a set of alternative futures is available can specific decisions be analyzed in terms of real uncertainties stemming from the range of relevant alternative futures.

In technical terms, the following points are to be noted:

The set of alternative futures should be prepared as a general systems-analysis staff paper, independent of specific decision issues under consideration. This is necessary to avoid biases and narrowing of perspectives resulting from fixation on a discrete item under consideration.

The alternative-futures-exploration exercise is an important systems-analysis output by itself, as an intelligence document enriching the decisionmakers' cognitive maps. It also serves as a basis for mapping uncertainties in specific decision issues by exposing underlying assumptions and their doubtful basis.

The alternative futures should be formulated in outline form, to be further elaborated as needed. The number of alternative main futures should be limited so as to preserve manageability. Each alternative future should be internally consistent. Their nonexclusive nature should be borne in mind, all of them never adding up to all possibilities. They should be supplemented with additional staff papers, such as lists of low-probability, high-impact surprise possibilities, as discussed later on. The alternative-futures set should be updated periodically.

An alternative future should be regarded, in the sense of a *futurible*—as suggested by Bertrand de Jouvenel (1967, p. 18)—as a future "that appears to the mind as a possible descendant from the present state of affairs." The generation of alternative futures requires combinational use of all prediction methods fitting the material, from extrapolation and modeling to Delphi and imagination. But, within the present protocol, no probabilities should be allocated to the alternative futures, their main purpose being to reduce qualitative uncertainty on decision contexts.

The set of alternative futures should be accompanied by an analysis of the dynamics of underlying processes, in terms ranging from stable to explosive. This is in order to explore the remaining qualitative uncertainty: The more underlying processes are turbulent the less will the set cover the real range of possibilities and the more are parts of the future in a state of hard qualitative uncertainty.

Special attention should be devoted to identifying declining curves, as possibilities or trends. This, as already mentioned, is as a basis for developing appropriate decision strategies, such as incrementalism versus search for breakthroughs, which, *inter alia*, will also be applied to discrete decision instances.

Chapter 9 offers a discussion of the issues relating to the scenario representation of alternative futures.

Mapping and Reducing Uncertainties

The second protocol, which in some senses is a sequel to the first one and is in an iterative relation with it, involves mapping and reducing uncertainties, both of the alternative-futures set as relevant to specific decision issues and of the variables endogenous to specific decisions (in the broad sense of that term, as already pointed out) up for systems analysis.

Relevant technical points can be summed up as follows:

Fixation on subjective probabilities must be avoided. As already mentioned, subjective probabilities are misleading unless an epistemological justification for their use exists, including both bases for assuming that uncertainties are not hard and availability of persons who can justifiably be assumed to have tacit knowledge serving as a grounding for their subjective probabilities. The Delphi method should be used sparingly, only when and insofar as the above two conditions are met.

Combinational use of all available prediction methods is needed in mixes fitting the nature of the subject matter.

Question marks, alternative assumptions, empty cells, ranges, and additional notations to signify hard and qualitative uncertainties, as well as ignorance, should be added to the languages used to denote various mappings of certainty and uncertainty in addition to available risk scales. Pressure to reduce uncertainty should be resisted when prediction methods reach the limits of legitimate use, as they often quickly do.

The end product will often be a map of certainties, risks, various forms of uncertainty, and ignorances. Adequate presentation of such mappings of uncertainty, after being reduced as much as correctly possible, raises problems of modeling as well as interface with decisionmakers, considered in what follows.

An additional, more complex possibility must be mentioned here, namely, the reduction of uncertainty as a goal. Reducing uncertainty may be an important policy goal by itself, as for instance, in legal and economic policies directed at assuring stable conditions to permit markets to operate rationally, and in international policies directed at reducing war probabilities by increasing stability. Reduction of uncertainty also increases possibilities for improving decisionmaking, because, after all is said and done, fuzzy decision gambles are often riskier than decisions under certainty, although often they may also offer chances of much larger benefits. Reducing uncertainty as a policy goal raises problems outside our concerns here, but two points should be mentioned. First, to reduce uncertainty it must first be mapped and its direct causes as well as its deeper grounds must be understood. Second, reducing uncertainty is often a major intervention with historic processes requiring, *inter alia*, critically large masses meeting impact thresholds (Schulman, 1980), a matter that suggests additional issues for advanced systems analysis.

Including Uncertainties Adequately in Models

Modeling is correctly considered as central to systems analysis (Miser and Quade, 1985, especially pp. 135–140; Quade, 1982, Chapters 9, 10). Models are essential for a justifiable process of systems analysis, but to fit reality they must reflect uncertainties fully. This requires significant changes in modeling, compared with much practice and most of the literature. In technical terms:

Qualitative models, including concept packages, may often be the best that can be formulated.

Models should reflect better the full range of uncertainties, which implies at the very least pluralistic parallel models to reflect main alternative futures, more use of stochastic modeling, integration of human gaming as an explorative modeling mode into the main body of systems analysis, and suitable notations of various forms of uncertainty, as already mentioned.

Models including a range of uncertainties are complex and usually require writing in the form of computer programs. Advanced artificial intelligence may be of help, but needs better handling of uncertainty than available at present (Kanal and Lemmer, 1986). Suitable heuristic models need development.

Full handling of uncertainties in complex issues requires combinational and integrated multiform models, including artificial intelligence and a variety of heuristic programs, stochastic and chance simulation, human elements, and a range of qualitative dimensions. Some experiences of the Rand Strategy Assessment Center are relevant (Davis and Winnefeld, 1983: Davis, Banker, and Kahan, 1986).

Handling Low-Probability, High-Impact Contingencies in a Special Way

A special problem is posed by the possibility of surprise events, some of which can be previewed qualitatively and some of which are in qualitative uncertainty, unless, somewhat by chance, imagined in advance. The general tendency is *a priori* to exclude surprise events from consideration as "esoteric," with the rationalization that in any case nothing can be done about them. But this is erroneous, as low-probability, but high-impact, possibilities may well deserve attention and preparatory action.

It should be noted that surprise impacts may be very positive, both by themselves and even more so as providing windows of opportunity, thus contradicting some professional prejudices of systems analysis against surprises as disruptive of certainty-assuming analysis and discrediting simplistic predictions. For instance, as further explored in Part 2, political feasibility may expand during the loosening up of stable realities caused by some kinds of surprises and consequent crisis situations. But, to avail oneself of rapidly passing opportunities provided by surprises, it is necessary to be well prepared for them, thus making prediction of surprise possibilities or, at least, prediction of the possibility of unforeseeable surprises, essential.

In this connection, a counterpoint theme should be mentioned, namely, the initiation of surprises and crisis instigation as a subject for systems analysis. In particular, when an important domain is on a declining curve and breakthrough trend-shifting interventions are needed, it may be necessary to destabilize negative momentum and create policy leeway by throwing surprises at history. This important subject goes beyond the confines of this chapter, but it should be kept in mind, both as a subject important for advanced systems analysis that requires attention, and as a source of uncertainty.

Returning to the systematic handling of surprise possibilities in systems analysis, I propose the following protocol:

1. A standard list of possible surprise events should be prepared and updated from time to time, as a general systems-analysis staff paper independent of any particular decision issue, similar to the set of alternative possible futures. A multiplicity of methods adjusted to the task should be used for preparing this list, such as introducing random variations into models, with special attention to imagination-stimulating techniques such as brainstorming.

2. The list should then be processed to analyze surprises in three dimensions, even if initially on the level of guesstimation: probability of occurrence, significance in terms of negative or positive impacts, and possibilities of influencing in advance the probability and/or cost-benefits of surprise.

3. The processed list should then be ordered by priority. Possibilities with high probability, which may still constitute "surprises" in the sense of contradicting subjective expectations, high costs, and ease of handling receive the highest priority, whereas possibilities with low probability, low impact, and difficult handling receive the lowest priority.
4. The cutoff line for further treatment is between low-probability, high-impact possibilities and low-probability, low-impact possibilities. The latter can be ignored in the main, whereas low-probability, high-impact possibilities need further handling. In particular, low-probability, high-impact possibilities that can be handled relatively easily should be taken up for appropriate treatment and put on the decision agenda. Low-probability, high-impact possibilities that are difficult to handle in advance should be monitored and serve as inputs into crisis-decisionmaking preparation.

The above protocol leaves many questions open, such as what to do about the possibilities of hard uncertainty or ignorance concerning probability, impact, and treatability. More important, whatever is done, many possibilities will not be identified in advance, including some with very high impact—thus making crisis-management upgrading essential in situations where significant surprises are likely, or at least cannot be excluded. Also, the number and range of low-probability, high-impact possibilities may overtax the capacities of even a very good decisionmaking system and of the systems-analysis unit itself.

Still, in my experience the suggested protocol can achieve significant improvements in actual decisionmaking behavior, which is the criterion by which systems analysis is to be evaluated. Well-prepared lists of surprise possibilities serve as important intelligence inputs to upgrade the world pictures of decisionmakers, including the important educational effect of learning to think in terms of surprise possibilities. Often, the protocol leads to identifying high-probability, high-impact possibilities that should not be "surprises" at all, but have been missed in routine-overloaded decisionmaking. Appropriate lists permit at least some advance preparation and efforts to influence the occurrence of a few main surprise possibilities. The use of surprise lists for upgrading crisis decisionmaking is also very important. Sensitivity testing of specific decisions to various surprise possibilities is an additional useful output.

Developing Robust, Resilient, and Uncertainty-Absorbing Options and Strategies

Standard decision-analysis methods, as usually used in systems analysis for handling uncertainty, suffer from two major types of weaknesses (see also Majone and Quade, 1980, especially pp. 32–33): many of the behavioral assumptions on which they are based are refuted by research (von Winterfeldt and Edwards, 1986), even when they are in principle applicable because no hard uncertainty exists; and, more important by far, most available decision-analysis methods are in principle inapplicable and misleading when used to handle hard uncertainties. The view of decisionmaking as fuzzy gambling developed in this chapter sharpens the second limitation, refuting in part the usefulness of most uncertainty-handling methods available nowadays in systems analysis.

Certainly sensitivity testing, search for less uncertainty-sensitive alternatives, contingency planning, various forms of insurance, parallel approaches, social experimentation, and the like, together with all available forecasting methods, continue to serve as the main tools of systems analysis in the face of uncertainty, especially in their more sophisticated forms (for example, Ascher and Overholt, 1983). But subjective probabilities in all their forms and uses must be rejected for most decisions, and consequently the modes of using all other methods and their relative importance change and additional methods must be developed.

The following comments illustrate appropriate approaches and methods:

Many of the available methods, such as sensitivity testing, must be used qualitatively, with alternative futures, scenarios, and surprise possibilities often unaccompanied by probabilities.

The idea of one-dimensional utility curves and preference functions is useless when value judgment has to be applied to qualitatively uncertain results. Choices become much harder with more complex and ambiguous unresolvable conflicts (going beyond Levi, 1986), and available ideas on decisions with multiple objectives (for example, Keeney and Raiffa, 1976), however interesting, are inadequate for handling complex mixes of uncertainties. Hence, the need to rely more on "acts of will" by decisionmakers, introducing all the complexities of *akrasia* (that is, weakness of will, see for instance Davidson, 1980, Chapter 2) and similar problematics of the interfaces between decisionmakers and systems-analysis units, and aggravating them, as discussed later in this chapter.

Upgrading of resilience rather than risk reduction may often be a preferable strategy in the face of hard uncertainties (Wildavsky, forthcoming).

The level of strategies for handling uncertainty as principles guiding specific decisions needs recognition as a subject matter for systems analysis and deliberate choice. This includes policies on attitudes to risk and uncertainty, on mixes between breakthrough attempts and incrementalism, on possibilities for using mixed strategies in the technical sense of utilizing loaded random devices as a decision mechanism, and more. Empiric studies of strategies actually developed by successful practitioners of various forms of gambling (for example, Hayano, 1982) may provide stimulation and relevant insights.

Accelerated and improved learning becomes a main response to hard uncertainties, with more emphasis on policy pilot testing, constant monitoring of ongoing decision implementation and improvement of underlying models (Holling, 1978), parallel approaches (Bendor, 1985), and upgrading of the learning abilities of decision systems, together with improvement of their long-range memories. Learning improvement involves both organizational aspects (for example, De Greene, 1982, and Mason and Mitroff, 1981; to be compared with Etheredge, 1985) and advancements in control theory (Beniger, 1986). More attention to learning from failures is also essential (compare Petroski, 1985), but with appropriate recognition that the fuzzy-gambling view of decisionmaking undermines reliance on results as tests of the quality of decisions, requiring more complex decision-process evaluation instead as a basis for learning.

Crisis-decisionmaking upgrading as the ultimate reserve against unpredictable high-impact events becomes an essential tool of fuzzy decision gambling that must be taken up, at least in part, within systems analysis.

Such changes and advances needed to handle fuzzy policy gambling must be supplemented with more radical innovations in the basic approaches of systems analysis, as illustrated by "debugging," to which I now turn.

Debugging the Handling of Uncertainty in Decisionmaking

The protocols just discussed move within the main body of the present state of the art of systems analysis, even if some radical changes are recommended in perceiving and mapping uncertainty. The debugging protocol brings us to a different mode of operation that needs to be added to systems analysis to handle uncertainty, in addition to other uses beyond the scope of this chapter. The usual main, underlying rationale of systems analysis is to try to approximate some conception of "rationality," whether mainly economic rationality (for example, Stokey and Zeckhauser, 1978; Amacher, Tollison, and Willett, 1976), or more complex optimal models, which include extrarational elements such as value judgment and creativity (Dror, 1983, Part 4). But, when hard qualitative uncertainties and ignorance pervade reality, available rational models are inapplicable and suitable models on a higher level of ultrarationality are not yet around, despite much ongoing work on the idea of rationality (for example, Elster, 1986), and they may be hard to apply to concrete decisionmaking once they are developed.

Therefore, as a partial substitute for nonavailable notions of rationality fitting stubborn uncertainties, and as a supplement to such models as they become available, a different decision-improvement strategy must be added to the repertoire of systems analysis, namely, "debugging" in the sense of offsetting, counteracting, balancing, and reducing actual error propensities in handling uncertainty during decisionmaking. This means that systems analysis must add three interrelated layers to its present main approaches:

1. Systems analysis must study and understand actual decisionmaking processes in all their complexity and identify the main widespread mishandlings of uncertainty that actually occur and characterize these processes.
2. Systems analysis must counteract these mishandlings in the context of specific decision issues constituting current subject matters.
3. Because of the difficulties of handling wrong decisionmaking features sporadically and in the narrow context of specific issues alone, and in order to make a broader contribution to the quality of decisionmaking processes as a whole, systems analysis must move in the direction of upgrading decisionmaking processes independent of specific issues under current consideration.

These proposals involve significant extensions in the theory and practice of systems analysis, although a number of works move in such directions (for example, Checkland, 1981; Mitroff, Quinton, and Mason, 1983). What is involved is an extension of systems analysis from problem handling to increasing the capacity of decisionmakers and of decisionmaking processes and structures to handle problems—which is quite a shift, although proposed some time ago within the idea of policy sciences (Lasswell, 1971; Dror, 1971a).

The following suggestions illustrate some of the dimensions of the debugging protocol in the context of handling uncertainty:

Decision psychology indicates serious fallacies in human handling of uncertainty (Kahneman, Slovic, and Tverski, 1982; Dörner et al., 1983). Also relevant are studies looking on uncertainty as a type of stress (Janis and Mann, 1977). Especially important are also so-called "motivated irrationalities" (Pears, 1984) and various aspects of "multiple self" (Elster, 1985), such as self-deception and weakness of will, which probably are reinforced by uncertainty and active in spoiling its handling.[7]

---

[7] Matters are more complex because self-delusion may sometimes be useful by avoiding the Hamlet syndrome, by activating the interesting process of self-fulfilling prophecies, and more. This leads to the already mentioned need for ultrarationality models, which adequately handle such paradoxical needs and other complexities, to serve as an advanced philosophic base for systems analysis.

Studies in group dynamics indicate widespread processes resulting in mishandling of uncertainty (Janis, 1982; Kahan et al., 1985).

Organizational research reveals many features of organizational behavior involving an incapacity to treat uncertainty correctly (March and Olson, 1976; Stinchcombe and Heimer, 1985; Sims et al., 1986, especially pp. 255–257).

Cultural studies show that contemporary cultures are hostile to risk (in the nontechnical sense, Douglas and Wildavsky, 1982), as are the social sciences (Douglas, 1986). This may impose blinders dominating all of our perception and thinking (Douglas, 1987), with pernicious consequences for the handling of uncertainty—and for systems analysis, which can hardly be expected to be able to "jump out of the system" (Hofstadter, 1979, Chapter 15). Also relevant are broad studies of societal and institutional irrationality (Elster, 1983).

Enough has been said to clarify the importance, as well as the difficulties, of debugging as a main contribution of systems analysis to handling uncertainty. Teamwork with professionals in decision psychology and organizational processes may be required, with the handling of cultural limitations as a whole being beyond the tasks of systems analysis. But systems-analysis units have to face the challenge of surmounting their own cultural limitations, which is posed starkly by the problems of handling uncertainty, where cultural biases are especially pronounced, but which is essential also for other systems-analysis problems, such as political feasibility estimation (as taken up in Part 2). Involving systems analysts with diverse cultural backgrounds and having some work experience in other cultures suggest operational possibilities for reducing cultural biases in systems analysis, however difficult.

With this and other problems, debugging constitutes a main essential protocol for handling uncertainty; we must include it within the scope of systems analysis, at least in part and in cooperation with other professionals.

Interfacing with Decisionmakers

The fuzzy decision-gambling perspective raises difficult issues on interfaces between systems-analysis units and decisionmakers, as already indicated. The first point to be noted is that viewing their decisions as largely fuzzy gambles is anathema to most decisionmakers. It is also dangerous, because it may result in recklessness and exclusive reliance on subjective feelings, as a result of regarding decisions in any case as "gambles." Alternatively, sophisticated systems analysis that clarifies the fuzzy-gambling aspects may be displaced by more primitive and "reassuring" practitioners who serve as modern versions of classical court astronomer-astrologers (Eberhard, 1957). Therefore, *a main function of systems-analysis units is to educate decisionmakers to understand correctly their function as fuzzy decision gamblers.* Conveying to senior decisionmakers a more sophisticated understanding of the natures of uncertainty and the correct modes for improving decision gambling should therefore be accepted as a main mission of systems analysts.

The debunking function leads to an additional dimension of interfaces between decisionmakers and systems-analysis units: the need for the latter to serve as constraints on the former against their error propensities. To fulfill this function, at least one and preferably two conditions must be met: either the decisionmakers must wish to bind themselves against their own failing, on the Ulysses and the Sirens metaphor (Elster, 1979; Schelling, 1984, Chapter 3); and/or the systems-analysis unit must be able to remonstrate with top decisionmakers without being punished or having its access to top decisionmakers curtailed. Some regimes have successfully met such requirements at least for some time, such as the censorial system in classical China (Hucker, 1966; Wu, 1970). But the closing down of the Central Policy Review Staff in the United Kingdom (Hennessy, 1986, pp. 111–112) illustrates the difficulties of meeting these requisites nowadays (see also Prince, 1983). Thus the decision-gambling perspective is very demanding on relations between systems analysis and top decisionmaking and may aggravate their interrelations, unless care is taken.

In the longer run, upgrading the qualifications of top decisionmakers is essential for correct utilization of more advanced systems-analysis knowledge (Dror, 1986a, pp. 291–294). Limiting myself here to salient improvements within the practice of systems analysis, an effort is needed to upgrade its communication with top decisionmakers. There is little hope of conveying the decision-gambling idea and applying it correctly, and in a form comprehensible and useful to nonprofessionals, without developing novel modes of presenting complex analyses to busy clients. Needed are multidimensional and interactive briefing and presentation systems that can present the main dimensions of decision gambles to top decisionmakers in a way that clearly explicates major uncertainties and the judgments required from them. Limits and main features of human information-processing capacities (for example, Lindsay and Norman, 1977; Minsky, 1968; Schank, 1975; Sternberg, 1985) must be fully taken into account.

The blocks for building such systems are available. But, in visits to a large number of offices of top governmental and corporate decisionmakers, and also to many crisis-management centers, I have failed to find a system adequately presenting uncertainties, or presenting operationally, correctly, and comprehensibly the decision issues and required judgments to the decisionmakers.[8] I am not sure whether decisionmakers will welcome such presentations, which clearly put on them the onus for hard choices; but it is up to systems analysis as a professional endeavor to do its share of the job and develop appropriate presentation and briefing systems, while carefully monitoring their use and preventing misuse.

7.7. Toward Systems Analysis Mark III

The leitmotif of this chapter is that most of decisionmaking constitutes a fuzzy-gambling activity. This view seems inescapable, however hard to accept. It has far-reaching implications, going well beyond the scope of this book, such as regarding history as between chance and necessity, casting doubt on most learning processes, and recognizing the occasional necessity of nations to gamble with history (Dror, 1986c). Within the narrower domains of systems analysis, three conclusions emerge.

The first conclusion is that far-reaching changes are required in systems-analysis praxis in order to handle hard uncertainties correctly and provide help with fuzzy decision gambles.

The second conclusion is that, within the present arsenal of systems analysis, much equipment is available that can be of service in improving decision gambling if correctly brought to bear.

The third conclusion is much more demanding: In order really to adjust systems analysis to the view of decisions as fuzzy gambling, many advancements in the theory and professional practice of systems analysis are needed, some of which require reconsideration of some fundamental issues, including for instance, the nature of systems analysis and its relations with neighboring and overlapping disciplines and professions. Also, relations between systems-analysis professionals and units and decisionmakers require considerable change in order to withstand the strains of regarding and handling decisions as fuzzy gambles.

When other needs of advancing systems analysis are also taken into account, such as the view of decisions as interventions with historic processes, the necessity for inventing breakthrough options when situations are on declining curves, and the requirement to improve complex value judgments by decisionmakers without encroaching on their prerogatives, then a fourth conclusion emerges: systems analysis as a whole must make a quantum jump, moving from its present Mark II, following the operations research version Mark I, forward to Mark III.

As Karl Popper has pointed out, present knowledge cannot logically

---

[8] This failure is related to the absence of workstations meeting the needs of top decisionmakers. Available workstations just do not fit the needs, in part because of their inability to handle ill-structured problems and hard uncertainties.

predict future knowledge. Therefore, the contents of Systems Analysis Mark III are shrouded largely in hard qualitative uncertainty. Nevertheless, my conjecture is that fuzzy decision gambling will constitute one of the main issues at which Systems Analysis Mark III will be directed. Systems-analysis professionals need not wait for Systems Analysis Mark III in order to improve their handling of decision gambling with available tools differently employed, as discussed in the proposed protocols. But advancing systems-analysis theory and practice by a big step forward—this is a main challenge posed, *inter alia*, by the decision-gambling concept.

PART 2. HANDLING POLITICAL FEASIBILITY

7.8. Systems Analysis and Political Feasibility

Political feasibility is one of the most vexing concepts facing systems analysis. It deals with the political resources needed for implementing an option, including the case of absolute nonfeasibility when needed resources are larger than available ones, and relative nonfeasibility when the opportunity costs in terms of political resources are too great. But, the vagueness of the very concept of political resources, its multidimensionality, its overlap with economic power, its complex dynamics including growing by being used and depreciating by being not used, its personal nature when attached to leaders, all these make the concept of political feasibility a vague one. It is not surprising, therefore, that even better texts in policy analysis handle it rather diffusely (for example, Brewer and deLeon, 1983), innovative books on the tools of government neglect it (Hood, 1983), and, more surprisingly and despite a strong tradition of in-part-realistic Mirrors-of-Princes literature, modern handbooks for politicians—as far as existing at all—often either ignore political feasibility issues or deal with them naively (for example, Heineman and Hessler, 1980; Walter and Choate, 1984).

When related uncertainties, as further discussed in this part, are added to the very vagueness of the concept of political feasibility itself, then it becomes clear that we are dealing with a domain that constitutes a main area of fuzzy gambling. *In its underlying perspectives and main recommended protocols the handling of political feasibility is therefore a special case of fuzzy decision gambling*, as explored in Part 1. The present part provides additional points that require clarification and treatment within the perspective of systems analysis, but does not elaborate detailed handling of political feasibility as a dimension of decision gambling, as resulting from direct application of Part 1 to the subject.

Handling political feasibility involves two separate, although interrelated, issues: (1) predicting the political feasibility of given sets of options, with or without mapping political feasibility domains going beyond a single specific decision, and (2) intervention strategies to change the political feasibility of options and enlarge political feasibility domains, including building up political resources and/or reducing the political costs of particular options. The first poses all the difficulties of mapping uncertainties. The second leads from systems analysis to political advice and political action, in the form of political marketing, Machiavellian (in the clinical, not perjorative, sense) tactics, the advancement of charismatic leadership, and so on.

A preliminary question that must be faced is how far systems analysis as a discipline and profession should get involved in such issues. The emergence of distinct professional specialization in political prediction (Ascher and Overholt, 1983, Chapter 5) and risk assessment on one hand (Kobrin, 1982; Simon, 1985) and in political marketing on the other (Sabato, 1981), in addition to all the classical practitioners of politics and power brokerage, adds to the problematics of the matter.

In many respects, it would be nice to exclude political feasibility from the domain of systems analysis: the basic modes of political reasoning and power handling are different from those of systems analysis; in some ways politics can be viewed as an "enemy" of the systems approach (Churchman, 1979, especially Chapter 8); clinical attitudes are particularly hard to maintain in the value-loaded and emotion-intense context of political power issues; the professional nature and credibility of systems analysis and its practitioners can easily be compromised by moving into political feasibility and its manipulation, all the more so as political career considerations of particular clients cannot be detached from more general considerations of political feasibility—posing harsh conflicts before advisers (Dror, 1987b, especially pp. 195–196); and more. Inbuilt incapacities of many of the frames of appreciation of systems analysis to comprehend the realities of politics that shape political feasibility add to the arguments for leaving this subject outside the scope of systems analysis.

But, in principle, an option that is not feasible is no good, and there exists no reason to regard political feasibility constraints as less real or less important than, say, economic and technological feasibility limitations (despite Ackoff, 1978, p. 72). Also, in practice, systems analysts who ignore the limits of political feasibility are soon excluded from the corridors of power. Therefore, there is no choice but somewhat to include political feasibility within the domain of systems analysis, comparable in part to considering organizational feasibility, for instance, but in a more limited way.

Empirically, my studies of advisory staffs to top decisionmakers show that usually two circles of advisors surround a top executive: political advisors, who share the political views of the top executive and look out for his political interests, and professionals, who look after the nonpolitical aspects of decision issues. But, unless some overlap between these two groups exists, the arrangement cannot work well. Political advisors must be familiar with professional knowledge and professional advisors must understand and be able to take into account political concerns. Therefore, *systems analysis must understand the basics of political feasibility and be able to take into account political considerations, as constraints and challenges. But, the main tasks of political intelligence, political feasibility estimation and mapping, and political feasibility changing should be outside the domain of systems analysis.* In particular, political manipulation (Goodin, 1980) should be kept separate from systems analysis.

According to this guideline, political strategies and tactics are not a main subject of systems analysis, but must be known to systems analysis and taken into account by it, especially when dealing with political feasibility. Further, to explore the relations between political feasibility shaping and systems analysis, the subject of entrepreneurship must be introduced. To put the matter into a broader frame, issues of changing political feasibility lead to major issues of innovativeness as a main challenge to policymaking and its relations with systems analysis.

Leaving the subject to more extensive treatment elsewhere (Dror, in print *a*), it can be summed up for the present limited purposes as follows: As many problems change or are at least in part novel, to develop and invent new options is a main requirement of improving decisions, up to the need for breakthrough options in the face of declining trends, as considered in Part 1. Empirically, it seems that often innovative ideas are scarce and come more from politicians than civil servants (Aberbach, Putnam, and Rockman, 1981), but inadequately so. Therefore, the question is posed of how to increase entrepreneurship, a problem well recognized in economic theory (Casson, 1982) and much discussed in the the business-sector context (for example, Peters and Waterman, 1983; Peters and Austin, 1985), but just beginning to be systematically considered in the domains of politics (for example, Polsby, 1984) and the public sector (for example, Lewis, 1980; Wholey and Nellavita, 1985; Merritt and Merritt, 1985).

This raises major issues on the nature of systems analysis and the roles of design, creativity, and innovativeness in it, including the danger that analytical parts of systems analysis may repress innovative ideas, which are harder to analyze. In the present context, three main implications deserve attention:

1. Innovative decisions face different and often harder problems of political feasibility and involve more uncertainty than incremental decisions. They require a more dynamic attitude to political feasibility and explicit treatment of the need to make novel options po-

litically feasible. Although tactical aspects should be left outside the main concern of systems analysis, the strategic aspects must be handled as an integral part of systems analysis, in cooperation with various political and other relevant professionals.

2. The likelihood of policy innovations themselves—whether by the clients of a systems analysis activity, and perhaps as a result of it, or whether in the external environment—and their political repercussions add to uncertainty, also in respect to political power and political feasibility domains. Therefore, a dynamic, and sometimes in part self-feeding, process of innovation facing political uncertainty and producing more of it adds to the fuzzy-gambling characteristics of decisionmaking in its political dimensions.

3. It is all the more essential for systems analysts to have a sophisticated understanding of political feasibility, and to pay attention to its dynamic consideration and handling, while, with due self-restraint, not entering the domain of political action.

7.9. Understanding Political Feasibility

A main recommendation above is that systems analysis must understand political feasibility. The trouble is that reading the main texts in political science hardly contributes to meeting this need. Therefore, systems analysis must break new ground to be able to fulfill its tasks with regard to political feasibility. In doing so, the following conjectures may serve as a useful framework and correct widespread misapprehensions and biases congenital to systems analysis because of its underlying modes of thinking.

In the very short run, political feasibility is often stable and moves in fixed cycles. In most industrialized countries, political feasibility is rather stable in the short run and many cycles follow a regular pattern, such as changes in political power and feasibility as correlated with electoral cycles (Frey, 1978; Tufte, 1978). This reduces uncertainty and permits systems analysis to rely on standard prediction methods for handling political feasibility, such as consulting experts in politics.

Once the short run is left behind—and frequently also within it—political feasibility is jumpy and shrouded in hard qualitative uncertainty. All contemporary societies are loaded with factors producing political instability (Sanders, 1981), including democracies (Powell, 1982). This is much more the case in many Third World countries. Recent developments in the USSR also demonstrate jumps in political feasibility in the Communist bloc. When a longer time perspective of, say, five to ten years and more, is taken into account, some aspects of political feasibility become jumpy and are shrouded in hard qualitative uncertainty in all countries. The general frame of political culture can be taken for granted for longer periods in most of the democracies, but changes in political priorities and in allocation of political power between main actors exhibit constant uncertainties. These are reinforced by uncertainty on longer-range political feasibility-shaping factors such as the economic situation, the overall legitimation of regimes, attitudes toward religion, and the like.

In many countries, far-reaching destabilization is assured with a high probability of many low-probability events. When single rulers are powerful, idiosyncratic action should be expected. When backward areas proceed through accelerated modernization, instability and political mutations are practically guaranteed. They may take forms strange to systems analysis in Western societies, such as religious fanaticism ranging up to "crazy states" (Dror, 1980; Dror, 1987a).

Factors further reducing the stability of political feasibility, and therefore its predictability, in Western democracies include, for instance:

The growing importance of political leadership (Tucker, 1981), including ideological rulers (Dror, in print b), which increases the political impacts of personalities that are dynamic, irregular, and hard to predict.

The propensities toward political violence (for example, Clutterbuck, 1986) and aggressive political participation (Muller, 1979), which in turn may influence political culture in erratic ways.

The political results of a variety of social tensions, such as products of ethnically different immigration and of large-scale endemic unemployment, which may take turbulent forms and are, in part at least, in hard qualitative uncertainty.

The cross-impacts of global developments, such as security tensions and economic surprises (Neu and Henry, 1986).

The unforeseeable, but perhaps major, implications for political dynamics of technological innovations, especially in the areas of information and communication technologies, as previewed by the political results of television, which too are still in the making.

As already discussed, the needs, possibilities, and likelihood of more political and policy innovativeness and entrepreneurship add further uncertainty, in part (and paradoxically also) as a result of better systems analysis itself.

If we aggregate such additional factors, we face some—and often a lot of—hard qualitative uncertainty in analyzing political feasibility in the long run, and even sometimes in the short run. This is the case in relatively stable democracies, and, even more so in other societies. Let me add that my own studies of relevant alternative futures (Dror, 1986a, Chapter 2) lead me to the conclusion that political turbulence must be expected to increase, resulting in growing hard uncertainties with respect to political feasibility.

Models of rational economic behavior are in part misleading with respect to political feasibility. A main danger in systems analysis is reliance on rational economic behavior as a basis for analyzing and predicting political feasibility. Although rational-economic-behavior models are somewhat useful as one of a range of ways of looking at political behavior, they ignore a number of main factors, such as mass psychology (Moscovici, 1985; Graumann and Moscovici, 1986), counter-economic deep motives of individuals (for example, Fromm, 1974), impacts of leaders (for example, Willner, 1984), changing ideologies (for example, Thompson, 1984), and shifts in social values (for example, Beckford, 1986). To these must be added various forms of motivated irrationality (Pears, 1984) and divided selves (Elster, 1985). It follows, to put it mildly, that Pareto optimum notions are a dangerous guide for considering political feasibility, since quite different motives control a lot of individual and collective human behavior.

Even better social choice theories (for example, Elster and Hylland, 1986; Coleman, 1986) do not adequately consider the noneconomically rational aspects of political-feasibility-shaping behavior. Economic theories applied to politics, as pioneered by Downs (1957) and Buchanan (Buchanan and Tullock, 1962), and collective choice theories shape much of systems-analysis thinking. Therefore, it is all the more important to correct and balance such biases by emphasizing the limitations of economic rationality as bases for understanding and analyzing political feasibility.[9]

Political feasibility is susceptible to deliberate change. A widespread danger is to regard political feasibility as a given factor to which analysis must adjust itself, rather than to regard feasibility as an elastic constraint and as an object that can be influenced and changed by appropriate operations. To take up a relevant popular saying, the aphorism that politics is the art of the possible has no more historic validity and much less prescriptive justification that my proposed substitute that politics is the art of making the necessary and desirable feasible.

Related is the danger of systems analysis having a "conservative" bias in the sense of taking present power structures as fixed and immutable. Although political reforms and transformations are not the usual subject

---

[9] I realize that it is not easy within the rationalist world of systems analysis really to understand in depth the operation of irrational forces in politics, but this is absolutely essential for realistic handling of politics and the potentials of political feasibility to jump. Good political novels may serve as a better road to understanding underlying realities of politics than most texts in political science. Broch (1987) and Llosa (1984) may serve as good starting points for systems analysts not familiar with this genre of literature.

of systems analysis, experienced politicians know that it is possible to change political feasibility domains by deliberate efforts, sometimes more so and sometimes less so. Often the very changeability of political feasibility depends on factors that cannot be predicted, adding to uncertainty. The needs and possibilities of policy entrepreneurship and political ventureship, as discussed above, add another dimension to the view of political feasibility as an object for shaping rather than as a rigid constraint.

Recognizing possibilities to influence political feasibility, systems analysis must also beware of the danger of overestimating the domains of policy leeway and underestimating the hard rigidities of political feasibility. Utopia writing is an important social function, but should be kept distinct from systems analysis. Accordingly, rigid limits of political feasibility must not be ignored and hopeful assumptions that what seems desirable from the perspective of "objective" systems analysis will somehow become possible must be resisted. For instance, one does not have to agree with all or even most of Gramsci's ideas to benefit from his analysis of rigid ideological limitations on feasible political dynamics within given societal (including ideological) infrastructures (Femia, 1987, Chapters 1, 2, 3; see also Abercrombie, Hill, and Turner, 1980, especially Chapters 1, 2).

The fact that some outstanding systems analysts have moved into what I regard as utopian, if not sometimes mystic, frames of thinking to overcome apparent contradictions between what they see as essential from their analysis of the human condition and what real politics can deliver (for example, Beer, 1975; Friedmann, 1979; Jantsch, 1975; Laszlo, 1983) reinforces the need for care in avoiding the pitfall of underestimating the rigidities of political feasibility constraints and the stubborness of political realities. At the same time, the opposite pitfall of underestimating the elasticities of political feasibility constraints and of possibilities to reshape them through deliberate intervention must also be avoided.

A variety of additional possibilities for handling political feasibility can serve systems analysis. It is important, for instance, to prepare contingency plans in case political feasibility shifts to open passing windows of opportunity after surprises and crises; it is even possible to consider planning crisis instigation in order to loosen rigidities and create political feasibilities, as mentioned in Part 1. The rebuilding of Paris under Napoleon III according to Haussmann's ideas offers an extreme but illuminating example of policies that seem infeasible under given conditions, but whose hour may strike (Pinkney, 1958). Such possibilities and their implications for systems analysis are further complicated by some paradoxical features of political feasibility and its images, to which I now turn.

Political feasibility is loaded with paradoxical dynamics requiring ultrarational models. Problems of political feasibility demonstrate in rather pure form some of the paradoxes of rationality mentioned in Part 1. Especially vexing is the striking function of self-illusions and true believers in political actions, many radical transformations in reality being achieved by engaging in the probably impossible, thanks to a deep illusionary feeling that it is possible and perhaps also historically inevitable. Zionism is a good illustration: Had a good systems-analysis unit been set up before the first Zionist Congress in 1897, it probably would have calculated the chances of setting up a Jewish state in Palestine within 50 years as approximating zero and as constituting a Messianic dream and not a political action program—and this very study, if taken seriously, could well have paralyzed action and prevented Zionism from taking off.

This example is dramatic, and in some respects singular, but in principle not exceptional: Most radical social transformations have *a priori* a very low success probability, but the very belief by human actors that they are feasible activates self-fulfilling prophecy effects and human efforts, and the devoted sometimes snatch success out of very low probabilities. This raises a fundamental problem, namely, under what conditions is systems analysis, in its conventional forms and standard political feasibility mappings, counterproductive because of its effects in diminishing motivations and thus fulfilling its own negative predictions? Needed is a more advanced conception of ultrarationality that takes into account such paradoxical effects of lower-level rationality and permits either deliberate self-limitation of systems analysis, or perhaps systems analysis on a higher level, which also fits the needs of, for instance, revolutionary situations.

Usually systems analysis does not run into these problems and I therefore leave their consideration for another context (Dror, 1986a, especially pp. 95–97). But recognition of the paradoxical dynamics of political feasibility, such as the usefulness of self-delusions, serves as a loud call for caution in evaluating the various methods used in systems analysis for considering political feasibility, and it raises questions about the usefulness of systems analysis as a whole under some conditions.

7.10. Methods for Handling Political Feasibility in Systems Analysis

Much work relevant to understanding and predicting political feasibility, including development of appropriate methodologies, takes place in security intelligence within efforts to understand and predict the politics of other countries. Turning the task around from "knowing one's enemies" (May, 1984) to knowing oneself involves many important differences. Thus, in analyzing domestic political futures (Dror, 1971b, Chapter 7), one has much better access to data, objective as well as subjective knowledge, and experts than when dealing with some external political future. But, leaving aside some technical differences, such as in clandestine data collection, handling political feasibility in one's own country involves harder problems of emotional involvement and motivated irrationalities (Pears, 1984), as well as many taboos.

Still, despite the differences, the basic task of analyzing political feasibility in one's own context and producing relevant "social intelligence" (Gross, 1964) is comparable to similar endeavors in defense intelligence (in addition to some overlaps in internal security matters) and in parts of external events forecasting as a whole. Therefore, methods can be transferred from parts of international relations forecasting (Choucri and Robinson, 1978) and from political intelligence studies in general (for example, Heuer, 1978). At the same time, studies of the failures and failings of security intelligence (for example, Laqueur, 1985) provide salutary lessons for political feasibility analysis and its inherent limitations—reinforcing the recommended fuzzy-gambling perspective and serving as an additional warning against simplistic models (for example, De Mesquita, Newman, and Rabushka, 1985) that do not meet the requirements of accommodating hard uncertainties according to the protocols in Part 1.

Subject to these caveats, the basic set of methods relevant to political feasibility analysis is in the main familiar to systems analysts. Three comments may help to adjust standard methods to the specifics of political feasibility materials:

Mapping political situations requires several layers of methods. Main actors should be analyzed in terms of their operational codes, cognitive maps, and psychological profiles (Axelrod, 1976; Falkowski, 1979). Historic contexts and anthropological perspectives should be added when complex situations are under consideration. To gain better insights, systems analysis should systematically adopt multiple perspectives (Linstone, 1984; see also Chapter 8).

Systematic interrogation of experts, such as with Delphi, is an essential method (see Chapter 3). But care must be taken in deciding who are experts on political feasibility. Also, experts may easily be emotionally biased when hot political issues are touched upon and may be overconservative in underestimating possibilities of changes and jumps in political situations.

I have become very skeptical about formal modeling in respect to political dynamics (including Dror, 1971b, Chapters 7, 8). Theoretically, the uniqueness of specific situations casts doubt on any general model of political dynamics that is detailed enough for the analysis of political feasibility (essential is Boudon, 1986). The available political science models are grossly inadequate for political feasibility analysis. Critiques of war gaming (Brewer and Shubik, 1979) apply all the more to political gaming, which presumes to deal with an even more complex and ill-structured subject. Still, as an explorative method providing some penetration into political feasibility, carefully designed games, in combination with stochastic quantitative models for subissues, if available, are useful.

The best approach is combinational: historic and anthropological studies of situations, careful monitoring of the opinions of experts, and systematic use of a multiplicity of formalized methods such as mentioned above provide entrance into political feasibility within systems analysis. All this is subject to the general protocols for handling hard uncertainties developed in Part 1, and to constantly bearing in mind the special features of political feasibility as between rigidity and changeability, as discussed in this part.

7.11. Implications for the Nature and Limits of Systems Analysis

The problematics of political feasibility analysis make it a test case for exploring the limits of systems analysis, as a discipline as well as morally. Morally, political feasibility poses the question of how far systems analysis should go into political intervention strategies and tactics as part of its concern with broadening feasibility domains. Entering political feasibility considerations and their related political intervention options also sharpens the fundamental question of professional ethics: how far to go in serving the values of clients, whether shared or anathema to the systems analysts (Miser and Quade, 1985, especially pp. 315–325).

Political feasibility brings out the dependance of systems analysis on other disciplines, such as history, psychology, and political science. Political feasibility issues expose some of the biases and limits of the modes of thinking underlying most of systems analysis, such as economic models of human behavior and choice. And the problems of changing political feasibility sharpen the systems-analysis dilemma between handling specific issues as discrete decision agenda items and moving into broad-gauge societal architecture to change the whole of the policymaking setting, in this case by broadening political feasibility domains.

Taken together with the uncertainty features of political feasibility, which return us to the fuzzy decision-gambling view, it seems that the conclusion of Part 1 on the need for advancing toward Systems Analysis Mark III is reinforced and radicalized by our exploration of some issues of political feasibility. The challenges posed by political feasibility as an essential but also quite prohibitive subject of systems analysis are, I think, even more difficult than those posed by the fuzzy decision-gambling perspective, going deeper into moral and professional dilemmas and, further, requiring from systems analysis the overall study and understanding of complex societies as a whole, thus casting doubt on the autonomy of systems analysis as an approach to improving the handling of complex issues.

Political feasibility issues also pose in aggravated form even more fundamental questions on the usefulness of systems analysis under some conditions, such as when "nonrationality" may be more useful. This leads again to the need for a higher level of ultrarationality, which encompasses also the question of when various modes of reasoning (expanding on Toulmin, Rieke, and Janik, 1979), of which systems analysis covers only a subset, are appropriate. These are existential problems of systems analysis, even less susceptible to satisfactory answers than hard uncertainties, and will continue to face Systems Analysis Mark III.

References

Aberbach, J. D., R. D. Putnam, and B. A. Rockman (1981). *Bureaucrats and Politicians in Western Democracies*. Cambridge, Massachusetts: Harvard University Press.

Abercrombie, Nicholas, S. Hill, and B. S. Turner (1980). *The Dominant Ideology Thesis*. London: George Allen and Unwin.

Ackoff, R. L. (1978). *The Art of Problem Solving*. New York: Wiley.

Amacher, R. C., R. D. Tollison, and T. D. Willett, eds. (1976). *The Economic Approach to Public Policy*. Ithaca, New York: Cornell University Press.

Ascher, William, and W. H. Overholt (1983). *Strategic Planning and Forecasting: Political Risk and Economic Opportunity*. New York: Wiley.

Axelrod, Robert, ed. (1976). *Structure of Decision*. Princeton, New Jersey: Princeton University Press.

Beckford, J. A., ed. (1986). *New Religious Movements and Rapid Social Change*. London: Sage.

Beer, Stafford (1975). *Platform for Change*. New York: Wiley.

Behn, R. D., and J. W. Vaupel (1982). *Quick Analysis for Busy Decision Makers*. New York: Basic Books.

Bendor, J. B. (1985). *Parallel Systems: Redundancy in Government*. Berkeley: California University Press.

Beniger, J. R. (1986). *The Control Revolution: Technological and Economic Origins of the Information Society*. Cambridge, Massachusetts: Harvard University Press.

Boudon, Raymond (1982). *The Unintended Consequences of Social Action*. London: Macmillan. (Translated from the French.)

———— (1986). *Theories of Social Change*. Cambridge, England: Polity Press. (Translated from the French.)

Braudel, Fernand (1980). *On History*. Chicago, Illinois: University of Chicago Press. (Translated from the French.)

Brewer, G. D., and P. deLeon (1983). *The Foundations of Policy Analysis*. Homewood, Illinois: Dorsey Press.

————, and M. Shubik (1979). *The War Game: A Critique of Military Problem Solving*. Cambridge, Massachusetts: Harvard University Press.

Broch, Hermann (1976). *The Spell*. New York: Farrar, Straus, and Giroux. (Translated from the German.)

Buchanan, J. M., and G. Tullock (1962). *The Calculus of Consent: Logical Foundation of Constitutional Democracy*. Ann Arbor: University of Michigan Press.

Casson, Mark (1982). *The Entrepreneur: An Economic Theory*. Towowa, New Jersey: Barnes and Noble.

Checkland, Peter (1981). *Systems Thinking, Systems Practice*. New York: Wiley.

Choucri, Nazli, and T. W. Robinson, eds. (1978). *Forecasting in International Relations: Theory, Methods, Problems, Prospects*. San Francisco, California: Freeman.

Churchman, C. W. (1979). *The Systems Approach and its Enemies*. New York: Basic Books.

Clutterbuck, Richard, ed. (1986). *The Future of Political Violence: Destabilization, Disorder, and Terrorism*. London: Macmillan.

Coleman, J. S. (1986). *Individual Interests and Collective Action: Selected Essays*. Cambridge, England: Cambridge University Press.

Davidson, Donald (1980). *Essays on Action and Events*. Oxford, England: Oxford University Press.

Davis, P. K., S. C. Banker, and J. P. Kahan (1986). *A New Methodology for Modeling National Command Level Decisionmaking in War Games Simulations*, R-3290-NA. Santa Monica, California: The Rand Corporation.

————, and J. A. Winnefeld (1983). *The Rand Strategy Assessment Center: An Overview and Interim Conclusions about Utility and Development Options*, R-2945-DNA. Santa Monica, California: The Rand Corporation.

De Greene, K. B. (1982). *The Adaptive Organization: Anticipation and Management of Crisis*. New York: Wiley.

De Mesquita, B. B., D. Newman, and A. Rabushka (1985). *Forecasting Political Events: The Future of Hong Kong*. New Haven, Connecticut: Yale University Press.

Dörner, Dietrich, et al., eds. (1983). *Lohhausen: Vom Umgang mit Unbestimmtheit und Komplexität*. Bern: Hans Huber.

Douglas, Mary (1986). *Risk Acceptability According to the Social Sciences*. London: Routledge and Kegan Paul.

———— (1987). *How Institutions Think*. London: Routledge and Kegan Paul.

————, and Aaron Wildavsky (1982). *Risk and Culture*. Berkeley: University of California Press.

Downs, Anthony (1957). *An Economic Theory of Democracy*. New York: Harper.

Dror, Yehezkel (1971a). *Design for Policy Sciences*. New York: Elsevier.

———— (1971b). *Ventures in Policy Sciences*. New York: Elsevier.

———— (1980). *Crazy States: A Counterconventional Strategic Problem*, revised edition. Millwood, New York: Kraus Reprints.

———— (1983). *Public Policymaking Reexamined*, revised edition. New Brunswick, New Jersey: Transaction Books.

———— (1984). Policy analysis for advising rulers. In *Rethinking the Process of Operational Research and Systems Analysis* (Rolfe Tomlinson and István Kiss, eds.). Oxford, England: Pergamon Press, pp. 79–123.

———— (1986a). *Policymaking under Adversity*. New Brunswick, New Jersey: Transaction Books.

———— (1986b). Planning as fuzzy gambling: A radical perspective on coping with uncertainty. In *Planning in Turbulence* (David Morley and Arie Shachar, eds.). Jerusalem: Magnes Press.

———— (1986c). "Gambling with history," however unpleasant, is normal to the human condition. *Technological Forecasting and Social Change* 29(1), 76–81.

———— (1987a). High-intensity aggressive ideologies as an international threat. *Jerusalem Journal of International Relations* 9(1), 153–172.

———— (1987b). Conclusions. In *Advising the Rulers* (William Plowden, ed.). Oxford, England: Basil Blackwell, pp. 185–215.

———— (in print *a*). Options for increasing innovativeness. In EIPA, *Innovativeness in Public Administration*. Maastricht, The Netherlands: European Institute of Public Administration.

———— (in print *b*). Visionary political leadership: On improving a risky requisite. *International Political Science Review*.

———— (in publication). *Breakthrough Policies for Israel: Memorandum to the Prime Minister*. Jerusalem: Academon. (In Hebrew.)

———— (in preparation). *Policy Gambling: A Radical Perspective on Decision Making*.

Eberhard, Wolfram (1957). The political function of astronomy in Han China. In *Chinese Thought and Institutions* (J. K. Fairbank, ed.). Chicago: University of Chicago Press.

Eggleston, Richard (1983). *Evidence, Proof, and Probability*, 2nd ed. London: Weidenfeld and Nicolson.

Elster, Jon (1979). *Ulysses and the Sirens: Studies in Rationality and Irrationality*. Cambridge, England: Cambridge University Press.

———— (1983). *Sour Grapes: Studies in the Subversion of Rationality*. Cambridge: England: Cambridge University Press.

————, ed. (1985). *The Multiple Self*. Cambridge, England: Cambridge University Press.

————, ed. (1986). *Rational Choice*. Oxford, England: Basil Blackwell.

————, and Aanund Hylland, eds. (1986). *Foundations of Social Choice Theory*. Cambridge, England: Cambridge University Press.

Etheredge, L. S. (1985). *Can Governments Learn? American Foreign Policy and Central American Revolutions*. New York: Pergamon.

Falkowski, L. S., ed. (1979). *Psychological Models in International Politics*. Boulder, Colorado: Westview Press.

Femia, J. V. (1987). *Gramsci's Political Thought: Hegemony, Consciousness, and the Revolutionary Process*. Oxford, England: Oxford University Press.

Fischhoff, Baruch (1981). *Acceptable Risk*. Cambridge, England: Cambridge University Press.

Frey, B. S. (1978). *Modern Political Economy*. New York: Wiley.

Friedman, John (1979). *The Good Society: A Primer for Social Practice*. Cambridge, Massachusetts: Harvard University Press.

Fromm, Erich (1974). *The Anatomy of Human Destructiveness*. London: Jonathan Cape.

Goodin, R. E. (1980). *Manipulative Politics*. New Haven, Connecticut: Yale University Press.

Graumann, D. F., and S. Moscovici, eds. (1986). *Changing Conceptions of Crowd Mind and Behavior*. New York: Springer.

Gross, Bertram, ed. (1964). *Social Intelligence for America's Future*. Boston: Allyn and Bacon.

Hayano, D. M. (1982). *Poker Faces: The Life and Work of Professional Card Players*. Berkeley: University of California Press.

Heimer, C. A. (1985). *Reactive Risk and Rational Action: Managing Moral Hazard in Insurance Contracts*. Berkeley: University of California Press.

Heineman, Ben, Jr., and C. A. Hessler (1980). *Memorandum for the President: A Strategic Approach to Domestic Affairs in the 1980s*. New York: Random House.

Hennessy, Peter (1986). *Cabinet*. Oxford, England: Basil Blackwell.

Heuer, R. J., Jr., ed. (1978). *Quantitative Approaches to Political Intelligence: The CIA Experience*. Boulder, Colorado: Westview Press.

Hofstadter, D. R. (1979). *Gödel, Escher, Bach: An Eternal Braid*. New York: Basic Books.

Holling, C. S., ed. (1978). *Adaptive Environmental Assessment and Management*. New York: Wiley.

Hood, C. C. (1983). *The Tools of Government*. London: Macmillan.

Hucker, C. O. (1966). *The Censorial System of Ming China*. Stanford, California: Stanford University Press.

IIASA (1985). Risk and policy analysis under conditions of uncertainty. *Options* (December), p. 5.

Janis, I. L. (1982). *Groupthink: Psychological Studies of Policy Decisions and Fiascoes*, revised edition. Boston: Houghton Mifflin.

————, and Leon Mann (1977). *Decision Making: A Psychological Analysis of Conflict*. New York: Free Press.

Jantsch, Erich (1975). *Design for Evolution: Self-Organization and Planning in the Life of Human Systems*. New York: Braziller.

Jouvenel, Bertrand de (1967). *The Art of Conjecture*. New York: Basic Books. (Translated from the French.)

Kahan, J. P., N. Webb, R. J. Shavelson, and R. M. Stolzenberg (1985). *Individual Characteristics and Unit Performance: A Review of Research and Method*, R-3194-MIL. Santa Monica, California: The Rand Corporation.

Kahneman, Daniel, P. Slovic, and A. Tversky, eds. (1982). *Judgment under Uncertainty: Heuristics and Biases*. Cambridge, England: Cambridge University Press.

Kanal, L. N., and J. F. Lemmer, eds. (1986). *Uncertainty in Artificial Intelligence*. New York: North-Holland.

Keeney, R. L., and H. Raiffa (1976). *Decisions with Multiple Objectives: Preferences and Value Tradeoffs*. New York: Wiley.

Kobrin, S. J. (1982). *Managing Political Risk Assessment: Strategic Response to Environmental Change*. Berkeley: University of California Press.

Laqueur, Walter (1985). *A World of Secrets: The Uses and Limits of Intelligence*. New York: Basic Books.

Lasswell, H. D. (1971). *A Pre-View of Policy Sciences*. New York: Elsevier.

Laszlo, Ervin (1983). *System Science and World Order: Selected Studies*. Oxford, England: Pergamon.

Levi, Isaac (1986). *Hard Choices: Decision Making under Unresolved Conflict*. Cambridge, England: Cambridge University Press.

Lewis, Eugene (1980). *Public Entrepreneurship: Towards a Theory of Bureaucratic Political Power—The Organizational Lives of Hyman Rickover, J. Edgar Hoover, and Robert Moses*. Bloomington: Indiana University Press.

Lindsay, P., and D. Norman (1977). *Human Information Processing*. London: International Academic Press.

Linstone, H. A. (1984). *Multiple Perspectives for Decision Making: Bridging the Gap between Analysis and Action*. New York: North-Holland.

———, and M. Turoff (1975). *The Delphi Method: Techniques and Applications*. Reading, Massachusetts: Addison-Wesley.

Llosa, M. V. (1984). *The War at the End of the World*. New York: Farrar, Straus, and Giroux. (Translated from the Spanish.)

Majone, Giandomenico, and E. S. Quade, eds. (1980). *Pitfalls of Analysis*. New York: Wiley.

March, J. G., and J. P. Olson (1976). *Ambiguity and Choice in Organizations*. Bergen, Norway: Universitetsforlaget.

Mason, R. O., and I. I. Mitroff (1981). *Challenging Strategic Planning Assumptions: Theory, Cases, and Techniques*. New York: Wiley.

May, E. R., ed. (1984). *Knowing One's Enemies: Intelligence Assessment before the Two World Wars*. Princeton, New Jersey: Princeton University Press.

Merkhofer, M. W. (1987). *Decision Science and Social Risk Management: A Comparative Evaluation of Cost-Benefit Analysis, Decision Analysis, and Other Formal Decision-Aiding Approaches*. Dortrecht, The Netherlands: D. Reidel.

Merritt, R. L., and A. J. Merritt, eds. (1985). *Innovation in the Public Sector*. Beverly Hills, California: Sage.

Miller, C. T., P. R. Kleindorfer, and R. E. Munn (1986). *Conceptual Trends and Implications for Risk Research*, CP-86-26. Laxenburg, Austria: International Institute for Applied Systems Analysis.

Minsky, M., ed. (1968). *Semantic Information Processing*. Cambridge, Massachusetts: The MIT Press.

Miser, H. J., and E. S. Quade, eds. (1985). *Handbook of Systems Analysis: Overview of Uses, Procedures, Applications, and Practice*. New York: North-Holland.

Mitroff, I. I., H. Quinton, and R. O. Mason (1983). Beyond contradictions and consistency: A design for a dialectic policy system. *Theory and Decision* 15, 107–120.

Monod, Jacques (1971). *Chance and Necessity: An Essay in the Natural Philosophy of Modern Biology.* New York: Alfred A. Knopf. (Translated from the French.)

Moscovici, Serge (1985). *The Age of the Crowd: A Historic Treatise on Mass Psychology.* Cambridge, England: Cambridge University Press. (Translated from the French.)

Muller, E. N. (1979). *Aggressive Political Participation.* Princeton, New Jersey: Princeton University Press.

Neu, C. R., and D. P. Henry (1986). *Surprises in the International Economy: Towards an Agenda for Planning and Research,* R-3401. Santa Monica, California: The Rand Corporation.

Neustadt, R. E., and E. R. May (1986). *Thinking in Time: The Uses of History for Decision Makers.* New York: Free Press.

Pears, David (1984). *Motivated Irrationality.* Oxford, England: Oxford University Press.

Peters, T. J., and N. Austin (1985). *A Passion for Excellence: The Leadership Difference.* New York: Random House.

———, and R. H. Waterman, Jr. (1983). *In Search of Excellence: The Lessons from America's Best-Run Companies.* New York: Harper & Row.

Petroski, Henry (1985). *To Engineer is Human: The Role of Failure in Successful Design.* New York: St. Martin's Press.

Pinkney, D. H. (1958). *Napoleon III and the Rebuilding of Paris.* Princeton, New Jersey: Princeton University Press.

Polsby, N. W. (1984). *Political Innovation in America: The Politics of Policy Initiation.* New Haven, Connecticut: Yale University Press.

Powell, B. G., Jr. (1982). *Contemporary Democracies: Participation, Stability, and Violence.* Cambridge, Massachusetts: Harvard University Press.

Prince, M. J. (1983). *Policy Advice and Organizational Survival.* Aldershot, England: Gower.

Quade, E. S. (1982). *Analysis for Public Decisions,* 2nd ed. New York: North-Holland.

Raiffa, Howard (1968). *Decision Analysis.* Reading, Massachusetts: Addison-Wesley.

Sabato, L. J. (1981). *The Rise of Political Consultants.* New York: Basic Books.

Sanders, David (1981). *Patterns of Political Instability.* New York: St. Martin's Press.

Schank, R., ed. (1975). *Conceptual Information Processing.* Amsterdam: North-Holland.

Schelling, T. C. (1984). *Choice and Consequence: Perspectives of an Errant Economist.* Cambridge, Massachusetts: Harvard University Press.

Schön, D. A. (1983). *The Reflective Practitioner: How Professionals Think in Action.* New York: Basic Books.

Schulman, P. R. (1980). *Large-Scale Policy Making.* New York: Elsevier.

Shackle, G. L. S. (1972). *Epistemics and Economics: A Critique of Economic Doctrines*. Cambridge, England: Cambridge University Press.

Sieber, S. D. (1981). *Fatal Remedies: The Ironies of Social Intervention*. New York: Plenum Press.

Simon, J. D. (1985). Political risk forecasting. *Futures* 17(2), 132–148.

Sims, H. P., Jr., D. A. Gioia, et al. (1986). *The Thinking Organization: Dynamics of Organizational Social Cognition*. San Francisco, California: Jossey-Bass.

Sternberg, R. J., ed. (1985). *Human Ability: An Information Processing Approach*. Princeton, New Jersey: Princeton University Press.

Stinchcombe, A. L., and C. A. Heimer (1985). *Organization Theory and Project Management: Administering Uncertainty in Norwegian Offshore Oil*. Oslo: Universitetsforlaget.

Stokey, Edith, and R. Zeckhauser (1978). *A Primer for Policy Analysis*. New York: W. W. Norton.

Suppes, Patrick (1984). *Probabilistic Metaphysics*. Stanford, California: Stanford University Press.

Thom, René (1983). *Mathematical Models of Morphogenesis*. New York: Halsted Press. (Translated from the French.)

Thompson, J. B. (1984). *Studies in the Theory of Ideology*. Berkeley: University of California Press.

Toulmin, Stephen, R. Rieke, and A. Janik (1979). *An Introduction to Reasoning*. London: Collier Macmillan.

Tucker, R. C. (1981). *Politics as Leadership*. Columbia: University of Missouri Press.

Tufte, E. R. (1978). *Political Control of the Economy*. Princeton, New Jersey: Princeton University Press.

von Winterfeldt, Detlof, and Ward Edwards (1986). *Decision Analysis and Behavioral Research*. Cambridge, England: Cambridge University Press.

Walter, Susan, and P. Choate (1984). *Thinking Strategically: A Primer for Public Leaders*. Washington, D.C.: The Council of State Planning Agencies.

Wholey, Joseph, and C. Nellavita, eds. (1985). *Promoting Excellence in Public Agencies*. Lexington, Massachusetts: Lexington Books (D. C. Heath).

Wildavsky, Aaron (forthcoming). *Searching for Safety*.

Willner, A. R. (1984). *The Spellbinders: Charismatic Political Leadership*. New Haven, Connecticut: Yale University Press.

Woodcock, Alexander, and M. Davis (1978). *Catastrophe Theory*. New York: Dutton.

Wu, S. H. L. (1970). *Communication and Imperial Control in China: Evolution of the Palace Memorial System, 1693–1735*. Cambridge, Massachusetts: Harvard University Press.

Young, R. J. (1977). *In Command of France: French Foreign Policy and Military Planning, 1933–1940*. Cambridge, Massachusetts: Harvard University Press.

Chapter 8

Quantitative Methods
Uses and Limitations

Edward S. Quade

8.1. Introduction

When first applied to the large-scale problems of business, industrial, and military organizations, operations research, and later systems analysis, owed their successes in large part to the introduction of quantitative methods and models that supplemented or replaced the superb craft skills of the first practitioners. The problems that were tackled arose in operations that man had designed or organized on the basis of logical principles; the underlying situations were highly technical and well structured, and their logic could be rediscovered and represented mathematically. Thus, the quantitative methods used by the early operations-research and systems-analysis workers were ideally suited to the type of problem to which they were applied.

Successes with these early problems led to demands for further applications of quantitative methods. Analysts responded by tackling an ever-widening class of problems, using increasingly more sophisticated mathematics. Few of us realized until later the extent to which the early successes had been made possible by the clearly structured and easily modeled nature of the original problem situations. The new problems involved many significant elements, such as human activities or political forces, that had been of less significance in earlier problems and that analysts did not know how to represent mathematically. The first approach was to leave these elements out of our models in the hope that we, or our clients, might be able to deal with them judgmentally after the conclusion of the quantitative work; however, the attempts to do this were often inadequate—and sometimes even nonexistent. The failures to take adequate account of the hard-to-quantify elements undermined the effectiveness of much of the work and led to the feeling that quantitative methods would be effective only for problems in which social and political factors were of little significance. Nevertheless, in recent years, experience has offered hope that logical analysis, even though it can at best be only partially quantitative when applied to soft and ill-defined problems, can be adjusted to help with them, thus reducing the extent to which the conclusions about them are based on subjective judgment and increasing the extent to which they are informed by an objectively rational view of reality.

This chapter discusses how quantitative methods must be modified and supplemented with judgment if they are to help with problems that are impossible to treat adequately without taking account in the analysis of factors that cannot be satisfactorily modeled quantitatively. The principal focus is on how the analyst can respond to situations for which quantification must be limited to certain aspects, because full quantification is either impossible or inappropriate. To this end, it first reviews briefly the development of quantification in operations research and systems analysis, and its advantages and disadvantages; then it considers the difficulties in measuring the attainment of goals and objectives and various ways of introducing useful quantification to such composite factors as comfort, health, the value of human life, and the "net output" or "net worth" of an alternative—as well as some of the pitfalls to guard against in so doing. The final sections contrast two ways of using models: as surrogates for, and perspectives on, a problem.

This chapter deals with some of the basic principles underlying the appropriate use of quantitative models; it does not describe specific models or techniques of quantitative analysis. These models and techniques are well covered in treatises emphasizing operations research methods, such as Mood's introduction to policy analysis (1983), Wagner's introductory treatise on methods (1975), or the Moder and Elmaghraby handbook survey of techniques and their applications (1978).

Scientists have long recognized the desirability of numbers and quantitative methods in their work, an attitude stated strongly more than a century ago by William Thomson, Lord Kelvin (1883):

> When you can measure what you are speaking about, and express it in numbers, you know something about it; but when you cannot measure it, when you cannot express it in numbers, your knowledge is of a meager and unsatisfactory kind: it may be the beginning of knowledge, but you have scarcely, in your thoughts, advanced to the stage of science.

Although Kelvin's view has dominated the physical sciences in the twentieth century, we now recognize that it is too extreme; one can do a great deal in science with methods that are not quantitative, or only partially so. Similarly, the early work in operations research and systems analysis (which, as we noted in the first chapter of this volume, consists only partially of scientific work) reflected the drive for quantification. In public affairs, however, problems and decisions almost always involve both quantitative and qualitative aspects, so that advice based solely on quantitative analysis alone is seldom satisfactory.[1] Therefore, systems analyses addressing broad social problems must make use of qualitative, as well as quantitative, reasoning. Quantifying to the extent possible without distortion is almost always advantageous, but systems analysts must always keep in mind that professional advice based on quantified experience and science is seldom fully comprehensive. Judgments, experience, political savvy, and just pure luck, as well as analysis and reasoning that are unquantified, are important and necessary complements to quantitative work—and, indeed, have often produced excellent results all by themselves.

8.2. Quantification in Operations Research and Early Systems Analysis

As Majone (1985, p. 40) points out, when operations research (OR) began, successful wartime applications "were not based on new theories or advanced technical tools, but on a sophisticated use of craft skills." In the decade after World War II, however, "a new generation of analysts was entering the OR scene—people primarily interested in the more formal aspects of scientific methodology and proficient in mathematical manipulations, but often lacking the craft skills and the mature critical judgment of the old masters" (loc. cit., p. 41). As one consequence, the nature of OR became more quantitative and mention in its literature of craft skills and judgmental analysis became rare.

Thus, in his 1957 keynote address to the first international conference on operations research, we find Morse, in spite of his position as the U.S. pioneer, saying that

> to make our subject truly quantitative, which I believe most of us desire, we must . . . develop a unitary theory about operations, parts of which can be expressed mathematically, and which can be checked by quantitative, operational experiments. . . . If operations research is to develop in the direction it has gone so far, we should look for general principles with a quantitative content. [Morse, 1957, pp. 5, 6.]

He saw no such theory or general principles in sight then, nor are they in view today. He reminded his audience, however, that the principles

---

[1] Or even possible. The very decision to use quantitative techniques (and the choice of techniques from among the possibilities) is judgmental, and dependent on craftsmanship.

of which he spoke "are quantitative ones, even though their general statement may have to include such non-mathematical phrases as 'tends to increase' or 'seldom gets less than'" (loc. cit. p. 6)—but it is worth noting that such relations are beginning to be partly mathematical through fuzzy set applications.

For the decades of the 1950s and 1960s operations-research and systems-analysis workers devoted almost all attention paid to methodology to mathematical models and computer methods. Quantitative methods brought results that were very well received for many problems in business, industry, and government (such as those of logistics, inventory control, and traffic congestion), but they turned out to be far less successful when, in the late 1960s, analysts applied these methods to questions of health, public welfare, education, ecology, and criminal justice operations. Indeed, solely quantitative approaches to these new questions not only were considered inadequate but also spawned an effective critical literature (of which Hoos, 1972; Brewer, 1973; Strauch, 1974; and Greenberger, Crenson, and Crissey, 1976, are representative examples). Although a good part of this criticism was justified, much of it missed the principal reason for lack of success: For social problems in the public sector, there is seldom a criterion of success that will satisfy all interested parties. Nevertheless, analysts generally lowered their expectations about the extent of their potential help with problems in which there were significant aspects that could not be included in their quantitative models. More importantly, however, some systems analysts began to search for ways of including nonquantitative factors in their analyses, and for better ways of handling the criterion problem and of supplementing their work with judgment, both their own and that of their clients and others.

8.3. Quantification: Advantages and Disadvantages

There are good reasons why quantification is an ideal toward which all analysts strive. Not only is it necessary if we are to measure and compare, particularly when different people are involved—and in this process to make full use of mathematics, statistics, and computers—but also it improves clarity, communication, and understanding. Too, since intuition is seldom an accurate guide to complicated and involved processes, the mathematical operations made possible by quantification bring out behavior that might otherwise not be understood. To become more quantitative is regarded—not always correctly—as a way of being more scientific and thus more rigorous.

In the effort to be more like physics and engineering, where prediction based on mathematical models is admittedly more accurate and reliable than in such behavioral disciplines as sociology, psychology, and political science, analysts hope that increasing quantification will yield better re-

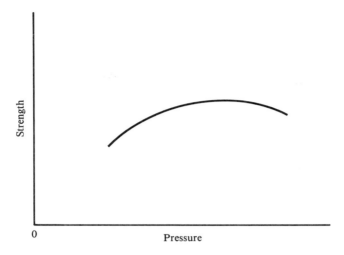

Figure 8.1. Strength as a function of clamping pressure in an industrial gluing process. (*From Quade, 1982, p. 165.*)

sults in their systems studies. That additional quantification will do so is often true to an extent, but it is not without its costs, and there are serious limitations on how far such a trend is useful—or can be carried. Moreover, a result obtained by quantitative methods is not necessarily better than one obtained by other means, particularly if the difficulties associated with quantification force the analyst to omit or drastically simplify too many aspects of the problem situation.

Two examples are useful in illustrating these points. The first, suggested by W. C. Randels, shows that even in high technology judgment must play an important, even decisive, role.

> Suppose we have a process for making a glue joint and wish to improve the strength characteristics. As a first approach we can consider variations in clamping pressure. This should produce a curve of strength versus pressure [Figure 8.1].
>
> This process gives rise to a further question. What temperature? To answer, suppose we generate a series of curves [Figure 8.2] of strength versus pressure for varying temperature.
>
> Fine, but for how long should it cure? We can run more tests, but we then have more answers than we can illustrate on a simple chart. Possibly a set of several overlays might be prepared. But there are still more questions. How much glue should be used? What viscosity should the glue be? How long should it set before joining? How smoothly should the surface be prepared? It is clear that we might as well give up on using conventional two-dimensional charts to illustrate all these relationships.

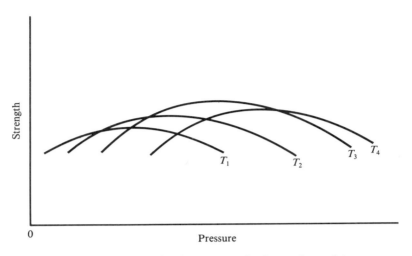

Figure 8.2. A sequential maximization process for four values of the temperature T of strength as a function of clamping pressure in an industrial gluing process. (*From Quade, 1982, p. 165.*)

There are things we can do, strategies we can adopt. For example, we can use a sequential strategy. After the first curve is computed take the pressure that yields maximum strength and, using this value, vary the temperature until strength is a maximum. Then, using these values for pressure and temperature, vary the curing time, etc. It is clear that each step represents an improvement upon the previously existing situation. Many strategies may be used, but the choice is subjective, based on the analyst's experience and judgment about the smoothness of the function involved.

Now suppose that to attain maximum strength a three-week cure is required . . . that involves tying up a good deal of equipment for the period. This may make the cost prohibitive; hence one must take cost into consideration. This might yield the relationship shown in Figure [8.3]. But to choose the point at which to operate we need to know how badly we need strength. In other words, what is the value of strength? But the value of strength is specific to a particular use, to a particular "scenario." If the uses are many and are to vary with future events, we must again depend on judgment.

Sometimes rules of thumb that have general applicability can be used. For example, when the cost-strength curve has a sharp bend (as in Figure [8.3]), one frequently operates close to the "knee" of the curve. But just where and how sharp is again a matter of judgment. [Quade, 1982, pp. 164–166.]

The lesson to be drawn from this example is much more general than just for the high-technology gluing process being described, or even just for high technology. Quade (1982, p. 166) puts it this way:

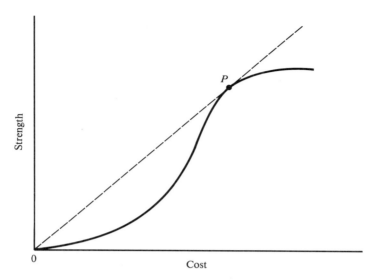

Figure 8.3. Finding an operating point *P* involving strength and cost for an industrial gluing process. (*From Quade, 1982, p. 166.*)

The point is that every so-called quantitative analysis, no matter how innocuous it appears, eventually passes into an area where pure analysis fails and subjective judgment enters. This is important; in applying this judgment the real decisions may be being made. In fact, judgment and intuition do not merely enter quantitative analyses when assumptions are made and when conclusions are drawn; they permeate every aspect of analysis in limiting its extent, in deciding what hypotheses and approaches are likely to be more fruitful, in determining what the "facts" are and what numerical values to use, and in finding the logical sequence of steps from assumption to conclusions.

The other example, from a letter by Leonard J. Savage criticizing the stress on quantification in analysis for policy decisions, makes the point that it is not quantification per se that is difficult to achieve, but rather quantification that is appropriately useful:

Consider an arctic station. Suppose, unrealistically, that the output of the station is clear and simple, a certain number of observations on a specified phenomenon, and that the objective of the expedition is to maximize the number of these observations for a fixed budget. Since the cost of fuel is serious, would it be best to have the temperature at 68° [20° Celsius] so everyone could work comfortably or would it be better to keep the temperature at 50° [10° Celsius] so that everyone would be miserable in sweaters and gloves but with facilities for two more men on the expedition? You might say that there is a serious problem in quantifying misery. But there are a million and one easy ways; we can measure misery as a function of temperature, for instance, by the amount per day that will induce a free workman to tolerate it. It then becomes a relatively easy matter to say how the number of men depends on the temperature within the fixed budget. What is difficult is to know output as a function of misery and the number of men. Merely knowing that output decreases with misery and increases with the number of men (if that be true!) is wholly inadequate in finding a solution. Also, we may not have measured the kind of misery on which performance depends. Extensive experimental work would reveal the unknown functions, but the experiments might be really prohibitively expensive and guessing unavoidable, in practice. What I have learned from this example is that difficulties that you at first sight might have attributed to the need for quantification are actually due to the presence of what you would call real uncertainty. [Savage, quoted in Quade, 1982, pp. 166–167.]

In addition to these limitations on quantification, there is another: Some situations involve such an unpredictable future, or one in which truly random phenomena dominate, that no quantitative analysis, no matter how perfectly it is carried out or how completely it covers the problem situation, could possibly predict the outcome of a single particular case, although it might offer an expected outcome, that is, an average of what will happen if the situation recurs many times. The study of the world's balance of energy supply and demand looking 50 years into the future (see Section 9.6) had to consider that the world's economic and political future could not be predicted with any precision over so long a time horizon. Similarly, a model for a situation containing random elements, such as game theory, may tell us how to behave in a given competitive situation to obtain an advantageous expected return, but it will do so only in a probabilistic sense, and will not predict the outcome for any single instance. (See also Chapter 7's discussion of uncertainty.)

In spite of these difficulties, systems analysts, reflecting the strong physical-science heritage that many of them bring to systems analysis, and knowing the advantages that quantification can offer, want to move as far in the direction of quantification as possible. There are, however, serious problems associated with the effort to quantify. Although a high degree of quantification, when achieved properly, invariably leads to improved analysis, the pressure and the effort to quantify in order to be able to use mathematical methods, and thus calculate results from mathematical models, can cause an analyst to introduce distortions into his analysis, such as losing information, neglecting intangible factors and influences, and adopting attractive methods that may be only marginally appropriate to the problem situation being studied.

Information can be lost, for example, when the analyst chooses a proxy measure to indicate the status of a relatively intangible or complex factor, such as using infant mortality rate as a measure of the health status of a population. Quade (1980, pp. 28–29) offers another example:

> One would like measures of effectiveness to reflect in a measurable way the ultimate good effect intended. For example, in a school health program aimed at detecting and correcting medical conditions that handicap a child's social adjustment or academic achievement, the ideal measure is the increase in the number of such children who, as a result of the program, engage in less socially aberrant behavior and make better grades. Unfortunately, improvement in behavior or even in grades due to a health program may be so difficult or expensive to measure that measurement of some intermediate goal may be operationally more significant. Thus, the number of health examinations given, or the number of children with handicapping conditions who are detected, or the number of those detected with handicapping conditions who receive appropriate medical treatment, or the number of those who receive appropriate treatment who have their handicap corrected, or even measurement of the number of handicapped children who are missed or not treated may have to serve as a proxy to indicate the success of the program. Actually, even our ideal measure is defective, for we should be talking about learning, not grades, if we remember what's important rather than what's measureable.

On the other hand, any systems analysis is an abstraction from reality that is intended to represent only the selected portions relevant to the problem in hand. Therefore, it is not unreasonable to expect the measures associated with it to represent only a portion of the available information; the issue is whether or not the portion of the information retained in the analysis is adequate for the problem being treated. Adequacy here is a matter of experience and judgment; various possibilities may have to be tried out and developed. Section 2.2 describes an analysis dealing with the problems of firehouse location in which proxy measures developed for fire problems were used successfully.

> Since a fire department's primary objectives are to protect lives and safeguard property, the most important measures of its performance are the number of fire fatalities and injuries and the amount of property loss. It is not possible, however, to use these measures to evaluate different deployment policies because there are as yet no reliable ways to estimate the effects that different policies have on them. . . .
>
> Therefore, in order to evaluate alternative firehouse configurations, three substitute, or proxy, measures were used, the first two of which are directly related to loss of life and property damage: (1) travel time to individual locations, (2) average travel time in a region, and (3) company workload. With these measures, the consequences of changes in firehouse location can be evaluated against the background of other considerations, such as hazards, fire incidence, costs, and political constraints. [*Overview*, p. 81.]

Many—and, indeed, probably most—of the issues investigated by systems analysis have major aspects that cannot be expressed satisfactorily in quantitative terms. An analyst may decide that he must have a completely quantitative analysis, but his work will be seriously incomplete if he does not still consider influences that can change the results, even if they cannot be quantified. He can, as has sometimes been done, postpone considering the unquantifiable factors until the end of his analysis, and then modify his conclusions to take them into account, but such a postponement too often neglects or underemphasizes these factors. The factors most commonly neglected are those treated in the social sciences, where few quantitative models of predictive ability comparable to those in the physical sciences and engineering have been developed. Whether or not it is due to the pressure to quantify, the most frequent—and possibly the most justified—criticism of systems analysis is that it overemphasizes the quantitative aspects of its problems and neglects the intangible ones; for some examples, see Hoos (1972), Brewer (1973), and Tribe (1972).

The pressure for quantitative analysis can lead to an infatuation with techniques, a bias leading to what is sometimes called *method-oriented* analysis—and some analysts start with such a predilection. Such an analyst is so well versed, or believes so strongly, in a certain modeling technique that he sees almost all problems as suitable subjects for his specialty, be it linear or dynamic programming, system dynamics, Markov processes, or whatever he's an expert in. He simply jams the problem into the model he understands well, making whatever assumptions, simplifications, or deletions are needed to achieve a fit. He then regards results calculated from the model as a solution to the problem. The observer, however, may have some difficulty in jumping to the conclusion that the calculated solution to the much modified problem provides much insight into the real problem from which the extrapolation began.

To someone without a broad knowledge of systems analysis practice over an extended period, this caricature may seem so blatant as to be unreal. Experienced analysts, however, often know informally of cases where the analysis was driven so strongly by a preferred paradigm as to ride roughshod over aspects of reality troublesome to the chosen approach. Although the literature, naturally, does not highlight such unfortunate cases, it is possible to find some there; for example, Churchman (1979, pp. 49–50) tells of an interaction with an analyst for whom "mathematical programming had become . . . a political ideology," and Meadows and Robinson (1985, especially pp. 75–86) describe how preferred modeling paradigms shaped the views of reality of a number of analysis teams.

On the other hand, a method-oriented analysis is not always a disaster, particularly if the omitted factors are reintroduced and handled judgmentally, say, by the decisionmaker. (Because it may not be easy to see how to amend quantitative results to make allowances for qualitative factors, the analyst should give careful thought to this problem before embarking on the analysis process.) Since no systems analysis can include everything that is relevant, its clients always have to make judgments about factors either omitted from or modified by an analysis; in fact, it is not unusual for such factors to dominate, and most problems can be modeled in more than one way. The real problem arises when method-oriented analysts neglect—and then fail to report candidly—whatever does not fit into their models.

Here are some of the pitfalls an analyst can stumble into in his effort to make his systems analysis quantitative:

Formulating or modifying the original real-world problem into another one that fits a particular model or method, rather than tailoring the approach to the original problem.

Becoming so fascinated by the model, particularly one implemented on a computer, as to pay insufficient attention to the original problem.

Deemphasizing—or even ignoring—the qualitative aspects or intangible factors in the original problem.

Losing important information by improper quantification.

Using technical, mathematical, and computer language to such an extent that communication fails, except with fellow specialists.

Tending to construct such a large central computer model to represent the problem situation that verification becomes difficult or impossible and validation (see Chapter 13) must remain inadequate, while costs in time and money soar.

There are a number of reasons for the pressure to quantify to be strong. Quantification is regarded, as a reflection of Kelvin's century-old view, to be essential to science, and thus automatically good. Analysts are trained to be quantitative. And experience tells us that intuition is seldom a good guide to prediction when quantitative intricacies are present. But the main reason, I think, is this: The analyst knows that, if his model is mathematical and quantitative, results from it can be deduced by mathematical means, with their attendant susceptibility to verification. It may be difficult and require considerable ingenuity to derive results, but once arrived at they can be proved to be correct—that is, correct with respect to the model (but not necessarily correct, of course, with respect to the problem situation). But, the quantitatively oriented analyst may argue, isn't it better to give the client something known to be correct, even for a modest distortion of the problem, than no result whatever, or something that may be based on a procedure without the logical rigor of analysis? And aren't those who have the responsibility for action likely to know more about how the intangibles will influence the outcomes than an analyst new to the context and trained largely in science and technology? And, finally, since the analyst must select the factors that can usefully be considered, isn't it reasonable for the partnership between decision-maker and analyst (as suggested, for example, in Section 12.4) to allot complementary responsibilities, deriving quantitative results to the analyst, and assessing less tangible factors to the decisionmaker? Indeed, there are clients who applaud this approach. Knowing that conclusions and decisions to act depend to a large extent on factors beyond the analyst's ability to quantify—or perhaps even to discover—they accept what the analyst provides and go on from there.

The craft of systems analysis provides no specific answers to these questions; rather, it suggests that each case must be approached and judged in terms of its own properties. But such a practice, although sometimes adequate, can be dangerous. Some policymakers are poorly informed about the limitations of quantitative analysis; they may not even understand that the analyst has made all sorts of assumptions in his analysis that they may not agree with, even if they understood what the assumptions were. Policymakers are thus at a disadvantage in trying to interpret the analyst's work. It follows that the analyst has the responsibility of trying to bridge the gap between his world and that of the policymaker—and not vice versa. It is vitally important, therefore, for all concerned to know what is being considered and what is not, and where the responsibility for judgment and analysis lies—and especially for the analyst to be sure that his client has a clear view of what the systems analysis includes and what it does not.

8.4. Measuring the Attainment of Goals or Objectives

One of the major difficulties in systems analysis lies in finding a satisfactory measure of effectiveness, that is, one allowing the extent to which competing alternatives attain desired goals or objectives to be measured in its terms. These difficulties are particularly acute in public policy analyses where the goals of public agencies are often multiple, conflicting, and not clearly defined. Even what seems to be a simple cost-effectiveness comparison of different educational programs can bring important complications.

Consider, for example, an elementary school district in the United States that includes many children who speak little or no English. To compare programs of various types that might be designed to teach such pupils more and better English, a scale measuring program effectiveness is needed. One such scale might be the increase in the average number of words in the pupils' English vocabulary at the end of a period of program application. But, even if this measure could be obtained with satisfactory accuracy, would it be a useful way to quantify English language achievement? Does an average count of words recognized measure the program's objectives?

Measures of Net Worth

To carry out a successful systems analysis, we would like to be able to compare and rank the alternatives by taking into account simultaneously the full range of consequences that could result from their implementation. The central problem arising from this desire is that, even when we can satisfactorily predict what the consequences will be, we find it impossible, except in very special circumstances, to devise a scheme to capture on a single scale the totality of consequences, or the net result, that would ensue from implementing each alternative. To make matters worse, we are not in many cases even able to conceptualize or identify comprehensively what consequences should make up such a "net output." Hence, in practice we must compromise and use some sort of approximation. In cost-effectiveness analysis, this is done by using only cost and the single most desired consequence—or a proxy for it—to represent the net (positive) output. This process, however, usually neglects not only the other desired consequences but also all costs other than money costs, such as pollution, discomfort, and loss of opportunity, that are imposed on people who have little or nothing to gain from attaining the objective. Cost-benefit analysis, on the other hand, uses, at least in theory, the full range of consequences, both positive and negative, but requires that they be measured in monetary terms, a task difficult to do in a way satisfactory to all interested parties. For scorecards (for examples, see Section 10.4 and the *Overview*, pp. 89–109 and 230–238), it is only the significant consequences (significance being judged by the analysts in consultation with their clients) that are displayed for judgment by the decisionmakers. Thus, there are defects in each of these approaches.

For a manufacturer, the number of physical units produced—automobiles, jet engines, washing machines, or loaves of bread—is often an acceptable measure of output for many purposes. But it is by no means the net output; other factors, such as pollution generated, taxes paid, capital invested, research and development investments, compliance with government regulations, safety records, community good will, and so on, are all part of the total output, and could be important for analysis, depending on the problem being addressed. The monetary value of the units produced might be a better measure of output, for it includes a measure of both the number of units produced and the value placed on them by the seller and the willing buying public. Similarly, profit might be a still better index of net output, for it includes both the monetary value of the output units as determined by the buying public and the costs incurred by some of the other factors mentioned. But profit, too, is likely to be an inadequate measure of the output of a business, since the business performs functions in society (such as, for example, offering employment and producing important goods and services) having values not usually comprehended in this measure (Drucker, 1973, pp. 58–61 and passim).

For a government agency it is usually even more difficult to conceptualize the output and its monetary value. What, for instance, is the output and monetary value of a strong military defense? Among other things, it may deter a potential enemy, but can such a factor be measured, since it exists only in minds, and the enemy's minds at that? For a municipal police department the situation is little better: The positive output is the protection of the public, the prevention and detection of crime, law enforcement, maintenance of order, traffic control, provision of emergency services, and so on. Each of these factors can be measured to some extent; for example, the Federal Bureau of Investigation in the United States has a crime index that takes some of them into account, but it has not worked out to be as useful as had been hoped—nor has any other single measure. The Urban Institute suggests (Hatry et al., 1977, pp. 85–108) that, to monitor adequately the performance of a municipal police department whose objective is "to promote the safety of the community and the feeling of security among the citizens, primarily through the deterrence/prevention of crime and the apprehension of offenders, [and] providing service in a fair, honest, prompt, courteous manner to the satisfaction of the citizens," 26 separate measures of effectiveness be used.[2]

A common—and often difficult—problem in quantification arises when one wants to compare the net worths or total benefits of a set of programs designed to alleviate a situation where the consequences of the various programs are made up of different packages of benefits and costs.

Consider, for example, what factors might have to be considered in an attempt to compare the net worths of various elementary-education programs. Such programs are commonly supposed to provide, increase, or improve, among other things: knowledge, skills, citizenship, social behavior, child care, mental ability, character, and health. Is there even a partially satisfactory proxy for this complicated set of outputs, as infant mortality is sometimes considered to be for the worth of a health system or transit time for the efficiency of a transportation system? The most frequently used proxy for education-system output is the amount of

---

[2] Although 26 may seem an excessively large number of measures to expect policymakers to deal with (and possibly one that could be reduced by eliminating measures of little significance or that are highly correlated with other listed measures), it is not an impossible one for them to handle. In the San Diego Clean Air Project (described in Section 10.4), they managed 29 successfully, and in the estuary-protection project (described in Section 2.2), they handled over 80.

money spent per pupil per year, but this proxy is an input rather than an output, and therefore can give no clear information about the quality of the system's operations. Two other proxies are also used from time to time: the pupil-teacher ratio (which reflects only the workload facing the teacher and thus gives no information about output or efficiency) and the percentage of former pupils entering college (a factor heavily influenced by such noneducational factors as the economic status of the family).

Quade (1982, pp. 85–86) calls attention to further difficulties in measuring the extent to which the goals of a school system are achieved:

> These different goals are to be achieved, if at all, at different times, some immediately, others—such as providing educated mothers who will in turn motivate their children to study—occurring in the next generation. A number of these goals are intangible and cannot be quantitatively measured; some even defy qualitative formulation. Others conflict directly or demand use of the same resources; education should promote both social stability and social change, both satisfaction and the fulfillment of social needs, both a desire to achieve (for the gifted) and contentment with self (for those of low ability).
>
> In addition, what might be regarded as a desirable goal or benefit by some people, say, changing the social structure, is certain to be regarded as a disadvantage by others. No policy output can be evaluated unless it is determined whether "more" or "less" of the given output is good; the percentage of criminals in jail, the number of highway miles paved, the number of individuals on the welfare rolls—all can have this indeterminacy.

To compare the net benefits of an educational program with that, say, of a health program involves further difficulties. For some thoughts on how this might be done, see Rivlin (1971).

Proxy Measures

The most important characteristic of a proxy is that it measures what it purports to represent; other desirable aspects are reliability and freedom from bias. To determine these properties is largely a matter of judgment (de Neufville, 1978).

In a proxy variable we seek something that is positively correlated with what it is supposed to represent, and usually something that can be measured directly or obtained from data that are not too difficult or costly to collect and process. Since the proxy assumes the role of the primary variable, its partial representation of the true measure of effectiveness can mislead if care is not taken to understand its limitations. For example, to measure the effectiveness of an industrial activity, any of these variables might serve as proxies: productivity, profits, client satisfaction, growth, or return on investment. But productivity might be achieved at the expense of building a skilled labor force for future growth, profits this year might be generated by postponing essential capital improvements for future operations, high client satisfaction might mask low quality and high expenses to correct faults, growth might occur at the cost of over-extension, and a low return on investment might deter important investments needed to strengthen the company's future. Thus, when a proxy is used in an analysis, the analyst needs to keep before him the larger picture that it represents and to examine what variations in the values of the proxy variable can mean for other important factors.

The usefulness of a proxy may also depend on the time and context in which it is used. Infant mortality can be a useful proxy for the general health of a population in which communicable disease is the dominant health problem and life expectancy is low. On the other hand, it is almost worthless today in the United States and Canada, where communicable diseases are under substantial control and life expectancies are long and increasing, and where the dominant health problems of young to middle age are trauma and those of old age are heart disease, cancer, and stroke.

Similarly, although the availability of full plumbing may serve the U.S. Census Bureau well as a proxy for the quality of housing in use by the general public, it may not be a valid indicator of the condition of the units in a public housing project. There is little reason for a complete set of plumbing to be linked with freedom from deterioration. In cases like this one in which a proposed proxy represents just one aspect of what we wish to measure, it is necessary to examine the behavior of other aspects to see if a combination of measures is needed.

It is clearly a mistake to select a proxy measure of effectiveness solely because it is readily measurable and can be handled easily in a mathematical model, even though it may be fairly well correlated with what properly needs to be measured. In addition to checking carefully on how its variations can affect, or be affected by, other important variables, as we have stressed above, it is also necessary to check on how the chosen proxy relates to various ethical, cultural, moral, and political factors that may bear on the results.

Proxy measures must also be watched carefully during the analysis, and especially carefully if they are used later to guide operations over time, because they can easily distort the desired situation. A classical example is the study of motor-pool operations that focused its analysis on hours per day of vehicle utilization, with a very low average result; by trading vehicle waiting time for executive waiting time, this measure could be driven up sharply, but at a serious cost to the higher-level operations of the organization operating the pool. Scioli and Cook (1975, p. 15) describe another example:

> Employing any single indicator of public agency output as the basis for structuring internal incentive systems can lead to pathologies. Jerome

Skolnick (1967) has graphically described one such pathology resulting from the excessive reliance upon the "clearance rate" by the Oakland Police Department as a measure of performance. Detectives in the department were led to treat the most active and confirmed burglars, once arrested on a charge, as a potential resource of great value for future promotions. If the detective were able to offer a suspected burglar a sufficient reward in the form of significantly reduced charges, he might convince him to confess to a large number of crimes—thus "clearing" them all with a single arrest and a reduced charge.

The knowledge that the quality of an arrest is possibly more important than the number of arrests has led many police departments to use multiple measures of effectiveness (Hatry et al., 1977, pp. 85–108).

Policy analysts can do just as badly if they rely on single components as proxies for measures of net effectiveness. Analyses of the effectiveness of public agencies have used total or per capita public expenditures as their yardsticks, measures that are not outputs at all, but inputs that may or may not be positively correlated with net effectiveness.

Excessive reliance on expenditure levels as the sole indicators of output may lead the courts into the position of finding the most wasteful (or most graft-ridden) cities providing the highest levels of output to their citizens.

Such methodological traps can be mitigated by the conscious development and reliance on multiple indicators of output derived wherever possible from multiple modes of data collection. [Scioli and Cook, 1975, p. 16.]

To use workload or operational measures as proxies for net effectiveness is another approach that is taken, with frequently unsatisfactory results. An Urban Institute study of solid waste collection (Blair et al., 1970) found many performance and efficiency measures to be in common use by local governments to measure the quality of service:

Tons or cubic yards of waste collected

Frequency of service

Number of special-request collections

Tons collected per man-hour

Miles of streets cleaned

Collections per crew

Tons collected per crew

Although each of these measures may be useful for operational management, none tells us anything about the quality or net effectiveness of the work. A later Urban Institute study (Hatry et al., 1977, pp. 10–11) suggested that the measures listed in Table 8.1 be adopted instead.

Table 8.1. The Objectives and Principal Effectiveness Measures for Solid Waste Collection

OVERALL OBJECTIVE: to promote the aesthetics of the community and the health and safety of the citizens by providing an environment free from the hazards and unpleasantness of uncollected refuse with the least possible citizen inconvenience.

Objective	Quality characteristic	Specific measure	Data collection source/procedure
Pleasing aesthetics	Street, alley, and neighborhood cleanliness	1. Percentage of (a) streets, (b) alleys the appearance of which is rated satisfactory (or unsatisfactory).	Trained observer ratings
		2. Percentage of (a) households, (b) businesses rating their neighborhood cleanliness as satisfactory (or unsatisfactory).	(a) Citizen survey (b) Business survey
	Offensive odors	3. Percentage of (a) households, (b) businesses reporting offensive odors from solid wastes.	(a) Citizen survey (b) Business survey
	Objectionable noise incidents	4. Percentage of (a) households, (b) businesses reporting objectionable noise from solid waste collection operations.	(a) Citizen survey (b) Business survey
Health and safety	Health hazards	5. Number and percentage of blocks with one or more health hazards.	Trained observer ratings
	Fire hazards	6. Number and percentage of blocks with one or more fire hazards.	Trained observer ratings

Fires involving uncollected waste	7. Number of fires involving uncollected solid waste.	Fire department records
Health hazards and unsightly appearance	8. Number of abandoned automobiles.	Trained observer ratings
Rodent hazard	9. Percentage of (a) households, (b) businesses reporting having seen rats on their blocks in the last year.	(a) Citizen survey (b) Business survey
Rodent bites	10. Number of rodent bites reported per 1000 population.	City or county health records
Minimum citizen inconvenience Missed or late collections	11. Number and percentage of collection routes not completed on schedule.	Sanitation department records
	12. Percentage of (a) households, (b) businesses reporting missed collections.	(a) Citizen survey (b) Business survey
Minimum citizen inconvenience (continued) Spillage of trash and garbage during collections	13. Percentage of (a) households, (b) businesses reporting spillage by collection crews.	(a) Citizen survey (b) Business survey
Damage to private property by collection crews	14. Percentage of (a) households, (b) businesses reporting damage to property by collection crews.	(a) Citizen survey (b) Business survey
General citizen satisfaction	15. Number of verified citizen complaints, by type, per 1000 households served.	Sanitation department records

Source: Hatry et al. (1977, pp. 10–11).

The Urban Institute has pioneered in developing practical ways for state and local governments to measure the performance of their programs. To a large extent, as Table 8.1 indicates, the information required to assess the efficiency, effectiveness, and quality of such programs is obtained from three sources: the government's written records, special observers trained to make ratings of the quality of government services, and local citizens surveyed about the performance of government services. The programs for which The Urban Institute developed performance measures ranged widely, including, in addition to solid waste collection, such diverse activities as programs for the short-term care of neglected and dependent children, the use of police patrol cars by off-duty officers, and treatment for hard drugs; Hatry et al. (1979) describe many performance measures for these and other programs.

Incommensurable Measures

In comparing the net worths of alternatives, there may be components in the spectra of consequences that cannot be expressed readily in the same units; they are said to be incommensurable. They may or may not be measurable in their own quantitative terms in a way useful for analysis; if not, they are also said to be intangible. If measures are to be assigned, they are selected subjectively, but it is not advisable for the analysts to make the assignments; they should be made by the responsible parties (but the analysts can assist by suggesting possibilities and helping to explore their implications). One systematic way to assign measure is to apply multiattribute decision theory (see the later subsection on indices). When decisionmakers use scorecards, as mentioned earlier in this section, to compare alternatives and make choices, they make subjective translations of the various incommensurables and intangibles into a common denominator bearing on their choices, thus avoiding the need for explicit assignments. Experience shows that they do this effectively and usually satisfactorily, but the process remains subjective, and without, so far, an objective description.

Efficiency Measures

Efficiency is often measured, particularly in government usage, as an output-input ratio with a workload measure as the output and a measure of resources (often cost) as the input. An increase in the number of tons of waste collected per dollar, however, if achieved at the expense of the quality of the result, cannot properly be considered an improvement in efficiency. A check on quality is always needed when output-input ratios are used with effectiveness data to measure efficiency. Thus, the number of clients improved (or helped) per unit of resource is a much better measure than the more common number of clients served per unit of resource, for the latter says nothing about the results.

In addition to the usual problems associated with ratios (see, for example, the *Overview*, pp. 221, 227–228), output-input measures are

> subject to the problems that affect all effectiveness measures. For example, some highly desirable effectiveness measures are virtually impossible to obtain; prime examples are those that attempt to assess success of "prevention" activities, such as prevention of crimes, fires, traffic accidents, or illnesses. Thus, the ideal efficiency measure, "number of such events (crimes, fires, accidents, or illnesses) prevented per dollar or per employee-hour," is rarely measurable. (An effort to estimate these, however, can and should be made periodically in special studies.) In effectiveness measurement, the approach generally substituted is to measure the number of incidents not prevented—that is, the number of crimes, fires, traffic accidents, or illnesses that do occur. These are useful and appropriate as measures of effectiveness, but constructing a ratio with these outputs related to dollars or employee-hours does not make sense. For example, the "number of crimes per dollar" or "the number of traffic accidents per dollar" are not meaningful efficiency ratios. If both the numerator (for example, the number of crimes) and the denominator (for example, the number of dollars) are reduced by 10 percent, the ratio will remain the same, implying that no improvement has occurred; in fact, there has obviously been improvement, because both costs and crimes are lower. [Hatry et al., 1977, pp. 235–236.]

Indices

Another approach is to bring all outputs (or as many as seem necessary or significant) into a common index of worth. For example, one might use a formula such as $(I - P)/C$ or I/P to compare the standards of living of various households, where I is income after taxes, P is the income required to maintain a standard of living at the poverty line, and C is the sum of the consumption weights of the members of the household (where, for instance, the weights assigned to children might vary with age). The commonest scheme is to use weights supplied by the analysts to combine the outcome indices (which may be in different units) into a single index.

A valid objection to a weighted index, however, is that the value judgments that assign weights should rightfully be made in the decisionmaking process and not by the analysts (Hatry, 1970). Many of the objections to using weights can be removed if the decisionmakers supply them; one systematic procedure for doing this with the aid of analysts uses decision analysis (Raiffa, 1968, and Keeney, 1980, the latter dealing with actual examples; see also von Winterfeldt and Edwards, 1986). Some experience, however, seems to confirm that it is better and more efficient (at least to me, as argued in the *Overview*, pp. 229ff.) to have the decisionmakers exercise their judgments (by means of scorecards, for example) based on the outcomes themselves, as in the case of choosing a protection plan for the Oosterschelde estuary in Holland (described in Section 2.2 and the *Overview*, pp. 89–109). For a comparison of cost-benefit analysis, decision analysis, and other formal decision-aiding approaches, see Merkhofer (1987).

Some outcomes or consequences may be difficult to evaluate usefully— state of health, the value of human life, the quality of life, the flexibility of a defense force—not only because of their intangible nature, but also because useful evaluation may depend on the willingness of the potential beneficiaries to pay for the programs aimed at producing the outcomes or consequences, or, indeed, to accept the programs at all. In some cases a proxy measure may help with the argument, in others a weighted index, but in still others, particularly where governmental programs are involved, the acceptance of the program and the amount set by the public's willingness to pay for the advocated benefit may be the proper and equitable choice.

Consider health, for instance. If the investigation is concerned with the entire population of a developing country where the chief threats to health are communicable diseases, the infant mortality rate might serve as a useful proxy (especially if it is the only reliable statistic that is available across the area of investigation); on the other hand, a health status index that could be readily applied to individuals (such as suggested by Fanshel and Bush, 1970) would permit more discriminating findings. If the concern is with the health of persons over 65 enrolled in a government health-care program such as Medicare in the United States, the number of days hospitalized per year per thousand enrollees might be a more satisfactory guide, provided hospitalization practices have been uniform over the period covered by the analysis.

Income might be used as a proxy for the quality of life, but income is just one dimension of a multidimensional objective, the other dimensions being such factors as health status, freedom, security, productive activity, creative leisure, and so on. In such a case, it may be that a weighted index in which all the important dimensions are represented will yield better results (Raiffa, 1968), provided the weighting is done by those involved and responsible.

In evaluating the worth of government programs aimed at such objectives as improving health, alleviating poverty, or equalizing the quality of life, it would be still better—or at least more logical—if the weights were assigned (or the proxy chosen) on the basis of what members of the public are willing to pay for the benefits that they would receive from the programs. On the other hand, evaluations in areas such as these have often employed cost-benefit calculations based on human capital considerations, which are unrelated to human preferences, and therefore irrelevant to political decisions. In any case, decisions about government programs of the sort being discussed in this paragraph, although they can be illuminated by policy decisions, remain largely political in the sense of responding to human desires, at least as perceived by political leaders. For a discussion of these points as they relate to government health activities, see Taylor (1970); for ways of determining willingness to pay, see Schelling (1968); for an example, see Acton (1973). Taylor's paper "analyzes the reasons for deficiencies in past analyses of government health programs and proposes remedies"; the introduction (Taylor, 1970, p. 50) argues the case in this way:

> Although the paper focuses specifically on health programs, the discussion is relevant to most government activity. In order to derive an appropriate approach to evaluating health programs, it was necessary to explore the general problem of the justification for government action. The concepts derived from this analysis are applied to evaluating both nonpoverty health programs and health programs for the poor. Much of the discussion is applicable to evaluating poverty and nonpoverty programs outside of health.
>
> This paper argues that cost-benefit analysis in health has been seriously deficient—not because the concept is bad but rather because the measures of benefit used have been based on a seriously inadequate understanding of how to value government activities. The paper takes as its starting point that the only legitimate justification for government action is to increase the satisfaction of individuals. There are no goals of government separate from the goals, wants, desires, demands of the individuals that comprise society. Although this should be an unremarkable premise, little attention has been paid to its implications for cost-benefit analysis. As a result, such analyses have often been seriously in error. . . .
>
> . . . the major conclusion [is that] . . . government health programs ought to be evaluated on the basis of their appeal and desirability to the public. The answer to the question [How much is good health worth?] . . . is . . . "good health is worth as much as people are willing to pay for it." At first glance, this answer may appear to be so obvious and uncontroversial that it is hardly worth making, but as . . . this paper . . . demonstrate[s], rigorous application of the principle implied in the answer leads to conclusions that are neither obvious nor uncontroversial.

Unfortunately, willingness-to-pay information probably must be collected by a questionnaire, a fact that brings problems, not the least of which are that (1) asking a subject what he believes or feels may not produce a response that reflects the subject's true beliefs and feelings, (2) a complete measure of beliefs or feelings requires a measure of their intensity, and (3) what a subject says he is willing to pay on a questionnaire may not be an accurate representation of his actual behavior in an actual event where a payment is required.

General Conclusion

The purpose of this section on measuring the attainment of goals and objectives has not been to prescribe how this should be done, or by what means, but rather to urge that the subject is a vitally important one demanding both careful thought and creativity. It is unlikely that there are any universal measures serving all purposes; thus, both analyst and client must choose the ones to be employed in the light of the problem situation and the uses to which results may be put. Some of the issues and difficulties in making these choices have been discussed here, but any important problem situation demanding analysis will surely bring up others; the important matter is for the analyst and the client to think carefully about them and derive appropriate measures before elaborate analysis is undertaken. This should include attention to interests not represented (such as future generations), not only because not to do so might come to frustrate the results during implementation or later, but also because it is the ethical thing to do. Time devoted to this important task will be well spent, and judicious, carefully chosen measures will do much to promote the acceptance of analysis results when they emerge.

8.5. Quantitative Analysis: Models as Surrogates

To obtain their results, most systems analysts depend almost exclusively on quantitative methods. When conclusions and recommendations are offered, however, they are frequently qualified, perhaps by adding a judgment or an uncertainty factor to the quantitative results, because there are usually one or more factors too important to neglect that cannot be dealt with quantitatively. Such a late modification of the quantitative results may not be necessary for certain operations-research studies of logistics or inventory problems, but it seldom happens that the larger, more complex problems addressed by systems analysis can be dealt with in a strictly quantitative way; social, political, and other "soft" factors are virtually always present in problem situations addressed by systems analysis, and they seldom can be fully characterized in strictly quantitative ways. In common practice, however, the term quantitative is applied loosely; a systems or policy analysis is said to be quantitative whenever a substantial part of the analytic work is carried out by quantitative means.

By the term *quantitative methods* here we mean the existing body of mathematical, statistical, and computational methods commonly used in operations research, management science, economics, and physics as well as other relevant branches of science. This body of methodology includes the theories, techniques, and tools of computers and computer science; probability theory; game theory; decision theory; linear and dynamic programming; simulation; econometrics; system dynamics; and so on. As Strauch (1974, pp. 5–6) explains:[3]

> This methodology has an abstract existence and meaning—as a body of mathematical knowledge about ways of dealing with theoretical problems defined in and examined by the supporting theory—that is independent of, but related to, its application to substantive problems. *The utility of the methodology as a tool for application to substantive problems depends on the existence of appropriate similarities between the theoretical and substantive problems.* . . .
>
> The supporting theory can be classed as mathematics, per se, in the sense that it is a study of mathematical systems and models as abstractly defined objects in their own right, independent of their relationships to real world substantive problems. The structure and behavior of mathematical models as such is determined fully by the defining premises and in no way by "real" considerations about any substantive process being modeled. [Emphasis added.]

By a substantive problem, Strauch means an actual real-world problem of the sort faced by a systems or policy analyst, or a practical problem in the sense of Ravetz (see Section 2.2).

The results obtained from mathematical models are based on their premises, and are arrived at by a sequence of logical inferences. The validity of the results from the model depends on the logical sequence alone, independent of the competence and judgment of the analyst who produced them, and can be verified entirely on the basis of the logical sequence leading from the starting point to the findings. Consequently, the analyst knows that, if his model is mathematical and his logic is error-free, the results obtained from the model are valid (as results from the model only, not as conclusions about the substantive problem the model might be designed to represent), and that any other analyst reasoning correctly from the same premises will obtain the same results.

This is not to imply that judgment and intuition play no role in determining the chain of logic leading from the model to the results. The opposite is true; good results in mathematics, like those of any science, depend on good craftsmanship. The nature of a mathematical model, however, is that, once the results have been reached, the analyst can go back over the intervening process that produced them, refine it, deleting references to the intuition or judgment that might have served to guide the analysis, and present the process as a sequence of logical steps. Regrettably, there is sometimes an exception to this general statement: for very large and complicated computer programs it may be a difficult, time-

---

[3] I am greatly indebted to R. E. Strauch for the ideas set forth in Sections 8.5 and 8.6, which follow closely the arguments of Strauch (1974), especially its Part II. Strauch (1975) is a shorter version that summarizes much of the earlier paper. Strauch (1980 and 1981) contain related material.

consuming, and almost impossible task to verify that the logic producing the results is indeed correct—and significant errors in findings are not uncommon (Meadows and Robinson, 1985, pp. 363–366).

To employ quantitative methodology[4] to help solve a substantive real-world problem requires the use of one or more mathematical models as simplified representations of the problem situation, or of important phenomena in this situation.

> The substantive problem or phenomenon is analyzed through the analysis of the model. The nature of the conclusions reached, and the amount of credence and confidence that can be placed in them, depend in part on the results produced by the mathematical analysis of the model. It also depends significantly on the relationship between the problem and the model—on what parts of the problem the model represents, and how well, and on what parts of the problem the model distorts or fails to represent, and how badly. [Strauch, 1974, pp. 8–9.]

The effect of the problem-model relationship, of course, falls entirely outside the scope of the mathematical theory on which the quantitative methodology is based, since this theory deals only with the analysis of problems defined by the models treated in the theory.

If a problem is strictly quantifiable, that is, if the logic, structure, and parameter values of the problem are identical with those of the model that represents it, the results produced by the mathematical analysis of the model can be given direct interpretation as substantive findings related to the problem. Strictly quantifiable problems arise in the investigations of physical systems obeying well-established laws; indeed, the relations between the phenomena and the models that represent them are so close that the literature of much of physics does not distinguish between them. Similarly, a strictly quantifiable problem can arise in an experiment in which the analyst introduces randomness to match that in the statistical model he plans to use in parallel with the experiment. In such cases, where the fit between model and problem situation is so close as to be unquestionable, the conclusions, like any mathematical result, are subject to validation or refutation on the basis of the logic of the mathematical analysis and the fit alone, without reference to the judgment of the analyst who produced them.

On the other hand, problems of the sort addressed by systems and policy analysts fall in a range of substantive problems that extends from what might be called reasonably quantifiable to soft and ill-structured, or "squishy" (Strauch, 1974). A reasonably quantifiable problem is one with

---

[4] Harold Linstone, a well-known analyst and teacher, estimates that 90 percent of the "tools" (of operations research and systems analysis) are suitable for only about 10 percent of the problems and subproblems.

a sufficient degree of natural structure to allow it to be represented reasonably well by a quantitative model that can be used to analyse it; its analysis is much like that of one that is strictly quantifiable. The model serves as a surrogate for the problem, as it did for the strictly quantifiable case, but now the question of the validity of the link between the problem and the model arises, although usually in mild form: The criterion is the reasonableness of the fit between the model and the problem situation. Reasonableness is subjective, however, and thus capable of yielding varying interpretations, depending on the person making the judgment. In fact, a critic of findings from an analysis using a model representing a reasonably quantifiable problem can usually find the problem-model relationship less reasonable than the analyst who did the work. When this possibility exists, it may be prudent for the analyst to devote some effort to buttressing his judgment of reasonableness with some work aimed at validating the model-problem link (see Chapter 13).

By a soft, or squishy, problem is meant one for which the analyst runs a very small chance of finding a clear and well-defined quantitative model that captures its essential nature unambiguously. Sometimes this case can arise when the analyst finds a good approximate model, but it turns out to be so intractable analytically or so computationally difficult that he is forced into some judgmental approach such as gaming (see Sections 2.8, 3.5, and 3.6, and Chapter 4) or the Delphi procedure (see Section 3.7) in order to produce any results. The desire for a workable analytic formulation may tempt an analyst facing a squishy problem to distort or restrict it by leaving out troublesome elements or making sweeping simplifications; in such a case, it is unlikely that the model can serve as a satisfactory surrogate for the problem.

To use a model as a surrogate for a substantive problem, we assume that the model captures its essence so satisfactorily that we are willing to take the results derived from the model and interpret them as conclusions with respect to the original problem. This way of using models lies at the heart of the physical sciences, as well as most areas where there are well-developed applications of mathematics, including much of operations research and some systems analyses. It is the one that students of science and engineering tend to accept as the way models ought to be used. Without question, it is a good way of using a model when the problem situation is well structured and its variables are quantifiable in a direct and natural way. But it is not necessarily adequate without thoughtful judgmental interpretation of the model results when the problem situation is at best only approximately quantifiable—and it may be well off the mark when the problem is soft, ill structured, and importantly affected by behavioral and political considerations.

Although it is clear that adding more realism—and thus more detail— to a model can make it a closer surrogate, such an effort, if carried too

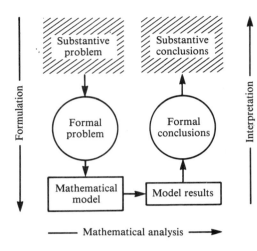

Figure 8.4. The general case of the application of the quantitative methodology. (*Adapted from Strauch, 1974, p. 14.*)

far, can be self-defeating. It can make calculations difficult and expensive and can even lead to error as the inaccuracies in the data pile up and the more the data are manipulated. The use of large-scale models in the policy process has produced very poor results (Ascher, 1978); such models involving social elements have, in addition, important costs and consequences that are often not recognized (Brewer, 1983).

Applying quantitative methodology to a substantive problem can be considered to involve three stages (Strauch, 1974, p. viii):

Formulation of a specific formal problem and the mathematical model used to represent it.

Mathematical analysis of the model within its context. This step produces results that are logically valid within this context.

Interpretation of the model results as formal conclusions (within the context of the problem's formal model) and then as substantive conclusions about the underlying substantive problem.

In the actual practice of systems analysis, these three steps are interwoven by iteration and feedback; however, after the process is over, it can be simplified so as to be pictured as in Figure 8.4. When a systems or policy analysis is called for, the substantive problem is seldom stated clearly enough, or sufficiently focused, to be modeled unambiguously. The first step in the formulation process is to define clearly a formal problem that can be modeled (at least in words). Then, the formulation component associates a mathematical model with the formal problem, mathematical analysis produces results from the model, and the interpretation component translates these results first into conclusions for the formal problem and then into conclusions related to the original substantive problem.

The formal problem links the substantive problem and the model by specifying, in substantive terms, exactly what parts of the original problem are being modeled: the parts allowed to vary and those held constant, the assumptions being made, the relations among the parts, the factors left out, and so on. Thus, the formal problem delineates the relationship of the model to the problem, often restating what is represented by the model in substantive terms. The formal conclusions act as a similar link, interpreting the model results in a context more closely tied to the substantive problem and less to the model formalities.

> Formulation and interpretation are essentially subjective activities—requiring and depending on careful and considered judgment on the part of the analyst doing them. Formulation, from the substantive to the formal problem and from the formal problem to the model, is a process of taking away—of removing pieces to make the problem smaller and more analytically tractable. With complex and squishy problems particularly, this requires an intuitive understanding of the substantive problem as well as of the methodology being used. It may also involve some adding on, in the form of assumptions that are questionable on substantive grounds, but that make analysis easier. The assumption of statistical independence of various types of events, for example, falls in this category, as do assumptions about the "rationality" of political decision-makers. Interpretation, conversely, involves putting things back—adding in the considerations removed to make the problem tractable and removing any distortions resulting from the added simplifying assumptions. [Strauch, 1974, p. 17.]

The activities of formulation and interpretation are judgmental in nature, and hence any conclusions reached are inherently subjective, even though the model may be quantitative (as in Figure 8.4) and the analysis producing the model results purely logical. If the model itself is not quantitative, the results will depend even more on the skill and judgment of the analyst.

In using the term judgment here we are not talking about an instantaneous expression of opinion or what is sometimes called unbuttoned or off-the-top-of-the-head judgment. We mean reasoned and considered judgment, offered, where possible, after analysis, research, and consultation with people steeped in the problem. It is opinion for which logical reasons can be given and documented, based on knowledge and experience.

Three of the examples of systems analysis described in Section 2.2 (and in the *Overview*, pp. 67–109)—blood supply, fire-station location, and protecting an estuary from flooding—are studies in which the models are used as surrogates. On the other hand, the fourth example (Sections 2.2

and 9.6, and the *Overview*, pp. 109–115)—global energy supply and demand—is one in which the conclusions were reached by using the model as a perspective on the problem, as discussed later in Section 8.6.

As a case exemplifying Figure 8.4, consider the estuary-protection problem. In 1953 a severe storm flooded much of the Delta region of the Netherlands, causing great losses of life and property. The next year the Dutch government embarked on a massive construction program for flood protection. In 1975, with the program nearly complete except for the largest estuary—that of the Oosterschelde—the project was interrupted by controversy, the goals other than security from flooding becoming increasingly more important.

The substantive problem facing the government and the analysts brought in to help was to find a politically acceptable scheme that would provide the needed security without doing excessive damage to the ecology; to fishing, shipping, and other industry; to recreation; and to the economy in general. After considerable debate, and with the help and advice from the government (in which certain alternatives were eliminated, political and defense considerations set aside, and the threat particularized), a formal problem was agreed on: to determine the consequences or impacts that would result from each of three radically different ways of providing the protection. The three alternatives were (1) a dam closing the mouth of the estuary, which would turn it into a fresh-water lake; (2) a storm-surge barrier with an opening to be closed by gates before an approaching storm; and (3) leaving the estuary open but building a series of high dams around its perimeter. There were also variations to be considered within each alternative, such as location, dam height, gate width, and so on.

The analysis team then designed—or adapted from the various disciplines involved—mathematical models to predict the various consequences that would follow from the implementation of each alternative. Several dozen models were required, as numerous impacts were considered important. (Thus, in this case, the mathematical model of Figure 8.4 consisted of a fairly large set of smaller models, each used to predict one of the many impacts considered.) Analysis of the models yielded model results, which were then translated into substantive impacts. These results, supplemented by analysis-team judgments, were used in various ways to screen out the inferior variations of the three alternatives. Then, for the chosen configurations of the alternatives, the analysts presented a selected set of these impacts on scorecards. Since it covered all of the major concerns of the government, the number of impacts presented was fairly large.

The Netherlands government then used these impact comparisons, supplemented by other studies and their judgments of the political climate and other factors not included in the formal models, to reach a conclusion about the substantive problem; the decision was to solve it by implementing a version of the storm-surge barrier alternative.

The validity of the conclusion produced when a mathematical model is used as a surrogate depends on two factors: the empircal validity of the linkages between the model and the substantive problem and the internal logical validity of the mathematical analysis. Quantifiable problems possessing a structure very close to that of the models used to represent them permit straightforward interpretations of model results as conclusions about the substantive problems themselves. For most problems to which the quantitative methodology of systems analysis is applied, however, the model is not a close mirror image of the problem, nor are the model results translatable directly into substantive conclusions about the problem. As Figure 8.4 illustrates, both the substantive problem (as is often its nature) and the conclusions (because they depend at least to some extent on judgment) may be far from sharply defined.

This is not to imply that quantitative methodology is of little or no value in dealing with squishy problems, for there are many examples where quantitative analysis has provided useful insights and acceptable solutions. But these successes have usually been due more to the skill and insight of the analyst—in sum, to the skillful application of his craft—than to the formal manipulation of the methodology.

> It stems not from the fact that he made calculations, but that he saw the *right* set of calculations to make, and was able to *interpret* them creatively. Nonetheless, the conventional dogma surrounding the application of quantitative methodology tends to attribute success to methodology per se, rather than to the wisdom and judgment of the analyst using it. It suggests that the appropriate standards by which to judge quantitative methodology should focus on the technical quality of the mathematical analysis, rather than on the quality of the judgment associated with the formulation and interpretation of the analysis. [Strauch, 1974, p. 21; emphasis in the original.]

8.6. Quantitative Analysis: Models as Perspectives

The idea of a model as a surrogate for a substantive problem—and a very close surrogate at that—is so thoroughly imbedded in our formal educational system that we may not realize that there are other ways of using models. We are taught to use models, particularly quantitative ones, in this way: for a substantive problem, build a model that captures its salient features, solve the problem as defined by this model, and then accept the model solution as directly applicable to the substantive problem; forget—and, indeed, classical scientific education has taught us to forget—that there is a significant difference between a problem and its model.

A model has a structure of its own. Although it may resemble closely the structure of the phenomena in the problem situation that it is supposed to represent, the model is independent of that structure, although based on it. The model defines its own problem, process, and solution that is related to the original substantive problem, but it is distinct from this problem.

This distinction is important to the way an analyst uses a model in problem solving. On the one hand, he may accept the structure of the model as a valid representation of the structure of the substantive problem and use the model as a surrogate for it; that is, he may adopt the problem defined by the model as the problem he wants to solve. On the other hand, he may supplement his use of the model with additional information and knowledge about the problem situation that is not reflected in the model's content or structure; in such a case he is using the model to furnish a perspective on the substantive problem.

To use a model as a perspective on a problem is so common in everyday thought with mental models as not to seem unusual. What is notable is the much less usual extension of this idea to formal models used to shed light on substantive problems of significant scale and complexity. There are problem situations that defy the construction of adequate surrogate models; here, if quantitative models are to be useful at all, it must be to yield perspectives. To use a model as a perspective on a complicated problem situation requires that the analyst carry along with the model— not cast away or subordinate—the full richness of the problem setting. He does this throughout the formulation process in deciding what models he wants to use, what to include in them and what to leave out, how to use them to exhibit the perspectives needed, and how to use them in combination with other knowledge to shed light on the problem and its possible solution.

Strauch (1975, pp. 180–181) offers an illustration of a model that has served as a perspective:

> [A] mathematical model that I, at least, think is an excellent perspective on a lot of problems, but a poor surrogate for any, is Prisoner's Dilemma. This is a well-known two-by-two game. Say two of us rob a bank and get caught. The local police have insufficient evidence to get a felony conviction, but enough evidence to hold both of us, say, six months for possession of concealed weapons. The police interrogator tells me that the District Attorney really wants a felony conviction, so if I turn state's evidence and let him convict my partner, he will let me off. At the same time another interrogator is making the same offer to my partner. If we both turn state's evidence, however, they cannot let us both off but will get both of us on a lesser felony conviction. We each get, say, five years as opposed to the ten if one confesses and the other does not.
>
> So, what's the situation I am in now, and what should I do? Suppose first that I do not know what my partner is going to do. If he holds fast,

if he does not confess, then by confessing I can reduce my own sentence, from six months for carrying a concealed weapon to getting off scot-free for turning state's evidence. So, I am much better off to confess, assuming he does not. Assuming he does confess, on the other hand, I reduce my own sentence from the ten years I will get if I remain silent and he confesses to the five years I will get if we both confess. Thus, whatever he does it is rational for me to turn state's evidence. I improve my situation either way. Now the situation is symmetric, so that the same considerations hold for him. Thus rationality would suggest that we are both better off if we turn state's evidence. However, we then both end up with five years in prison, rather than the six months we would have had if we both held fast.

The dilemma captured in that game is one that tends to occur throughout human affairs. It is present in many of the environmental problems that are making a lot of news now. It is always cheaper for a riverfront town to dump raw sewage into the river than to treat it, for example, but if all the towns do this, the river will become polluted for all. The dilemma is present in our daily lives. Regardless of what others do, it would be to my advantage, if I could get away with it, not to pay taxes. If no one paid them, however, organized society would collapse. It is also present in the affairs of nations. Each nation in a defense alliance may find it cheaper to let others carry the defensive load. If all try this, however, the alliance may not serve its purpose.

The usefulness of Prisoner's Dilemma as a model of these situations lies not in the fact that it tells us how to "solve" them. It clearly does not. The solution depends on more context than the model captures. Depending on the context, it may contain elements of criminal sanction, coercion, mutual trust, and living with degraded outcomes. This last we find particularly in the environmental area. The value of Prisoner's Dilemma as a model of these situations lies in the fact that it provides a context-free integrating perspective on the core elements common to all of them. The value of this perspective is not diminished by the fact that it does not prescribe a "solution," since these core elements do not provide enough structure for that.

The questions that systems and policy analysis address frequently arise from problem situations that preclude constructing an adequate surrogate model in the sense in which such models exist in the physical sciences. When this is the case, a model, or each of several models, can still be useful as a perspective—one way, but not a unique or even a completely valid way, of viewing the problem. Considerable additional substantive knowledge is usually required to determine how the model should be used and how its results should be interpreted with respect to the substantive problem.

The argument can be carried a step further by suggesting that a richly complex problem situation can usefully be illuminated by more than one perspective, each embodied in a model, and all surrounded by knowledge of the problem situation not embodied in any perspective. Linstone (1984) has carried this idea into research on a variety of retrospective case studies, but he limits himself to three perspectives (technical, organizational, and personal), rather than such perspectives as might emerge naturally from the problem itself; too, in most cases, his perspectives are ways of organizing and focusing the analyst's contextual knowledge, rather than models. Nevertheless, his work, and the cases it examines, sheds a good deal of light on the argument of this section, the net effect of which is to urge the further exploration of the uses of models as perspectives, since most complex policy problems demand multiple perspectives for their appropriate solution. Linstone et al. (1981) provide a brief introduction to the work reported more fully in Linstone (1984).

To use a model as a perspective, the process of analysis begins just as it would in the case where the model is to be a surrogate. The analyst starts with the substantive problem and whatever information he has or can gather about it and possible ways of solving it. Some of this information is—or will be made—internal, and stored in his mental models; some is external, existing in such forms as written material, data banks, and other people's opinions.

After absorbing the external information about the problem situation, the analyst combines it with what he already has about similar problems and contexts and promising tools and techniques for solving them. He generates and examines possible alternatives, screening out the clearly inferior ones. He develops a formal model, probably mostly verbal at this stage, to give the substantive problem a structure and to define more clearly just what he will try to model mathematically. If the problem is not reasonably quantifiable, he may have to try several modeling approaches before accepting one that, although unlikely to yield a surrogate model, will develop a useful perspective, one that will serve as an external aid to developing the the logical implications of the premises that have emerged relevant to the problem.

These activities make up the formulation process indicated in Figure 8.4; they are largely judgmental and take place in the analyst's intellect.

After producing results from the models, mathematically if the models are sufficiently quantitative, the analyst begins the interpretation process, which integrates the model results with whatever other information he has that is relevant and not included in the models. He uses his judgment to interpret the results, first at the formal and then at the substantive level. When the models are intended as perspectives rather than surrogates, some portions of their inputs are likely to be attributable to simplifying assumptions chosen to make the calculations tractable, rather than as close reflections of the real world. The results largely attributable to these simplifications, although valid in the mathematical sense, are unlikely to have substantive value. Finally, integrating the model results with the analyst's further knowledge produces three types of findings: acceptance of some results as valid substantive conclusions, qualification of others to account for differences between the model and reality, and the rejection of still other results as the products of discrepancies between the models and the problem situation.

Each perspective on a complex problem, like the resulting models, is always simplified and somewhat distorted, and may be far less complex than the original problem. Several, often many, different perspectives on the same problem, all equally valid, are possible. An intelligent human— say, the person with the responsibility for action—has the potential ability to use multiple perspectives by switching back and forth among them and adding his own to get a good picture of the full situation.

The tale of the blind men and the elephant is used by Strauch (1981, p. 7) to illustrate this possibility:

> Each blind man sees not the elephant, but his own simplified and dis-
> torted image of the elephant. Each of these images is different and comes
> from a different perspective. Each is in fact valid, though none is com-
> plete. The only thing missing is the multiple levels of perception, and
> the overall model which ties the individual partial images together. None
> of the blind men have that, but it is supplied by the reader. That is what
> gives the story its humor as well as its insight.

To aid their understanding—and to learn how to improve the manage-ment—of a complex ecological process, Holling and his colleagues (Holl-ing, 1978, pp. 57–105) used a similar scheme. They examined and tested a series of alternative models, each a simplified version of a more com-plete model of the process, to gain perspectives on the larger model, and thus insight into the process it represented.

Documenting these analyses—producing a permanent external version of the conclusions and the grounds on which they were reached—is im-portant, for several reasons.

Documentation allows the analyst to examine and verify what he has done after details may have slipped from memory. It gives him a permanent structure that will sit still while he makes sure that he is satisfied with what was done, and it will do this in a way that the internal working of his head will not.

Documentation provides a basis for communicating both the conclu-sions and the basis for them.

Documentation provides a basis for judging the validity of the conclu-sions. It exposes two structural characteristics of the analysis that are especially important for judging its validity: the comprehensive-ness of the work and the extent to which the conclusions are rational consequences of the inputs rather than prejudices of the analyst.

Comprehensiveness has two aspects: (1) the extent to which the analyst covered the ground needed to give an adequate basis for his findings and their relevance to the substantive problem and (2) whether the documentation articulates the grounds for the conclusions so as to allow another analyst to trace the logic without adding bridging inferences.

It is doubtful whether any analysis can be completely objective, but professional history shows that an experienced analyst can usually judge when findings arise from logic and well-established fact and when they have been dominated by subjective judgment, although the middle ground between these two extremes may present a tangle difficult to unravel. But to do this adequate documentation is required; without it another analyst cannot reach his own conclusions in order to see if they match those of the original workers.

The characteristic of being a surrogate for or a perspective on a problem does not lie within the model itself. Rather, as an interpretation of the evidence bearing on the problems as it relates to the model, it lies in the mind of the analyst—in the way that he thinks about the model and its relationship to the problem, in the extent to which he accepts the model as representing the problem and sets aside what he knows that is not embodied in the model or keeps it in mind, and whether or not his conclusions about the substantive problem embody not only model results but also the rest of the body of relevant knowledge about the problem situation.

From a technical point of view, building a perspective model is less difficult than building one to be used as a surrogate. A perspective need only reflect the parts of the problem the analyst wishes to investigate; it can be simpler than the real problem because it does not need to reflect all of the problem situation's intricacies. It may, for instance, be a sub-optimization or not be concerned with certain feasibility considerations. On the other hand, because the model does not capture the entire structure of the problem, the analyst cannot depend entirely on mathematical rules and the validity of the logic within the model as the basis for translating the model results into substantive findings about the problem. Too, an analyst working with a perspective model need not restrict himself to working within the model, but can also bring to bear other factors that he knows about. Thus, in arriving at conclusions about the substantive problem, he can use a variety of material, and must use it all to defend his judgments about it and how it was used to reach conclusions about the substantive problem.

As mentioned earlier, the global analysis of energy supply and demand looking 50 years into the future described in Sections 2.2 and 9.6 provided a perspective on this problem situation. The most persuasive fact about the analysis that supports this statement is that it did not—and indeed could not—consider all possible futures looking so far ahead, but confined its analysis principally to two plausible surprise-free scenarios. Too, co-operation, rather than competition, between nations was assumed tacitly, and all considerations of political feasibility within nations were omitted from consideration. Although results based on these scenarios offered a very valuable perspective, they could in no way be regarded as expected futures, since myriad possibilities exist for the world to move in directions other than those stipulated in the scenarios, and in fact the possibilities can easily be imagined. Nevertheless, if variant futures come to pass without major discontinuities, extrapolations from the scenario results will often be possible, thus allowing the study results to continue to offer information as a perspective on the problem of world energy supply and demand for the long-range future.

The development and use of models, both surrogate and perspective, enhance an analyst's judgment about the situation modeled. A good illustration is the case of Jan Leendertse, a fluid flow expert, who designed a detailed computer simulation model of the water and pollution flow in New York's Jamaica Bay (Greenberger, Crenson, and Crissey, 1976). In developing and using his models, he acquired such a thorough knowledge of the bay that New York City's water authorities were able to rely on his judgment, even when no opportunity to apply his modeling techniques existed (loc. cit., pp. 79–84).

Quantitative methods have a great potential for helping with policy analyses, both as surrogates, when possible, and as perspectives for a much larger class of problems, provided they are used in conjunction with adequately based and carefully considered judgment. Unfortunately, the use of models as perspectives, although it may be fairly common in practice, is not yet well developed, nor is it yet dealt with by supporting systems-analysis theory or more than barely recognized, under this or any other name,[5] in the conventions that exist concerning the use of quantitative methods.

An analysis that gives no more than one or two important perspectives on a problem may constitute only a partial analysis of the problem situation as a whole, but users may find it more helpful than an analysis that carries results to the point of choice by using models and methods that strain to fit the problem beyond what common sense and the methodology can justify.

8.7. Concluding Remarks

Much of the pressure for quantification—using numerical inputs and outputs to processes in the problem situation, measuring quantitatively the extent to which goals and objectives can be met, developing rigorous

---

[5] Tomlinson (1981, Near Truth No. 4) uses the term "partial representation" to mean something very close to what we mean here by the term perspective.

mathematical surrogate models—comes from the belief among analysts that the only payoff from systems analysis lies in problem solving. But this belief is in error: problem solving may not even be the most important payoff. Although operations research and systems analysis were started with problem solving in mind—in fact, most of the early work was very successful at problem solving—and we write about these subjects often as if problem solving were their single goal, it is so less frequently than one might expect. How frequently is not known definitely, but Vickers (1981, p. 28) cites a 1975–1976 review by the British National Coal Board of the work of its operational research unit:

> It was found that over a period of years only 15 percent of its activities could be clearly identified as "problem-solving." The rest and the most prized result of its activities was to increase "understanding" relevant to some "concern." Nor could this understanding be expressed in wholly mathematical terms. The "problem-solving" had more than paid for the whole exercise; but the deepened understanding was at least as greatly prized.

For further discussion of these points, and other misconceptions of operations research and systems analysis, see Tomlinson (1981).

At The Rand Corporation in the United States, where much of the pioneering work devoted to systems-analysis models as surrogates was done, the same point was being appreciated as early as the late 1950s. In fact, a review of the history of a very large and influential mid-1950s systems analysis done there makes this statement (Smith, 1966, p. 230):

> [I]n many cases the advisor's most important contribution is simply to force a rethinking of some complex problem. The end product of most planning and research [i.e., systems-analysis] activities is not an agenda of mechanical policy moves for every contingency—plainly an impossible task—but rather a more sophisticated map of reality carried in the minds of the policy makers.

Our view, stated in the *Overview* volume (p. 2), is that:

> The central purpose of systems analysis is to help public and private decision- and policymakers to ameliorate the problems and manage the policy issues that they face. It does this by improving the basis for their judgment by generating information and marshaling evidence bearing on their problems and, in particular, on possible actions that may be suggested to alleviate them.

Analysts may feel they need surrogate models that can be analyzed mathematically to obtain problem solutions, because the best help to decision- and policymakers emerges from quantitative models. But this is not the universal view:

> As a tool for solving problems which can be specified in mathematical

terms, systems modeling is an important contribution to the tool subjects of those responsible for governance, though a limited one because of its limitation to situations which can be mathematically modeled. As a contribution to human understanding, systems thinking is an immeasurable boon and need for every individual now living in our interdependent and unstable world. But to try to corral the human understanding within the bounds of what can be explicitly expressed in mathematical terms is an act of arrogance so outrageous as to merit the nemesis which will certainly attend it. It is nonetheless happening now. [Vickers, 1981, p. 28.]

Many policy problems to which analysis can be applied involve social, behavioral, and political factors as well as considerations of equity and reasonable treatment, and so do not always have unambiguous answers that can be reached by purely logical means. Quantitative methods and models, however, tend to produce such answers. Nevertheless, such answers, whether from surrogate or perspective models, can be valuable provided their results are thought of as opinions—and well-informed and properly considered ones if the analyst has done his work thoroughly. They are his opinions because they rest on his craftsmanship regarding how to choose a model, how to use it properly, how to test it to find out how far to trust it, and how to interpret its results in the light of the factors that it does not include as well as those it does contain. Strauch (1975, p. 184) sums the matter up:

> This is not to denigrate the value of quantitative methodology. It has great potential value, but as an aid to careful and considered human judgment, and not as a replacement for such judgment. When we forget that and focus our attention too strongly on our methods and our computations as the source of our answers, we not only fail to realize that potential but we run the risk of being seriously misled. The source of knowledge and understanding about squishy problems has always been, and will continue to be, wise people more than sophisticated methods. This is a subtle distinction, perhaps, but an important one. And if we fail to make it, there is no way we can realize the real potential of the people or of the methodology we do have available to us. I am not suggesting that we give up quantitative methods, but rather that we use them with a goodly grain of salt, and that we grant "the computer" no more *automatic* and unquestioned authority than we would grant a shaman or a fortune teller. [Emphasis in the original.]

In the light of the arguments of this chapter, it is useful for the systems analyst to think of the problem situations that he may face as lying somewhere along a spectrum stretching from highly structured ones on the left to very soft and squishy ones on the right. Problems on the left—where operations research and systems analysis began—are susceptible to being represented by surrogate models that ape the behavior of the real situation very closely. Here, since human intuitions are frequently seriously inadequate about the potential behavior of very complicated structures, results emerging from analyses based on surrogate models can serve as "bulwarks against the fallibilities and limitations of human judgment" (Strauch, 1980, abstract). On the other hand, for problems on the right of the spectrum, where surrogate models are not possible and where quantitative models are limited to roles as perspectives, rigorous analysis is likely to be much more partial. Here, since the models are inadequate predictors of the full range of potential behaviors, we see "judgment and intuition as bulwarks against the fallibilities and limitations of formal methodology" (loc. cit., abstract). The important lesson is this: In neither case does the prudent analyst rely solely either on quantitative analysis or on intuition and judgment; rather, he considers an appropriate mixture of both, the relative emphasis depending on where his problem situation is on the spectrum.

References

Acton, J. P. (1973). *Evaluating Public Programs to Save Lives: The Case of Heart Attacks*, R-950-RC. Santa Monica, California: The Rand Corporation.

Ascher, W. (1978). *Forecasting: An Appraisal for Policy-Makers and Planners*. Baltimore, Maryland: The Johns Hopkins University Press.

Blair, L. H., H. P. Hatry, and P. A. Don Vito (1970). *Measuring the Effectiveness of Local Government Services: Solid Waste Collection*. Washington, D.C.: The Urban Institute.

Brewer, G. D. (1973). *The Politician, the Bureaucrat, and the Consultant*. New York: Basic Books.

——— (1983). Some costs and consequences of large-scale social systems modeling. *Behavioral Science* 28(2), 166–185.

Churchman, C. W. (1979). *The Systems Approach and Its Enemies*. New York: Basic Books.

de Neufville, J. I. (1978). Validating policy indicators. *Policy Sciences* 10, 171–188.

Drucker, P. F. (1973). *Management: Tasks, Responsibilities, Practices*. New York: Harper and Row.

Fanshel, S., and J. W. Bush (1970). A health-status index and its application to health-services outcomes. *Operations Research* 18, 1021–1066.

Greenberger, M., M. A. Crenson, and B. L. Crissey (1976). *Models in the Policy Process: Public Decision Making in the Computer Era*. New York: Russell Sage Foundation.

Hatry, H. P. (1970). Measuring the effectiveness of non-defense public programs. *Operations Research* 18, 772–784.

———, L. H. Blair, D. M. Fisk, J. M. Greiner, J. R. Hall, Jr., and P. S. Schaenman (1977). *How Effective Are Your Community Services? Procedures for Monitoring the Effectiveness of Municipal Services*. Washington, D.C.: The Urban Institute.

————, S. N. Clarren, T. van Houten, et al. (1979). *Efficiency Measurement for Local Government Services: Some Initial Suggestions*. Washington, D.C.: The Urban Institute.

Holling, C. S., ed. (1978). *Adaptive Environmental Assessment and Management*. Chichester, England: Wiley.

Hoos, Ida R. (1972). *Systems Analysis in Public Policy: A Critique*. Berkeley: University of California Press.

Keeney, R. L. (1980). *Siting Energy Facilities*. New York: Academic Press.

Kelvin, Lord (1883). Lecture to the Institution of Civil Engineers, May 3, 1883.

Linstone, H. A. (1984). *Multiple Perspectives for Decision Making: Bridging the Gap between Analysis and Action*. New York: North-Holland.

————, et al. (1981). The multiple perspective concept: With applications to technology assessment and other decision areas. *Technological Forecasting and Social Change* 20, 275–325.

Majone, G. (1985). Systems analysis: A genetic approach. In Miser and Quade (1985), pp. 33–66.

————, and E. S. Quade, eds. (1980). *Pitfalls of Analysis*. Chichester, England: Wiley.

Meadows, D. H., and J. M. Robinson (1985). *The Electronic Oracle: Computer Models and Social Decisions*. Chichester, England: Wiley.

Merkhofer, M. W. (1987). *Decision Science and Social Risk Management: A Comparative Evaluation of Cost-Benefit Analysis, Decision Analysis, and Other Formal Decision-Aiding Approaches*. Dortrecht, The Netherlands: D. Reidel.

Miser, H. J., and E. S. Quade (1985). *Handbook of Systems Analysis: Overview of Uses, Procedures, Applications, and Practice*. New York: North-Holland. Cited as the *Overview*.

Moder, J. J., and S. E. Elmaghraby (1978). *Handbook of Operations Research* (2 vols.). New York: Van Nostrand Reinhold.

Mood, A. M. (1983). *Introduction to Policy Analysis*. New York: North-Holland.

Morse, P. M. (1957). Operations research is also research. In *Proceedings of the First International Conference on Operational Research* (M. Davies, R. T. Eddison, and Thornton Page, eds.). Baltimore, Maryland: Operations Research Society of America, pp. 1–8.

Overview. See Miser and Quade (1985).

Quade, E. S. (1980). Pitfalls in formulation and modeling. In Majone and Quade (1980), pp. 23–43.

———— (1982). *Analysis for Public Decisions*, 2nd ed. New York: North-Holland.

Raiffa, Howard (1968). *Decision Analysis*. Reading, Massachusetts: Addison-Wesley.

Rivlin, Alice (1971). *Systematic Thinking for Social Action*. Washington, D.C.: The Brookings Institution.

Schelling, T. C. (1968). The life you save may be your own. In *Problems in Public Expenditure Analysis* (Stuart Chase, ed.). Washington, D.C.: The Brookings Institution, pp. 127–176.

Scioli, F. P., Jr., and T. J. Cook (1975). *Methodologies for Analyzing Public Policies*. Lexington, Massachusetts: D. C. Heath.

Skolnick, Jerome (1967). *Justice without Trial: Law Enforcement in Democratic Society*. New York: Wiley.

Smith, B. L. R. (1966). *The Rand Corporation: Case Study of a Nonprofit Advisory Corporation*. Cambridge, Massachusetts: Harvard University Press.

Strauch, R. E. (1974). *A Critical Assessment of Quantitative Methodology as a Policy Analysis Tool*, P-5282. Santa Monica, California: The Rand Corporation. Also published later (1976): A critical look at quantitative methodology, *Policy Analysis* 2, 121–144.

——— (1975). "Squishy" problems and quantitative methods. *Policy Sciences* 6, 173–184.

——— (1980). *Risk Assessment as a Subjective Process*, P-6460. Santa Monica, California: The Rand Corporation.

——— (1981). *Strategic Planning as a Perceptual Process*, P-6595. Santa Monica, California: The Rand Corporation.

Taylor, V. (1970). How much is good health worth? *Policy Sciences* 1, 49–72.

Tomlinson, R. (1981). Some dangerous misconceptions concerning operational research and systems analysis. *European Journal of Operational Research* 7, 203–212.

Tribe, L. H. (1972). Policy science: Analysis or ideology? *Philosophy and Public Affairs* 12, 66–110.

Vickers, Geoffrey (1981). Systems analysis: A tool subject or judgment demystified? *Policy Sciences* 14, 23–29.

von Winterfeldt, D., and W. Edwards (1986). *Decision Analysis and Behavioral Research*. Cambridge, England: Cambridge University Press.

Wagner, H. M. (1975). *Principles of Operations Research with Applications to Managerial Decisions*, 2nd ed. Englewood Cliffs, New Jersey: Prentice-Hall.

PART IV
SOME COMPONENTS
OF ANALYSIS

Chapter 9
Forecasting and Scenarios

Brita Schwarz

9.1. Introduction

Every systems-analysis study is oriented toward the future, since it deals with the consequences of decisions not yet taken and alternatives not yet adopted. Thus, assessing these consequences is a kind of forecasting. The discussion of forecasting in this chapter, however, is confined to forecasting the environments in which the various decisions and alternatives will be implemented and have their effects.

The environment for a problem situation is the portion of the surrounding world that interacts with it. For example, in the Dutch study of alternative ways of protecting the Oosterschelde estuary from flooding (see Section 2.2 and the *Overview,* pp. 89–109), an obvious part of the environment was the possibility of superstorms and their severity and frequency. Other environmental factors were the fishing and shipping industries, the recreational use of the estuary area, the human activities potentially protected from flooding, and the general economic activity of the Netherlands.

Possible changes in the environment may affect all stages of a systems-analysis study: the conception of the problem, the design of decision alternatives, the operational interpretations of the objectives, and the estimates of the impacts of the alternatives. Forecasts of the environment are therefore always an important part of the work.

Explicit forecasts, however, are not always necessary. For example, there may be no need for environmental forecasts when the time period is fairly short, when the environment is stable and can be expected to remain so, or when the flexibility of the decision alternatives either allows them to adjust readily to changing environmental conditions or makes them insensitive to such changes. For example, in the study of blood availability and utilization in the Greater New York area (see Section 2.2 and the *Overview*, pp. 68–79), the environment was fairly stable and the alternative adopted was designed to respond promptly to changes in it, whether large or small.

Forecasting techniques have developed rapidly since the 1950s and their results have become widely used. Recently, many comparative evaluation studies have contributed important insights into the accuracy and adequacy of the various approaches (see, for instance, Armstrong, 1985, and Ascher, 1979). Too, the field of futures research has entered the scene and with it a broader interest in possible, probable, and desirable futures. An overview of these developments, however, is outside the scope of this chapter. Instead, the main interest here is in approaches, experiences, and pitfalls in dealing with environmental changes from the viewpoint of the systems analyst.

The chapter first discusses the possibilities and limitations of future-oriented knowledge, and some controversies regarding it. Then it takes up some categories of development in forecasting and scenario writing. The next-to-last section uses a major study of world energy supply and demand to illustrate some of the main points, and the last section discusses the strengths and weaknesses of scenarios.

9.2. Future-Oriented Knowledge

Opinions differ on the "knowability" of the future. In systems analysis this issue is crucial on several levels. On one level, there are the concerns about knowledge of future changes in the environment of the problem area being investigated as well as about the future impacts of the decision alternatives. On a methodological level, there is the problem of what knowledge can be gained about the adequacy and reliability of various techniques and approaches for dealing with future environmental changes.

The idea that the future is unknowable is usually connected with the view that knowledge must consist of well-founded statements about observable phenomena. As future events cannot be observed as long as they remain in the future, we cannot have knowledge about them in this sense. When futurists argue that the future is to some extent knowable, however, they are either using a different concept of knowledge or relaxing the requirement for a direct and close empirical connection, or both. For instance, Edward Cornish, a president of the World Future Society, claims that some statements about the future may be at least as well founded and certain, if not more so, than a statement such as "Christopher Columbus discovered America in 1492" (Cornish, 1977, p. 95).

For many of the phenomena that have been studied by natural science we have theories allowing us to make good predictions. For instance, we can make accurate predictions of the future movements of the planets and of the outcomes of many chemical reactions. On the other hand, our knowledge of the future of complex social systems, even in the near future, is much less accurate and more conditional—but our knowledge of the past and present of such systems is often similarly incomplete.

Another dimension of controversy concerns the extent to which the future is predetermined. Are there laws governing the development of all aspects of social evolution? A belief in a completely predetermined future is not uncommon, but it is not the majority view among forecasters or futurists:

> Regrettably the discipline of forecasting is widely misunderstood, not only by the public but by many practitioners. The idea still lurks both in our culture and in other cultures that the future is preordained—it exists, though hidden from ordinary mortals.
>
> Since the future is conceived as indeterminate to some degree, a forecaster deals with possibilities and probabilities, not certainties. [Ayres, 1979, preface.]

Similar messages come from the futures field. It has, for instance, been argued that the term "futures research" is better than "future research" because "it conveys a crucial point: the aim is not to prophesy what the actual future will be but rather to rehearse futures accessible to political choice" (Calder, 1967). And Roy Amara, president of the Institute for the Future in the United States, considers as basic premises of the futures field the existence of alternative futures and the possibilities of influencing future outcomes by individual choices (Amara, 1980). These are also among the basic premises of systems analysis: if the future is completely predetermined, systems-analysis efforts to improve decisionmaking are futile.

Let us now look at some consequences of these premises for the possibility of gaining knowledge about the future development of the environments of the alternatives studied by systems analysis.

Obviously, in some cases—a stable environment, a short time period—it is possible to make fairly accurate predictions. In other cases, only very conditional forecasts, or forecasts with unknown accuracy, may be made. In the latter cases, future developments are sometimes described in the form of a set of scenarios that outline possible, but not necessarily the most probable, developments. Descriptions of future events and developments, however, can be of many different kinds. A study that compared the methods of three futures study projects found the following different types of "futures" (Schwarz et al., 1982):

Developments that are probable under certain conditions.

Desirable developments (normative projections).

Boundary developments (delimiting a space of probable or possible developments).

Reference cases, obtained, for example, with the help of trend extrapolations.

Chains of events, designed, for example, in order to illustrate possible causal relations.

This leads us to the second type of knowledge problem: What knowledge do we have, or what knowledge can be gained, about the adequacy and reliability of various techniques and approaches for dealing with future environmental changes in systems analysis?

Comparative evaluations are difficult to make because of the heterogeneity of systems-analysis projects, and the *ad hoc* nature of this work. There is one field, however, with somewhat more homogeneity and repetitiveness: strategic business planning. Here the problem situations are embedded in a natural system, the firm, and the environment consists of various factors influencing the development of markets, of the behavior of competitors, and so on. Strategic planning and management have been introduced in many firms since the beginning of the 1970s, and an extensive amount of relevant research has been carried out. So, in this field some comparisons can be made about the adequacy and reliability of techniques, and thereby some guidance may be offered for other areas of systems analysis. Apart from this, heterogeneous systems analysis must rely on rough assessments obtained by looking at the approaches that have been used and identifying such pitfalls as they may have exhibited.

9.3. Forecasting One Variable

The use of formal forecasting methods has a long tradition in some fields, such as demography, energy, and econometrics. More recently, we have seen the development of quite sophisticated techniques, some available as computer software packages that facilitate their widespread use. Developments such as the Delphi technique for using expert judgments (see Section 3.7 and 3.8, and Linstone and Turoff, 1975) and the Box-Jenkins time-series-analysis approaches (Makridakis, 1976) have had major stimulating impacts on the forecasting field.

To some extent, the developments in different areas have followed parallel streams without much intercommunication; for example, economic, social, technological, and political forecasting have developed as more or less separate fields of knowledge, as is evident from existing handbooks and overviews: *The Handbook of Forecasting: A Manager's Guide* (Makridakis and Wheelwright, 1982) deals mainly with economic forecasting; *Technological Forecasting for Decision Making* (Martino, 1983) is a major work on technology forecasting; and *Strategic Planning and Forecasting—Political Risk and Economic Opportunity* (Ascher and Overholt, 1983) focuses on political forecasting, but it also includes perspectives and approaches of interest for other fields.

Short-term forecasts are, naturally, often more accurate than long-term ones. However, the time period over which (fairly) reliable forecasts can be made varies considerably among different fields. National demographic forecasts, for instance, are usually made for 10 to 20 years; they have, historically, often been accurate enough for many purposes, although during some periods less accurate than the forecasters have expected (Ascher, 1979). In some other areas, long-term forecasting is inevitably more difficult. Macroeconomic forecasting is usually limited to a few years. Forecasting the time interval within which a specific technological breakthrough can be expected to occur is inherently difficult in case most of the knowledge needed is not already available; if this knowledge were available, the breakthrough would be likely to occur in the near future and forecasting would become an easy matter. There are also areas where accurate long-term forecasts are hardly possible because the future largely depends on individual and societal free choices, which—fortunately—are not predetermined.

A systems-analysis study is usually carried out by a team that includes workers from a number of different fields, but its expertise may not include forecasting environmental changes. Sometimes existing forecasts are used, or experts outside the analysis team are called on. In such cases, it is useful to have existing forecasting methods divided into categories in a way that facilitates choosing external forecasts or outside experts and clarifies the assumptions on which the forecasts are based. Such a classification is difficult, however, since forecasting methods differ in a number of different dimensions. The discussion that follows in this section examines some categories and takes up some issues of importance when external forecasts are used in systems-analysis work.

Judgmental Methods

All forecasting methods include elements of judgment, as regards, for instance, the choices of data base, procedures, time intervals to be covered, and so on. The method is called judgmental, however, only when the values forecasted are obtained from individual opinions. Such a forecast may be made by one person or by a group—for example, by a committee, by a survey, or by the Delphi method.

There are a number of advantages in using judgmental methods: They offer the possibility of gaining access to the "right" experts, they are flexible, and usually inexpensive. On the other hand, a great deal is also known about the disadvantages: various types of bias, sensitivity to the choice of experts, effects of the way questions are worded, and so on (Evans, 1982; Hogarth and Makridakis, 1981). For the systems analyst, judgmental methods may be adequate when it is important to use outside expertise and when the need for accuracy is not very great. But to use the results of judgmental forecasts made elsewhere may be quite risky; the contexts, hidden assumptions, and biases of such forecasts may not be easy to ferret out.

Extrapolative Methods

Extrapolative methods use data about the past values of the forecasted variable as the basis for future forecasts; and they are based on the assumption that the future will evolve in accordance with the pattern of historical behavior. This class of methods includes the frequently used trend curves, as well as decomposition methods, exponential smoothing, and the Box-Jenkins (or ARIMA) models. There are several surveys of extrapolative methods (Fildes, 1979; Makridakis, 1976, 1978). According to a number of comparative evaluations, they often give good accuracy in short-term forecasting, but less so in medium- or long-term forecasts. And the more sophisticated techniques do not necessarily give better accuracy than simpler approaches (Armstrong, 1985).

The presentation of an extrapolative forecast usually includes estimates of its accuracy, but the underlying assumption—that the pattern of the past will continue into the future—is less often clearly spelled out. When new causes of change enter the picture, which is often the case in long-range forecasting, an extrapolative forecast and its estimates of accuracy can become badly wrong, and thus can easily mislead its user. The systems analyst may, however, still find it useful when interpreted merely as a reference projection.

Theoretically, the extrapolative forecast answers this question: If this data series continues into the future for a certain number of years in accordance with the pattern identified for the past, what values will the variable attain at different points of time during this period?

Causal and Mathematical Models

The linear single-equation regression model is one of the most frequently used types. The variable to be forecasted, the dependent variable, is assumed to be determined by its own previous values, as well as by a number of "causes" or exogenous factors. A regression analysis determines the relation and it is assumed to be invariant over time. In this process the dependent variable is forecast by making assumptions about, or by forecasting, the values of exogenous factors in the future.

A well-known pitfall here is to infer causality from covariation over some previous period of time (see, for example, the discussion of correlation and causality under Action 11 of Section 11.4). Stumbling into this pitfall can lead to large forecasting errors, particularly in long-term forecasts. There are many other forms of "causal" models—such as nonlinear equations, sets of equations, complex econometric models, and so on—but there is always the risk that a mistake about true causality, or the invariance of its impact, will lead to serious forecasting errors.

The types of quantitative forecasting models are so numerous and varied that they cannot be easily divided into homogeneous classes. For example, there is a broad range of models called mathematical models. Sometimes this is just a synonym for quantitative models, but usually it refers to models of specific mathematical structure, such as system-dynamics models, input-output models, mathematical-programming models, and so on. An obvious pitfall here is to assume that the forecasting accuracy is determined solely by the accuracy of the related data and to disregard errors that stem from inadequacies in the assumed structure and how it relates to the reality that it represents. This is, of course, a special case of a problem that exists for all models in systems analysis.

Discontinuities and Breaking Trends

We know from history that discontinuities and sharp changes in trends sometimes occur, but most classes of forecasting models do not allow for such possibilities. One approach is to use expert judgments to identify possible new events, such as technological breakthroughs, natural disasters, or important policy turning points. The class of cross-impact models consists of approaches in which probabilities of new events and the interdependencies of these probabilities are estimated subjectively (see Section 3.6). The impact of these events on existing trends may also be estimated. Little is known about the accuracies of cross-impact models, but they are fairly widely used because of the insights they yield about interdependencies and dimensions of uncertainty.

Purpose and Responsibility

It is often taken for granted that the purpose of a forecast is to predict the development of some future aspect of reality as accurately as possible. On the other hand, the phenomena of self-fulfilling and self-defeating predictions are well known. The former might occur if facilities were provided to meet an optimistic demand forecast, since the existence of the facilities might create further demand. On the other hand, suppose, for example, that an analyst forecasts an undesirable situation, and, on the basis of this forecast, the decisionmaker acts to prevent the situation from arising. If he succeeds, the forecast does not come true—which, indeed, may have been the purpose of the forecast. That is why, for instance, Martino (1983) argues that the usefulness of a forecast for decisionmaking, and not necessarily its accuracy, should be the criterion of quality for it.

The varying purposes served by forecasting may create problems for the systems analyst who looks for existing forecasts that he can use for taking account of possible changes in environmental factors. A forecast may, for instance, be purposely on the side of optimism or pessimism to be useful in its original decisionmaking context. This sort of intended bias may not be easy to discern.

For the external user of a forecast it is sometimes difficult to know what the responsibility of the forecaster has been. Has his task been to use a certain well-established exploratory method that has been successful earlier? Or has he also tried to look out for new causes of change or used a forecasting method that is able to capture part of an emerging new pattern (such as exponential smoothing, see Carbone and Makridakis, 1986)? If there is a new cause of change that is outside the field of expertise of the forecaster, he may not feel obliged to consider it. Nanus (1979, pp. 288–289) discusses an example of this type of problem; in an essay on interdisciplinary policy analysis in economic forecasting, he quotes an economist as having argued that economists cannot be held responsible for not having foreseen the part of the double-digit inflation of 1974 to 1975 that depended on political factors (such as the Middle East crisis and the oil-price policy of the OPEC cartel).

9.4. Scenarios

If someone happened to overhear the word scenario in a conversation thirty years ago, his first thought would probably have been that the conversation was about a film strip or theater play. But today the associations might be quite different. The discussion could be about a hypothetical future development that might cause a threat or create new opportunities, or, perhaps, about a desirable future for the world, a nation, or some sector of society.

Today, the term *scenario* belongs to the professional vocabulary in such areas as systems analysis, planning, and futures research—and as such it has become a loan-word in several languages other than English. It has a variety of meanings, however, a fact that poses a semantic problem for anyone trying to get a grip on the use of scenarios as a methodological approach. The various forms and roles of scenarios can probably best be understood in the light of the historical development of the term, beginning with its first connections with systems analysis and futures studies.

According to the 1955 *Oxford Universal Dictionary,* a *scenario* is "a sketch of the plot of a play; giving particulars of the scenes, situations, etc." or "the detailed directions for a cinema film." From the starting point of this definition, it may seem far-fetched to use the word scenario to mean a description of future developments. There is, however, a connection. It arose in the increasing use of gaming in operations research and planning during the decades of the 1950s and 1960s. Long-range defense planning, in particular, made use of gaming as an important tool; new weapon-system concepts were "tested" in map games before final decisions were taken about development and acquisition. These games were therefore simulations of hypothetical conflicts 10 to 20 years ahead, and the scenarios were the "sketch of the plot" made before a gaming exercise was started. Such scenarios usually consisted of a verbal description of a hypothetical situation at a future point in time; but they could also include a sketch of the main changes and events that were assumed to have taken place in the intervening period.

In the 1950s, defense analysts at The Rand Corporation used the term scenarios in connection with their work, and, as one of the products of this stream of work, Herman Kahn later made the term familiar to a large audience. The book *The Year 2000* (Kahn and Wiener, 1967) defines a scenario as "a hypothetical sequence of events constructed for the purpose of focusing attention on causal processes and decision points." It presents a number of different scenarios, consisting of macrolevel descriptions of alternative developments of society toward the year 2000. An interesting methodological development here was the design of the so-called "surprise-free" scenario, a starting point for constructing alternative scenarios. As Kahn and Wiener point out, a surprise-free stream of development is not very likely; history has produced few periods of any length without various breakthroughs or turning points. Nevertheless, the design of a surprise-free scenario as a reference projection has sometimes proved to be a convenient step when developing a set of scenarios to explore possible futures.

The Kahn and Wiener book was not uncontroversial, but it gave a noticeable impetus to the growth of the new field of futures research. A few years later the Club of Rome's study *The Limits to Growth* (Meadows et al., 1972) appeared, not less controversial, but stimulating its critics to develop other views of global futures and models to support them. Since these authors did not use the term scenario for the futures they presented (which had been developed by computer models), in the field of futures research mathematical models have come to be regarded as the opposite (or main alternatives) to scenarios in the methodological debate. Although this view—or should we say this terminology—is not accepted by all futurists, it is discernible in the literature.

In a survey of scenarios, Wilson (1978, pp. 226–227), discusses the meaning of the term in futures research; among other things, he emphasizes the holistic principle, arguing as follows:

Scenarios are (or should be) multifaceted and holistic in their approach to the future. In the early days of futures research, and still to a great extent in popular literature, it is the isolated event, the specific prediction that rivets attention; also there is a fatal fascination with particular dates in the future, e.g., 1984, 2000, or the [U.S.] tricentennial 2076. However, history is a "booming buzzing confusion of events," trends and discontinuities; it is constantly in motion and so is more accurately represented by a motion picture than by a snapshot. Scenarios have a special ability to represent this multifaceted, interacting flow process, combining (when appropriate) demographic changes, social trends, political events, economic variables, and technological developments.

There has been an extremely rapid growth in the use of scenarios since the early 1970s, and the meanings and uses of scenarios have become increasingly varied. As a consequence, however, misunderstandings and communication problems about scenarios can arise easily. To sort out this confusion somewhat, it can be useful, for example, to make distinctions:

between the traditional conception of scenarios in systems analysis and the more recent broader scenario concepts in systems analysis, strategic planning, and futures studies;

between the forms and roles of the scenarios; and

between the various methodological ideas associated with the use of scenarios.

Traditional Systems Analysis and New Developments

In a 1968 survey of scenarios in systems analysis, Brown (1968, p. 299) makes these observations:

There are many notions floating around of what a scenario is or ought to be. More often than not these notions, or attempted definitions, are the product of the specialist's acquaintance with those things which are called scenarios in his special field of work and exclude those things which other specialists choose to label scenarios in their own fields.

Some of the notions of a scenario that he mentions are:

an outline of a sequence of hypothetical events;

a record of the actions and counteractions taken by parties to a conflict;

a specific set of parametric values selected for a given computer run.

Brown notes that they all—disparate as they are—refer to descriptions of the conditions under which the system they are analyzing, designing, or operating are assumed to be performing. He concluded that "whatever the scope and properties of the specific system, a scenario—in systems analysis—can be defined as a statement of assumptions about the operating environment of the particular system we are analyzing'' (op. cit., p. 300). This is still a common conception of a scenario in systems analysis, where the terms context, setting, situation, or environment are often used loosely as equivalents.

Lately, however, systems analysts have become increasingly involved in studies of more complex problem areas. Also, the fields of systems analysis and futures studies have come to overlap to some extent. Scenarios are no longer used solely for descriptions of the environment of the problem situation; they are also used in studies of interactions between the system and its environment and as descriptions of alternative developments of the problem area itself. The scenarios in the International Institute for Applied Systems Analysis (IIASA) energy study (see Section 9.6) are examples of this latter type of use.

Forms and Roles

As can be seen in the examples of scenarios that Brown mentions, a scenario may or may not be in verbal form, may or may not consist of a set of data, and may or may not include assumptions about how various actors act and react.

Scenarios may take on a variety of forms. In fact, this freedom of form is probably one of the reasons for their widespread use. It has become legitimate to use a short narrative or a vividly described chain of events to draw attention, for instance, to threats or opportunities that otherwise might be neglected in a systems-analysis study. In some cases, forecasts from several areas, or knowledge from different disciplines and experts, need to be integrated in a description of, for instance, a possible future development. Freedom of form is here an obvious advantage, as it helps to avoid an unnecessary constraint on this type of knowledge-integration process. A disadvantage is that it may not be easy to assess what knowledge the end product actually represents.

There are a number of different possibilities, not only as to the form of a scenario, but also as to its role or type of future orientation. A scenario may, for example, be a description of:

a hypothetical, likely or unlikely, development or situation;

a development that is described as caused to some extent by the actions and reactions of various actors; or

a desirable or undesirable development or situation.

To some a scenario is, by definition, an outline of a probable or desirable development or situation (see, for example, Julien et al., 1975). Such a scenario may serve, sometimes more or less directly, as a basis for policy conclusions, but it can also lead to identifying problems to be addressed by systems analysis.

Scenarios have also been used to explore developments that are undesirable or dangerous but possibly unlikely. Examples are conflict scenarios for defense planning, and scenarios providing the context for studies of various types of risks.

Scenarios as a Methodological Approach

The situation is still very much the one that Brown described in the 1960s; everyone who has been engaged in scenario work is quite certain about what a scenario is and about its form and role—but these ideas differ. Sometimes a recommendation is made to use "the scenario method." But, if a scenario may be nothing but a verbal description of some hypothetical future development, how can it be regarded as a method?

In their study of the use of scenarios, Schwarz et al. (1982, pp. 28–29) found that the scenario method could be interpreted as usually referring to the application of one or several of a number of methodological ideas:

It is not usually possible to predict the future with precision, but, by designing a number of plausible and consistent descriptions of hypothetical future developments or situations (scenarios), we can delimit the uncertainty space that we want to take into account in studying the problem at hand.

If we are concerned with the future development of a specific system or planning object (for example, a business firm or energy system), it is useful to make explicit assumptions (scenarios) about the future development of its environment. (Otherwise the implicit assumption is often made that there will be no change at all.)

It may be useful to try to synthesize fragmented, dispersed, and sometimes vague knowledge into a holistic and consistent picture (or scenario) of a future development or situation. This is particularly important when there are a considerable number of cross-impact effects between developments in various fields that cannot be taken into account if they are considered separately.

To get a realistic picture of possible future developments in an area where discontinuities and changing trends may emerge as the results of specific events and actions, it can be useful to try to describe a number of hypothetical developments as resulting from the decisions and actions of various actors.

When certain future developments are considered unlikely but dangerous, we may be able to make preparations to render them less probable or less dangerous by making an effort to imagine in some detail how these developments might arise.

Scenarios may take on quite different forms and roles, partly as a consequence of which set of methodological ideas is being applied. Section 9.6 describes a major systems analysis study that used the first of the ideas listed above as the basis for its scenarios. The description points out how a misunderstanding of the role scenarios played in the work played an important role in a debate that followed publication of the study's results.

Methodological ideas that are the reasons for developing and using scenarios should not be confused with methods for developing them. Various such methods, or guidelines, have been developed, but their applicability seems often to be partly confined to the problem area for which they were designed. For an overview of considerations and guidelines for scenario writing that focuses on scenarios for free-form political-military games, but is more general in its orientation, see deLeon (1975).

9.5. The Problem Environment: Approaches and Pitfalls

There are a number of questions that need to be considered in developing forecasts or scenarios for the future environment of a system or problem situation, such as:

How can the external (environmental) variables be identified?

Are there interdependencies among the different variables? If so, how should they be dealt with?

In case uncertainties require several alternatives to be considered, how should the set of scenarios—or alternative futures—of the environment be developed and considered?

What criteria of adequacy should be applied?

Each of these questions is discussed in what follows.

Identifying the External Variables

Various guidelines for the process of identifying external variables have been developed. Experience indicates, however, that a direct copying of an approach developed for one problem situation is seldom appropriate for another; even using an approach developed for an earlier time for the same problem area may be inappropriate for a later time. Approaches used in past analyses can be used as sources of learning and inspiration—and as warnings of pitfalls to be avoided—but they will almost surely need to be extensively tailored if they are to be applied to a new problem situation.

In corporate planning, identifying external variables is usually labeled environmental scanning. This process needs to be adapted to, for instance, the size of the company:

> Environmental scanning in *small companies* is a strategically oriented task that can go far beyond the mere collection of data about markets, competitors, and technological changes. . . . The task of monitoring detailed environmental changes in *large companies* is too difficult to be performed by top management alone. Division managers, therefore, are expected to study the external environment that may be relevant to their particular business. In these circumstances, headquarters typically provides only a few environmental assumptions—mainly economic forecasts. [Lorange and Vancil, 1976, p. 78; emphasis in the original.]

Amara and Lipinski (1983, pp. 49–50) offer an example of an environmental scanning approach in corporate planning:

> To obtain the data from which the important scenario components will be drawn, we recommend that knowledgeable managers in a particular strategic unit (it could be a corporation as a whole or one of its strategic business units) be interviewed. To create a sufficient numerical base for analysis, about 20 interviews are needed. The purpose of the interviews is to establish the relative importance of key external factors needed for evaluating the consequences of options or candidate strategies. . . . To make the data from different interviews comparable, a common interview plan must be followed. Each interview consists of:
>
> The interrogation of the "clairvoyant."
> A report on an unfavorable future environment.
> A report on a favorable future environment.
> The identification of pivotal decisions to be made.
> The identification of current management assumptions about the future environment.

In a study of the airline insurance industry, Vavrin (1980) has reported experiences from a somewhat different interview procedure. In this case, the analyst used a modified Delphi approach, but obtaining adequate material was found to be more difficult than expected, owing to "a lack of planning and future oriented thinking" among the managers. In the public sector, systems analysis problems sometimes cut across the areas of responsibility of several governmental bodies, which may limit the possibility of identifying all important environmental variables by using interview procedures.

Ideas about what kinds of external variables are important have changed over time. During the 1970s it began to be a generally accepted view that there was a need to look out, not only for economic and technological changes, but also for social and political ones (see, for instance, Taylor, 1975).

When systems analysis is applied to public-sector problems, the possibility of environmental changes sometimes poses problems of a somewhat different character than those usually encountered in corporate planning. Such questions as these may arise:

Do the assumptions on which the models in use are based presuppose an unchanging environment?

Is the way the problem has been posed or the objectives operationalized based on implicit assumptions about the environment?

As examples, we can mention the models and approaches used in the fields of energy and transportation in the early 1970s. A frequent, implicit assumption was that the future energy prices were of no importance, or would continue to decline (Asher, 1979).

A study by Mitroff (1979) found that advocates of different policies or decision alternatives sometimes implicitly base their preferences on differing assumptions about the environment. Identifying and challenging such assumptions can thus be an important step for improving decision-making. More generally, the policies to be compared can sometimes be a useful starting point for identifying environmental factors and the assumptions being made about them, whether explicit or implicit (see also Section 7.6).

Interdependencies

If only a few important external variables have been identified, and if they can be adequately forecast, the analysis may have no further problem. But, as an example, let us assume that only high and low estimates can be given for each variable, because of various uncertainties; but, in such a case a low estimate for one variable may rule out the possibility of a development according to the high estimate of another if the variables are interdependent in this way. Thus, unless the analysts are careful to explore such matters, incompatible estimates could lead to impossible environmental futures, which obviously is not what is wanted.

This problem of interdependence was, in fact, the origin of the practice of scenario writing in the 1950s at The Rand Corporation. When new weapon systems were tested in map games, the context was described by a set of assumptions about the politico-military situation. To examine the consistency of these assumptions it was found to be useful to try to write down one—or several—coherent development paths that would lead to the assumed future situation. If no plausible path could be imagined, the conclusion could be drawn that the postulated situation was in fact quite unlikely.

It is sometimes argued that scenarios should be holistic, not only because of the need for consistency, but also for other reasons: to provide information about first warning signals for future branching points, to highlight the cross-impacts of events and trends, and to help bracket the future by means of a set of realistic scenarios.

In a systems-analysis study, there may be uncertainties and interdependencies among some, but not necessarily all, of the external variables. In such a case, it could be useful to limit the scenario writing to the part of the environment that encompasses the interdependent variables, if this is possible.

The problem situation being studied, and therefore the system that it implies, is often thought of as the part of the real world that is more or less under the control of the decisionmakers for whom the study is being made. Although the environment is, by definition, external to this system, and therefore not under the control of the decisionmakers, it is not necessarily unaffected by the decisions that are made, and it may in fact respond to them. Such an interdependence between the problem-situation system and its environment is sometimes studied by using gaming, or by developing scenarios encompassing alternative future developments of both the system and its environment.

A Set of Scenarios

When scenarios are being used in an analysis, one way to delimit the uncertainty space that needs to be taken into account is to select an appropriate set of scenarios, perhaps by varying one or several parameters. The problem here is that the number of combinations increases rapidly with a growing number of parameters. For practical reasons, the number of scenarios—and consequently the number of parameters—must be kept down. As few as one to two parameters and three to six scenarios seems usual in practice (Wilson, 1978). The choice of parameter or parameters is thus of crucial importance, particularly when only one is varied.

One of the possible pitfalls is to choose just one parameter, and not the most important one. Experience indicates that the parameter to be varied is sometimes chosen automatically, without special attention, at least in noninterdisciplinary groups. If, for instance, a technological, economic, social, or political parameter is used, it often seems to be linked to the disciplinary background or to the special interests of the scenario writer, and is in this case not the result of deliberate choice based on a wide-ranging survey. Ideally, the factor to be varied in the scenario should be determined by what kind of uncertainty is the most important for the problem being studied. If the task of the set of scenarios is to delimit a space of probable developments, and if an important and very uncertain parameter has been omitted and not varied, the scenarios may lead to a bias in the study's outcomes.

As Brown (1968) pointed out long ago, the function, form, and content of scenarios have to be determined by the specific research task at hand. This statement also applies to choosing the set of scenarios to be used. Developing a relevant set of scenarios may be accomplished most easily when the needed competence is available in the analysis team. But this happy situation is not always—perhaps not even very often—the case; the team will then need to consult expertise outside its own members (see, for instance, Levien, 1968). If a special scenario group is set up, the composition of this group and good communication between it and the systems-analysis team can be of crucial importance.

A possible shortcut, similar to the case already discussed of forecasts of one variable, is to use existing scenarios, developed for some other problem, but covering roughly the same environment. Such scenarios must be carefully reviewed, and will usually need adaptation to the new problem situation. The obvious pitfalls of borrowing existing scenarios relate to the differing needs in the new study regarding the composition of the set of scenarios, as well as their roles, forms, and content.

Traditionally, systems analysis has been regarded as a problem-oriented inquiry with the systems-analysis team primarily concerned with policy and decision alternatives while relying on outside sources for information about the environment. Increasingly, however, systems analysts have become involved in developing environmental forecasts and scenarios for use not only in their own work but also by planning bodies in business and the public sector. Many journal articles about scenarios—for example, in *Long-Range Planning*—testify to this development.

The increasing use of scenarios—instead of forecasts—since the early 1970s can be interpreted as a response to more unstable environments and to an increasing awareness of the possibility of discontinuities and breaking trends, particularly after the 1973–1974 oil crisis. Consequently, a set of scenarios may consist of a pair bracketing probable developments in combination with others that depict other futures, possibly not very likely but with important implications for the problem being studied. Thus, the various scenarios in a set may describe quite different types of futures. Theoretically, the possibility of this diversity is an advantage of the scenario approach; but, in practice, it may sometimes cause communication problems with the users (see the first protocol in Section 7.6).

Criteria of Adequacy

There is no general theory that allows us to assess scenario adequacy or quality. In the practice of scenario writing, however, it is possible to find a number of criteria that are often referred to. Discussions of criteria are most likely to be found in more elaborate scenario works, or when scenarios are used over longer periods of time in formal planning systems. Such criteria cannot be expected to be universally applicable, but can serve as useful items on a checklist. Here are some of the—partly overlapping—criteria that are often mentioned:

Consistency
Plausibility

Credibility

Rationality

Relevance

Utility

Probability

As already mentioned, the consistency requirement was, in a way, the origin of the first uses of scenarios in systems analysis. Coherent paths of development leading to a hypothesized future situation were sought in order to test whether all the assumed future conditions could be imagined to occur for the same path of development; if not, the assumptions were regarded as nonconsistent.

The need for, or usefulness of, internally consistent scenarios is emphasized in many applications, as, for instance, in the IIASA energy study discussed in the next section. Obviously, if a scenario is perceived as clearly inconsistent, this fact will tend to block its intended integration into the user's pattern of thought or scheme of analysis. But overall consistency may not always be feasible, or even necessary. For instance, a scenario that is perceived as technically consistent may facilitate communication between experts from different technical areas and serve as a tool for knowledge integration, but it may be perceived as inconsistent by, for instance, a social scientist. Such a conflict limits its area of use, but may not necessarily exclude all uses.

It is often stated that a scenario must be "plausible," "reasonable," or "credible." In this context, these concepts are more or less synonymous. Plausibility may seem as obvious a requirement as consistency, but it should not be given too limited an interpretation. As Kahn and Wiener (1967, p. 264) put it:

> Since plausibility is a great virtue in a scenario, one should, subject to other considerations, try to achieve it. But it is important not to limit oneself to the *most* plausible, conventional, or probable situations and behavior. History is likely to write scenarios that most observers would find implausible not only prospectively but sometimes, even, in retrospect. [Emphasis in the original.]

A "rationality" criterion is sometimes used for testing scenarios describing the actions and reactions of various actors. Rationality, however, has different connotations. In the sense of consistent preferences it might not be a useful criterion, as scenarios often do not include sets of decisions of such a similar nature that preferences can be compared. And rationality, in the sense of optimal decisions under perfect information, is not applicable when the information available in the scenario is described as being imperfect. In fact, it may be "reasonable"—one sense of rational—

to allow what is usually regarded as irrational behavior if, for instance, an actor is described as a fanatic.

In sum, rationality, plausibility, and consistency can all be regarded as different facets of a more general credibility criterion. Relevance and utility are criteria of a different kind, but, like credibility, hardly controversial. Certainly a scenario must be useful, and its form, role, and content should be relevant to the problem at hand.

However, the question of whether or not a scenario depicts, or should depict, a probable development is often a source of misunderstanding. As already mentioned, it is not always the most probable future that is of interest; in some problem areas, scenarios are used to describe environmental developments that may be unlikely but have important, possibly dangerous or highly desirable, impacts. Even when the "probable" future environment is sought, each scenario may represent a more or less extreme and unlikely case, the intention being to make the probability high that the actual future will lie somewhere in between. Thus probability may be irrelevant as a criterion of adequacy for a single scenario but relevant to judging the adequacy of the span of a set.

A kind of relevance criterion may also be needed for selecting a set of scenarios. Do they, taken together, represent adequately enough the environmental uncertainties, possible first warning signals, and so on? A relevance or comprehensiveness criterion for the set may here come into conflict with the seemingly obvious criteria mentioned above for the individual ones. A scenario might have to include a certain amount of detail to be easy to use—but these details can endanger the comprehensiveness of the set of scenarios. (For additional discussion of the use of scenarios, called alternative futures there, see Section 7.6, particularly protocols one and four.)

9.6. An Example of Scenarios: A Global Energy Study

The international oil crisis of 1973–1974 raised questions about the security of oil imports and about how the world's growing energy needs could be supplied. Furthermore, these questions arose in a climate of public conflict over the long-term safety and potential of nuclear energy and over whether or not energy consumption could grow without seriously degrading the environment. And behind all these concerns there lay the realization that the world's supplies of oil, gas, and coal are limited—and could well be seriously depleted within the foreseeable future, thus forcing the world to turn to renewable resources for its energy needs, but on a scale and schedule that had not yet been foreseen, or even comprehensively studied.

Thus, it was natural for the International Institute for Applied Systems Analysis (IIASA) in Laxenburg, Austria (founded in 1972) to turn its early attention to this subject, and it began what emerged as a comprehensive inquiry into the issues related to the problem of meeting the world's energy needs over the half-century from 1980 to 2030. The analysis was completed by the end of the decade, and its findings were reported in a series of three publications: Energy Systems Program Group (1981a, 1981b, 1981c). In what follows, these reports will be referred to as ESPGa, ESPGb, and ESPGc, respectively. The first is a volume of 250 pages aimed at the general reader interested in the questions addressed, the outline of the approach, and the major findings; the second, *Energy in a Finite World: A Global Systems Analysis,* is a volume of 830 pages covering the work and its results in considerable detail (and giving complete references to the large number of supporting reports and working papers published both at IIASA and elsewhere); the third is a shorter executive report of 74 pages that offers an overview of the study and its principal findings.

Scenarios lie at the core of the IIASA energy study and form an important part of the structure underlying its findings. Since this case exemplifies many of the important problems and issues relating to scenario development and use, the purpose of this section is to describe the scenario work and how it supported the results.

In order to establish the connections that will support the discussion of this use of scenarios, it is necessary to offer a brief description of the aspects of the study in which the scenarios are embedded; references to appropriate sections of the Energy Systems Program Group publications will tell the reader where more comprehensive information can be found.

Before proceeding further, some special characteristics of the IIASA energy study should be noted. In comparison with how systems analysis is often described, it is atypical in some respects. The future development of the world's energy system is not under the control of a single decisionmaker, or even a small group of decisionmakers, so the alternative developments that were studied were not clearly defined decision alternatives. On the other hand, the study was typical as regards the approaches used and the clear problem orientation. If the yardstick of comparison is systems analysis as it is actually practiced, however, the energy study is hardly atypical; the relationship to decisionmakers and clearly defined decision alternatives usually becomes less close the broader the problem being addressed.

As Section 9.3 mentions, scenarios in systems analysis are usually descriptions of the environment of the problem situation that is being studied. The IIASA energy study includes macroscenarios of this type; they describe population and economic growth in different world regions. In addition, several scenarios were also developed for the world energy system itself, and they were linked to the macroscenarios by iterative procedures.

The Focus and Approach of the IIASA Energy Study

Since the IIASA energy study was conceived and its design laid out about a decade ago, to understand its focus, some comments on the problem conceptions and the level of knowledge during the mid-1970s are useful.

A major criticism of the global model used in *The Limits to Growth* (Meadows et al., 1972) was its disregard of the large differences between the developed and developing countries. The later global modeling projects increasingly focused their attention on the disparities between the development patterns and resource uses of different countries and world regions. The Bariloche model, for instance, used as its point of departure a "desirable future" with an equitable allocation of resources within and between different countries (Herrera et al., 1976); the Leontief study analyzed the economic growth and international aid needed to diminish the gaps between the developed and developing countries (Leontief et al., 1977).

At that time, the usual approach in energy forecasting was to make separate energy supply and demand projections (see, for example, WEC, 1978). Usually the demand was found to exceed the supply. In the early 1970s the gap was interpreted as a need to increase oil production in the Middle East, but by the mid-1970s the conclusions were different; limitations of supply could lead to lower economic growth and cause disturbances in the energy market. Suggested remedies were increases in energy supply through increased investment, decreased reliance on oil imports, and energy conservation. However, the analysts rarely addressed questions of how various measures would affect the economy or what implications could be drawn from the long-term limits on the supplies of fossil fuels.

The IIASA core team was comparatively small, but many researchers from different parts of the world participated in the energy program for various periods of time. Contacts were also established with a number of national and international energy research groups. The team's intent was to complement other energy studies, particularly by providing a long-range global view.

Several reasons prompted the search for a long-range global perspective. Estimates of world population and economic growth, especially possible and desirable economic growth in the developing countries, indicated a significant growth in energy demand. The limited supplies of oil and natural gas, which in the mid-1970s supplied two-thirds of the world's primary energy, will therefore force a conversion to other sources, a transition seen as slow on the basis of the historical experience that in the past transitions to other mixes of primary energy carriers have been slow. As the life of a power plant averages 30 years, the time span needed to consider for major innovations may well be 50 years or more. The IIASA team chose to study the period from 1980 to 2030, and focused their work on whether available physical resources will restrain economic growth during this period and whether feasible transition paths could be found to more sustainable, equitable, and resilient energy systems.

Since the world's energy situation is characterized by wide regional disparities in demand, supply capabilities, economic development, and natural energy resources, the analysis team wanted to take account of these differences. To consider all such factors would complicate the work beyond the bounds of feasibility; therefore, the team simplified these considerations by dividing the world into seven regions, chosen so as to be relatively homogeneous with respect to energy consumption, energy resources, and economic structure (and not necessarily with respect to geographic proximity).

This division of the world into seven regions allowed intraregional variables to be considered separately (a very important property in view of their great differences in value) and interregional trade (an essential feature of the world's future energy postures) to be factored into the analysis.

The analysis team's approach consisted of several phases:

Reviewing in detail the world's resources, fossil as well as nuclear, solar, and other renewables.

Studying the constraints that limit changes in the world energy system.

Balancing energy supply and demand.

Considering the future perspectives illuminated by previous steps and formulating their implications.

The scenario approach entered explicitly at the third phase, but work on the different phases proceeded partly simultaneously, as one would expect.

At the time the work was done, the general energy debate was often confused by unclear distinctions between reserves and various resource concepts. Reserves are resources that are known explicitly and that can be extracted with economic feasibility. Resources include reserves, but also sources that are presumed to exist with a certain probability based on geological evidence. The IIASA study surveyed the available information about resources; it considered fossil resources (coal, gas, and oil, with the oil resources divided into three different production-cost categories), nuclear options, large-scale solar energy capture, and other renewable sources (biomass, hydropower, wind, ocean currents and waves, ocean thermal electric conversion, heat pumps, geothermal energy sources, and tides). For details, see ESPGb (Chapters 2–7).

The constraints considered in the second phase fell into these areas: market penetration; resource use (water, energy, land, materials, and manpower); climatic impacts of energy systems (from releases of waste heat, gases, particulates, and carbon dioxide, and from the effects produced by deploying large-scale solar energy systems); and risk. For details, see ESPGb (Chapters 8–12).

The Approach to Scenarios

In order to define the scope of the IIASA energy study sufficiently narrowly to make it feasible, the analysis team made certain important basic assumptions that had an important bearing not only on the study as a whole, but also on how the scenarios used in it were constructed (ESPGb, p. 8):

> A completely comprehensive study of energy problems of the future is impossible. One would have to list all conceivable eventualities and trace out their structures and consequences. This is manifestly an infinite task. [Therefore,] . . . we made a number of assumptions, which helped to reduce the job to a finite one.
>
> First, we limited the constraints on solutions to the energy problem to those that are physical or structural. Political and social constraints are recognized, but were not applied explicitly. This allowed us to explore the entire range of the possible.
>
> Second, our study assumed a surprise-free future. No major catastrophes such as wars or large local upheavals were assumed to take place. We also assumed no positive technological breakthroughs of a nature that cannot be expected today. This was not meant to exclude such positive unforeseen breakthroughs, but we did not want to rely on them.
>
> Third, we assumed in general only a modest population and economic growth. To keep the energy problem manageable, we assumed major energy conservation and aggressive exploration for additional energy resources. We also assumed a functioning world trade in oil, gas, and coal such that the needs of the various parts of the world can be taken care of.
>
> Fourth, we assumed in all evaluations of the study that the U.S. dollar, and any other monetary unit, has a constant value. This is a more sweeping assumption; it amounts to decoupling the terms of trade from the side effects of inflation. In effect, we note that these problems are of a social and political nature, so that ignoring them is consistent with the first assumption.
>
> We emphasized as much as possible the economic and energy growth of less industrialized as compared with more industrialized countries. This was done on a per capita basis as well as on a national or regional basis.

This narrowed scope still allowed an infeasibly large set of possibilities, so that further choices were necessary in order to bring the prospective work into the range of feasibility, a remark that is particularly relevant to the question of how many scenarios could be considered, and how they were to be chosen. The analysis team adopted this view of using scenarios in their work (ESPGb, pp. 423–424):

> A scenario, as used here, is a logically consistent statement or characterization of a possible future state of the world. Often the scenario statement also specifies a logical sequence of events that could transform the reference or base year state into the postulated future state. The postulated future state can represent the consensus of many experts or be outrageously absurd, provided that it is internally consistent and follows from the assumptions made. A scenario, therefore, is not a prediction, but simply one of infinitely many future states that might happen.
>
> Scenario development and analysis can serve many varied purposes. Depending on one's purpose, one can be interested in scenarios that are most unlikely to happen—as, for example, in contingency planning or developing emergency preparedness. In contrast, for the purpose here, scenarios are used to provide synthesis for conceptions and projections about a wide variety of global energy issues in the long term. This is above all a learning process; the scenarios are an aid to this process. As the primary goal of this quantitative analysis is to learn about the smooth evolution of energy demand and supply, discontinuities, abrupt changes, and radically changed lifestyles have been excluded from the scenario set.
>
> The development of scenarios is necessarily subjective. Certain assumptions must be made that cannot be defended rigorously. Choosing what to make assumptions about is even more subjective. While acknowledging this subjective input to the scenario definitions of this [study], we offer no apology; it is necessary in order to serve the purposes of synthesis and learning. It should also be understood that these projections do not necessarily represent either the most preferred future state of the world or the most probable.

Developing the Scenarios

Having gained insights into and understandings of the supply possibilities and their constraints, as well as a clear idea of the role that scenarios should play in the analysis, the team could turn its attention to the demand for energy that would have to be contemplated over the half century under consideration. The main factors determining energy demand can be divided into four categories:

Population growth, differing markedly in different regions.

Economic growth, also varying from region to region.

Technological progress in the processes and machines involved in energy conversion.

Structural changes within the regional economies.

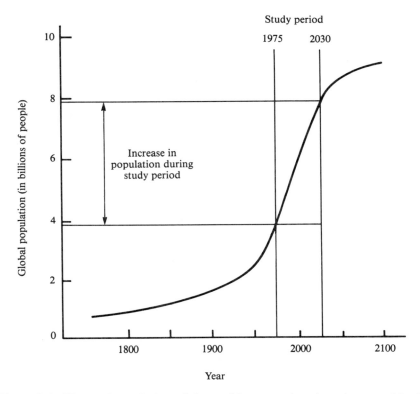

Figure 9.1. The total population of the world, past and projected, as used in the IIASA energy study. (*Note:* One billion here is 10^9.) (*From ESPGc, p. 4*).

The first of these factors was treated somewhat separately, but the others are so intertwined that an interactive treatment was found to be necessary.

World population. The assumptions about population growth were based on studies by Keyfitz (1977), who envisaged a doubling of the world's population from four billion in 1975 to eight billion by the year 2030, but a leveling off in succeeding years, as Figure 9.1 shows. Population growth estimates were also made separately for the seven world regions, and these estimates were used in all of the scenarios used in the analysis. These observations about the population estimates have important bearings on the analysis.

The population projections are based on the assumption that a bare replacement level of fertility will be achieved in developed regions by 2015.

As a stable population is approached, an important change occurs in the age distribution: the older portion of the population increases in size relative to the younger portion. This change decreases the number of persons in the population who must produce, in an economic sense, for their children and for one older person in the population who has retired from economic production.

The rapidly increasing populations in developing regions imply increasing urbanization, a fact that increases energy use and concentrates environmental pollution.

It follows that, although population is an important driver of energy demand, the status of the population segments is also important.

Economic growth. With estimates of the populations in each region in hand, the team's next step was to estimate the economic development each region would experience, as measured by the growth in its gross domestic product (GDP). In making projections of economic activity, a range of values was selected for examination in order to reflect the large uncertainties associated with making such projections and to avoid having a single estimated projection being mistaken for a prediction. A high and a low alternative for the growth rates of GDP per capita were used to define two different scenarios. The basic criterion, however, for selecting the set of scenarios was not economic growth differences; rather, it was related to energy demand. The process of arriving at the preliminary scenario assumptions was a complicated one demanding a variety of input information:

> The assumptions made range from those concerning near-term energy demand in economies where data are voluminous, if not always directly useful, to such things as the appropriate room temperatures in India in 2010, where available data prove less helpful. The object in each instance is, first, to make the best use of available data, analytic techniques, and collective human wisdom to produce reasonable numerical values and, second, to organize the resulting numbers systematically. Our basic method for meeting these objectives involved the writing of two quantitative scenarios. These two benchmark scenarios are labeled "High" and "Low," the former referring to a situation in which the demand for energy is relatively high, the latter to one in which demand is relatively low. [ESPGa, p. 19.]

Later on, the analysis team also considered three variations of these two scenarios, although in less detail: a worldwide nuclear moratorium case, an enhanced nuclear development case, and a very-low-energy-demand case.

Energy supply and demand. One might naively imagine that the population estimates, combined with the high and low GDP estimates, would establish two scenarios in which energy supply and demand considerations could be fitted easily to calculate results on which to base broad findings. The actual situation, however, was much more complicated. Many factors, such as technical possibilities and constraints mentioned earlier, the costs, maximum build-up rates, resource and production limits, interregional trade, and so on, must be considered carefully if a feasible dynamic scenario is to be evolved—that is, one that not only represents a calculated result but also one that can be achieved by a practical path of decision and action. Thus, the preliminary estimates of economic growth and energy demand had to be subjected iteratively to a process of adjustment, as outlined in Figure 9.2.

This process, as the figure indicates, combined formal mathematical models implemented on a computer with intermediate steps in which the analysts could view the results, consult appropriate experts, and decide how to pursue the adjustment process. (They were thus in line with the advice against very large, comprehensive computer models and in favor of a series of more partial computer models offered in Section 2.8.)

To understand the different steps outlined in Figure 9.2 it is important to be familiar with the distinctions relating to energy at various stages of conversion and use (ESPGb, pp. 5–7):

> *Primary energy* is the energy recovered from nature—water flowing over a dam, coal freshly mined, oil, natural gas, natural uranium. Only rarely can primary energy be used to supply *final energy*—energy used to supply energy services. One of the few forms of primary energy that can be used as final energy is natural gas, which is why it is a fuel of preference whenever it is available.
>
> For the most part, primary energy is converted into *secondary energy*—defined as an energy form that can be used over a broad spectrum of applications. Examples are electricity, gasoline, . . .
>
> Primary energy is converted into secondary energy in several ways. Central power plants produce electricity and sometimes district heat. Refineries convert petroleum, which is not an easy fuel to use at the end point, to more convenient liquid fuels—gasoline, jet fuel, diesel oil, and naphtha. When gasoline is not available, coal conversion plants can make liquid fuels. Sometimes the conversion plant is the end point of a system, as with nuclear fission energy. . . . [The] final steps are the conversion of secondary energy into final energy—the energy in a motor, a stove, a computer, or a lightbulb—and of final energy into useful energy—the energy actually stored in a product or used for a service.

The calculations in the model called MEDEE-2 lead to estimates of final energy consumptions in three macrosectors: transportation, household service, and industry (agriculture, construction, mining, and manufacturing). A further step involves the calculations of secondary energy

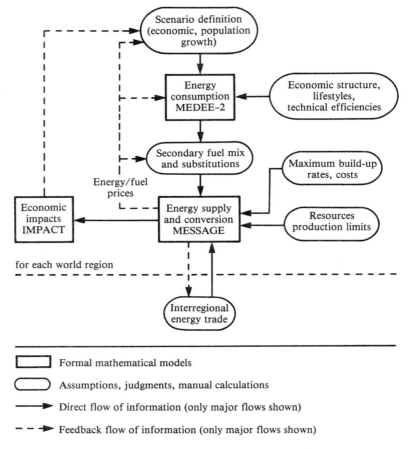

Figure 9.2. The structure of the adjustment procedure used for the scenarios in the IIASA energy study. (*From ESPGb, p. 401.*)

demand required as input to the energy supply and conversion model MESSAGE. This model (a multiperiod linear programming model) calculates the supplies of primary energy required to meet the secondary energy demands at lowest cost and within various constraints on resource availabilities, energy production, and the rates of buildup of new energy facilities.

Some flavor of the scenario development process can be gained from this very brief overview of it (ESPGb, pp. 396–397):

> The analysis begins with postulates of basic development variables for each of the seven world regions to the year 2030. The two basic development variables are population growth and economic development (as

measured by growth in gross domestic product—GDP). Two sets of estimates are made in order to bound from above and below the set of likely developments over the next fifty years. These estimates may be (and are, in practice) modified as further energy-related insights are gained. . . .

Given the initial scenario-defining estimates of population growth and economic development, consequent energy consumption patterns are calculated for each of the regions. While the objective here is aggregate energy demand consistent with the postulated economic development over fifty years, each region's energy consumption is examined in considerable detail. Key assumptions about technology, lifestyle, and penetration of specialized energy sources (e.g., solar, district heat, heat pumps) for each region and for each time interval drive the estimation procedure. The details enable a more careful examination of the linkages of energy and the economy and help to keep straight the many and complex end uses of energy. In this way, some notion can be gained of the requirements for reducing energy consumption growth. . . .

These aggregate energy consumption projections provide one kind of input to energy supply system analysis. Adding constraints that act upon the total energy supply system—estimates of the amount, rate, cost, and time of availability of each prospective energy source—enables calculation of alternative energy supply mixes for the future. . . . Water availability, hostile local environment, difficult and costly transportation requirements, skilled labor limitations, and other local or regional conditions must also be incorporated in some way. . . . [I]n many cases it is necessary to select some quantitative constraints (e.g., an upper limit on annual coal production within a region) to represent, generally, the aggregate of a number of restrictive factors—water supply, suitable transportation, and so forth. . . .

The general approach is to consider the energy system from resource extraction through end use with sufficient detail so as to reflect the most important constraints and to produce the "best" energy supply mix within such constraints. This must be done in order to ensure overall global consistency, while not unduly sacrificing regional specialness, for a plausible scenario range (High or Low). The specific approach is to use an optimization program to find the combination of energy sources that satisfies the specific temporal sequence of regional final energy demands, subject to (often very dominant) given constraints, while minimizing the discounted total cost.

This procedure is iterated from demand estimation to supply optimization until the demand and supply projections are judged to be "reasonable" and consistent. An important part of the iteration and global consistency is the treatment of energy trade. . . ; the goal is to produce balanced, consistent world scenarios without energy "gaps." . . .

Once two globally consistent energy scenarios are produced in this way, the results can be evaluated in terms of their consequences for cumulative resource consumption. . . , the direct and indirect investments. . . , and other criteria of interest. It is important, in this highly

iterative analytical approach, to evaluate the complete scenario specification and results in as many ways as are useful for determining reasonableness and identifying misfits. . . . Then one can iterate wherever
necessary and/or appropriate in order to alter unacceptable results. Once
acceptable scenarios are generated in this way, variations can be
examined.

Scenario Criteria

As Section 9.5 mentions, criteria of adequacy are usually discussed explicitly in more elaborate scenario work. The IIASA analysis team discussed the guiding criteria that they used as follows (ESPGb, pp. 424–
425):

> The most important criterion for developing a scenario is consistency.
> Consistency was assured among the details through the use of detailed
> accounting systems and models. This procedure ensured that the supply
> of each fuel and energy type was consistent with the projected demand
> for energy of each fuel and type for each region. Consistency evaluations
> were made among and across regions with respect to energy availabil-
> ity and supply costs and utilization of globally traded energy commod-
> ities, with respect to energy use by economic sectors in various stages
> of development, and with respect to energy price response and
> conservation.
>
> A second important criterion is that the scenario projections in their
> totality must not be unreasonable. Application of this criterion required
> judgment—judgment made collectively by accepting and soliciting com-
> ments and reactions from experts. . . .
>
> A third criterion is the degree of continuity or smoothness of the sce-
> nario projection. Not only are the values of key parameters and variables
> expected to be evolving smoothly during the projection period, they also
> must be connected to the past. The inertia of accumulated capital cannot
> be ignored. . . .
>
> The fourth criterion is the degree of variation between scenarios. An
> important determinant of the long-term transition in energy use away
> from petroleum is the level of energy use. At least two scenarios are
> required, and these must span a sufficiently wide range in order to in-
> corporate the unavoidable uncertainties. Thus, iterations originally af-
> fecting only one scenario often resulted in related changes to the other
> scenario so as to preserve the necessary range between the two.

Quantitative Characteristics of the Scenarios

The scenario-development work involved processes of synthesis and
learning and gave important insights, for instance, regarding the orders
of magnitude of different aspects of possible evolutions of the global en-

Table 9.1. The Historical and Projected Per Capita Final Energy Consumption
Calculated for the High and Low Scenarios in the IIASA
Energy Study

Region	Base year 1975	High scenario		Low scenario	
		2000	2030	2000	2030
I (NA)	7.89	9.25	11.63	7.95	8.37
II (SU/EE)	3.52	5.47	8.57	4.98	6.15
III (WE/JANZ)	2.84	4.46	5.70	3.52	3.90
IV (LA)	0.80	1.75	3.31	1.28	2.08
V (Af/SEA)	0.18	0.42	0.89	0.32	0.53
VI (ME/NAf)	0.80	2.34	4.64	1.76	2.46
VII (C/CPA)	0.43	0.93	1.87	0.64	0.93
World	1.46	1.96	2.86	1.58	1.83

Source: ESPG, 1981c, p. 39.

Note: The entries in the table are in kilowatt-years per year (kWyr/yr). A kilowatt is 1000 watts and a kilowatt-year is one kilowatt flowing for a year.

The seven world regions are defined as follows:

Region I (NA): North America (developed, market economies rich in resources)

Region II (SU/EE): Soviet Union and Eastern Europe (developed, centrally planned economies rich in resources)

Region III (WE/JANZ): Western Europe, Australia, Israel, Japan, New Zealand, and South Africa (developed, market economies, but poorer in resources than the other developed regions)

Region IV (LA): Latin America (a developing region with market economies and many resources)

Region V (Af/SEA): South and Southeast Asia and sub-Sahara Africa excluding South Africa (developing regions with mostly market economies, but with relatively few resources, except for some notable exceptions, such as Nigeria and Indonesia)

Region VI (ME/NAf): The Middle East and North Africa (a special case with their economies in transition and with rich oil and gas resources)

Region VII (C/CPA): China and other Asian countries (centrally planned economies in developing regions with only modest resources)

ergy system. Tables 9.1 and 9.2 illustrate some of the quantitative characteristics of the high and low scenarios, and Figure 9.3 shows some aggregated values.

Although the results shown in these tables and the figure could easily be taken as some of the findings of the study, they only exhibit selected characteristics of the scenarios produced by the iterative adjustment process. Therefore, the numerical results must be thought of merely as illustrative, with the principal findings of the systems analysis to emerge later, not only from the work on the two main scenarios, and the scenario variants, but also from the study as a whole.

The IIASA project team distinguished three different types of mathematical models in future-oriented studies in order to explain how the quantitative aspects of the scenarios should be interpreted (ESPGb, p. 27), and they emphasized that the models they used were of the third kind:

The first comprises all laws of nature as incorporated in the natural

Table 9.2. The Historical and Projected Total Primary Energy Requirements
Calculated for the High and Low Scenarios in the IIASA
Energy Study

Region	Base year 1975	High scenario		Low scenario	
		2000	2030	2000	2030
I (NA)	2.65	3.89	6.02	3.31	4.37
II (SU/EE)	1.84	3.69	7.33	3.31	5.00
III (WE/JANZ)	2.26	4.29	7.14	3.39	4.54
IV (LA)	0.34	1.34	3.68	0.97	2.31
V (Af/SEA)	0.33	1.43	4.65	1.07	2.66
VI (ME/NAf)	0.13	0.77	2.38	0.56	1.23
VII (C/CPA)	0.46	1.44	4.45	0.98	2.29
World total[a]	8.21[b]	16.84	35.65	13.59	22.39

Source: ESPG, 1981c, p. 39.

Note: The entries in the table are in terawatt years per year (TWyr/yr). One terawatt is 10^{12} watts; one TWyr is 10^9 kWyr.
[a] The columns may not sum to the totals owing to rounding.
[b] This total includes 0.21 TWyr/yr of bunker fuel used in international fuel shipment.

sciences (as in physics). These laws are meant to represent nature fairly precisely within a given scope.

The second kind of quantitative model is based largely on statistical or other experimental data, without (necessarily, at least) implying the existence of rigorous laws of nature. An example is an econometric model that may forecast the development of an economy for a fairly short period, such as two or three years. It is conceded that there can be errors in these forecasts, but the aim is to project real developments.

The third kind of model is meant to conceptualize a complex and conceivable development. Although such models are also quantitative, here such quantification is meant to deal with an otherwise unmanageable complexity by providing inherent consistency and an explicit identification of assumptions and results. In other words, the third kind of model provides synthesis. The results of such models are therefore not forecasts or predictions.

These distinctions have an important bearing on a debate that followed the publication of the IIASA energy study and they will be referred to later in this section, after a brief summary of some of the more important findings of the work.

The Findings of the IIASA Energy Study

The analysis yielded many findings that could be considered reasonably robust (as distinguished from the numerical scenario results, which must be considered exemplary and not predictive). Although it is inappropriate

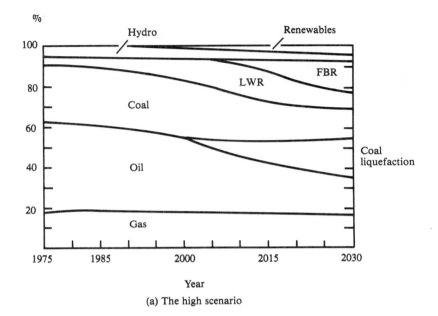

(a) The high scenario

(b) The low scenario

Figure 9.3. The shares of the historical and projected primary energy supplies allocated to sources by calculations for the high and low scenarios in the IIASA energy study. (LWR = light water reactors; FBR = fast breeder reactors.) (*From ESPGc, p. 40.*)

to set all these findings forth here, it is important to offer at least enough of them to give a flavor of the sort of results that the scenario work and the studies of supply options and their constraints led to and supported; for a fuller discussion, see, for example, ESPGa (pp. 169–201).

The most central and important result of the work is that, with prudent planning and well-based decisionmaking, it is possible for the world to meet a reasonable level of energy needs while sustaining a moderate growth of the standard of living over the half century from 1980 to 2030. The continuing heavy demand for liquid fuels, however, dictates that venturesome ways of supplying this need must be explored if the succeeding half century is to offer a promise of continuing the growth of the half century studied. Here are some of the other conclusions of the work:

> Balancing energy demand and supply means striking a balance among energy sources, capital, labor, skill, and social desires in order to provide both a feasible and desirable energy service.

> Only radical changes in lifestyles could lead to very low energy demands such as those that could feasibly be supplied over the next half century by renewable sources.

> The demand for liquid fuels is a principal driving force in the energy problem for the half century studied, and is expected to remain so for the half century beginning in 2030.

> Under any set of circumstances, the world's economic growth rates are likely to be limited by energy supply considerations.

> Fossil fuels will continue to be available in the half century studied, but they will become increasingly unconventional and expensive.

> Coal liquefaction will increasingly be needed to maintain an adequate supply of liquid fuels.

> The carbon dioxide buildup caused by fossil fuel combustion is probably the most severe climate issue generated by the scenario views of the world's energy future.

It is worth noting that the IIASA study—and particularly its scenario core—established a base on which variant, and possibly more local, analyses can be built against a reasonable world background; for one such extension to the European Communities countries, see Sassin et al. (1983).

A Critique

In 1984 *Policy Sciences* (volume 17, number 3) published three critiques of aspects of the IIASA energy study and a later issue (volume 17, number 4) published a response from the leader of the project and one of his principal assistants. Some of the points made by Keepin (1984) will be discussed here, together with how they were viewed by the analysis-team leaders. The intention is not to summarize this extended debate, but only to distill from it some issues related to the use of scenarios that are of general importance, such as the strengths and weaknesses of the scenario approach, the differences in model conceptions, and the role of assumptions.

The main basis for Keepin's paper is the study he made of the use of MESSAGE, one of the models used in the scenario development process (see Figure 9.3). His work, which was most detailed for one of the seven world regions (Region III), compared the set of input data with the model outcomes. The following quotations illustrate the general direction of his critique:

> The principal argument developed in this article is that the quantitative analysis behind the scenarios does not support the conclusions drawn from them, and that these conclusions are more accurately described as opinions rather than findings. Two major analytical results are established here that support this claim. First, the complex computer models used in the quantitative analysis do not play a significant role in determining the final numerical results of the scenarios. . . . Second, the scenarios are seriously lacking in robustness with respect to minor variations in certain input data. . . .
>
> . . . [I]t cannot be overstated that this article addresses only one aspect of the IIASA energy study, and it is definitely not a general criticism of the entire program. In fact, the program contributed in many important ways to a more complete understanding of many aspects of the global energy system. It was the first serious attempt to systematically account for, and gather consistent data from all regions of the world with roughly equal emphasis. Given the magnitude and complexity of the global energy system, this was no simple task. A genuine attempt was made to properly incorporate all nations on earth, which required painstaking analysis and aggregation of masses of detailed economic, geographical, demographic, and resource data from countless sources. Valuable results were obtained from systematic studies of the global potential of each major source of energy. In addition, the program produced several important contributions, including pioneering work on atmospheric carbon dioxide (CO_2) and the global "greenhouse" problem. . . , solar energy. . . , technological risk. . . , etc. . . . Finally, perhaps the most important contribution has been the innumerable personal and working relationships, interactions, and professional discourses that developed at IIASA, and as a result of the many conferences and workshops that were held. [Keepin, 1984, pp. 201–202.]

Despite the assertion that "models should be designed for gaining insight and understanding" [ESPGb, p. 399], the basic conclusion reached here is that the dynamic and analytic contents of the IIASA energy scenarios are directly attributable to assumptions and quantitative judgments that are specified outside of the set of mathematical models. . . . In fact, the models were used primarily in the capacity of accounting aids. . . . [Loc. cit., pp. 221, 230.]

[R]obustness is the property that an analysis must have if it is expected to be of some validity in the face of uncertainties in the underlying assumptions. . . . The major finding is that the supply scenarios are not robust with respect to variations in several input data. One example concerns the structure of the nuclear contribution, which is found to be sensitive to minor variations in the assumptions concerning the availability and cost of uranium. [Loc. cit., pp. 221, 222.]

Although scenarios are not presented as decisive forecasts, there is an inevitable tendency to view them as such, even by their authors. [Loc. cit., p. 233.]

In their response to the Keepin critique, Haefele and Rogner (1984) made these points:

The IIASA study consisted of a number of "strata" (that is, the four phases listed in an earlier subsection) and "the essential and complex task of synthesizing the study findings was undertaken only after the completion of the analyses at the various strata" (loc. cit., p. 343).

The research activities of the first two strata were completed before the modeling work began. Thus, the inputs to the modeling exercise were determined within the activities of these strata, and not, as Keepin claims, "tentative projections and arbitrary assumptions that have not been carefully substantiated or tested" (Keepin, 1984, p. 232).

On the relations between the assumptions and the outputs, Haefele and Rogner (1974, p. 347) say that: "Within the context of the IIASA study an important purpose of the mathematical modeling was to ensure the consistency of the calculations. Keepin himself has demonstrated the need for deploying mathematical models for quantitative consistency." Then, after discussing an example showing the complications involved in keeping solutions within the boundaries of the constraints—and the careful treatment needed to do this—Haefele and Rogner conclude: "Once this painstaking exercise was completed, the calculations were straightforward. Keepin calls this a simplistic transformation of assumptions into outputs. But he can claim this only after others have completed the formidable task of quantifying the inputs for him. Indeed, such modeling forces the systematic organizations of otherwise overwhelming amounts of data" (loc. cit., p. 347).

For the purposes of this discussion it is useful to distinguish three types of models (as described in an earlier subsection): those used for pre-diction and forecasting, those used to describe complex and short-range trends when a descriptive theory of the phenomena is not known, and those "used to describe in a consistent way scenarios of evolution—that is, not the one future that is to come but rather several conceivable futures—[and] . . . used not for forecasting but for the maintenance of consistency and thereby for consistent dis-aggregation" (loc. cit., p. 344). Against this background, Haefele and Rogner say: "Keepin's misunderstanding is therefore colossal as he seems to reach out for a mixture of modeling of the first and second type, while in the IIASA study the modeling was meant to be of the third type" (loc. cit., p. 344). "It was simply not the intent to make the energy modeling an image of the study results; energy modeling was just one test besides the others employed" (loc. cit., p. 352).

Keepin's conclusion that the models were not robust in their responses to "minor" changes in uncertain input data arose from the fact that the changes he introduced to test them were not minor but quite large (Haefele and Rogner, 1984, pp. 349–352).

9.7. Scenarios Revisited

As a tool for systems analysis, scenarios have both strengths and weak-nesses. The IIASA energy study and the brief glimpse of one of the de-bates that followed its publication serve to shed light on them—and, al-though these strengths and weaknesses are likely to differ depending on the problem situation being explored, a discussion against the energy-study background suggests some of them, and how they can be regarded.

Here are some of the strengths of the scenarios as they were used in the IIASA energy study:

They played the role in the work that is usually expected of scenarios: that of bringing together a coherent picture of a limited number of feasible futures as the basis for further analysis.

Since any systems-analysis study involves the analysis team in making many important judgments, the scenarios forced the judgments about the future environment to be brought together so that they could be tested for feasibility and consistency.

In developing the scenarios the combination of various models and approaches as summarized in Figure 9.2 played an important role. It allowed for a synthesis of knowledge from many different areas. It also enforced essential disciplines: the same resources could not be used twice; the supply had to balance demand; only available resources could be tapped to yield energy supplies; constraints were applied quantitatively over time; and so on.

The scenarios forced the analysis team to face the realities in each of the seven world regions and to produce calculations for them that could be seen to represent a reasonably expectable reality. They also forced the analysis to totals that lay within the values of feasibility for world variables.

The scenarios that the IIASA analysis team developed can be characterized as semiquantitative. As Figure 9.2 shows, the scenarios emerged from a combination of results from quantitative models and from findings and judgments in various forms from other studies and experts. The freedom of choice as regards the scenario form and the process of its development made it possible to synthesize knowledge from many different areas and sources, and to gain important insights and understandings related to many aspects of the global energy system and its future. This freedom of form, however, as well as, in this case, the complexity of the development process, made it difficult to assess the findings: just what knowledge they represent and how it can be used.

A clear documentation of the scenario development process and the role of the scenarios in the analysis is, of course, an essential foundation for clarifying the interpretations of the findings emerging from work based on them. But this detailed documentation may sometimes enlarge the risk—as it did in the IIASA energy study—that details in the scenario descriptions are mistaken for end results. Although this seems to be a dilemma that is inherent, clear explanations and further experience in the systems-analysis community may help to overcome it in the future.

The IIASA energy-project team emphasized the importance of making clear distinctions among different types of future-oriented mathematical models. Others (Ayres, 1984; Schwarz and Hoag, 1982) have discussed similar distinctions. For example, Schwarz and Hoag introduce typologies implying distinctions between forecasting and such nonforecasting models as "what if?" and learning models, and between different "levels" of results, such as model outcomes, model inferences, and policy-issue oriented interpretations. However, Keepin's "misunderstanding" of the type of models used in the energy study and of the role of scenarios raises some issues of general importance:

Different model conceptions indicate the existence of different schools of thought. Communication difficulties between these schools of thought, as well as between modelers and policymakers, will undoubtedly remain until the differences are clarified and explained effectively.

Scenarios are based on assumptions; therefore, the crucial role of the choice of assumptions should not be underestimated, nor should it be glossed over in presenting study findings.

Any complex systems-analysis inquiry, and the scenarios that may be used in it, rests on the knowledge and judgments available at the time it is carried through—and all these elements may be affected by the passage of time and the new influences and knowledge that emerge. Thus, most studies and their findings experience diminishing relevance with the passage of time and the emergence of new influences that would prompt changed assumptions and judgments if the study were done anew. For example, the IIASA energy project as a whole, as well as its scenarios, was based on the assumption that the highest possible economic growth in the developing countries is desirable, as well as on a number of assumptions aimed at making the analysis feasible. Political and social constraints were recognized, but not applied explicitly. One of the conclusions was that, under any set of circumstances, the world's economic growth rates are likely to be limited by energy-supply constraints. Now, ten years later, evidence supports the validity of this finding. But now, from the IIASA energy work, from other studies, and from the intervening developments, we have a deeper understanding of the global energy problem. Keepin's critique could be reinterpreted as saying that in 1984 he would have found a global energy study based on somewhat different assumptions more interesting than the earlier IIASA study; to judge from the rebuttal, the IIASA team would probably agree.

Since to consider adopting any policy or action alternative means to look to the future and its consequences in that future, systems analyses intended to support such considerations must incorporate such projections of the future as present knowledge will support. Single-variable forecasts are useful, but, when many variables interact, scenarios that consider their potential interactions are often used. The energy-study example shows that considering these interactions in developing feasible views of the future can be a demanding and complicated—but essential—task. Although technical methods can aid this task, and often act to reduce the uncertainties in the views of the future, judgment remains an indispensable ingredient in developing such views. Scenarios are one device that is being used to bring these judgments together with other knowledge to produce coherent views of possible futures. With its strengths and weaknesses properly understood and interpreted, it can be an effective tool— and one that will be much further developed in the future.

References

Amara, Roy (1980). *The Futures Field*. Menlo Park, California: Institute for the Future.

———, and A. J. Lipinski (1983). *Business Planning for an Uncertain Future—Scenarios and Strategies*. New York: Pergamon.

Armstrong, J. S. (1985). *Long-Range Forecasting—From Crystal Ball to Computer.* New York: 1985.

Ascher, W. (1979). *Forecasting—An Appraisal for Policy-Makers and Planners.* Baltimore, Maryland: The Johns Hopkins University Press.

———, and W. H. Overholt (1983). *Strategic Planning and Forecasting—Political Risk and Economic Opportunity.* New York: Wiley.

Ayres, R. (1979). *Uncertain Futures—Challenges for Decision-Makers.* New York: Wiley.

——— (1984). Limits and possibilities of large-scale long-range societal models. *Technological Forecasting and Social Change* 25, 297–308.

Brown, S. (1968). Scenarios in systems analysis. In Quade and Boucher (1968), pp. 298–310.

Calder, N. (1967). What is futures research? *New Scientist* 36(570), 354–355.

Carbone, R., and S. Makridakis (1986). Forecasting when pattern changes occur beyond the historical data. *Management Science* 32, 257–271.

Cornish, E. (1977). *The Study of the Future.* Washington, D.C.: World Future Society.

deLeon, P. (1975). Scenario designs: An overview. *Simulation and Games* 6(1), 39–60.

Energy Systems Program Group, Wolf Haefele, Program Director (1981a). *Energy in a Finite World: Paths to a Sustainable Future.* Cambridge, Massachusetts: Ballinger.

——— (1981b). *Energy in a Finite World: A Global Systems Analysis.* Cambridge, Massachusetts: Ballinger.

——— (1981c). *Energy in a Finite World: Executive Summary,* Executive Report 4. Laxenburg, Austria: International Institute for Applied Systems Analysis.

ESPG. See Energy Systems Program Group.

Evans, J. St. B. T. (1982). Psychological pitfalls in forecasting. *Futures* 14(4), 258–265.

Fildes, R. (1979). Quantitative forecasting—the state of the art: Extrapolative models. *Journal of the Operational Research Society* 30, 691–710.

Fowles, J., ed. (1978). *Handbook of Futures Research.* Westport, Connecticut: Greenwood Press.

Haefele, W., and H.-H. Rogner (1984). A technical appraisal of the IIASA energy scenarios?—A rebuttal. *Policy Sciences* 17, 341–365.

Herrera, A. D., et al. (1976). *Catastrophe or New Society? A Latin American World Model.* Ottawa, Canada: International Development Research Centre.

Hogarth, R. M., and S. Makridakis (1981). Forecasting and planning: An evaluation. *Management Science* 27, 115–138.

Julien, P.-A., P. Lamonde, and D. Latouche (1975). *La Méthode des Scénarios.* Paris, France: La Documentation Française.

Kahn, H., and A. Wiener (1967). *The Year 2000—A Framework for Speculation on the Next 33 Years.* New York: Macmillan.

Keepin, B. (1984). A technical appraisal of the IIASA energy scenarios. *Policy Sciences* 17, 199–275.

Keyfitz, N. (1977). *Population of the World and its Regions, 1975–2050*, WP-77-7. Laxenburg, Austria: International Institute for Applied Systems Analysis.

Leontief, W., et al. (1977). *The Future of the World Economy*. New York: Oxford University Press.

Levien, R. (1968). The analysis of force policy and posture. In Quade and Boucher (1968), pp. 279–297.

Linstone, H. A., and M. Turoff, eds. (1975). *The Delphi Method: Techniques and Applications*. Reading, Massachusetts: Addison-Wesley.

Lorange, P., and R. F. Vancil (1976). How to design a strategic planning system. *Harvard Business Review* 54(5), 75–81.

Makridakis, S. (1976). A survey of time series. *International Statistical Review* 46, 29–70.

——— (1978). Time-series analysis and forecasting: An update and evaluation. *International Statistical Review* 46, 255–278.

———, and S. C. Wheelwright, eds. (1982). *The Handbook of Forecasting: A Manager's Guide*. New York: Wiley.

Martino, J. P. (1983). *Technological Forecasting for Decision Making*, 2nd ed. New York: North-Holland.

Meadows, Donella, et al. (1979). *The Limits to Growth*. London: Potomac Associates, Earth Island Limited.

Miser, H. J., and E. S. Quade, eds. (1985). *Handbook of Systems Analysis: Overview of Uses, Procedures, Applications, and Practice*. New York: North-Holland.

Mitroff, I. (1979). On strategic assumption-making: A dialectical approach to policy and planning. *Academy of Management Review* 4, 1–12.

Nanus, B. (1979). Interdisciplinary policy analysis in economic forecasting. *Technological Forecasting and Social Change* 13, 285–295.

Overview. See Miser and Quade (1985).

Quade, E. S., and W. I. Boucher, eds. (1968). *Systems Analysis and Policy Planning*. New York: Elsevier.

Sassin, W., A. Hölzl, H.-H. Rogner, and L. Schrattenholzer (1983). *Fueling Europe in the Future—The Long-Term Energy Problem in the EC Countries: Alternative R&D Strategies*, RR-83-9. Laxenburg, Austria: International Institute for Applied Systems Analysis.

Schwarz, B., and J. Hoag (1982). Interpreting model results—examples from an energy model. *Policy Sciences* 15, 167–181.

———, U. Svedin, and B. Wittrock (1982). *Methods in Future Studies: Problems and Applications*. Boulder, Colorado: Westview Press.

Taylor, B. (1975). Strategies for planning. *Long-Range Planning* 8(4), 27–40.

Vavrin, J. (1980). *The Airline Insurance Industry—A Future Study*, Ph.D. thesis. Stockholm, Sweden: Department of Business Administration, University of Stockholm.

WEC (1978). *World Energy: Looking Ahead to 2020*, Report by the Conservation Commission of the World Energy Conference. Guildford, United Kingdom: IPC Science and Technology Press.

Wilson, I. (1978). Scenarios. In J. Fowles, ed. (1978), pp. 225–247.

Chapter 10
Cost Considerations

H. G. Massey

10.1. The Importance of Cost

Systems analysis, as discussed in this handbook, addresses problems that emerge from interactions among people, the natural environment, and artifacts of man and his technology. It aims to understand these problems, and to use this understanding to help decisionmakers adopt ameliorative solutions. As Section 1.1 states:

> In practice, analysis of this type clarifies and defines objectives, searches out alternative courses of action that are both feasible and promising, gathers data about them and their environments, and generates reliable information about the costs, benefits, and other consequences that might ensue from their adoption and implementation.

The purpose of this chapter is to discuss the cost aspects of systems analysis.

Because decisionmakers must make choices under resource constraints, the cost elements of systems analysis are fundamentally important. If society's resources were unlimited, problems of choice would be much easier than they are; but resources are always limited. Therefore, decisionmakers are faced with questions like these:

What way of achieving a desired objective will require the least cost (that is, the least use of resources)?

For a specified level of cost, and a given type of objective, what alternative means will provide the highest level of achievement (that is, the greatest effectiveness) toward these objectives?

For a given objective, is there some combination of cost and level of achievement that can be viewed as preferred?

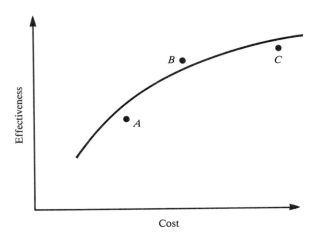

Figure 10.1. Comparing the cost and effectiveness of alternatives. The curve represents an estimated average relation between cost and effectiveness for a spectrum of alternatives; the points A, B, and C represent three specific alternatives for examination.

To help decisionmakers with questions of this type, it is clear that analysts must deal with cost as a central issue.

The first question poses an equal-effectiveness problem: A set of alternatives must be designed that satisfy the objective and their costs estimated so that the one (or ones) with the least cost can be identified.

The second question poses an equal-cost problem: A set of alternatives with equal costs must be designed and the effectiveness of each must be estimated so that the one (or ones) that do the most to achieve the objective can be identified.

Unfortunately for analysis, few problems fit into such simple frameworks. For example, consider the situation depicted in Figure 10.1, where average cost and effectiveness relations have been derived for a spectrum of alternatives, but where three have been selected as particularly interesting. Here the issue has the form of the third question: Which of the three alternatives has a preferred combination of cost and effectiveness? To help answer this question, the decisionmaker might have to choose, for example, whether the incremental increase in effectiveness in moving from alternative B to alternative C is worth the incremental increase in cost. Again, as in the first two questions, cost considerations are central to the analysis.

In any case, no matter how the question is posed, cost is an essential ingredient in evolving a preferred course of action. Furthermore, the nature and definition of cost in a systems-analysis problem are fundamentally connected with the characteristics of the problem and of the alternatives chosen for consideration. These statements especially hold true as the situation becomes—as it usually does—more complicated than the relatively simple formulations implied by the three questions we have considered.

10.2. What Cost Is

Like many common words, the word *cost* is used differently by different people, and vaguely by most. Therefore, it is important to be alert to the various modifications of the concept that will, unless we watch out for them, confuse or mislead.

For the purposes of this chapter, costs will be divided into two major categories: economic and noneconomic. From a systems-analysis point of view both categories may be considered to represent *undesirable* consequences of the alternative courses of action, the desirable consequences being, of course, the anticipated benefits.

Economic Cost

Let us start with a simple example. The Jones family, like most of us, has limited resources. Suppose that one day they consider taking a trip around the world, for which they will have to pay x monetary units (MUs). The next best item on their scale of preferences that could be purchased for x MUs is a new automobile. What is the *real* cost of the trip to the Joneses? Although the monetary cost is x MUs, the real cost is represented by the pleasures and satisfactions (that is, the utility) they give up by not having the new automobile. This is why economists refer to economic costs as benefits lost or opportunities foregone.

Since institutional decisionmakers, both public and private, must operate within budget constraints, the opportunity-cost concept applies in their cases as well. Thus, if public-works administrators are contemplating construction of a new bridge, they may have to consider that the real (economic) costs of such a decision are best measured by the benefits the public must forego because alternative projects, such as a new park or sports arena, could no longer be provided for in the public-works budget.

How do cost analysts on systems-analysis teams show the economic-cost consequences of alternative courses of action? The most direct way is to exhibit these costs separately, factor by factor. For an example where this was done, see the discussion of finding the preferred way of closing a Dutch estuary to avoid flooding in Section 2.2 and the *Overview* (pp. 89–109; see also Goeller et al., 1977). This procedure has the advantage of showing decisionmakers in detail the costs that a proposed program will incur, but the disadvantage that it forces the decisionmakers themselves to bring the costs together into a synthesis, a procedure that is sometimes desirable and sometimes not. The commonest way for cost analysts to make this synthesis is to translate the physical and performance characteristics of the alternatives, together with their operational concepts, into monetary measures.

This translation of economic costs into monetary measures raises an important question: Are money costs likely to be an acceptable proxy for economic costs? The correct answer is sometimes yes and sometimes no, but it is frequently positive when the alternatives under consideration pertain to the more distant future, as is often the case in systems-analysis studies; here is where money costs are most likely to serve as rough approximations of the real economic costs. As a general rule, the more distant the future alternatives are—keeping in mind that a longer time permits almost all resources to be substitutable and therefore fungible— the better money costs are as proxies for economic costs (Hitch and McKean, 1960, pp. 25–26).

When monetary cost estimates are used instead of direct measures of foregone benefits, there is an indirection: Detailed information about outcomes is being sacrificed in order to increase the ease of evaluating alternatives. On the other hand, in some cases it may be appropriate and desirable for the analyst to provide more direct information for the decisionmaker about the benefits lost when resources are employed in one way rather than another. For example, if land or facilities owned by a public body are included among the resources required for a program, the most appropriate measure of cost is the benefits that could have accrued if the land or facilities had been used by other programs—or the cost could be measured by the costs that would be incurred to obtain substitute land and facilities for the other programs.

Noneconomic Cost

In the early years of systems-analysis practice, analysts tended to ignore noneconomic costs, that is, negative impacts on the factors associated with the quality of life. Such impacts were considered to be external effects, and were therefore not included in the analyses. More recently, however, these impacts have been recognized as important costs, and systems analysts have endeavored to take them into account in their work and its findings.

For instance, early studies of alternative future transportation systems focused primarily on transportation benefits (such as travel time, low costs to individuals, relative comfort, and other such measures of worth) and on the monetary costs of achieving these benefits. They did not consider such noneconomic costs as increased noise pollution, increased congestion in some areas, the displacement of some households, or increases in air-pollution emissions. More recently, such work has taken such external costs (that is, external negative benefits) into account, thus bringing them into the analysis, or internalizing them (see, for example Chesler and Goeller, 1973). In the case of noise pollution, for example, analysts have developed noise-impact assessment models for calculating noise exposure contours expressed in appropriate metrics: noise pollution level, noise exposure forecast, and decibels on the A scale (Chesler and Goeller, 1973, pp. 52–64). The number of households likely to be subjected to critical exposure levels near airports and along ground routes were also computed.

10.3. Principles and Methods of Cost Analysis

This section considers how cost analysts on a systems-analysis team go about estimating the cost implications of alternative future systems or programs; to this end, it stresses general principles and methods. Section 10.4 illustrates them in an example.

Defining the Problem

The first stage of a systems analysis is problem formulation, in which the problem situation is transformed in order to identify and clarify the desired objectives, define the boundaries and constraints (limit the scope), and establish good criteria for choosing among alternative courses of action (see the *Overview*, Chapter 4, especially Figure 4.1).

If the cost analyst or analysts are to do a good job of working out the cost considerations that are involved in the problem, it is essential for them to work as an integral part of the systems-analysis team from the beginning. Participating in the formulation stage of the work forms an important background for their work—and it is especially important for the cost analysts to participate in the development of the alternatives that will be given detailed consideration in the analysis. A good cost analysis is impossible unless the alternatives are described and specified in considerable detail: their physical characteristics, operational concepts, impacts on the objectives, and other factors relating to cost and effectiveness.

The formulation stage must answer many key questions as it prepares the ground for the later analysis. Several of them are particularly important for the cost-analysis work. First, although the formulation stage must identify the primary objectives that are sought, it is also important to establish the form of those objectives in order to define the measures of effectiveness by which alternatives will be judged. For example, is an absolute or threshold level of effectiveness to be sought; is it desired to obtain as high a level of effectiveness as possible, subject to budget or other constraints; or will the analysis be expected to consider tradeoffs between cost and effectiveness?

Second, the time frame of the problem must be identified. Is there a starting point and an end point between which the specified objective is desired, or is an infinite time horizon appropriate? The example problem that is discussed later in this chapter has an objective goal that is to be reached at a particular future date and then maintained indefinitely. These time considerations form an essential basis for considering how the analysis will deal with costs that are incurred over time.

Third, it is necessary to understand where the costs of the alternatives fall. Are the only costs of concern those that are to be financed from the decisionmaker's budget, or should costs that are incurred elsewhere be considered as well? What are the boundaries of the problem? The answers to these questions will help to define the costs to be considered as internal and those that are external (that is, outside the problem's domain of concern). Eventually, it may become desirable to consider internalizing some external costs, and the boundary definitions will help with the focus on this issue.

Fourth, to provide useful cost information to the decisionmaker, the cost analysts must attach costs to corresponding benefits; in other words, a fundamental part of the framework of cost analysis is its foundation on output-oriented units. Attaching costs to outputs (or benefits) allows the analysis to link resource inputs to the benefits that accrue from their expenditure, and therefore to estimate the extent to which the objectives are achieved.

An Analytical Framework: The Input Side

The practical problem for the cost analyst is to estimate the cost of the system for each of the alternatives. What does the cost of the system mean? What items should be included in this cost? Generally, the cost of the system means the sum of the costs of everything required to develop the alternative over a period of years: equipment, new facilities, manpower, supplies, and so on. It includes everything *directly related to the decision* to achieve the capability embodied in the alternative; but it excludes the cost of items not so related, such as the costs of administrative and support activities that would go on regardless of the decision under consideration.

One can rarely, if ever, estimate the cost of the system in the aggregate; it must be broken down into meaningful components—an analytical framework. Over a wide range of systems (transportation, health care delivery, water resources, and the like) experience has shown that a useful first cut identifies three life-cycle cost categories:

Research and development (R&D) costs—the resources required to develop the new capability to the point where it can be put into operation at some desired level of reliability. (In some cases R&D will already have taken place, in which case R&D costs are sunk, that is, already paid, hence not a consequence of the decision, and thus should not be included in the cost of the system.)

Investment costs—the one-time outlays required to place the new capability in an operational mode.

Operating costs—the recurring outlays required year by year to operate and maintain the capability in service, usually for a specific period of years. (The period of operation chosen for the analysis may affect what costs need to be included; for example, some system components may last five years and require added costs for repair and replacement to sustain a ten-year operational life.)

In most cases, the three life-cycle categories must be broken down further into resource and/or functional categories; Table 10.1 shows such a more detailed cost-category structure. This input structure is still general enough to apply over a wide range of systems, but how much detail may be needed depends on the problem and alternative being dealt with. For example, in an analysis dealing with new aircraft transportation systems, the major equipment category would probably be broken down into aircraft subsystems, such as airframe, propulsion, and avionics. On the other hand, this category would be quite different for a flood-control or a health-care-delivery system.

Regardless of what set of input categories is chosen, it is vitally important to define carefully what each includes. This step is a fundamental prerequisite to developing estimating relationships (to be discussed in following material) and to working out consistent estimates of the cost implications of the various alternatives under consideration. It should be noted here that, in a given case, it may be that not all the categories will be assigned positive numbers; for example, if an alternative is able to utilize facilities made available from phasing out an existing system, the facilities-investment category cost may be zero, or close to it, unless there are competing uses for the facilities.

Some points are worth making about the appropriate level of detail in input structures. Trying to structure the input in great detail for long-range planning is usually not rewarding; indeed, in most cases it is impossible. It is important, however, to have input structures that are specific enough to distinguish between aspects of an alternative that are really new and those that are not; for example, for a new piece of equipment,

Table 10.1. An Example Cost-Category Structure

	Cost of the alternative	
	System A	System B
Research and development costs		
Preliminary design and engineering		
Fabrication of test equipment		
Test operations		
Miscellaneous	————	————
Total R&D costs	═════	═════
Investment costs		
Facilities		
Major equipment		
Initial inventories		
Initial training		
Miscellaneous	————	————
Total investment costs	═════	═════
Annual operating costs (assuming a ten-year operational life)		
Maintenance		
Personnel pay and allowances		
Replacement training		
Fuels and lubricants		
Miscellaneous	————	————
Total annual operating costs	═════	═════
Total system cost		
R&D + investment + 10 × (annual operating costs)	═════	═════

some parts may represent the current state of the manufacturing art, whereas others may call for new developments. It is usually the case that even the most advanced systems contain elements that are not significantly new; and these should be separated in the analysis from those that require development so that appropriate attention can be devoted to the more difficult new elements. For items of hardware, this principle usually means going down at least to the subsystem level, and often to individual components.

Estimating Relationships

The cost analysts must develop a cost-estimating procedure for every category and subcategory in the input structure (for a detailed treatment, see Fisher, 1971, Chapter 6). Often these estimating procedures can be embodied in a mathematical equation, which may be simple or complex. Such estimating relationships (ERs) are at the heart of the cost analyst's work. They enable him to estimate the costs of components and subcomponents of an alternative as functions of their physical characteristics, performance levels, and operational concepts; for example:

The equipment-maintenance cost for a bus system as a function of the annual bus-miles traveled and the number of seats per bus.

The unit investment cost of an aircraft airframe as a function of aircraft speed, weight, and the cumulative number produced.

The capital cost of a light-water-reactor nuclear power plant as a function of size, location, type of cooling system, and the cumulative number built.

The unit investment cost of unmanned spacecraft as a function of accuracy and stabilization requirements, weight, and annual production rate.

Where possible, cost analysts determine the forms of ERs and the estimates of the values of the parameters contained in them empirically from data on past, present, or very-near-future systems similar to the ones being considered (for a discussion of deriving ERs, see Fisher, 1971, pp. 143–156).

Even though ERs usually have an empirical basis, care must be taken in how they are used. Projections into the distant future may imply a universe of relevant supporting facts and estimates that is quite different from the one used in formulating the ERs and estimating their parameters. ERs must never be used mechanically. Often they must be modified to take account of supplemental information about a new context of both quantitative and qualitative factors (for a discussion on "how to be careful in using ERs," see Fisher, 1971, pp. 157–163).

Cost Models

The basic ingredients for building a cost model are at hand when the cost analyst has developed a definitive input structure and has derived ERs for all the categories and subcategories in it. A cost model is an input-output device that enables the cost analyst to go from system characteristics (physical characteristics, performance levels, operational concept) to estimates for input categories to total system cost. For example, such a total system cost is often the R&D cost plus the investment cost plus a number of years of operating costs expressed in monetary units. In addition, in some cases it is desirable to include a treatment of noneconomic costs in the cost model (as discussed in the example presented in the following section).

Cost models may be simple or complex, depending on the problem being considered. The usual practice is to computerize the more complex models in order to facilitate their use. Frequently the cost analyst must do sensitivity testing (as will be discussed further in following material) to see how total system cost varies with variations in system characteristics; this process is greatly facilitated by automation. Although automation is desirable, it is a secondary consideration; the primary one is to build a representation of reality appropriate to the problem and the alternatives being considered.

The output of a cost model can take various forms, depending on what questions are being asked in the systems analysis. To illustrate some of the possibilities, let us assume a very simple situation where the model generates total system cost (TSC) as a function of only three variables: (1) a single system performance characteristic P over the range (a, b); (2) system size or scale of operation S over the range (c, d); and system activity rate R over the range (e, f). Assume further that the base case of interest in the analysis is characterized by P^*, S^*, and R^*, and for an operating period of $N = 10$ years. The analysts are interested in the TSC for the base-case system configuration and for variations around it.

By exercising the model, the cost analyst obtains the results shown in Figure 10.2, in which the TSC for the base case is TSC* and variations around it are shown for P and S over their ranges, the variable R being held constant at R^*. The figure shows that TSC is not very sensitive to variations in P but quite sensitive to variations in S; it also shows that the marginal cost with respect to S declines as S increases.[1] Facts like these are usually of great interest to the systems analysts. For example, this figure tells them that, for a given R^* and for S varied over its entire range, they can get a significant increase in P without incurring severe penalties in the values of TSC. Similarly, the results for variations in R and N can be derived and studied.

Treatment of Time

So far we have considered models that do not include explicit treatment of time, and which therefore may be considered static. Such a course may be appropriate for some purposes, but for most purposes time must be treated explicitly, and the cost models have to be constructed to do so.

A static TSC can give a rough preliminary indication of the economic impact of each of the alternatives being considered. In fact, in the early stages of an analysis, when many alternatives are being considered and screened in order to arrive at the small set to be considered in detail,

---

[1] The marginal cost with respect to S is the slope of the curves in Figure 10.2; that is, if TSC is considered for the moment to be a function of S alone, the marginal cost is the derivative of TSC with respect to S, $d(\text{TSC})/dS$.

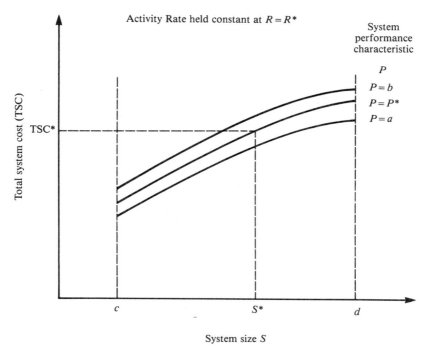

Figure 10.2. An example of results from a cost model in which the total system cost (TSC) in constant monetary units (MUs) depends on system performance characteristic P (ranging from a to b), system size S (ranging from c to d), and system activity rate R (ranging from e to f). The graph shows the value TSC* of TSC for the base case with $P = P^*$, $S = S^*$, and $R = R^*$. It also shows the values of TSC for R held constant at R^* and with P and S varying over their ranges. TSC = R&D costs + investment costs + operating costs for N years, with N in this case taken to be 10; it is assumed that these costs are expressed in constant-price-level MUs.

static system-cost estimates are usually sufficient. These unrefined estimates can help the analysts with the screening process.

Later in the analysis, when a preferred set of alternatives begins to emerge, the static estimates should be converted to time-phased MU cost streams (that is, a vector showing the distribution of costs or expenditures by year, month, or other time period) for each of the alternatives in the set. Since decisionmakers must always be concerned about budgetary matters, these cost streams should be expressed in terms appropriate to the problem and its administrative setting, such as obligational authority, annual expenditures, or whatever budgetary or monetary element is commonly used in the setting. Time-phased estimates are important, because they serve as proxies for the timing of the economic impacts of the various alternatives; and timing is almost always a critical consideration in policy- and decisionmaking.

Decisions about expending resources nearly always involve time preferences: in one case, it may be desirable to limit immediate expenditures and accept larger ones later; whereas in another case, it may be preferable to make a major capital expenditure right away in order to limit future costs. Presenting cost estimates in time-phased streams helps a decisionmaker to consider the time-preference problem. On the other hand, it is sometimes helpful to condense these time streams into single estimates representing the present, discounted values of the future annual expenditures. (The present value of a stream of future monetary amounts is analogous to the amount one would need to invest today in order to receive a future income stream, assuming a constant interest or "discount" rate. For further discussion of the concept and the mechanics of discount calculations, see Fisher, 1971, pp. 51–52 and pp. 224–226.)

Given that committing resources to any alternative is necessarily at the expense of the benefits these resources would bring in other uses, and that this expense is heavier if near-year costs are large relative to far-year costs, it follows that, in converting cost streams to a common present value measurement, future costs should be discounted at a rate that reflects the marginal productivity of capital. On the other hand, since this rate is always controversial, a range of rates should be used to test the sensitivities of the outcomes to assumptions about the discount rate. Above all, the cost analyst must be explicit about his discount assumptions and their basis, so that the decisionmakers can consider them explicitly in arriving at their preferences for a course of action.

In the past, it was not uncommon for cost analysts to make a single arbitrary assumption about discounting without disclosing it to the decisionmakers. This rate was usually zero for the time period of interest and infinity thereafter, a very useful assumption for some purposes (such as some program budget contexts) but one that is inappropriate for others. Both decisionmakers and analysts should join in considering the appropriateness of discount rates, so these assumptions should always be made explicit, with alternative assumptions and their consequences also displayed.

Treating time explicitly, including discounting, can create a heavy computational burden if it must be done late in the analysis. Therefore, in analyses where a large volume of time-phasing calculations must be made, the cost analysts usually build subroutines into their cost models to facilitate generating discounted cost streams over time.

Treatment of Uncertainty

Most important decision problems involve major elements of uncertainty, and so systems analyses addressing them must treat uncertainty explicitly, as far as possible. Thus, cost analysis, as an integral part of systems analysis, must also pay attention to the problems posed by uncertainty. (This is a complex subject, about which space constraints force us to limit our discussion to a few main points. For a more detailed treatment, see Fisher, 1971, pp. 201–217.)

Uncertainty affects the work of cost analysts in two ways. First, their estimating relationships and cost models are never exactly correct, owing to statistical uncertainty. Second, the world of the future may turn out to be different from what they are assuming it to be—because, for example, certain technological developments may progress more rapidly or less rapidly than they assumed—owing to uncertainty in the future state of the world.

Statistical uncertainty can often be handled in a fairly straightforward way, particularly if estimating relationships have been derived statistically and are accompanied by standard errors, prediction intervals, and other standard statistical measures. These measures can contribute to a decision about whether or not statistical uncertainty is worth considering explicitly in the overall analysis, whereas other uncertainties may dwarf the statistical errors in the ERs. Also, these measures can readily show which cost categories in the input structure are subject to the greatest uncertainties, and sensitivity testing can be done for these to see what effects the uncertainties have on the TSC.

State-of-the-world uncertainty[2] usually dominates statistical uncertainty in systems analyses, but state-of-the-world uncertainty is more difficult to deal with, largely owing to the paucity of reliable information and well-supported assumptions about the future. With this fact in mind, one of the most important things that cost analysts can do is to look critically at system descriptions (such as performance and physical characteristics, operational concepts, assumed operating environments, and so on) for each of the alternatives to see if these factors are being stated too optimistically. This occurs frequently, because proponents of new systems and programs tend to be optimistic about what their new developments can do, how soon they can be acquired, and what they will cost. One of the usual duties of cost analysts, when they detect such possibilities, is to consult the relevant technical experts and work with them to evolve more realistic specifications and assumptions; the cost analysts can then calculate the effects on TSC for each of the alternatives to see if this affects their rankings in the analysis. This kind of sensitivity testing can be very useful in helping decisionmakers with their policy and program choices.

It often happens in systems-analysis work that the ranking of the alternatives is sensitive to the ways the major uncertainties have been as-

---

[2] Uncertainties that involve events about which we have some conception, but for which neither the full range of outcomes nor the probabilities of occurrence are known. Dror (Section 7.3) calls these "qualitative" uncertainties.

sumed to be resolved in the future. Sometimes it even happens that no significant dominating alternative can be found independent of such future uncertainties. Then systems analysts must ask questions like: Can we think of ways for the decisionmakers to buy either more time or more information in order to reduce the uncertainties? Are there ways of hedging against some of the more important uncertainties? Since decisionmakers usually want to adopt positive answers to such questions, provided they do not cost too much, cost implications of such measures are important subjects of investigation. To help, cost analysts can undertake special analyses aimed at questions such as these:

> How much would it cost and how long would it take to conduct exploratory development work on critical, long-lead-time subsystems or components, with a view to preserving a wide range of options for the future?

> How much would it cost to conduct experiments in critical technical areas in order to reduce technological uncertainties?

> How much would it cost to conduct field exercises designed to resolve uncertainties about novel operational concepts associated with new alternatives?

Presenting Results

How results are presented varies with the nature of the problem being analyzed and the type of analytical approach adopted for the systems analysis as a whole. Let us consider some illustrative cases.

One of the simplest situations—one that is becoming rarer in practice—is where both costs and benefits can appropriately be measured in MUs. Here, for each policy alternative, the cost analyst estimates the costs and benefits in streams over time. The alternatives are then compared on the basis of discounted net benefits (benefits less costs), internal rate of return (that is, the discount rate that makes the present value of the net benefits equal to zero), or some other convenient metric. Thus, if there are six alternatives, the cost-benefit study results for the nominal case can be summarized in six numbers; other sets of six numbers could be presented to show the results of sensitivity analyses or variants from the nominal case.

Another relatively simple case is one in which the systems analysis adopts an equal-effectiveness framework, in which all the alternatives are designed to achieve a desired level of effectiveness. Since the alternatives are normalized with respect to benefits, the focus of attention in the results is on cost considerations. (This case, like the cost-benefit one, is also becoming rarer in practice, as the problems considered involve multidi-

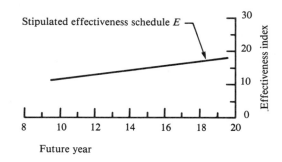

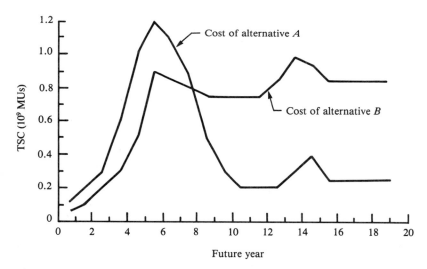

Figure 10.3. An equal-effectiveness case: The time streams of cost for an example in which alternatives *A* and *B* are both required to meet the effectiveness schedule *E* starting ten years after the program begins. (*From Fisher, 1970, p. 229.*)

mensional benefits for which equal-effectiveness alternatives are diffi-cult—and usually impossible—to design.)

Figure 10.3 shows an equal-effectiveness example. Here, alternatives *A* and *B* are required to satisfy the effectiveness schedule *E* over a period of years beginning 10 years from the present. Because of their develop-ment and production lead times, both alternatives have substantial costs (mostly development and investment) occurring during the first ten years, before their operational capabilities can be applied. On the other hand, the two alternatives have markedly different levels of operating costs, and both have to incur modifications in years 12 to 15 to enable them to

Table 10.2. The Total System Costs (TSCs) for the Example
Shown in Figure 10.3 Assuming Four Different
Discount Rates

	The present value TSC (10^9 MUs)	
Discount rate (percent)	Alternative A	Alternative B
0.0	9	14
4.75	6	8
7.0	5	6
10.0	4	5

meet the requirements of increasing effectiveness as stipulated in schedule
E. (The cost streams are expressed in terms of constant year $-$ 1 MUs.)

If the time preference assumption is a zero discount rate for the first
20 years and an infinite rate thereafter, the total system costs for A and
B are 9×10^9 and 14×10^9 MUs, respectively, so that alternative A is
the least-cost choice to meet effectiveness schedule E over a distant period
of the future. The obvious question—and one that must be investigated
by the cost analyst—is whether or not the choice is sensitive to the discount rate for time preference. To explore this question, let us assume a
minimum rate of 4.75 percent and two higher rates of 7 and 10 percent.
Table 10.2 gives the results. Over a wide range of assumptions about time
preference, the ranking of the alternatives does not change, even in this
rather extreme case involving marked differences in the ratios of acquisition and operating costs for the two alternatives. On the other hand, for
the higher rates of discount the differences in total system cost become
less significant—so much so that the decisionmakers would probably be
indifferent in choosing between A and B on the basis of their discounted
total system costs.

The opposite of the equal-effectiveness framework is the equal-cost
framework, in which a decisionmaker aims to get the greatest capability
or benefit from a given budget—or cost. Here the main function of the
cost analyst is to assist in designing the equal-cost alternatives. Thus,
since TSCs are fixed in advance, the final results to be compared are the
benefits and noneconomic costs or disbenefits associated with each alternative. For example, for a study comparing transportation-system alternatives, a format such as that shown in Table 10.3 could be used. It
is usually the case that neither an equal-cost nor an equal-effectiveness
framework can be employed, because most systems analyses ask broader
questions than these frameworks are suited to answering, and because
decisionmakers are focusing more on external costs that were formerly
ignored both in their decisions and in the systems analyses supporting
them. Thus analysts today and in the future face problems in how to

Table 10.3. An Example Format for Presenting the Results of an Equal-Budget Analysis of the Benefits and Disbenefits of Four Alternative Transportation Systems

| Attribute | Equal-budget (X million MU) alternatives | | | |
	A	B	C	D
Travel time				
Comfort				
Passengers per year				
Households impacted by noise pollution[a]				
Air-quality impacts[a]				
.				
.				
.				
etc.				

[a] Noneconomic costs, quality-of-life disbenefits.

conduct analyses over these broader contexts and how best to present their results to decisionmakers in meaningful and effective ways.

It would obviously be desirable to be able to combine the results of an analysis into a single index—as was done in the equal-effectiveness example by using total system cost—and considerable effort has been devoted to devising schemes aimed at this goal (for a brief overview, see Schwarz et al., 1985, pp. 229–239). Although a detailed description of such possible approaches is beyond the scope of this chapter, some general conclusions from the point of view of applied systems analysis are worth noting here (Schwarz et al., 1985, p. 230):

> Any aggregate approach . . . has two serious disadvantages. One is that a great deal of information is lost by aggregation; the fact that alternative A has environmental problems whereas alternative B has political implementation problems is suppressed. The second is that any single measure of value depends on the relative weights assigned by the analyst and the assumptions he used to get them into commensurate units.
>
> To produce anything resembling a valid value function is clearly difficult, and perhaps impossible in many situations. . . .
>
> Many analysts believe that, while such value functions are clearly useful for screening the alternatives, the final designation of a preferred alternative must be made by other means. Particularly when the decision concerns the public sector, and the preferences depend on basic values, the decision thus being essentially political, more disaggregated information needs to be communicated to decisionmakers. Nevertheless, the analyst may, while developing and using value functions for his own initial inquiry, find that his understanding of the complexity of the prob-

lem (and consequently the quality of the advice he finally offers) has been enhanced.

From this point of view, the problem of presenting the results of a broad systems analysis becomes one of how to summarize, organize, and present a complex array of information so as to make its meaning clearly understood by the persons to whom it is addressed. Experience shows that there is no universal solution to this problem; rather, each analysis and its audience dictates a solution tailored to its unique situation. One notable and useful approach is the scorecard technique developed by Goeller in 1972 (Chesler and Goeller, 1973). This technique is part of an overall approach called *system impact assessment*, which aims to evaluate spillover effects on the environment and other segments of society, along with the financial costs and service benefits normally addressed in analysis. Each of these impacts is presented in terms of its natural units, rather than being converted into a single measure of worth, such as monetary units. The comparative ranking of the alternatives for a single impact is indicated by color coding, shading, or some other symbology. A scorecard for a set of alternatives is a table showing both the rank-coding and numeric outcomes for the different impacts.

Table 10.4 is an example summary scorecard that was used to present the results of a study of future transportation alternatives. Here, each row deals with a single impact or attribute (note that some of these are benefits and some are disbenefits, or costs) and shows the numeric results for each alternative. Color codes are used (in the original) to indicate the relative ranking of the alternatives for each impact. When the results from several impacts are combined in a scorecard, the decisionmakers have a comprehensive picture of the multiple attributes of the alternatives. For two other examples of the use of this scorecard technique, see the *Overview* (pp. 99–108 and 230–238).

The results shown in Table 10.4 are for a single, nominal set of assumptions about statistical uncertainties and about the state of the world at the time when the alternative(s) are to be implemented. The scorecard lends itself to presenting results of sensitivity analyses through the use of overlays to show the changes in rankings that result from different sets of assumptions. This can help to find situations where the *rankings* of alternatives are insensitive to certain assumptions whether or not the *absolute* levels of costs or benefits are sensitive to them. (Note that this arises largely from the advantages of *color* coding, which can convey an overall impression of how rankings change more readily than can other rank-coding schemes.)

One important principle of this approach is that it provides factual information about the costs, benefits, and disbenefits of complex choices without imposing external views about the relative importance of the at-

Table 10.4. A Summary Scorecard for a Systems Analysis of
Transportation Alteratives

System impacts		Transportation system alternatives				
		CTOL	STOL	VTOL	TACV	IPT
Annual passenger volume (millions)	(+)	6.5	5.3	4.9	9.1	5.8
Annual passenger miles (millions)	(+)	1956	1709	1463	2440	1625
Cost/passenger mile (cents)	(−)	4.34	5.79	6.00	15.41	8.20
Capital investment required ($ millions)	(−)	135	305	190	1856	434
Annual subsidy required ($ millions)	(−)	0	0	0	277	74
Travel time: Los Angeles/ San Francisco (min.)	(−)	171	145	145	199	314
Travel cost: Los Angeles/ San Francisco ($)	(−)	17.17	21.66	21.69	18.46	18.49
Net change in employment- peak year (000s)	(+)	0	14	5	52	11
Urban land required (acres)	(−)	0	135	12	514	4
Noise impacts (no. of households exposed)	(−)	64029	31521	21523	4989	8463
Minority noise exposure (% of all exposed)	(−)	8.8	11.9	7.8	5.7	9.9
Low-income users (% of all users)	(+)	13.0	12.9	12.9	13.0	24.3

Source: Chesler and Goeller (1973).

Transportation system abbreviations:
    CTOL = Conventional takeoff and landing aircraft (base case)
    STOL = Short takeoff and landing aircraft
    VTOL = Vertical takeoff and landing aircraft
    TACV = Tracked air-cushion vehicle
     IPT = Improved passenger train

Symbols:
    (+): Larger numbers are better outcomes (benefits)
    (−): Smaller numbers are better outcomes (costs or disbenefits)
    ☐ : Indicates an alternative with best outcome for this impact (green)
    ▨ : Indicates an alternative with an intermediate outcome (orange)
    ■ : Indicates an alternative with worst outcome for this impact (red)

tributes. The decisionmakers can then add to this information their feelings for human values and the values of the society they represent, the intangibles that underlie all human decisionmaking. When public policy choices are at stake, the decisionmakers and those who are affected by the choices may have many diverse (and often conflicting) values. Hence, it is doubly important that policy analysis results retain this multi-attribute aspect.

The scorecard technique is a means of expanding the number of separate attributes readily addressable by decisionmakers in making choices among a *limited* number of alternatives. This is of obvious benefit in presenting results for complex, multiattribute policy issues. But other means must be used to screen the usually larger number of alternatives needing examination in the earlier stages of a study and thereby reduce to a manageable level the number to be evaluated in more detail. (See Chapter 6 and Walker, 1985, for discussions of screening methods.) Although it may be feasible to use some sort of weighting scheme to combine multiple attributes into a single index for this screening, the more common practice is to choose one major index for costs and another index for benefits for an overall cost-effectiveness index by which all alternatives can be examined and major sensitivity analyses done. This was the approach taken in the systems analysis study discussed in the next section.

Conclusion

The discussion and examples in this section have summarized the basic principles and approaches of cost analysis as a part of systems analysis; for further detail the reader should consult treatises focusing on cost analysis. Fisher (1971) is a definitive text on cost considerations in systems analysis. More recent, shorter essays on cost issues appear in Quade (1982) and Fisher (1977). Although not specifically devoted to cost analysis, Mood (1983) discusses several techniques of applied mathematics for policy analysis that are also fundamental to cost-analysis practices. A number of works on cost estimating have been published in recent years, most of them with a cost-engineering focus, that is, cost analysis for project management. Two books with a somewhat broader focus are Blanchard (1978) and Seldon (1979). The cost-analysis techniques discussed in these works are useful for analysis applications, although the focus of the books is primarily on cost analysis for the management of government procurement programs.

For examples of cost-analysis applications in studies of electrical power production, see Mooz (1978), Hess et al. (1983), and Stucker (1985). A variety of cost-analysis methods are demonstrated in reports documenting a policy analysis of water-management techniques for the Netherlands (see especially the summary volume, Goeller et al., 1983, and the two volumes treating major cost estimates and models, Petruschell et al., 1981 and 1982).

The next section of this chapter deals with a specific application of cost-analysis techniques to the problem of reducing air pollution in a large metropolitan area. The solutions proposed for this problem require several different kinds of cost-estimating techniques, and their evaluation involves both economic costs and noneconomic costs and disbenefits.

10.4. An Illustrative Example: The San Diego Clean Air Project

Air pollution is a significant problem in many areas of the world. The air basins surrounding urban areas are especially subject to pollution from industry, electric power generation, and transportation systems. In 1970 the U.S. Congress passed the Clean Air Act Amendments setting various air-quality standards for air basins in the United States and requiring that the states prepare and promulgate air-quality implementation plans for each region to meet the standards.

The implementation of air-quality control strategies to meet strict standards can have both positive and negative effects on the quality of life in regions where air pollution levels are high. These, of course, include the desired benefits (improvements in air quality) and the obvious economic costs (expenditures for control devices). But other disbenefits may accrue from the control strategies, including changes in travel volumes, times, and costs; changes in employment, taxes, and energy consumption; and changes in the distribution of such effects among different locations and social groups (such as the poor). Different control strategies will produce quite different regional effects: rationing gasoline, for example, reduces vehicle miles and auto-related employment but produces savings in energy consumption; adding emission-control devices to older cars increases their costs and energy consumption but produces employment opportunities for mechanics; increasing the use of buses increases average travel times for individuals but produces energy savings, employment, and travel opportunities for those without cars.

Concern over these issues led the government of San Diego County, California, to ask The Rand Corporation to undertake a systems analysis, the San Diego Clean Air Project, to analyze alternative strategies in terms of a comprehensive set of effects on the quality of life in San Diego and to identify the most promising strategies for implementation. (Goeller et al., 1973, summarize the results of the analysis and the methods used.)

Analytical Approach

Pollutant emissions come from a variety of both mobile sources, such as automobiles, and fixed sources, such as smokestacks. In the San Diego Clean Air Project, a variety of control *tactics*, such as adding emission-control devices to automobiles or providing incentives to carpool, were identified for each source. Control *strategies*, comprising mixes of tactics for the different emission sources, were devised to achieve air-quality improvements over the entire region. The number of potential strategies was so large that it was impractical to evaluate all of them in terms of detailed effects; hence, an initial screening was done to identify the most promising strategies. The criterion by which this screening was accomplished was the cost-effectiveness of a strategy—cost being measured by the monetary cost of procurement for and operation of the control tactics used in the strategy and effectiveness being measured by the amount by which the strategy would reduce emissions, primarily of reactive hydrocarbons, which yield oxidant, the most troublesome pollutant.

After screening, the next stage of analysis was a more extensive evaluation of the strategies identified as superior. This required that detailed predictions be made for each strategy of a variety of effects or *system impacts*. These were grouped as follows:

Environmental impacts, such as air quality and energy consumption.

Transportation-service impacts, such as travel times and volumes.

Economic impacts, such as investment costs, taxes, and employment.

Distributional impacts, or how selected impacts are distributed by social group (costs by income group, for example) or by location (air quality by geographic region, for example).

The strategies were evaluated under alternative sets of assumptions about technology and about the future population growth of San Diego.

Finally, a comparative analysis of the alternative strategies was prepared using scorecards composed of tables that displayed the various impacts and showed, by color code, each strategy's ranking for the most significant impacts. The scorecards were used to convey the research findings to the decisionmakers and to the public to aid in the selection of the preferred strategy.

Components of the regional strategies. It is convenient to consider an overall strategy for the region as made up of three components, termed *pure strategies,* where each applies a certain class of tactics to a particular class of sources:

Fixed-source controls. These include both technological controls (such as emission-control devices), and managerial tactics (such as the use of tractor towing to move aircraft while they are on the ground at airports). Aircraft were classified as fixed sources of pollutant emissions because their significant emissions occur at fixed locations: at or near airports.

Retrofit-I/M. These involve applying retrofit control devices and inspection/maintenance (I/M) policies to in-use vehicles. Vehicles are categorized as (1) light-duty motor vehicles (LDMVs), such as automobiles and light trucks; (2) heavy-duty vehicles, such as heavy trucks (over 6000 pounds) and diesel buses; or (3) motorcycles.

Transportation management. These strategies involve tactics whose purpose is either to reduce LDMV miles (the origin of most mobile-source emissions) or to reduce congestion. They include such tactics as bus-system improvements, mileage surcharges and/or gasoline rationing, incentives to promote carpooling, and traffic controls (expressway on-ramp metering, for example).

An overall strategy for the region, involving a particular mixture of pure strategies, is termed a *mixed strategy.*

Components of the study methodology. The San Diego Clean Air Project presents a rich assortment of analysis problems. The strategies for reducing pollutant emissions vary both in the agencies or individuals who must pay to implement them and in the kinds of costs that are incurred. The importance of potential quality-of-life disbenefits in the region required that the framework used for evaluating the strategies provide visibility to these effects in addition to economic costs and air-quality improvements. A single optimization model was out of the question under these circumstances (see Section 2.6).

Several models and methods were adopted and tailored to San Diego's particular characteristics for the study. At the same time the models were purposely made general enough to be easily adaptable to studies of other regions. Four of these methodology components pertain particularly to the cost-analysis aspects of the study:

The fixed-source analysis techniques (DeHaven and Woodfill, 1973) evaluate the cost and effectiveness of controls on fixed sources, including aircraft. They also estimate the distribution of costs among various industries and governmental agencies.

The motor-vehicle emission and cost (MOVEC) model (Mikolowsky, 1973) evaluates the cost and effectiveness of retrofit and inspection/maintenance strategies for in-use vehicles. MOVEC also reports detailed effects for each retrofit strategy considered, including the increase in fuel consumption and the distribution of retrofit costs among different income groups.

The bus-system cost model (Petruschell and Kirkwood, 1973) estimates the costs of providing a particular quantity and quality of bus service to a region. It also estimates a variety of economic and financial effects, such as employment, fuel consumption, and taxes.

The transportation model (Bigelow et al., 1973) evaluates the costs and other effects of transportation-management strategies, such as added gasoline taxes or improved bus service. It incorporates an aggregate version of the bus cost model to estimate costs for the bus system included in a strategy. In computing transportation demands, it considers induced changes in the number of trips made by individuals, amount of carpooling, and the degree of traffic congestion. It reports detailed transportation-service effects (both in the aggregate and as distributed among income groups and travel purposes) and various economic effects.

Another component of the methodology, the *tradeoff model* (Mikolowsky et al., 1973), brings the other components together in order to find the best mixed strategy, given a single fixed-source control strategy and menus of vehicle-retrofit and transportation-management strategies.

Cost Concepts

In most problems of public policy analysis the principal monetary costs of the alternatives are borne by a single governmental body. But in this example, the costs could be borne either by government agencies, by private businesses and industries, by individuals, or by some combination of these. Who bears the costs depends on the control strategy employed and the type of financing used, and real costs would be understated if the only ones considered were those borne by the governmental agencies. For the initial screening of strategies, the study employed a *regional* cost concept: the total monetary cost of procuring, installing, and operating the control devices, control techniques, or alternative transportation means used in the San Diego region. Under this concept, certain transportation payments made by individuals—such as bus fares, regional transportation taxes, or mileage surcharges borne by motorists—are transfer payments within the region (assuming there are no interactions with other regions). These affect the *distribution* of costs but do not change the overall regional cost of a strategy. (The distribution of the costs is, of course, an important consideration in the final evaluation of strategies.)

Two factors raise problems in dealing with the time-incidence of costs in this study. First, the costs are distributed irregularly over time, and second, the strategies are mostly open-ended (that is, the time horizon is indefinite). The approach that was adopted was to express all costs as *annualized* values in constant dollars (with 1972 as the base year). This is readily done for annual operating costs, but procurement costs (for purchase and installation of equipment and facilities) usually occur as lump-sum costs at the point in time when the strategy is implemented.

To transform procurement costs into annualized figures, a capital-recovery approach was used. A simplified view of this approach is that of amortizing a loan over a given number of years by making equal annual payments that include both payment on the principal and interest on the unpaid balance, as one typically pays off a home mortgage or a car loan. (The concept is described fully in Grant and Ireson, 1970, pp. 35–36.) The annualized procurement cost of an item needed for one of the control strategies is calculated using a capital-recovery period that approximates the useful life of the item and a discount rate that is analogous to the interest rate in amortizing a loan.[3] A before-tax discount rate of 15 percent was used in the study for all investment costs borne by nongovernmental bodies. This approximates an after-tax rate of 7.5 percent, which was the rate used in discounting investment costs borne by governmental agencies.

Using these cost concepts, the alternative control strategies can be evaluated in simple cost-effectiveness terms: Cost is the annualized cost of procuring and operating the strategy in the San Diego region, and effectiveness is the amount by which the strategy reduces emissions. These unitary measures of the cost-effectiveness of a strategy are useful for screening a large number of alternatives to identify the most promising strategies. Those so identified can then be examined further in terms of their full cost-effectiveness, including other regional impacts on quality of life and distributional effects in addition to monetary costs and air-quality effects.

The models and methods used for each of the major types of control strategies employ different cost analysis approaches, reflecting their differing analysis requirements.

Fixed-Source Strategies

The fixed-source strategies employ separate, individual control tactics to reduce reactive hydrocarbon (RHC) emissions from specific sources. These include: petroleum storage facilities and gasoline service stations, organic solvent users (such as commercial dry cleaners), aircraft operations on and near airports, electric power plants, and metallurgical industries. The diversity of control tactics applicable to these sources precludes the use of a general model to evaluate them. Rather, a menu of tactics for each source must be devised and the cost and effectiveness of each must be estimated individually.

---

[3] In its basic form, the annualized payment R required to recover the capital cost P of an investment over a period of n years at a discount rate i is given by the formula $R = P\{i/[(1 + i)^n - 1] + i\}$. The expression in braces is termed the *capital recovery factor*.

For example, the Western Oil and Gas bulk terminal, which handles finished petroleum products, had storage tanks that were a source of substantial hydrocarbon emissions (in 1970). The control tactic for these emissions involved the installation of loading mechanisms and holding tanks to save vapors, which would otherwise be released to the atmosphere, and permit them to be reabsorbed by the fuel remaining in the storage tanks. The investment cost for this tactic was based on the cost of similar installations at other fuel storage facilities in the country, scaled on the capacities of the storage tanks. For the Western Oil and Gas storage tanks, the cost was estimated at $135,000. Using the capital recovery technique discussed above, and assuming a ten-year life and a discount rate of 15 percent (before taxes), the annualized cost would be $26,900. No additional operating and maintenance costs were expected for this control tactic, but some monetary savings were expected because most of the entrapped vapors could be recovered as gasoline that would be lost in the absence of the control. The annual saving in recovered gasoline was estimated at $51,000. Hence net annual cost (saving) is:

Annualized investment	$ 26,900
Operating and maintenance	0
Saving from gasoline recovered	(51,000)
Total annual cost (saving)	$(24,100)

It was estimated that this control tactic would reduce total hydrocarbon emissions by 1976 tons per year, and RHC emissions are about 55 percent of these. Hence the saving per ton of RHC removed would be $22.18.

Similar estimates of costs and of emission reductions were made for a number of other fixed sources of RHC emissions in the San Diego region. Ranking these by the gross cost-effectiveness measure of cost per ton of RHC removed permits construction of the graph in Figure 10.4, which shows the amounts by which RHC emissions are reduced and the cumulative annualized cost to achieve these reductions as the various control tactics are introduced successively.

Retrofit and Inspection/Maintenance Strategies

The vehicles responsible for most of the hydrocarbon emissions in urban areas of California are light-duty motor vehicles (LDMV—vehicles with a gross weight of fewer than 6000 pounds). Because of the increasingly stringent emission standards imposed on new vehicles sold in California since 1962 (and nationwide since 1968), new vehicles already have reduced emission levels. But older vehicles remaining in use can contribute significantly to the pollution problem for years, and improper maintenance

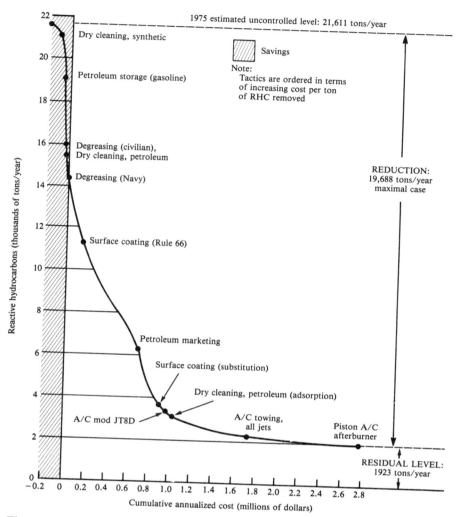

Figure 10.4. The relation between the reduction of hydrocarbons and cost as control tactics are introduced successively. (*From Goeller et al., 1973, p. 14.*)

on vehicles sold after the stricter emission standards came into being can erode the effectiveness of their emission controls. Hence, an important set of control tactics involved retrofit installation of emission control devices on used vehicles and mandatory vehicle inspection and/or maintenance schemes to assure that in-use vehicles are kept operating at low emission levels.

The number of candidate retrofit devices and inspection/maintenance schemes is not in itself so large as to cause any great difficulty in estimating either the cost of installing and operating them or the effectiveness of each in reducing emissions from vehicles of a given category. But the number of possible combinations of control tactics, applied to different model years under varying vehicle usage rates and average trip speeds, is quite large. Therefore the MOVEC model was built to ease the computational load. For a given combination of retrofit-I/M tactics and a specific calendar year, the model calculates: (1) vehicle emissions for all types of vehicles; (2) total annualized costs to the region for purchase, installation, and operation of the devices and I/M policies; and (3) the distribution of costs by income group under three alternative financing schemes. Characteristics of the vehicles in use (such as age distribution and mileage distribution by age of vehicle), average speed factors, and other regional characteristics are used in the model to specify the basic environment within which the candidate tactics are evaluated.

The cost calculations for the retrofit-I/M tactics are straightforward, but of course they must be applied to various combinations of model years. Each of the vehicle exhaust emission control tactics is characterized by four cost components: (1) purchase price of the retrofit device, (2) installation charge for the retrofit device, (3) annual increase in maintenance cost incurred by the device or I/M policy, and (4) increase in fuel consumption experienced by vehicles to which the device or I/M policy is applied. The first two of these are investment costs, and the second two represent operating costs. The incremental investment cost per vehicle for tactic n applied to vehicles of model year i can be expressed as

$$C_i^N = C_{1i}^n + C_{2i}^n,$$

where

C_{1i}^n = purchase price of device n for model year i vehicles, and

C_{2i}^n = installation charge for device n on model year i vehicles.

To calculate operating costs, the cost analyst must have factors representing the average annual miles traveled V by each vehicle, and average fuel cost f per vehicle-mile. Vehicles typically are driven fewer miles each year as they grow older, hence the value of V varies by vehicle model year. (The mileage variation by year is used explicitly in the MOVEC model for emission calculations, but for the annual operating cost calculation, the value V for vehicles of a given model year is simply the sum of the declining annual-miles-traveled figures divided by the expected remaining vehicle life.) The expression for the annual incremental operating cost per vehicle for tactic n applied to vehicles of model year i is

$$C_i^R = C_{3i}^n + fV_iC_{4i}^n ,$$

where

C_{3i}^n = increase in annual maintenance cost for tactic n applied to vehicles of model year i, and

C_{4i}^n = fractional increase in annual fuel consumption for tactic n applied to vehicles of model year i.

For total-annual-cost calculations, the investment costs are annualized over the expected remaining life for model year i vehicles, using the appropriate capital-recovery factor CRF, calculated according to the formula given earlier. In actual application, such costs may be financed through any of several user-paid or tax-subsidized schemes, depending upon the importance attached to the distributional effects of the various payment schemes. In all cases, however, the financing schemes represent costs borne by individuals and government bodies; hence, the public/government discount rate (7.5%) rather than the business rate is used in the capital-recovery formula.

For a given regional strategy, which is composed of various combinations of tactics applied to specific mixes of model years, the total annualized regional cost is calculated simply by (1) multiplying the annualized cost per vehicle for each tactic/model-year combination by the number of vehicles of that model year in the population and (2) summing over the combinations that are included in the strategy. Hence, the total annualized cost C^T for the strategy can be expressed as

$$C^T = \sum_{i=y}^{i=Y} [C_i^N - \text{CRF}_iC_i^R]N_i,$$

where

CRF_i = the capital-recovery factor for vehicles of model year i,

N_i = the number of model year i vehicles in the total population,

y through Y are the earliest through the latest model years of vehicles included in the population, and

C_i^N and C_i^R include the costs of all retrofit devices and/or I/M schemes applied to the vehicles of model year i.

In its most basic application, the MOVEC model is applied to a large number of retrofit-I/M strategies, producing for each an estimate of the total RHC emissions per year from the vehicle population and an estimate of the annualized cost of implementing the strategy. Figure 10.5 shows the results for 28 promising strategies evaluated for the San Diego Clean

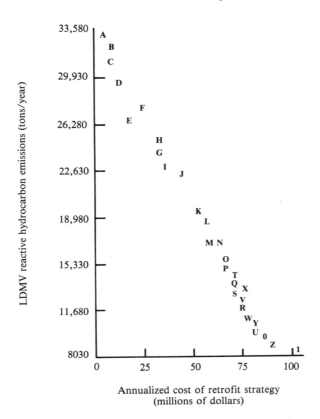

Figure 10.5. For light-duty motor vehicles (LDMVs), the total annual reactive hydrocarbon emissions versus the total annualized cost of retrofit strategies for 28 selected strategies. The strategies are labeled A through Z, 0, and 1. (*From Goeller et al., 1973, p. 19.*)

Air Project. Each strategy represents a particular combination of tactics; that is, unlike Figure 10.4, they do not represent individual tactics that are to be added together successively. For example, strategies L, M, and R, which later were determined to be appropriate components of some of the most cost-effective mixed strategies, consisted of the tactics and applicable model years shown in Table 10.5.

Transportation Management Strategies and the Generalized Bus-System Cost Model

Transportation management strategies mainly involve tactics that reduce light-duty motor vehicle miles traveled by lessening the number or length of LDMV trips that would occur in the uncontrolled situation. In the San

Table 10.5. The Components of Selected Retrofit-I/M Strategies

Strategy	Tactic	Applicable model-years
L	Crankcase blowby control	1960–1955
	Catalytic exhaust gas converter and state inspection/maintenance and NOx control kit and/or VSAD	1974–1968
	State inspection/maintenance and pre-1966 exhaust retrofit kit and NOx control kit and/or VSAD	1967–1955
M	Crankcase blowby control	1960–1955
	Evaporative loss control	1969–1968
	Catalytic exhaust gas converter and state inspection/maintenance and NOx control kit and/or VSAD	1974–1968
	State inspection/maintenance and pre-1966 exhaust retrofit kit and NOx control kit and/or VSAD	1967–1955
R	Crankcase blowby control	1960–1955
	Evaporative loss control	1969–1964
	Catalytic exhaust gas converter and state inspection/maintenance and pre-1966 exhaust retrofit kit and NOx control kit and/or VSAD	1974–1968
	State inspection/maintenance and pre-1966 exhaust retrofit kit and NOx control kit and/or VSAD	1963–1955

Source: Goeller et al. (1973, pp. 83–84).

Abbreviations:
   NOx = oxides of nitrogen
  VSAD = vacuum spark advanced disconnect (reduces NOx and hydrocarbon emissions)

Diego region, the most promising tactics appear to be bus-system improvements, taxes, LDMV mileage surcharges, gasoline rationing, and other incentives that induce people to make fewer trips, form carpools (for trips between work and home), or otherwise reduce the number of LDMV miles traveled in the region. The analysis of these strategies makes use of both the transportation model and the generalized bus-system cost model.

The transportation model is the primary tool for examining the transportation-service effects of mixtures of tactics. It employs an aggregate version of the bus-system cost model (a single, nonlinear function) that calculates a total annualized bus-system cost as a function of three variables generated in the transportation model: buses required, bus-miles, and annual bus-hours. Of course, the cost of a bus system is affected by many other variables as well. The function used in the transportation model is an approximation of the generalized bus-system cost model, which provides the capability to make in-depth analyses of particularly interesting alternatives.

The bus-system cost model was designed to be capable of examining a wide range of possible bus systems. However, it uses only a small number of high-level design and policy inputs to generate estimates suitable for use in the analysis and to examine the sensitivity of those estimates to variations in major system characteristics. The model was designed to meet these requirements: (1) an ability to handle system configurations ranging from demand-responsive systems using small vehicles to regularly scheduled systems employing the standard 50-passenger bus; (2) an ability to yield analyses of financial and economic effects, such as labor costs, taxes by recipient, and major investment items by kind; (3) an ability to reflect equally well the cost implications of public or of private ownership of bus-operating companies; and (4) a limitation to design inputs that could be generated within the framework of the transportation model.

Cost-analysis methods. The approach to cost classification and estimation used in the bus-system cost model is quite conventional for models of this kind. Initial investment costs are estimated first and include the total costs of vehicles, yards and shops, other equipment, other facilities, and land. Recurring annual operating costs such as drivers' and helpers' wages, fuel expenses, equipment maintenance and garaging are then estimated. Operating cost elements closely parallel those used by the American Transit Association (American Transit Association, 1971).

Initial investment costs are annualized using a capital-recovery factor like that used in the analysis of other strategies. The model allows the users to choose between public or private financing; tax considerations are accounted for in calculations for the latter. Total annualized costs combine after-tax return on investment, corporate profit taxes, bond interest, and depreciation along with other recurring annual costs. If the bus fares collected are insufficient to cover the total annualized costs, then a publicly-financed subsidy is used to cover the deficiency.

All cost estimates are made using cost estimating relationships (ERs) that were empirically derived from a cross-section of data on approximately 100 bus companies operating in the United States and Canada and from information supplied directly by bus manufacturers. (For example, the purchase cost of a bus is estimated as a function of the number of seats.) The inputs to the model are classified as:

System design specifications, such as number of seats per bus, fuel type (gasoline or diesel), fraction of buses in maintenance, and extra buses to cover downtime from scheduling and maintenance.

Inputs related to form of ownership, that is, public or private.

Inputs generated in the transportation model, such as number of buses required in the peak period, annual bus-miles driven, and annual bus-hours of operation.

Financial inputs, including the required after-tax return on investment, debt/equity ratio, bond interest rates, and relevant corporate profits tax rates.

Parameter values for the cost ERs, such as fuel cost per gallon, fuel-tax rates, land cost per square foot, and useful lifetimes and salvage values for each kind of asset.

The output information consists of:

Total system cost, split between initial investment and annual operating costs.

Total annualized cost, which includes investment costs annualized by the capital-recovery approach, and both direct and indirect annual operating costs.

Other financial information, such as after-tax return on investment, corporate profits taxes paid, bond interest paid, and subsidies if required.

Cost per trip and cost per passenger-mile for break-even operation, that is, the fares that would be needed to make revenues equal to costs.

Information required for analyses of economic effects, including costs of labor, buses, other equipment, structures, land, fuel, interest, and fuel taxes.

Coefficients of the aggregate form of the bus-system cost model for use in the transportation model.

Model overview. Figure 10.6 shows the major elements of the cost model and the flow of information within it. This flow chart is used as a guide for the discussion below. Arrayed to the left of the dashed line are the external inputs required by the model. Their sources are indicated at the top of the boxes (the supply model is part of the transportation model), and representative examples of input variables are shown below.

As the first step in the model, annual operating and initial investment costs by cost element are estimated using inputs from the supply model, system design specifications, and the relevant cost ERs. Total investment costs by cost element are printed out from these calculations. The next step in the model uses the financial analysis parameters (representing

402

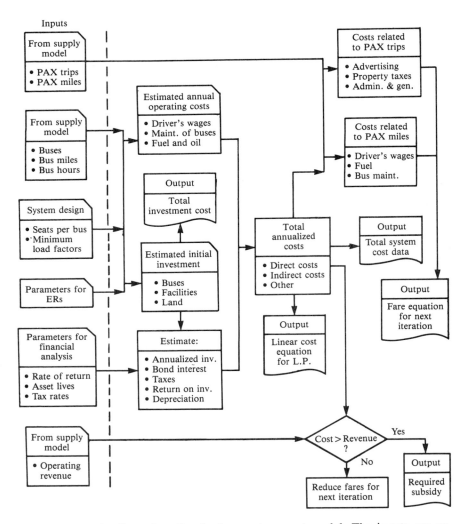

Figure 10.6. The flow chart for the bus-system cost model. The inputs are enclosed by punch-card symbols, outputs by paper-copy symbols, data processing and calculations by rectangles. Inputs to the left of the dashed line are external inputs. The diamond-shaped figure is a branch point. PAX is an abbreviation for passengers. (*From Petruschell and Kirkwood, 1973, p. 8.*)

different forms of ownership and financing arrangements) to estimate annualized investment cost, bond interest cost, various taxes, return on investment, depreciation, and so on. These cost estimates are combined with the annual operating cost estimates and divided, by cost element, into direct, indirect, and "other" cost categories.

Total annualized costs (annual operating, plus annualized investment, plus corporate profits taxes and bond interest), are next divided by the model into those that vary with total passenger miles (such as fuel, drivers' wages, and bus maintenance expenses), and those that may vary with the number of passenger trips but are relatively insensitive to trip length (such as advertising costs, property taxes, and general and administrative expenses). A total cost is calculated for each group and divided by the total passenger trips or passenger miles (input from the supply model) to obtain the coefficients for the break-even fare equation. The total annualized costs are also printed out at this point for use in analyses of other economic effects.

The supply model uses bus fares as one of its inputs in generating bus-usage rates (passsenger demand varies inversely with fare levels) and other supply model outputs, which are used in turn as inputs to the bus-system model. The fares produced by the break-even equation in the bus-system model can be used iteratively as inputs to the supply model to seek break-even fares that balance bus transportation demands with bus-system costs. Alternatively, fares can be specified exogenously for both models, and the bus-system cost model will calculate the subsidy required (if any) to bring total revenues up to the calculated annual cost to purchase and maintain the required bus system.

Cost element framework. The bus-system cost model is assumed to be applied to systems that required no new development; hence, only initial-investment and annual-operating-cost categories are required. Initial-investment-cost elements are listed in Table 10.6. These cost elements cover the one-time outlays necessary to provide the equipment, structures, land, and so on required to operate any bus system. The categories for buses and for yards and shops are self-explanatory. The special-control-facilities cost elements are included to pick up the costs of dispatching, routing, and control facilities and equipment for demand-responsive systems. The bus-system cost model includes cost-estimating relationships for most of these cost elements. Where any or all of these items represented in a category are already available, or inapplicable, the calculations can be adjusted or zeroed out by the user.

Table 10.7 lists the annual operating cost and annualized investment cost elements. These cost elements follow those of the American Transit Association, thereby providing comparability with other transit-system cost studies and helping to ensure that data in the proper categories are

Table 10.6. Initial Investment Cost Elements

Buses
Yards and shops
   Structures
   Equipment
   Paving and parking areas
   Land
Special control facilities
   Structures
   Equipment
   Land

Source: Petruschell and Kirkwood (1973, p. 20).

available to construct ERs for the operating-cost elements. Bond interest, depreciation, taxes on corporate profits, and net return on investment would be adjusted by input choice to reflect the desired form of public or corporate ownership and financial structure.

Cost estimating relationships. The primary investment and annual operating costs are calculated in the bus-system cost model as functions of

Table 10.7. Annual Operating Cost and Annualized Investment Cost Elements

---

Equipment maintenance and garaging	Other operating taxes and licenses
Labor	Property taxes
Materials and other	Bond interest costs
Drivers' and helpers' wages	Buses
Fuel and oil (not including taxes)	Equipment
Other transportation cost	Structures
Labor	Land
Materials and other	Depreciation costs
Administration and general costs	Buses
Labor	Equipment
Materials and other	Structures
Maintenance and operation of special	Corporate profits taxes
control facilities	Buses
Labor	Equipment
Materials and other	Structures
Insurance and safety cost	Land
Labor	Profit (after-tax return on investment)
Materials and other	Buses
Traffic and advertising cost	Equipment
Labor	Structures
Materials and other	Land
User taxes on fuel	
Federal	
State	

Source: Petruschell and Goeller (1973, pp. 23–26).

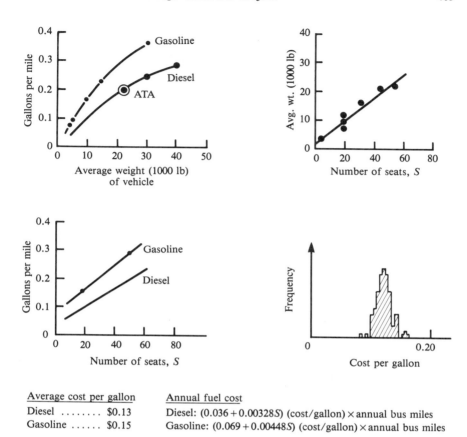

Figure 10.7. The development of a cost-estimating relationship for fuel cost (less tax). (*From Petruschell and Kirkwood, 1973, p. 11.*)

the inputs generated by the transportation model (buses required in peak period, bus-miles driven, and so on) and inputs describing the bus system's design (number of seats in bus, fuel type, and so on).

Cost-estimating relationships were developed for these calculations from data on existing buses, equipment, and transit systems. The development of an ER for fuel cost typifies the process used for most of the ERs. Annual fuel cost is to be estimated as a function of bus size (number of seats), annual miles driven, type of fuel used (gasoline or diesel), and the cost (less taxes) of a gallon of fuel. The major steps in developing the ER from the data collected for the study are shown in Figure 10.7.

The first step was to relate fuel consumption measured in gallons per mile to average vehicle operating weight. Data on vehicles ranging from passenger cars to buses and large trucks—both diesel- and gasoline-powered—were collected from various sources. The data were adjusted for driving patterns and plotted as shown in the upper left graph in Figure 10.7. The overall average (American Transit Association, 1971) for the typical diesel bus (indicated by the dot marked ATA) falls nicely on the lower curve, indicating that the data are in agreement with another authoritative source (at least in that range). In addition, the results shown were checked against current theory regarding diesel versus gasoline fuel efficiency. Diesel is more efficient than gasoline for two reasons: (1) it contains more BTUs per gallon than gasoline and (2) diesel engines operate more efficiently than gasoline engines. When the theoretical values of these differences are combined, they account quite well for the difference between the two curves.

The next step was to use data on various sizes of buses obtained from bus manufacturers to relate average operational weight to number of seats. The result is shown in the upper right graph of Figure 10.7. Each of the curves shown in the upper graphs were first drawn free-hand and then expressed mathematically. The mathematical expressions were combined to obtain the final cost-estimating relationship shown in the graph on the lower left of Figure 10.7.

Fuel costs per gallon are left to be specified when running the bus-system cost model so that price variations (due, for example, to geographic price differences, or to quantity purchase discounts) can be taken specifically into account. The lower part of Figure 10.7 shows the mathematical form of the ERs developed for fuel cost and the lower right graph shows some typical fuel prices and fuel-price distributions (in 1971).

Applications and Sensitivity Analyses Using the Bus-System Cost Model

Figure 10.8 presents a graph of output from the bus-system cost model showing the relationship between annual cost and annual bus-miles assuming a nominal 50-passenger diesel bus operating at two average velocities. Two facts are apparent from these curves: (1) increased velocity results in decreased cost and (2) diseconomies of scale exist (doubling the size of a bus system, as measured by annual bus-miles traveled, results in more than double the cost). Increases in average velocity reduce total cost because—other things being equal—a given annual volume of passenger-miles can be accomplished in a shorter period of daily bus-hours of operation. A reduction in bus-hours leads to a similar reduction in driver-hours and, because drivers' wages account for a large part of annual cost, a significant reduction in total annual cost.

The diseconomies of scale were first observed in the data and, as the fact was somewhat surprising, the results were checked with persons

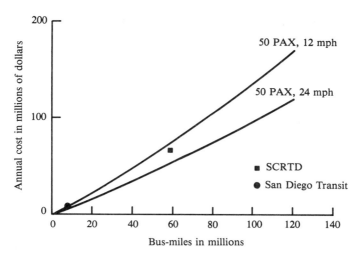

Figure 10.8. The bus cost model output: Annual cost versus total bus-miles for two average velocities for 50-passenger diesel buses. SCRTD is an abbreviation for Southern California Rapid Transit District. (*From Petruschell and Kirkwood, 1973, p. 42.*)

involved in transit-company operations. The explanation of this phenomenon was that, as transit companies become larger, they make less efficient use of drivers' time, spend proportionately more on advertising, have a proportionately larger and more expensive administrative structure, and so on. The less efficient use of drivers' time seemed to be the main factor.

As a further validity check of the model, data on bus-system experience were examined for both the San Diego Transit Corporation and the Southern California Rapid Transit District (SCRTD) (as reported in American Transit Association, 1971). The results are plotted on the graph in Figure 10.8. San Diego Transit, which has an average bus speed of about 12 miles per hour (mph), falls right on the 12-mile-per-hour curve. SCRTD probably achieves a somewhat greater average speed than San Diego, because average trip distances are longer (data on SCRTD bus speeds were not available), and its plot point falls somewhat below the 12-mile-per-hour curve.

Figure 10.9 shows graphs of the cost per bus-mile for systems operating at 12- and 24-mph average velocities with 50-passenger diesel buses. In addition, it shows cost per bus-mile curves for 20-passenger gasoline-powered buses. Here the diseconomies of scale are more evident. The curves also show that doubling the average velocity of the 50-passenger bus has a much greater effect on cost per bus-mile than does using the smaller-sized buses.

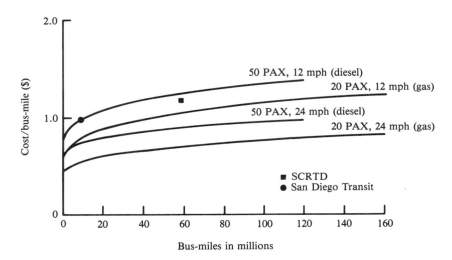

Figure 10.9. The bus cost model output: Cost per bus-mile versus total miles for two average velocities and two bus types and sizes. (*From Petruschell and Kirkwood, 1973, p. 43.*)

Figure 10.10 shows the results obtained from the bus-system cost model in a more extensive sensitivity analysis of factors that influence the costs of a bus system. Here, total annual cost is plotted against annual passenger miles. The major purpose of this analysis was to generate estimates of the cost of providing bus systems ranging between two benchmark volumes of passenger usage: (1) 100 percent of the passenger miles now traveled in the San Diego region and (2) 80 percent of the passenger miles now traveled in the Los Angeles region. Figure 10.10 makes clear how sensitive costs are to the assumptions one makes about load factors and velocities, both at peak hours and average over the day. The figures spotted along several of the curves are the number of buses necessary to handle the peak-hour load. Note that the number of buses required is more dependent on the peak-hour operating factors. Other analyses with the model indicate that neither the type of bus-company ownership (private versus public) nor the type of fuel used (diesel versus gasoline) has as much influence on total bus system cost as do load factors and velocities.

Overall Cost-Effectiveness of the Transportation Management Strategies

The transportation model and the bus-system cost model were used together to examine the costs and effectiveness of a variety of bus-system improvements and mileage surcharge levels. The transportation model,

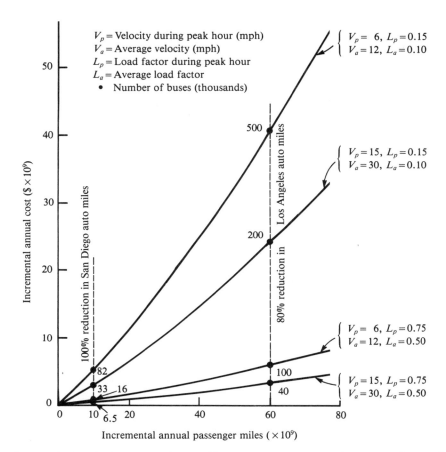

Figure 10.10. The incremental annual bus-system cost versus incremental annual passenger-miles for 50-passenger buses operated by a public corporation. (*From Petruschell and Kirkwood, 1973, p. 46.*)

with the aggregate form of the bus-system model it incorporates, was run for a range of strategies to produce the graph of cost versus effectiveness in Figure 10.11. The primary direct measure of effectiveness in this case is the percent reduction in annual LDMV miles traveled. The reduction of RHC emissions that results from the mileage reduction is a function of the retrofit and inspection/maintenance strategy employed. The right-hand ordinates of the graph in Figure 10.11 show the reduction in RHC under two of the alternative retrofit strategies. The bus-system costs shown are solely for bus-system improvements, that is, for incremental investment and operating costs above those for the nominal bus system. The surcharge expenditures indicate the amount of surcharge collected

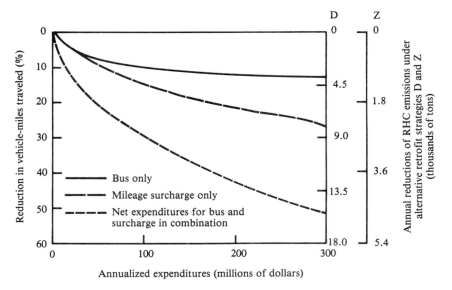

Figure 10.11. The cost-effectiveness of transportation management strategies. (*From Goeller et al., 1973, p. 27.*)

on the residual LDMV mileage (that is, the LDMV mileage remaining after curtailment and/or diversion of travel to buses and other means). Note that the surcharge is also an incremental cost (to the normal cost of owning and operating a car) that affects trip-making behavior by individuals.

The uppermost two curves in Figure 10.11 show that for an annualized expenditure of about $20 million, a 5% reduction in LDMV mileage could be obtained by either bus improvements or a mileage surcharge alone. For greater levels of reduction in LDMV mileage, the surcharge is more cost-effective than bus improvements because it induces people to reduce their mileage in three ways: by forming carpools, by taking the bus, or by foregoing some trips entirely. Bus improvements only provide incentives to use buses for travel. The curves clearly indicate that bus-system improvements alone, no matter how substantial, cannot alone produce mileage reductions greater than about 10 percent.

The lower curve in Figure 10.11 shows that it is most productive to use a combination of bus-system improvements along with mileage surcharges. This is in part due to the fact that mileage surcharge revenues can be used to offset the costs of bus improvements. The net annual costs used to plot the curve is the net cost after this transfer of revenues. We observe, as an illustration of the mutually reinforcing effects, that a $20

million net expenditure on bus improvements in combination with a surcharge would produce a more than 15 percent mileage reduction. In contrast, the same expenditure on either bus improvements or a surcharge
alone would produce only a 5 percent reduction.

Cost-Effectiveness Ranking of the Three Strategy Classes

The three strategy classes can now be evaluated in comparable cost-effectiveness terms (cost per ton of RHC removed). Figure 10.12 portrays
the cost-effectiveness of the most promising pure strategies (under 1975
conditions). Except for the most extreme strategies in each class, fixed-
source tactics are more cost-effective than retrofit tactics, and retrofit
tactics are more cost-effective than transportation management. This is
only a rough indication of the relative desirability of the strategies, however, as these are average cost figures. The maximum *total* amounts of
RHC emissions removable by the various strategies is quite different for
the three categories, and combinations of strategies may be required to
reduce emission levels to the point required by air-pollution regulations.
It is notable, however, that the maximum fixed-source strategy achieves
a reduction of over 19,000 tons for about $2.67 million annually, whereas
the retrofit and transportation management strategies require much larger
expenditures to achieve comparable reductions. The general conclusion
from this is that fixed-source strategies should be applied to their maximum extent before other strategies are employed.

Synthesis: Evaluation of Mixed Strategies

The three kinds of pure strategies—fixed-source, retrofit-I/M, and transportation management—can be combined in a variety of ways to form
mixed strategies for reducing RHC emissions. But there still are many
attractive alternatives remaining at this point in the analysis. Before proceeding to a detailed assessment of these (looking at all the economic and
noneconomic costs and benefits), it is necessary to identify a few superior
mixed strategies by applying a somewhat broader cost concept—but one
that still will be readily usable in screening a large number of alternatives.
This broadening of the cost concept involves an economic cost (opportunity cost) that has not to this point been included in the monetary cost
measures used for screening pure strategies. Its inclusion is important
when evaluating mixed strategies that include transportation management
with other strategies.

One of the impacts of transportation management is that the quality
and quantity of travel may be adversely affected by the strategy employed.
This is a social cost that needs to be included in the cost-effectiveness

412

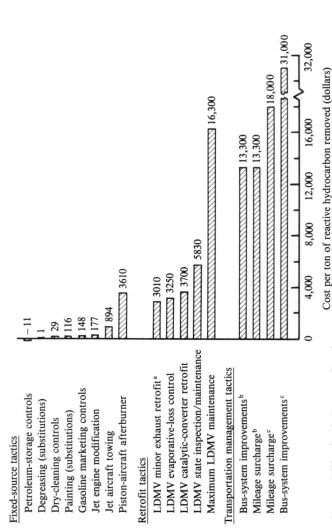

Figure 10.12. The cost-effectiveness of selected tactics for reducing reactive hydrocarbon emissions under 1975 conditions. LDMV is the abbreviation for light duty motor vehicles. (*From Goeller et al., 1973, p. 30.*)

[a] Current California Air Resource Board program, retrofit strategy D

[b] For 5% under retrofit strategy D

[c] For 10% reduction in vehicle-miles travelled under retrofit strategy D

evaluation. Otherwise a mixed strategy that produced a large amount of foregone travel would appear unreasonably attractive if the annual, regional monetary costs required were small; or a strategy that produced a particular mileage reduction with a large social cost (such as a large reduction in the number of trips taken) would be unreasonably favored over another that produced the reduction at a smaller social cost (for example, one that shifted more trips to buses without reducing the total number of trips substantially).

Fortunately, the analysis provides a convenient means of identifying the monetary cost of the foregone travel that transportation-management strategies produce. The transportation model permits us to infer the value of foregone trips by looking at the changes induced in travel behavior (number of trips taken) as the mileage surcharge rate is varied. The cost-per-trip-foregone estimate generated in this way can be used to add a monetary cost for foregone travel to the other annualized investment costs of each strategy and thereby account for this important social cost in the overall cost-effectiveness evaluation of mixed strategies.

The method used to identify superior mixed strategies involves use of the *tradeoff model* (Mikolowsky et al., 1973). The model seeks the best combination, that is, the superior strategy, for a given fixed-source control strategy and menus of retrofit-I/M and transportation-management strategies. The best combination is defined as the one that meets the desired air-quality standard for the minimum total cost (including the cost of foregone travel, as discussed in the preceding material). The model also is useful for sensitivity analyses; for example, it can select the superior strategy for varying levels of air-quality standards for a particular pollutant, thereby showing how strategy composition and cost vary with the level of the standard. The model also identifies a variety of economic, environmental, and transportation-service effects for the superior strategy it selects.

Impact Summary for Superior Strategies

The tradeoff model is used to identify a number of superior strategies under varying sets of conditions. When the detailed effects of the various cases are examined we usually find mixed results—that is, a strategy that has a better transportation service effect than another (fewer trips foregone, for example) may have a less desirable economic effect (more short-term employment disruptions, for example). The presentation of these mixed impacts is made clearer by the use of a scorecard, a table that uses color-coding to highlight rankings among strategies for the most significant effects. For a particular effect, green indicates the best ranking, red the worst, blue the second best, and orange the second worst. Comparable values are given the same color, and neutral effects are shown without

Table 10.8. The Impact Summary for the 1977 Reference Cases for the San Diego Clean Air Project

Strategy components	Case					
	Nominal	A	B	C	D	E
Fixed source tactics	Nominal	Nominal	Maximal	Medium	Maximal	Medium
Retrofit[a]	D	R	L	M	L	M
Mileage surcharge, cents/mile	0	0	0	0	0	0
Bus-eligible population, %[b]	70	70	70	70	80	80
Bus headways, peak/off-peak, minutes	40/40	40/40	20/40	20/40	20/40	20/40
Bus fare, cents/trip	25	Free	Free	Free	Free	Free
Environmental impacts						
Worst-day oxidant concentration, parts per million	*(0.13)*	[0.08]	[0.08]	[0.08]	[0.08]	[0.08]
Number of days oxidant above standard	*(15)*	[1]	[1]	[1]	[1]	[1]
Lead standard achieved?	*(No)*	Yes	Yes	Yes	Yes	Yes
Annual gasoline consumption, millions of gallons	615	618	609	607	604	603
Annual diesel-fuel consumption, millions of gallons	2.4	2.4	3.4	3.4	4.8	4.8
Transportation-service impacts						
LDMV mileage reduction, % of uncontrolled	0.0	2.9	4.3	4.3	5.0	5.0
Trips foregone, % of uncontrolled	[0.1]	*(1.0)*	(0.9)	(0.9)	(0.9)	(0.9)
Trips by bus, % of all trips	1.1	1.4	2.6	2.6	3.1	3.1
Carpooling, % of home–work trips	5	5	5	5	5	5
Economic impacts						
Annual strategy expenditures, $ millions						
Fixed-source control	2.8	2.8	5.1	3.2	5.1	3.2
Retrofit	7.7	67.6	51.5	55.2	51.3	55.0
Mileage surcharge	0	0	0	0	0	0
Bus	10.0	10.0	15.0	15.0	23.0	23.0
Net expenditures after transfers, $ millions[c]	[20.5]	(80.4)	[71.6]	[73.4]	(79.4)	*(81.2)*
Unallocated income from surcharge, $ millions	0	0	0	0	0	0
Incremental recurring employment	*(77)*	(815)	[1041]	[1048]	[1536]	[1543]
Incremental nonrecurring employment	[212]	*(2071)*	[1144]	(1483)	[1144]	(1483)
Number of buses	[175]	[175]	(348)	(348)	*(490)*	*(490)*

(continued)

Table 10.8. *(continued)*

Strategy components	Nominal	A	B	C	D	E
			Case			
Distributional impacts						
General-aviation annual cost, $/aircraft	[0]	[0]	*(500)*	[0]	*(500)*	[0]
Retrofit procurement cost per household, % of income[d]						
Under $500						
User-pays financing	**[0.8]**	*(8.9)*	[5.4]	(6.5)	[5.4]	(6.5)
Income-proportional financing	0.2	2.6	1.6	1.9	1.6	1.9
Over $15,000						
User-pays financing	0.1	1.2	0.8	0.9	0.8	0.9
Income-proportional financing	0.2	2.6	1.6	1.9	1.2	1.9

Source: Goeller et al. (1973, p. 38).

Note: Rank coding for each impacts:

Best case (green):	**[Best case]**
Next best case (blue):	[Next best]
Next worse case (orange):	(Next worse)
Worst case (red):	*(Worst case)*

[a] See Table 10.5 for the compositions of strategies L, M, and R. Strategy D, the nominal case, is the current California Air Resources Board requirement.

[b] As a percent of the total county population.

[c] The surcharge income is used to offset the bus subsidy to the extent possible.

[d] The percentage of the average annual income for the group when the retrofit–procurement expenditure is made in a single year.

color-coding. (Since we cannot show color-coding here, typographic distinctions are used to indicate rankings in the summary table.)

Table 10.8 shows the composition and impact summary (costs, benefits, and disbenefits) for the superior strategies that constitute the 1977 reference cases for the San Diego Clean Air Project. The compositions of the nominal case (which consists only of programs that were already mandated at the time of the study) and of five alternative cases are shown at the top of the scorecard. Case A emphasizes LDMV retrofit. Case B introduces maximal fixed-source control and improved bus-system headways. Case C backs off from maximal to medium fixed-source control and keeps the same improved bus headways. Cases D and E investigate the tradeoff between maximal and medium fixed-source control while at the same time increasing the bus-eligible population by 10 percent. All of the alternative cases provide free bus service. The impacts are classified as environmental, transportation-service, economic, and distributional, and are presented numerically (in units appropriate to each) in the main or summary part of the scorecard.

The nominal case is, of course, the least costly, both in terms of economic effects and in terms of distributional and transportation-service effects. But it fails to meet the legally mandated air-quality standard. All of the other cases shown in the summary scorecard meet the standard and have mixed positive and negative effects in other areas. Cases B and C have the lowest overall expenditure requirements of the five alternatives, and they also have relatively good attributes in terms of transportation service, other economic effects, and distributional effects. In net terms, case B appears to be the best balanced in its effects, but its major drawback is the $500-per-aircraft annual cost for general aviation aircraft. If the added emission controls for these aircraft proved not to be feasible, then Case C, which employs somewhat more extensive retrofit and inspection/maintenance controls on automobiles to make up for the lack of general aviation aircraft controls, would appear to be the best alternative.

10.5. Summary

This chapter has described and illustrated the major concepts and some of the techniques of cost analysis, an important aspect of systems analysis. In general, systems analysis is concerned with complex problems of choice under conditions of uncertainty, and cost considerations are central to these problems because policymakers are subject to resource constraints. For any policy area, the general approach of systems analysis involves (1) identifying systematically the desired objectives and alternative courses of action to achieve them; (2) examining the consequences of the alternatives in terms of their costs, benefits, and risks; and (3) presenting the results of this examination to decision- and policymakers in a framework that permits them to make comparative judgments about the worths of the alternatives. (For a fuller outline of this process, see the Appendix to Chapter 1.)

Cost analysis focuses largely on the identification of resources and *monetary* costs associated with alternative policies and systems. From a broader view of systems analysis, however, cost includes *all undesirable consequences* of an alternative—monetary and economic costs and non-economic costs, such as noise, air pollution, and other discomforts or inconveniences that may be forced on individuals or organizations as a consequence of the installation and operation of new policies and systems. Hence, although economic cost may be a primary factor in identifying the comparative cost-effectiveness of policy alternatives, a broader view, bringing in the multiple dimensions of cost and effectiveness, may be useful to policymakers in some areas.

To be effective, cost analysts should work as integral members of the systems-analysis team. In particular, they should participate in the study design phase and in the identification and description of the alternatives, which should usually be framed in terms of benefit or output-oriented packages in order for all costs to be attached to corresponding sets of benefits. The relevant costs include everything that is the consequence of selecting one alternative instead of another, but they exclude those costs that would be incurred regardless of the decision under consideration.

It is useful for analysis to divide economic costs into three life-cycle cost components: (1) research and development, (2) investment, and (3) operations. These categories may be further divided into subcategories that reflect separate types of resources or sets of resource requirements that are differently affected by different aspects of the systems being considered. In all cases, it is essential that what is included in each category be defined carefully. Great detail is usually not desirable (and often not possible) for studying long-range, future alternatives, but the cost category structure must be specific enough to determine which aspects of a proposal are really new and which are not.

Estimating relationships enable the analyst to relate the cost of a component or part of a total system to the proposed system's performance and physical characteristics and to the operational concept specified for it. Where possible, these should be determined empirically from data on past, current, and near-future systems similar to the ones being considered. However, care must be taken when projecting estimates into the distant future, where the universe may differ from that from which the sample data were taken.

Cost models combine the estimating relationships for all of the cost categories being used in an analysis into an input-output structure, which enables the analyst to go from system characteristics (performance, physical, and operational) to estimates of total system cost. They may be simple or complex, depending upon the context of the problem. More complex models may be computerized, which greatly facilitates sensitivity testing. Outputs may be detailed or aggregate, static or time-phased. Aggregate, static cost estimates are often useful in screening alternatives to reduce them to a more manageable number suitable for more detailed treatment. Time-phased cost streams can be useful to decisionmakers who must fit the alternative selected within budgetary constraints, and they also provide the basis for treating the time-preference problem when the alternatives under consideration differ in the mix of near-term versus future costs.

Uncertainties may affect cost estimates in two ways: (1) the estimating relationships may contain statistical uncertainties and (2) there may be uncertainties in the future state of the world. The former often may be handled by using standard statistical measures (standard errors, prediction intervals, and so on). The latter type of uncertainty often dominates in systems-analysis problems, and is usually best handled by the use of sensitivity analysis, whereby the costs of the alternatives are calculated under less (or more) optimistic assumptions about system performance, operating environment, and other important variables. Such testing of sensitivities can sometimes identify dominant solutions (alternatives that are preferred regardless of the assumptions) and can help to identify key uncertainties, which might be resolved by further study or test programs.

The presentation of results can be one of the most important aspects of a systems or cost analysis. Where costs and benefits can both be measured appropriately in monetary units, the alternatives can be compared by unitary measures, such as discounted net benefits. An equal-effectiveness framework, where all alternatives are designed to meet a specified objective, also permits ready comparison of alternatives by showing only the costs of each at the stated level of effectiveness. Sensitivity analyses of assumptions, desired level of effectiveness, discount rates, and so on should also be included in results presented to decisionmakers when these sensitivities are critical to the outcomes. In an equal-cost framework, the presentation should emphasize the effectiveness level of each alternative at specified budget levels. Again, selected sensitivity analyses should be included in results when the specified budget level or other factors strongly affect either the rankings of alternatives or the absolute levels of cost and effectiveness outcomes.

When either costs or benefits are multidimensional, neither a single cost-effectiveness measure nor the equal-cost or equal-effectiveness frameworks may be effective ways to present results. When no dominant alternative exists—one which has higher benefits and lower costs for all types of costs, benefits, and disbenefits—some presentation method is needed that displays the multiple attributes of the alternatives and lets the decisionmaker apply his own judgments about the importance of each attribute.

The scorecard approach does this by displaying results in an array, showing the various benefits and costs (measured in appropriate units for each) that result when one or another of the alternatives is chosen. The relative ranking of the alternatives for each cost or benefit can be shown by means of color coding, shading, or other means to show which alternatives, if any, tend to have the most positive attributes (higher benefits or lower costs). Sensitivity analyses in this type of presentation emphasize the way in which *rankings* are altered as assumptions or constraints are changed. This format supplies a considerable amount of information in one table or picture and permits the decisionmaker to apply his own judgments as to the relative importance of each effect.

All presentation methods that show outcomes as multiple, incommensurable attributes carry with them the danger that the analyst may provide too many individual pieces of information for decisionmakers to consolidate and apply their judgment. But when public policy decisions are at stake, the decisionmakers and the various parties affected by the decisions may have different goals and preferences for the multiple effects produced. In such cases, no unitary measure or weighting scheme can guarantee that all preferences will be properly represented. Therefore, analysts must take care to assure that some costs and benefits are not obscured or ignored in favor of others. The purpose of analysis is to clarify the costs, benefits, and risks of the alternatives. Thus, it is important that analysts seek a balance between the desire to present all of the available information that may be relevant to the decision and the need to simplify and consolidate results[4] to reduce confusion and highlight the most important aspects of the decisionmaker's problem of choice.

References

American Transit Association (1971). *Transit Operating Report.* Washington, D.C.: American Transit Association.

Bigelow, J. H., B. F. Goeller, and R. L. Petruschell (1973). *A Policy-Oriented Urban Transportation Model: The San Diego Version,* R-1366-SD. Santa Monica, California: The Rand Corporation.

Blanchard, B. S. (1978). *Life Cycle Cost.* Portland, Oregon: M/A Press.

Chesler, L. G., and B. F. Goeller (1973). *The STAR Methodology for Short-Haul Transportation: Transportation System Impact Assessment,* R-1359-DOT. Santa Monica, California: The Rand Corporation.

DeHaven, J. C., and B. M. Woodfill (1973). *Cost and Effectiveness of Strategies for Reducing Emissions from Fixed Sources: The San Diego Air Basin,* R-1363-SD. Santa Monica, California: The Rand Corporation.

Fisher, G. H. (1971). *Cost Considerations in Systems Analysis.* New York: American Elsevier.

――― (1977). Cost considerations in policy analysis. *Policy Analysis* 3, 107–114.

Goeller, B. F., et al. (1973). *San Diego Clean Air Project: Executive Summary.* R-1362/1-SD. Santa Monica, California: The Rand Corporation.

―――, et al. (1977). *Protecting an Estuary from Floods—A Policy Analysis of the Oosterschelde: Vol. I. Summary Report,* R-2121-NETH. Santa Monica, California: The Rand Corporation.

―――, et al. (1983). *Policy Analysis of Water Management for the Netherlands: Vol. I. Summary Report,* R-2500/1-NETH. Santa Monica, California: The Rand Corporation.

Grant, E. L., and W. G. Ireson (1970). *Principles of Engineering Economy.* New York: The Ronald Press.

Hess, R. W., et al. (1983). *An Analysis of the Cost, Schedule, and Performance*

---

[4] For instance, by setting aside impacts that are highly correlated with impacts already displayed, that are comparable in size for all cases, or that can be shown to be of little significance to the choice.

of the Baseline SRC-1 Commercial Demonstration Plant, N-1952-DOE. Santa Monica, California: The Rand Corporation.

Hitch, C. J., and R. N. McKean (1960). *The Economics of Defense in the Nuclear Age.* Boston, Massachusetts: Harvard University Press.

Mikolowsky, W. T. (1973). *The Motor Vehicle Emission and Cost Model (MOVEC): Model Description and Illustrative Applications,* R-1364-SD. Santa Monica, California: The Rand Corporation.

————, et al. (1973). *A Tradeoff Model for Selecting and Evaluating Regional Air Quality Control Strategies,* R-1367-SD. Santa Monica, California: The Rand Corporation.

Miser, H. J., and E. S. Quade, eds. (1985). *Handbook of Systems Analysis: Overview of Uses, Procedures, Applications, and Practice.* New York: North-Holland. Cited in text as the *Overview.*

Mood, A. M. (1983). *Introduction to Policy Analysis.* New York: North-Holland.

Mooz, W. E. (1978). *Cost Analysis of Light Water Reactor Power Plants,* R-2304-DOE. Santa Monica, California: The Rand Corporation.

Overview. See Miser and Quade (1985).

Petruschell, R. L., and T. F. Kirkwood (1973). *A Generalized Bus System Cost Model,* R-1365-SD. Santa Monica, California: The Rand Corporation.

————, T. Repnau, and G. Baarse (1982). *Policy Analysis of Water Management for the Netherlands: Vol. XIII. Models for Sprinkler Irrigation System Design, Cost, and Operation,* N-1500/14-NETH. Santa Monica, California: The Rand Corporation.

————, ————, J. W. Pulles, and M. A. Veen (1981). *Policy Analysis of Water Management for the Netherlands: Vol. XIV. Costs for Infrastructure Tactics,* N-1500/14-NETH. Santa Monica, California: The Rand Corporation.

Quade, E. S. (1982). *Analysis for Public Decisions,* 2nd ed. New York: North-Holland.

Schwarz, Brita, K. C. Bowen, István Kiss, and E. S. Quade (1985). Guidance for decision. In Miser and Quade (1985), pp. 219–247.

Seldon, M. R. (1979). *Life Cycle Costing: A Better Method of Government Procurement.* Boulder, Colorado: Westview Press.

Stucker, J. P. (1985). *The Costs of Closing Nuclear Power Plants,* N-2179-RC. Santa Monica, California: The Rand Corporation.

Walker, W. E. (1985). *The Use of Screening in Policy Analysis,* P-6932-2. Santa Monica, California: The Rand Corporation.

Chapter 11
Program Evaluation

Harry P. Hatry

11.1. Introduction

This chapter offers a basic introduction to the major principles of program evaluation. It does not examine its theoretical underpinnings or detailed methods (many works, such as those cited in the references, cover these methods well), nor does it describe how a program evaluation process might be established (see, for example, Hatry et al., 1981, for how to do this for public organizations). Chapter 5 discusses social experimentation, an important tool of evaluation.

Webster's dictionary defines *to evaluate* as "to determine the significance or worth by careful appraisal and study. . . . Evaluate suggests an attempt to determine either the relative or intrinsic worth of something in terms other than monetary."

Evaluation as a recent distinguishable activity draws its roots from laboratory experiments in past decades, leading to expansion of their principles to out-of-the-laboratory trials of programs involving human subjects, such as the clients of mental-health treatment programs. In the past two decades, evaluation has been applied to many types of activities, from human-services programs to programs with substantial technological content.

Program evaluations provide information on how well existing programs have been working, and thus their focus is largely retrospective. On the other hand, especially if they are designed for this additional purpose, they can offer clues as to how the program can be improved. Thus, a program evaluation can play any of three roles in systems analysis:

It can help target candidates for analysis by identifying existing programs that have not been working well.

It can provide baseline data on the performance of an existing program, whether or not it is the subject of a systems analysis and whether or not the evaluation may be used to identify the need for systems analysis. These baseline data can be used as a basis for predicting future performance if the same, or a similar, program is continued. Presumably, proposed alternatives need to be expected to outperform the current program.

It can assess the progress of a program that has been implemented as a result of a systems analysis. This will have the added benefit of helping the systems analysts verify their methodologies or of indicating the need to modify their procedures (if the evaluation findings differ significantly from the performance predicted by the systems analysis results).

Program evaluations can be heavily quantitative, but in recent years, as data problems and the difficulties of controlling for numerous confounding variables have surfaced in many settings, there has been a trend toward using more qualitative approaches, including mixtures of quantitative and qualitative techniques.

Program evaluations can be quite complex or, in some situations, the analyst can use simple statistical conversions to indices of operational data that are easily kept. For example, analysts of the system for supplying blood to hospitals in the Greater New York area (see Section 2.2) developed a revised system that continued to meet demands while reducing the discard rate for blood from 20 to 4 percent and lowering the average number of deliveries per hospital per week by 46 percent. These simple numerical results that emerged from the new system's implementation have two implications. Since they were close to what the analysis had predicted, they confirmed its relevance, and, since they exhibited obvious improvements in the operations of the blood-supply system, they indicated that implementing the new system was worthwhile. On the other hand, if other aspects of the new delivery system (such as its dollar costs or adherence to safety standards) had needed to be investigated, a more demanding program evaluation would have been required. In either case, the evaluation would have explored the results arising from putting a new program into place, and possibly suggested desirable program changes.

Program evaluation can be applied to just about any activity defined as a program, extending from relatively narrow activities to large-scale undertakings. In particular, it can be applied to such governmental services as police and fire protection, solid waste management, the delivery of physical and mental health services, transportation services, water treatment and supply, recreation facilities and activities, library services, employment and training programs, education, and so on.

11.2. The Objectives of Program Evaluation

The term *program evaluation* is commonly used in the United States to refer to an assessment of a government program's past performance. Evaluation is aimed at determining the program's past impacts—its effectiveness. Some evaluations also include estimating the dollar costs of the program, but this is rare. The private sector does not usually use the term program evaluation. Private firms routinely examine past performance by studying such evidence as sales, profits, and market shares. Nevertheless, analysts in the private sector face many of the issues and choices that this chapter discusses and apply many of the procedures used in the public sector, as, for example, in marketing research.

The key objectives of program evaluation are:

1. To determine what the effects of the program have been over a specified previous period of time.
2. To estimate the extent to which the program, rather than other factors, has caused the observed effects.
3. To identify, where possible, program characteristics and other factors associated with its success or failure in order to guide future improvements.
4. Ultimately, to identify future actions that should be taken regarding the program, such as whether it should be dropped, continued as is, expanded, or modified.

Evaluators have long debated whether evaluations should solely assess past performance ("summative" evaluations) or should focus on uncovering factors associated with success and offering suggestions for improving the program ("formative" evaluations). Both purposes are praiseworthy. I suspect, however, that, to be sufficiently useful, most evaluations need to attempt to serve both purposes, even if this means lessening the evaluator's ability to do either in as much depth as is ideally desirable.

The objectives listed above are typically difficult to achieve with high confidence, as few programs can be set into operation under the sort of controlled conditions that one expects in a laboratory experiment. Program effects are often unclear, sometimes even ill defined, and they can offer tricky measurement problems. Separating the influences of key program variables from those of extraneous factors in the program's environment usually offers important difficulties. Nevertheless, the need for program evaluation is great, and even evaluation work of limited success can offer decisionmakers important insights about policy and future decisions.

A great deal of program-evaluation work has been done in past years, and a substantial literature has grown up in this field.[1] This literature has important things to say to the systems analyst facing the need to conduct a program evaluation. Its tools and principles are quite similar to, and compatible with, those used by systems analysts. It is difficult to define the fine line between program evaluators and systems analysts; many professionals, in effect, do both.

If the analyst is in the fortunate situation of evaluating a program emerging from a systems analysis in which he participated, he has important advantages: His exploratory work at the beginning of his systems analysis undoubtedly had many of the elements of a program evaluation in it as he sought to understand how the existing system was operating, and his work throughout the analysis has given him a deep understanding of the situation in which the new program is embedded. Thus, he already has a running start on the work that he must do in his prospective evaluation. Too, the findings from his program evaluation can serve as a check on the accuracy of the systems-analysis results.

11.3. Types of Program Evaluation

The most conceptually desirable form of evaluation, and therefore the one most written about, is the controlled—or true—experiment. It is also, however, the form of evaluation undertaken least often because of the difficulties, in practice, of meeting its special requirements. In a controlled experiment, members of the target population (or a random sample of it) are assigned to groups at random; one group receives no treatment relating to the program being evaluated, and the others receive varying treatments, one to each group. Ideally, each member of the population should have the same probability of being assigned to any particular group. In practice, however, pure random assignment and maintenance of full control over the treatment applied to each group are seldom possible and have to be approximated. Such experiments also generally require substantial numbers of personnel to undertake and supervise them, and they usually take long periods of time, often several years.

The population the analyst is concerned with consists of the entities being subjected to study, such as people, communities, organizations, equipment, operating systems, and so on; the treatment is some variant

---

[1] A selection of useful references: Campbell and Stanley (1966); Cook and Campbell (1979); Fitz-Gibbon and Morris (1978); Glaser (1973); Guttentag and Struening (1975); Hatry et al. (1981); Herzog (1959); Riecken and Boruch (1974); Rossi and Freeman (1982); Rossi, Wright, and Anderson (1985); Rutman (1977); Struening and Brewer (1983); Suchman (1967); Weiss (1972a, 1972b); and Wholey (1983). In addition, *New Directions for Program Evaluation,* a quarterly publication of the American Evaluation Association, provides coverage of major program evaluation developments.

of the policy or program that has been adopted. The random assignment is intended to assure that the groups have similar characteristics at the beginning of the experiment and that extraneous variables, insofar as possible, have the same effects in each group, so that the differing outcomes observed can properly be attributed to the varying treatments. When these conditions are met or closely approximated, controlled experimentation can be expected to yield discriminating results. (The situations in which this form of evaluation is likely to be appropriate are discussed later in Section 11.5; see also Chapter 5.)

Less Demanding Types of Evaluation

Analysts will most often need to use less demanding and less time-consuming approaches to program evaluation. The price they must pay for meeting less stringent demands is that, in general, their results will be less conclusive both as to the magnitudes of the effects attributable to program treatments and whether or not observed differences between treatment groups can be attributed unambiguously to the treatments.

The less demanding and less conclusive forms of program evaluation include approaches such as these (perhaps used in combination):

Looking merely at the values of the evaluation criteria before the program interventions are made and then at a later time seeing what changes have occurred—a simple pre versus post design.

Deriving an historical time series based on several years of past data, extrapolating the trend into the future, and then comparing what actually happens with the extrapolation (this procedure controls for certain types of underlying long-term trends)—a time-series analysis.

Using natural comparison groups that already exist, when it appears that they have properties that will allow reasonable inferences to be drawn from the comparisons—a comparison-group design.

The literature contains many examples of successful evaluations that were not controlled experiments; for example, an evaluation of the impacts of community alcohol treatment centers (Armor, Polich, and Stambul, 1976), an evaluation of automatic vehicle-monitoring equipment systems for police use (Larson, Colton, and Larson, 1977), and an evaluation of the State of Connecticut's crackdown on speeding in the late 1950s (Campbell and Ross, 1968). The first example shows what can be done for a human-service program, the second for a technology program, and the third for a combination.

The simplified discussion of this section is aimed at persuading the reader that the major difficulty in program evaluation is usually less to determine the outcomes of the program than to determine whether or not the observed outcomes were caused by the new program or should be attributed to other factors. To the extent that the evaluators can rule out other plausible explanations, they can offer support for the conclusion that the new program has caused the observed effects.

A Simple Illustration

To illustrate these approaches—and some of the principles of program evaluation—it is useful to discuss a simple example. Although not a typical topic for program evaluation, this example is easy to describe, and it presents issues similar to those arising in the more usual and more complex social programs that are subjected to evaluation.

Let us consider an activity common to many large organizations, both public and private: maintaining the vehicles in the organization's fleet. (This example was inspired by evaluation work done several years ago by the city of Wilmington, Delaware.) Our scenario is that a year ago a government unit introduced a preventive-maintenance plan to reduce total vehicle maintenance and operating costs. Now, a year later, with budget time at hand, the issue of vehicle costs arises again, with the budget analysts challenging the economy of the new vehicle-maintenance policy. The evaluation office is called in to shed light on the issue.

The analysts, with limited time available, decide to collect data relating to the total cost per vehicle-mile. Although they are aware of the limitations of such a ratio criterion (see the *Overview* volume, pp. 221 and 227–228), the analysts nevertheless feel that it will be useful and appropriate in this case. The initial goal is to compare the value of this criterion for the year since the new maintenance program was instituted with the value for the previous year under the old maintenance policy. They hope that the difference between the two values can be used to indicate whether or not the new maintenance program has been worthwhile. One of the simplest evaluation designs, this procedure is called a *one-group pretest, posttest design*. Its design is conceptually simple and involves low cost; it is probably the evaluation design that is most commonly used.

To the surprise and chagrin of the program manager, the analysts find that the total cost per vehicle-mile has risen sharply under the new policy, as the budget analysts had suspected. The maintenance shop supervisor then steps in and points out that labor and material costs have been rising sharply during recent years. He suggests that the increasing costs are extensions of a trend and expresses the opinion that cost would have risen even more than it did if the new preventive-maintenance policy had not been put into effect. To test these contentions, the analysts then obtain data for several past years in order to determine the trend with time. With this done and the trend extrapolated to the recent year with the preventive-maintenance policy in effect, a comparison with the actual results now shows that the new policy was only a bit more expensive than the old one. However, the analysts are quick to point out that a firm conclusion is not warranted, as the difference is within the boundary of reasonable error for the extrapolation from the past years. They also point out that, if the program is retained for several more years, the comparison with the time trend will be more likely to be decisive. This type of approach is sometimes called a *discontinuous time-series design*.

During these considerations the maintenance supervisor points out further that recent articles in trade journals have reported significant increases in vehicle-maintenance costs in other organizations, suggesting that external factors are at work throughout the maintenance community to drive costs up significantly. If so, these forces may be the principal sources of increasing costs, rather than the new preventive maintenance program. To explore this contention, the analysts will have to obtain results from other organizations to compare with their own unit's experience. Such an approach is called a *pretest-posttest comparison group design*.

The situation is still further complicated by a fact pointed out by one of the mechanics: During the year being studied the mix of vehicles changed substantially from previous years. He claims that the newer vehicles are more complex and tend to have many more difficulties than those used by the organization in previous years. He contends that higher costs per vehicle-mile should be expected for these vehicles, and that they probably have dominated the net vehicle-mile cost for the fleet as a whole, thus masking the effects of the preventive-maintenance program.

To explore the consequences of the change in the vehicle mix, the analysts can isolate the costs per vehicle-mile for each type of vehicle, both in the recent year and, when available, for the previous year. An appropriate marshaling of this evidence would then indicate to all concerned what the influences of the new vehicle types have been and how they have affected the cost per vehicle-mile index as a whole.

Fleet users, however, may suggest that a criterion of worth chosen solely on the basis of the fleet's economics may be inadequate: They want the amount of vehicle downtime and frequency of road breakdowns considered. Data should then be obtained on these elements. Such an analysis, with its lessons presented carefully to the management and budget analysts, could well head off the classic problem with large-organization motor fleets created by a narrow view of their operations and costs: A very low-cost operation that renders poor service to its users is not adequate.

This scenario is intended to suggest two lessons:

The analysts should have first explored the situation thoroughly in order to discover such difficulties as those described in the scenario before any analysis was undertaken. This approach would have enabled them to plan and carry out a discriminating evaluation, thus avoiding the embarrassing experience of discovering important factors after the analysis was complete. (Incidentally, this more comprehensive approach would also have avoided the continuing controversy that their early results fueled.)

The analysts should have been aware a year earlier that the new maintenance policy was being introduced, designed an appropriate analysis to explore its effectiveness, and recommended that an experimental design be incorporated into the policy and that records appropriate to it be kept. With this careful preparation, the analysts would have been able to complete a discriminating evaluation shortly after the end of the year. Its results would then have been available in time to assist policy and budget decisions for the following year. In sum, the new maintenance policy should have contained appropriate provisions for evaluation. (If the new policy had emerged from some sort of systems analysis, the systems analysts should have recommended that an evaluation be undertaken to test the new policy after it had been in place for a suitable period.)

The evaluation scheme incorporated into the maintenance policy at the time it was introduced could well have been a controlled experimental design. A first step in evolving such a design would be to survey the factors that could affect the outcome, such as those described earlier and others (changes in use patterns, altered fleet size, different types and makes of vehicles, and so on). Such an experimental design would arrange to assign each vehicle at random either to an experimental group of vehicles subject to the new preventive-maintenance program or to a control group where the older maintenance policy would be continued. This assignment, probably by stratifying the random assignment, should ensure that similar proportions of different types and ages of vehicles are in each group. Such a design, if monitored carefully during the experimental year to be sure that its provisions were carefully observed in actual practice, would give the analysts a confident basis for inferring that differences in costs per vehicle-mile and in vehicle reliability are attributable to the different maintenance policies.

The controlled-experiment approach is generally the most desirable evaluation concept, since it offers a good prospect of achieving sound evaluation results. On the other hand, it can seldom be used unless it is incorporated in the plan for the new program at its inception, it tends to be the most costly approach, and it is often difficult—or even impossible—to control in practice (for example, a city in the United States considered such an experiment in connection with a vehicle-maintenance program but had to fall back on a different and less satisfactory evaluation design when it met resistance from the agencies being served by the vehicles). It is unlikely that use of the controlled experiment in connection with the scenario of a vehicle pool would add much to the cost of the experiment. For other types of programs, such as large-scale social service programs, planning and carrying out a controlled experiment can be very difficult and costly indeed. Section 11.5 will discuss conditions under which the experimental approach is the most likely to be appropriate for program evaluation.

More Qualitative Types of Evaluation

Some evaluators, frustrated at being unable to determine program effects quantitatively (for example, by statistical means), have turned to "qualitative" evaluation.[2] Qualitative procedures tend to emphasize case studies, using unstructured (that is, open-ended) or semi-structured interviews, and using detailed descriptions and direct quotations. These procedures can provide more depth and richness of detail but not statistical generalizations. Qualitative procedures can be used alone or to complement quantitative techniques. They can also be used as a preliminary procedure to help develop structured questionnaires. Perhaps most importantly, these procedures can provide insights as to why a program has not worked well and on ways to improve it.

For example, a recent evaluation of a local government's program to improve nutrition for elderly citizens included qualitative interviews of a sample of potential clients to identify a number of likely reasons elderly citizens were not using the government's congregate meals program. The findings of these interviews did not yield statistically reliable data on program impacts or on reasons for nonuse, but did give a number of insights on likely reasons, and thus on possible program improvements.

Where there are a number of different facilities (sites) providing the service, the use of common information collection instruments for all can provide useful data and insights into how the program is working, how well, and its problems. This information can be useful even if the number and choice of sites examined are not a representative random sample of all the facilities. The evaluators will, of course, need to be careful in drawing conclusions and generalizing their findings.

11.4. Actions to Improve the Usefulness of Program Evaluations

To improve the quality of their work and the likelihood that its results will be used, program evaluators should take a number of actions. Some are needed before the program's implementation, some during it, and

---

[2] Some references on qualitative evaluation are Miles (1979), Miles and Huberman (1984), Patton (1980), and Yin (1984).

some after the data have been collected (as part of data analysis and report preparation). Table 11.1 lists a number of such actions; each is discussed further in what follows.

Ongoing Outcome Monitoring

Although this chapter focuses primarily on in-depth, ad-hoc evaluations, agencies may also find a lesser form of evaluation useful: They can establish procedures to provide regular (for example, quarterly) feedback on principal indicators of service quality, usually the same ones used for ad hoc evaluations.

Such evaluation information has the considerable advantage that it gives public managers timely information on performance. Effort is not spent (as it is in in-depth evaluations) on determining the extent to which the outcomes are being caused by the program rather than by other factors. Because much less effort is required only to estimate outcomes, an agency can generally afford to monitor regularly many—and probably most—of its programs. With in-depth evaluations, any one program can probably be evaluated at most every few years.

Outcome monitoring has been the subject of increasing attention in the United States as pressure to increase the accountability of public managers has increased. (For discussions of the need for such performance monitoring, see Carter, 1983; Epstein, 1984; and Wholey, 1983. For procedural details, see Austin and Associates, 1982; Hatry et al., 1977; and Magura and Moses, 1986.)

Actions Relating to the Design of the Evaluation

The actions in this category are designed to assure that an effective evaluation will be possible and that all concerned will not face the embarrassment of an operating program whose effects cannot be measured.

Action 1. Wherever possible, plan the evaluation before the new program begins. With the flurry of activity following a decision to implement a program and the primary focus of attention on planning the implementation and getting it started, it is hard to look forward to the time when the results—predicted by the program's advocates (possibly supported by a systems-analysis study)—will need to have been evaluated. Nevertheless, this is the time to formulate preliminary plans for the evaluation, and to see to it that needed baseline data are obtained before implementation begins and that the needed data-collection procedures are in place as the program proceeds. If not, the evaluation analysts will have to depend on whatever outcome-related data are developed as part of program

Table 11.1. Actions to Improve the Usefulness of Program Evaluation

Actions relating to the design of the evaluation
1. Wherever possible, plan the evaluation before the new program begins.
2. Undertake an assessment of the program's evaluability to assure that it can be evaluated within available resources and in time for the findings to be useful.
3. Prepare a detailed plan for the evaluation.
4. Schedule the major data-collection effort (on new programs) to begin only after the program has shaken down. Schedule an evaluation period long enough to be sure that a stable period of program observation is being observed and evaluated.
5. Plan the evaluation to help the program manager to improve the program as well as to measure its impacts.
6. Communicate clearly to all concerned what can—and what cannot—be expected from the evaluation, so that the expectations will be realistic.
7. After reviewing both intended and actual effects, select adequate measures of program impacts.

Actions during the period of observation, data collection, and early analysis
8. Monitor the operations of the program closely during the evaluation period to determine what really has taken place, as contrasted with what may have been planned for the implementation.
9. Identify, and adjust for, changes in data-collection procedures that occur between pre- and postprogram data gathering.
10. Avoid using persons whose work is being evaluated as evaluators of their own work.
11. Look for plausible explanations for the outcomes other than the program treatments.
12. Avoid pitfalls in undertaking sample surveys.
13. Allow enough time to analyze the data that are collected; avoid collecting too much data.

Actions relating to deriving the results and formulating the findings
14. Avoid overdependence on statistical significance, particularly at the expense of practical significance.
15. Consider the component effects as well as overall or average results.
16. Identify study limitations; do not overgeneralize the results.
17. Provide an opportunity for the program personnel to review the tentative findings before they are put into final form and reported.

Actions relating to presenting the findings
18. When presenting the findings of an evaluation, be at pains to describe its conditions, assumptions, limitations, and uncertainties.
19. Separate facts from opinions.
20. Be at considerable pains to present the findings to users comprehensively, clearly, concisely, and effectively.

implementation (often programs collect little outcome-related data) and do strictly after-the-fact data collection.

The evaluation can be greatly strengthened if provision can be made to obtain special evaluation data prior to the program's implementation. Evaluations are handicapped if they are forced to rely solely on data collected months, or years, after the program has begun.

Action 2. Undertake an assessment of the program's evaluability to assure that it can be evaluated within available resources and in time for the findings to be useful. If Action 1 has been taken, the question of evaluability may already have been settled and this step will be unnecessary. Nevertheless, prudence dictates that the situation be carefully reviewed to confirm this judgment. On the other hand, if Action 1 has not been taken, Action 2 should be. Also, as Majone (1985, pp. 61–62) remarks,

> the social and institutional context in which systems analysis is done has changed dramatically in the last two decades. In the early days the relationship between decisionmaker and advisor, between producer and user of analysis was much clearer than it is today. . . . Now it is quite common for policy research to be sponsored by one organization, carried out by another, used by a third organization, and perhaps evaluated by yet another agency (which, in turn, may commission the evaluation to an independent research group).

Thus, if the prospective evaluation team is new to the program, this step constitutes an important beginning for its work.

Failure to assess the potential utility and possibility of carrying out an evaluation effectively has been the cause of considerable frustration with program evaluation in the United States (see, for example, Horst et al., 1974). Program evaluation efforts will face major obstacles if the programs have properties such as these: the program has vague objectives; even though the objectives are reasonably clear, the current state of the art of measurement does not permit reasonable measurements of the program's impacts; or the program's major impacts cannot be expected to appear until many years into the future, or until after the results would be useful (for example, after any decisions potentially based on them would already have been made). Table 11.2 suggests some criteria for assessing evaluability and the merit of an evaluation effort.

In many cases, evaluability problems can be alleviated, so that the evaluability assessment should not overreact negatively to apparent difficulties. For example, with persistence it is sometimes possible to identify major specific objectives for programs that seem vague at first; it is often possible to obtain rough but adequate impact information about characteristics that, at first glance, appear to be too subjective (such as by using structured interviews of systematic samples of clients on various aspects of program services); and sometimes evaluations conducted over prolonged periods turn out to be useful for decisions in later years even though they are not useful during a current administration.

Proper evaluation requires adequate staff, money, and time, and the evaluation plan clearly needs to be compatible with the resources available. Although some corner cutting and less sophisticated approaches can

Table 11.2. Criteria for Selecting Issues for Program Evaluations

Can the results of the evaluation influence decisions regarding the program?
Programs for which a decision regarding continuation, modification, or termination needs to be made are obvious candidates.

Programs are poor candidates if decisionmakers have strong preconceptions of program value or if there is considerable support by influential vested-interest groups—if these circumstances make it unlikely that the program will be altered regardless of evaluation findings. On the other hand, there are cases where the program may be so important to a community that government officials will proceed with an evaluation and be prepared to seek changes despite the political risks involved.

A program is not a good candidate unless the evaluation can be completed in time to be helpful to decisionmakers. Evaluations finished after the relevant decisions have been made not only are not likely to be useful but are also likely to make evaluators look foolish.

Can the evaluation be done?
To be a good candidate for evaluation, a program's innovation must be sufficiently defined, both as to what it consists of and what its purposes are, and to some extent there must be linkages between the innovation and its desired impacts. It is not necessary for everything about the project to be spelled out in advance, only for its details to be sufficiently specific for the evaluators to determine what is actually being implemented and what adequate criteria for judging its worth might be.

For a program to be evaluable, sufficient data must be obtained to yield reliable and valid indices of the program's important effects. Evaluations can seldom, if ever, cover all program effects, but it should be possible to gather meaningful data on significant aspects.

To be doable, the evaluation must have sufficient resources available to meet the time schedule and technical requirements.

To be evaluable, a program must be stable enough to yield data that can be converted to information characterizing its properties. If the program shows continual change—or is about to undergo major change—in significant ways, it is not likely to be a good candidate for evaluation.

Is the program significant enough to merit the evaluation effort?
Programs meet the significance test if they involve large amounts of resources relative to the cost of the evaluation or if they have important positive or negative consequences for major segments of the public. The likely cost of an evaluation effort should be compared to possible benefits from its results, either decreased costs or improved effectiveness.

A program can meet the significance test if it is suspected of being marginal in its performance, and if an evaluation promises to identify important improvements.

A program can meet the significant test if it is a major one whose potential benefits and costs are highly uncertain, but important to measure.

New programs and programs to be substantially expanded are often particularly good candidates for evaluation, since there is likely to be more uncertainty about their effects, and evaluation requirements can probably be more easily instituted.

Source: Adapted from Hatry, Winnie, and Fisk (1981, Exhibit 15).

sometimes be used when resources are scarce, too many compromises can weaken an evaluation to the point where it is not worth doing—or worse, could yield misleading results.

The literature seldom discusses the need to distinguish whether the program to be evaluated is under development or is operational: Evaluations of projects in development stages, in general, seek less definite information on impacts, since such evaluations are more concerned with determining feasibility and preferred program characteristics. Ignoring this distinction appears to have led to inappropriate expectations and improper evaluation designs in some instances. Programs dominated by technology usually go through a formal development stage before being put into full operation, with evaluation activities tailored to the needs of each of these stages. This precedent could well be considered for human-service projects by including a development test phase, though at the cost of delaying full implementation somewhat.

Action 3. Prepare a detailed plan for the evaluation. This action extends to program evaluation a similar one for analysis in general (see the *Overview,* pp. 294–297), and has the same importance. With the first two actions behind him, the evaluation analyst may feel that he knows what to do and that this step is unnecessary—but he could hardly be more mistaken. All the arguments for a plan for systems analysis apply here, and the content will, in fact, consist of similar elements. It is an essential professional step in evaluation.

Action 4. Schedule the major data-collection effort (on new programs) to begin only after the program has shaken down. Schedule an evaluation period long enough to be sure that a stable period of program observation is being observed and evaluated. These admonitions apply particularly when evaluating new programs or major revisions of old ones. Sometimes there is the temptation to begin evaluating the program as soon as the new implementation activity begins. For many new or revised programs, however, the shakedown period may last many months, during which program procedures stabilize, new people become adjusted to new procedures, and the program settles into a stable state. Thus, a proper evaluation should be based on observations of the stable state, should not begin until the stable state is reached, and should continue long enough to constitute an adequate test of the program's performance.

It is also important for the evaluation period to be long enough to consider effects associated with time. For example, programs affected by seasonal differences require an evaluation period of at least a year in order to encompass the four seasons; a road-repair program evaluated over six months not including the special problems of the spring thaw would miss perhaps the most important difficulties facing such a program, whereas a six-month period that includes the spring thaw might overemphasize this period because the less demanding six-month period was omitted from the evaluation. Although appropriate timing depends, of course, on the nature of the program and its setting, it is not uncommon for six months to be allowed for the program to settle into a stable state and for the evaluation to cover an ensuing year or more.

A variety of factors may make an early evaluation based on results impossible. For example, a proposal to close a Dutch estuary, which required choosing among three major engineering works designed to protect the nearby land from flooding, emerged from a large systems analysis (see Section 2.2). A full evaluation of the implemented program by observing results, however, must await not only completing the barrier but also the passage of enough time for the operational, economic, ecological, social, and safety results to develop (and the major storm that is most feared and from which the barrier is designed as protection may not occur for as long into the future as several centuries). Operating policies, such as those for opening and closing the barrier gates, may require only a few seasons for evaluation. Evaluating most economic, social, and ecological results will likely take longer; but the most important of the safety results cannot be expected to be known firmly for a long time. Many of these aspects of the program can be assessed by partial program evaluations, which may shed light on both the quality of the analysis that led to the proposal and the ultimate value of what was done.

Similarly, a study of world energy published in 1982 (see Sections 2.2 and 9.6) looked at supply and demand 50 years into the future; the possible impacts it estimated are quite unlikely to be verifiable until the twenty-first century, and then only if significant portions of its implications are translated into programs. In a case such as this one, the quality of the findings must be judged largely on the basis of criteria related to the process that produced the results. Issues related to this form of evaluation are discussed elsewhere in this book (see Chapters 1, 13, and 15).

Action 5. Plan the evaluation to help the program manager to improve the program as well as to measure its impacts. Evaluations should, whenever possible, not only assess the program impacts, but also aim to identify constructive ways of improving the program being evaluated. Although such an effort seems obviously desirable, it is not always undertaken, since, with limited resources for evaluation, it will dilute the impact-evaluation component.

A large number of factors can affect program success in addition to the new approach and procedures, such as the quantity and quality of the staff operating the program; the organizational structure under which the program is being managed; and external factors such as economic and weather conditions, and so on. Evaluators from outside the program are in a particularly advantageous position to obtain insights about the program and its problems, and how the problems might be overcome. The central evaluation itself should be able to shed light on program problems by examining the relations between program outcomes and selected program characteristics (for further discussion, see Hatry et al., 1981, Chapter 5).

It is to the mutual advantage of the managers and evaluators for these perspectives to be shared, so that the managers can make use of them in their management decisions. In the long run such a helpful posture by the evaluators will not only increase their credibility—and the usefulness of their findings—but also help reduce the instinctive hostility that most program managers have to being evaluated.

Action 6. Communicate clearly to all concerned what can—and what cannot—be expected from the evaluation, so that expectations will be realistic. On the one hand, decisionmakers sometimes have unrealistic expectations about what program evaluation can accomplish; on the other, evaluators may be tempted to promise more than they can deliver. An evaluation seldom yields a completely definitive, conclusive, and unambiguous result defining program success or failure. Even under the best circumstances with a controlled experimental design, numerous problems inevitably arise in keeping the experiment uncontaminated by unexpected influences. Therefore, in almost all instances, the results of an evaluation will not be definitive. Evaluators should be at considerable pains to tell decisionmakers beforehand what to expect from an evaluation and what limitations it will have, as well as issue warnings about possible contaminations that will have to be guarded against—but that may nevertheless occur.

Unrealistic expectations that are not met can destroy confidence, not only in the evaluation of concern, but also in evaluation activity in general.

Action 7. After reviewing both intended and actual effects, select adequate measures of program impacts. Choosing adequate measures of effectiveness is a central issue faced by systems analysts, and it has been discussed in the *Overview*. It is also a central issue in program evaluation. If the program to be evaluated is one that was adopted after being recommended by a systems-analysis study, it is reasonable to expect that the systems-analysis work identified appropriate measures of effectiveness; therefore, it is reasonable for the evaluation to explore the extent to which these criteria have been met by what actually occurred during the program implementation.

On the other hand, it would be imprudent for the evaluation team to presume that the criterion question was entirely settled by the prior systems analysis. New problems and issues may have arisen during the implementation, and new surrounding circumstances may have raised new questions. Therefore, as the evaluation is being planned and carried out, the evaluation team should review the issue of what measures of effectiveness should be used.

Multiple evaluation criteria will generally be needed. The credibility and usefulness of an evaluation can be called into considerable doubt if it uses inadequate measures of program effects: if it limits its attention to only one criterion when others are clearly relevant for the decisionmakers (even if they have agreed in advance to limit the evaluation to this single criterion); or if the evaluation neglects important consequences of the program, whether or not intended, and whether or not beneficial. For example, an evaluation of a mental health program aimed at leaving patients in their homes and communities rather than placing them in large institutions should consider not only the welfare of the patients but also the impacts on families and communities; evaluations of transportation programs should consider not only safety, travel-cost, and convenience criteria but also environmental effects; and so on.

Before establishing the final evaluation criteria, the evaluation team should review the objectives of the program from the viewpoint not only of the agency installing the program but also of the program's clients. The team should also look for significant impacts that were not anticipated at the beginning. The danger that evaluations will be overly affected by the evaluation team's own outlooks and preferences (for example, Shaver and Staines, 1972) suggests that considerable effort should be devoted to obtaining various perspectives on both the explicit and implicit objectives of the program: for example, perspectives from supporters and opponents, from program operators and clients, from budget officials and program managers, from the legislative and administrative branches, and from special-interest groups and the general public.

Two aspects of program evaluation deserve to be singled out here. The first is the importance of considering separately the effects on significant groups within the client population. Inevitably, programs have different effects on different groups, helping some more than others, perhaps even being detrimental to some. Thus, the measures of program effects should be disaggregated to show the effects on each group. Too much data aggregation can hide such differential effects, thus preventing the users of the evaluation from considering differences among client groups that the program has caused. It is important for the evaluation to seek and measure these differential effects so that future program revisers can include such considerations.

The second special aspect of program evaluation is to assess program financial costs. Most evaluations do not estimate past program costs, but such information is likely to be of vital interest and considerable use to officials concerned with the program.

In spite of the desire for comprehensiveness, as stressed in this discussion, the pursuit of multiple, comprehensive measures of effectiveness must be tempered by the practicalities of time, effort, and usefulness. Marginally useful criteria requiring costly data collection and complicated analysis will have to be dropped from the evaluation work (see Action 13 for further discussion). In general, the evaluation will have to focus its effort on important issues and leave marginal ones aside.

Actions during the Period of Observation, Data Collection, and Early Analysis

These actions are designed to move the evaluation safely from the plan through to the stage of formulating the findings.

Action 8. Monitor the operations of the program closely during the evaluation period to determine what really has taken place, as contrasted with what may have been planned for the implementation. This action stipulates that the analysis team should carefully identify and describe the program that has actually been implemented. It is not uncommon for the best-laid plans of program managers—and evaluators—to go awry when they face reality. The possibility of significant deviations from the program implementation plan is always present, and becomes more likely as the period of evaluation grows longer. For example, in an evaluation of neighborhood police teams, the consistent assignment of teams to specified neighborhoods may degenerate if the dispatchers alter assignments to meet unusual needs or convenience. If evaluators watch what is going on carefully, they may be able to compensate for such deviations from plan if they are not widespread and act to ensure that further deviations are not made.

Similarly, maintaining the practice of random assignments is often difficult when the assignments have to be made throughout a prolonged period by persons other than the evaluation team. For example, in an experiment to test the effects of requiring appearances before a judge for moving traffic violations, the court clerk had the legally mandated responsibility for making assignments; however, he had fixed ideas about the need for court appearances for young drivers and consequently did not adhere to the random assignment procedure for them (Conner, 1977, which also gives other examples). Prejudices, heavy workloads, personnel not familiar with the imperatives of the research design, and many other influences can alter assignment patterns to the detriment of an evaluation effort. To make sure that the evaluation plan is followed and to spot important deviations from the plan, prudent evaluators cannot afford to relax their vigilance.

In spite of great care by program managers and evaluators, serious changes do occur in mid-evaluation, particularly when the external environment thrusts uncontrollable changes on the surrounding situation. Sometimes a revised evaluation, perhaps of degraded quality, can be patched together after such a change. On the other hand, if such an adjustment cannot be made to a reasonable degree of satisfaction, the evaluators should face the situation realistically and recommend that the evaluation be terminated.

Action 9. Identify, and adjust for, changes in data-collection procedures that occur between pre- and postprogram data gathering. This problem typically occurs when the evaluation depends in part on data collected before the evaluation began, for example, on past government records. Evaluators should be aware of this warning (Riecken and Boruch, 1974): "Too often a new program is accompanied by changes in record-keeping." Data collected by others may have been guided by standards other than those desired in the evaluation and government data definitions and collection procedures often change. Either situation can cause important differences in what the resulting data mean and how they can be interpreted. Thus, evaluators who must use such data must take pains to verify their relevance and, when appropriate, make required adjustments to the data.

Action 10. Avoid using persons whose work is being evaluated as evaluators of their own work. In spite of the obvious objections to self-evaluation, there are sometimes temptations to ignore them. Government agencies with tight resources are especially tempted. For example, some mental-health and social-service agencies have used caseworkers' ratings to assess the progress of clients being served by these caseworkers. Such a procedure may be reasonable when the information serves solely internal purposes, such as casework management, but, for evaluations on which to base decisions about program continuation and budgets, it seems to be asking more of human nature than is reasonable.

Action 11. Look for plausible explanations for the outcomes other than the program treatments. After outcomes have been measured, the key dilemma of program evaluation is to determine the extent to which the effects can be attributed to the program itself rather than to other influences. Controlled-experiment designs are aimed at reducing this problem, but they do not eliminate it altogether. For other evaluation designs this problem assumes major importance.

Not looking hard for other plausible explanations is a pitfall for evaluators using nonexperimental evaluation designs. A number of variations of this pitfall have been discussed in the literature.[3] Some are discussed below.

1. *The influence of initial conditions.* The state of affairs at the beginning of the evaluation should be considered when interpreting the findings. These initial conditions might have been sufficiently extreme that, even without the program's intervention, there would have been a natural shift toward improvement without the program;[4] or the opposite could have occurred because the initial conditions were so bad that no improvement could reasonably have been expected. Clients that, at the outset, have the greatest need (who, for example, are in crisis situations) are capable of showing much more improvement than others. Conversely, the less needy clients can show only small improvements regardless of the program (Conner, 1977). If, as often happens, a program is initiated because of a rash of recent problem occurrences, a remedial program is likely to be accompanied by improvement, simply because the probability of such an unfortunate cluster recurring is small; a "rash" of traffic fatalities is a case in point (Campbell and Ross, 1968).

2. *Maturation.* In some cases, there is a normal maturing process, which is accompanied by a gradual improvement in client behavior without any sort of programmatic intervention. For example, as persons with criminal, alcoholic, or drug-addiction backgrounds age, there often appears, over a period of many years, to be a reduction in these adverse behaviors without the influence of treatment programs. An evaluation of community alcohol treatment centers included as a comparison group a sample of persons that had an intake record but for whom only nominal treatment was provided. A follow-up 18 months after the intake showed that 54 percent of this group were in remission (Armor et al., 1976).

3. *Dropouts.* It is important not to neglect dropouts when formulating evaluation findings. Usually, clients that drop out of a program reflect negative results. Basing the findings only on the clients who complete the program can make the reported results too positive. Dropouts could also be used as a comparison group, to see if dropouts "succeeded" as much as the treatment group. In any case, to report results only for clients who complete the program is to present a misleading characterization of the program's performance.

4. *Comparison groups.* The lack of a comparison group, or, on the other hand, the use of one that is inappropriate, can lead to findings that distort the true situation. Even when a controlled experiment is impos-

---

[3] Such circumstances, sometimes called *threats to validity*, are discussed in Campbell (1969); Campbell and Stanley (1966); Bernstein, Bohrnstedt, and Borgatta (1975); Nunnally and Wilson (1975); and Fischer (1976). Not all of the threats to validity described in these references are described here.

[4] This effect is sometimes called *regression toward the mean.*

sible, examining groups that received no treatment—or reduced treatment (such as dropouts)—can help to provide evidence relating to whether or not the program yielded the observed effects. In the evaluation of the Connecticut highway speeding crackdown (Campbell and Ross, 1968), comparisons with nearby states that did not have similar programs helped to screen out other causes of reduced fatalities, such as unusual weather conditions and the introduction of safer automobiles.

A study of the treatment of arthritis (Boruch, 1976) provides a case in which the effects of a program could have been misunderstood if there had not been a comparison group. The results for the group subjected to the treatment program were that the severity of the arthritic conditions increased; however, the condition of the untreated control group worsened more than those in the treated group. This indicated that the treatment program had enjoyed some success, rather than being the failure that it might have been judged to be if the comparison group had not been studied.

The evaluation of community alcoholism treatment centers (Armor et al., 1976) mentioned earlier followed up not only those who had received significant amounts of treatment but also samples from two other comparison groups: (a) persons who had made only one visit to a treatment center and had received no further treatment and (b) clients who had received minimal services (usually detoxification) and then left the center and had not returned. The evaluators found that, 18 months after treated clients had entered the program, 67 percent were in remission. But, as already noted, 54 percent of the group receiving only nominal treatment was in also remission. Thus, it appears possible that a significant portion of the treated group might have attained and kept remission with only nominal help from the program. Considering only the treatment-group remissions would probably lead one to overstate the effects of the treatment program.

When comparison groups are used, considerable care should be taken to avoid misinterpretations. If a program's clients differ in some potentially critical characteristic from the comparison group (for example, if persons entering the program have greater motivations than those with similar problems who do not), differences in impact could be due to this characteristic rather than the program's treatment.[5] A useful procedure when using comparison groups (even when the experiment is controlled) is to check for similarity on key characteristics, where this is possible.

---

[5] Programs whose procedures allow clients to determine for themselves whether or not they will enter a treatment program raise the question of whether or not the motivation to join is an important contributor to program success. Highly motivated persons might make more progress than others even without the program. Volunteers for experiments tend to be more innovative or self-confident than the average eligible participant (Hamrin, 1974).

Table 11.3. Data from a Hypothetical Evaluation of a Program Aimed
at Reducing the Dependency of Clients Presenting
Two Levels of Difficulty

	Number of cases	Number and percentage of cases where dependency was reduced
Group 1		
High client difficulty	400	100 (25%)
Low client difficulty	100	100 (100%)
Total	500	200 (40%)
Group 2		
High client difficulty	100	0 (0%)
Low client difficulty	400	300 (75%)
Total	500	300 (60%)

Source: Millar, Koss, and Hatry (1977).

(Edwards et al., 1975, discuss the problems with comparison groups further.)

5. *Hawthorne effects.* These effects can occur when program personnel and receivers of services alter their behavior because they are in a special program or become aware that an evaluation is under way (see, for example, Tannenbaum, 1970). Since such behavior may distort evaluation results, and thus restrict their generalizability, it is well to plan the evaluation so as to avoid the possibility of such effects if possible. For example, as Roos (1975) points out, if the evaluation can be done after the fact without requiring new data to be collected or new observations made, the evaluation results will not occasion such modified behavior.

6. *Handpicked personnel.* Using handpicked personnel to operate the new program can make the outcomes unrepresentative and ungeneralizable. Using specially chosen workers may be appropriate in a program's developmental stages, but it is to be avoided when generalizable results are wanted. To achieve representative results, personnel of the type that would ordinarily operate the program should be used. Otherwise, any observed advantage to the treatment group could be attributed to the use of the special personnel.

7. *The difficulty of the workload.* When assessing results, changes or differences in the difficulty of the program's workload should be considered explicitly. Not to control for this important variable is to risk misinterpreting what has occurred. A program with a lower apparent success rate because it dealt with a mix of clients who were more difficult to help is not necessarily less effective than a program with a higher success rate but a less difficult mix of clients. Consider the hypothetical outcomes shown in Table 11.3 Based on the totals alone, the second group appears to be superior because success was achieved in 60 percent of the cases,

as contrasted with 40 percent for the first group. But Group 1 had a higher success rate for both high-difficulty clients (25 percent as compared to 0 percent for Group 2) and low-difficulty clients (100 percent as compared to 75 percent for Group 2). The overall higher success rate for Group 2 stems from its having a large proportion of clients with low difficulty.

Thus, the difficulty of the incoming workload can be a significant reason for the observed effects.

When random-assignment, controlled experiments can be used, they should greatly reduce the likelihood of this problem arising. Even if workload difficulty is not controlled explicitly in making random assignments (such as by stratifying the sample), randomization will likely lead to assigning similar proportions of difficult-to-help clients to each group. Nevertheless, even in controlled experiments, the control and treated groups should be examined after they are chosen to determine if they are indeed sufficiently similar in difficulty.

Note that this point applies, not only to programs whose workload consists of human clients, but also to most, if not all, types of programs (such as, for example, different or changing quality of water entering a water-treatment program, roads with different traffic or different underlying soil conditions in a road-maintenance program, different ages and models of vehicles in a vehicle-maintenance program, and so on).

8. *Correlation and causality.* In spite of the standard warnings in statistics textbooks, it is tempting to assume that, when two variables are closely correlated, variation in one of the variables causes the variation in the other. In the absence of other information, all that can be said is that the two variables are associated.

What is needed to strengthen the minimal conclusion of association? Two eminent statisticians respond in this way (Mosteller and Tukey, 1977, pp. 260–261):

> Two or three sorts of ideas usually are required to support the notion of "cause":
>
> 1. Consistency: that when other things are equal in the population we examine, the relation between x and y is consistent across populations in direction—perhaps even in amount.
>
> 2. Responsiveness: that if we can intervene and change x for some individuals, their y's will respond accordingly.
>
> 3. A mechanism: that there is a mechanism, that someone might sometime understand, through which the "cause" is related, often step by step, with the "effect"—the sort of mechanism where, at each step, it would be natural to say "this causes that.". . .
>
> Of these three, only one—consistency—can be confirmed by observation alone. We might look at a variety of different populations and see whether the relationship between x and y is consistent in direction or in direction and amount.

The second—responsiveness—can be confirmed by experiment, if an experiment is feasible. If we can intervene and change x, we can see if y then changes. Occasionally natural experiments as well as manmade ones offer some evidence here, the natural ones being almost always less valuable. (Beware especially of "natural experiments" where no change has been made.)

The third—the mechanism—can be confirmed only by constructing a detailed mechanism, and supporting the correspondence between each step in the mechanism and that in the process under study.

Causation, though often our major concern, is usually not settled by statistical arguments (indeed, is usually not settled at all in social problems), though statistics has many inputs that can help. Causation is made more plausible by considerations drawn from data, most effectively perhaps when three essentials are provided:

a clear and consistent association between x and y;

a showing that either there are no plausible common causes of x and y or that the quantitative relationships between the plausible common causes and x and y are inadequate to explain an observed clear and consistent association. . . .

a showing that it is unreasonable for y to cause x, which is often far from easy. . . .

To this overview, two points can be added:

Conclusions about causality or lack of it do not arise from statistical evidence; they must rest on what the analyst knows about the situation as a whole. The statistical evidence then can be viewed as supporting or denying the hypotheses generated by the knowledge of the total situation.

Since a program evaluation is only one specialized test within the broad sweep of inquiry—summarized in the quotation above—that must be mounted to arrive at firmly confident conclusions about causation, the findings of the evaluation should be qualified appropriately.

Summary of Action 11. In general, it is important to search widely for plausible explanations for observed differences other than the program being evaluated. A wide variety of possible circumstances or factors can cause unrepresentative or misleading findings. For example, changes in the employment status of persons given training programs may occur because of improved general economic conditions even in the absence of training programs (Boruch, 1976). A correlation found between the amount of scotch whiskey imported into New York and the growth of the number of ministers there depended more on the growth of the population during the Industrial Revolution than on the tippling propensities of the divines (Mosteller and Tukey, 1977, p. 261). Using a control group in the first example could shed light on the influence of economic conditions, and observing that the amount of whiskey imported was far more than could possibly be drunk by the ministers would confirm the conclusion stated in the second example.

Action 12. Avoid pitfalls in undertaking sample surveys. Evaluations frequently use surveys of sample groups of program clients or comparison groups to obtain information about aspects of the program. Because of their prevalence and importance as an evaluation tool, sample surveys should be designed and administered carefully. To this end, evaluators should avoid pitfalls by taking actions such as these:[6]

Provide adequate training for interviewers and test their performance.

Pretest the wordings and order of the questions; correct the difficulties discovered by the pretest, and, if necessary, pretest again.

Use considerable care in selecting the sample; avoid sampling in a way that omits or undersamples a key group (such as persons without telephones; those with unlisted numbers; those living in trailers; or those living in multiple dwelling units, where the list from which the sample is drawn includes only individual structures and not the individual households in each structure). Low-income, disadvantaged persons often are particularly likely to be underrepresented in sample surveys.

Provide for an adequate number of callbacks so that persons not generally at home are heard from; provide for interviews at various times throughout the week so that persons not at home during certain hours (such as during the day) are not excluded systematically from the sample.

Schedule enough effort to achieve high completion rates, so that the number of persons that cannot be located or refuse to be interviewed will be small. This goal is particularly hard to achieve with mail surveys, especially those without provision for following up nonrespondents, but is difficult in all modes of interviewing. The number of mail returns may be large, but mail-return percentages can be as low as 25 percent, or even less. The temptation to claim significance because the *number* of returns is large is to be shunned, as such a low response *rate* cannot be said to represent the population—and therefore it is not a reliable basis for deriving generalizations about the population.

---

[6] There is a large literature discussing survey procedures in more detail. Here are some useful entry items: Abramson (1974); Babbie (1973); Campbell et al. (1969); Duckworth (1973); Rossi, Wright, and Anderson (1985); Survey Research Center (1976); Turner and Martin (1984); Webb and Hatry (1973); Weiss and Hatry (1971). In addition, the *Public Opinion Quarterly* frequently contains articles on the problems of surveys. The U.S. National Center for Health Statistics Series 2 reports ("data evaluation and methods research") provide occasional studies on the problems of survey research, such as Gleeson (1972).

Action 13. Allow enough time to analyze the data that are collected; avoid collecting too much data. These two concerns go hand in hand. Evaluators are tempted to collect data on any characteristic of the client or situation that could conceivably be relevant and then allow too little time for adequate analysis. The urge to collect data is difficult to overcome, particularly in evaluations, since it is often difficult at the beginning to know which data will turn out to be important. The argument can be advanced that some data can be excluded later, but, once collected, data pile up and have a pyramiding effect on the data-processing and analysis effort (not to mention the added collection costs).

Interview questionnaires are particularly good places to exercise restraint. They can become longer and longer as members of the evaluation team consider what information could possibly be valuable. This tendency can be restrained by requiring that each question be justified by identifying the use to which the resulting data will be put in the analysis and the light that the information can shed on the relevant hypotheses.

Allowing enough time for data analysis is complicated by the tendency to schedule too short a time period for the evaluation as a whole. Implementation difficulties can delay the start of the program. The new-program shakedown period can be longer than expected. Data can come in later than anticipated. Computer processing can take longer than planned. Any or all of these delays can severely shorten the amount of time available for completing the analysis.

Limiting the data to be gathered to what is clearly essential for the analysis and adopting a schedule that allows time to cope with delays can help to alleviate these problems.

Actions Relating to Deriving the Results and Formulating the Findings

Although the four actions in this category are all part of the broad skills of systems analysis, it is worth considering the special issues and choices faced by analysts involved in program evaluation.

Action 14. Avoid overdependence on statistical significance, particularly at the expense of practical significance. A narrow focus on precision and on statistical significance can lead to excessive evaluation costs (perhaps by requiring samples larger than needed) and findings that are misleading from a practical point of view. Statistical significance at the traditional academic levels of 95 to 99 percent often are too demanding for programs other than those involving health or safety elements. For most government decisions such precision is neither appropriate nor needed. ''It doesn't pay to lavish time and money on being extremely precise in one feature if this is out of proportion with the exactness of the rest" (Herzog, 1959, p. 82). Lower significance levels, such as 90 percent, or even 80 percent, are often likely to be more appropriate and efficient.

Statistical significance levels alone can be a misleading guide to officials using the results. What may be statistically significant can, particularly when large samples have been involved, suggest that important differences exist when in fact the actual differences are too small to have practical significance. Differences of two or three percentage points arising from large samples can be statistically significant, but may not have any practical significance for officials charged with making decisions. Good practice is to present users with both the actual differences and the statistical significance levels so that the users can judge the practical significance for themselves.

Action 15. Consider the component effects as well as overall or average results. Although examining aggregate data is useful for assessing a program's aggregate results, the analysis should seldom be limited to the aggregate data. Usually it will be highly useful, if not critical, to explore disaggregations.

When a number of different projects, facilities, sites, and so on, are included in a program that is being evaluated, the evaluation team should explore whether or not the program affected certain segments differently from others. Also, variations in the conditions underlying different projects may differ, and examining these may shed light on possible reasons for the effects, thus possibly suggesting program variations that should be applied to different types of projects. Thus, examining component effects and underlying conditions often will yield useful findings and identify program successes, even if the overall program did not emerge as successful.

In addition, some types of clients served by the program may exhibit better outcomes than others, even though such outcomes were not anticipated in the original evaluation design. Therefore, in general, as discussed under Action 7, the evaluation team should examine various subgroups to detect whether or not there were some subgroups exhibiting better outcomes than suggested by the aggregate figures. For example, a program may work better with severe cases than with mild ones, with older clients rather than younger ones, with females rather than males. Sometimes the samples can be stratified this way at the beginning of the work, but if not, an after-the-fact analysis is better than not considering potential differences at all.

Action 16. Identify study limitations; do not overgeneralize the results. Even when the evaluation is well done and very well controlled, there are numerous pitfalls to be avoided in trying to generalize the results to the future, possibly more widespread, applications. "Too many social

scientists expect single experiments to settle issues once and for all"
(Campbell, 1969). "The particular sample from which control and experimental group members are drawn . . . may be idiosyncratic in that
other potential target populations are not represented. If the conditions
. . . in the experiment . . . differ markedly from conditions which prevail
in other populations, then it is reasonable to believe that additional testing
of the program is required . . . " (Riecken and Boruch, 1974).

There are several variations of this pitfall, four of which are discussed
below. Recognizing them should temper statements about the generalizability of findings.

1. *The limitations of a single trial.* The results of one trial under one
set of conditions may represent only a single sample point. Before one
can make confident general statements about the general effectiveness of
the program it may be necessary to replicate the program at other sites
and times. Of course, if the initial trial covers the entire target population
of clients and a lengthy period of time, this limitation will be of less concern, but it is more usual for the trial to have limitations on its size and
coverage. Stating these limitations clearly is an important part of reporting
the findings. An example, the New Jersey Graduated Work Incentive
experiment, dealt with only one type of geographical location (the U.S.
urban east coast), covered only male-headed households, and varied only
the level of guaranteed income and the tax rate (Roos, 1975). The generalizability of the findings had to be judged accordingly. Since this first
experiment was completed, it has been replicated elsewhere in order to
test the generality of the results.

2. *The characteristics of the intervention.* As already discussed under
Action 8 it is vital for the evaluators to know in detail what was actually
implemented, and to be alert to features of the trial that appear to be
significant factors in its outcomes. For example, in social service casework the outcomes can be affected by such characteristics as the techniques used, the caseworkers' personalities, the amounts of time spent
with clients, and the caseworkers' styles (Fischer, 1976). Unless the evaluation procedures isolate these characteristics, the evaluators will be unable to understand how they may have affected the outcomes. Without
considering such possibilities, the findings should not be overgeneralized,
and should include appropriate qualifying conditions.

3. *Declining novelty.* Behavior may alter when the novelty of a new
program wears off. Both program operators and their clients may be affected. Similarly, clients in a program adopted for an entire population
may behave differently from those in a trial regarded as novel and especially interesting. For example, consider a program at one or two locations to explore the effects of using group homes rather than large
institutions to care for children with records of delinquency. The findings
might not be representative of what would happen if the program were
substantially expanded, especially when expanded to homes where the

eagerness to participate in a promising experiment is not present. In fact, the community attitude to having a larger number of group homes could well be quite different from its attitude to having one or two as part of an experiment. Thus, it would be unwise to generalize too broadly from the test experience.

4. *Not-included groups*. The evaluators should be at considerable pains to determine which types of clients were included in the experiment and which were not. It may turn out that some groups were not included, even though this was not intended. In any case, the test findings should not be attributed to groups not included unless there is evidence that the findings can be so attributed. There are, of course, a variety of reasons why an initial test program would restrict its attention to a carefully selected group of clients. When such restrictions occur, either intentionally or unintentionally, the findings should be appropriately qualified. The New Jersey Graduated Work Incentive experiment cited earlier was such a program.

Action 17. Provide an opportunity for the program personnel to review the tentative findings before they are put into final form and reported. It is a matter of courtesy and good practice to permit program personnel to review the findings before they are put into final form and reported. There is also, however, an important technical purpose. Such a review provides a critical consideration of the evaluation's validity—its strengths and weaknesses—from a well-informed point of view. The persons involved in the program's operation may spot situations and factors that the evaluation team has missed. They can often add considerable insight into the interpretations of data and results. They will sometimes correct misinterpretations and misunderstandings of the evaluators (who may have come on the scene late in the life of the program and observed only part of its operations). The program managers may be defensive and hostile, but their review can subject the work and its outcomes to a severely critical test—something that can be viewed as an important advantage. In an evaluation of a county drug treatment program, program personnel reviewing the draft evaluation report pointed out that a group of program clients had been omitted from the evaluation—requiring the evaluation team to follow up this neglected group before issuing the final report. The opportunity for program personnel to suggest modifications may also reduce their defensiveness, and thus serve to encourage them to accept the evaluation findings and implement changes arising from them.

Actions Relating to Presenting the Findings

The three actions in this category are aimed at creating in the minds of the users of the evaluation's findings correct impressions of their implications and range of validity.

Action 18. When presenting the findings of an evaluation, be at pains to describe its conditions, assumptions, limitations, and uncertainties. This admonition emphasizes for evaluation an important principle of systems analysis (see, for example, Sections 13.6, 13.7, and 15.9). The evaluation report should make clear what was done and, more importantly, what was not done. It should indicate the boundaries of validity of the evaluation criteria, the limitations of the measuring instruments, the uncertainties underlying the findings (not only those associated with statistical analysis but any others as well), and any other qualifications that should attend the results (see Action 16). This background is essential if decisionmakers are to arrive at fully informed judgments about the implications of the findings. As a result, the users of the evaluation should be able to reach judgments that factor in both the findings and their limitations.

Action 19. Separate facts from opinions. Since it may be easy for nonscientific users of evaluation findings to be misled as to the nature and basis for findings, especially when they are numerous and complicated, it is important for evaluators, in discussing their results, to distinguish carefully between their opinions and reasonably objective findings and conclusions. Similarly, they should delineate carefully the strength of the evidence underlying each recommendation.

Government decisionmakers seldom enjoy the luxury of having completely objective and decisive information on which to base decisions—nor do they usually expect such an ideal commodity from an evaluation. Rather, they are accustomed to making decisions on the basis of an amalgam of fact, opinion, and even hunch. Knowing this, evaluators may choose to accompany their objective findings with insights and opinions that arise naturally from their work, because they are aware that this will be expected and may be useful to the users of the evaluations. This fact makes it even more important for the facts and opinions to be clearly delineated, so that users know with confidence and clarity the bases for the conclusions and any recommendations.

Action 20. Be at considerable pains to present the findings to users comprehensively, clearly, concisely, and effectively. Unfortunately, poor presentation of findings is a chronic problem of program evaluation. The casual, time-consuming, academic blackboard lecture to which many researchers have become accustomed in universities simply will not do in this context. This phase of the evaluation work can make all of the previous effort either worthwhile or worthless. How to present the findings deserves a great deal of careful thought and preparation, as in the case of any systems analysis.

The principles of effective presentation for systems analysis, as discussed in Sections 13.7 and 15.9, also apply to reporting the results of evaluations. Summary presentations, whether oral or written, should be clear, concise, intelligible to the users, and couched in language they are familiar with. The supporting technical evidence should be available in writing, so that technical staffs and reviewers can examine the bases for the findings.

11.5. Conditions under Which Controlled Experiments Are Likely to Be Appropriate

The controlled experiment, as Section 11.3 says, is conceptually the most desirable form of program evaluation because it is the type most likely to provide clear and relatively unambiguous results. The purpose of this section is to explore the questions that must be answered in the process of adapting this form of inquiry in a practical way to a program evaluation that an agency might employ.

Chapter 5 of this book discusses social experimentation as a tool for measuring the impacts of alternatives; therefore, what it says also applies to program evaluations that take the form of social experiments. The purpose here is to extend the discussion of social experiments to program evaluations by emphasizing the important issues and choices that evaluation analysts must face if they are to adopt this mode of analysis.

Special Requirements for Controlled Experiments

There are two special requirements that must be met if controlled experimentation is to be successful: random assignment and control of implementation so that the experiment is carried out properly. These requirements dictate that the experimental procedures be planned ahead of the introduction of the new program, and that the plan specify the procedures for making the random assignments appropriately and for assuring that prescribed treatment procedures be followed throughout the evaluation.

Random assignment. A controlled experiment involves some groups who are administered different treatments, one to each group, and one or more control groups, which are given no treatment, less treatment, and/or different types of treatment. Each unit to be subjected to the experiment is assigned to a group at random. That is, the process of assignment has the property that a unit has the same probability of being assigned to a given group as to any other group. The units to be assigned are most commonly people, as in experiments involving human-service programs, such as health, correctional, training, or social-service programs. The units can also be geographical areas, administrative units, facilities, equipment (subjected, for example, to different maintenance procedures), and so on, depending on the subject of the experiment. In some cases, random assignments are made only at the beginning of the experiment, with the assignments remaining in effect throughout. In other cases, where clients may come and go during an experiment, such assignments may continue throughout the evaluation.

Control. Not only must the principle of random assignment be maintained, but also the integrity of the treatment protocols must be preserved throughout, so that the groups receive the prescribed treatments. No group should receive treatment significantly different from that prescribed in the evaluation plan. This discipline is needed to avoid contaminating the experiment with other extraneous factors that might affect the outcomes.

Violating either of these experimental conditions will weaken the ability of the experiment to provide definitive findings.

Questions to Be Dealt with in Planning Controlled-Experiment Evaluations

In approaching the evaluability of a program—Action 3 of Section 11.4— there are a number of questions that must be answered satisfactorily if a controlled experiment is desired. Table 11.4 lists a number of these questions; the discussion that follows addresses them.

Question 1. Can some units (for example, citizens) be given treatments different from those given others without the potential for significant danger or harm? If there is a significant chance that withholding a treatment or service to the control group would cause substantial personal harm or would put them in recognizable danger, the experimental approach is not likely to be appropriate.

The converse also applies: If there is a possibility that the treatment will be harmful or dangerous to the experimental group, the experiment becomes questionable, except in the relatively rare case where it is possible to proceed with volunteers who are fully aware of the potential hazards and ramifications. This condition also applies, of course, to any experiment, whether or not controlled.

As an example, consider the withdrawal or reduction of police or fire-protection services in certain areas as part of a test of certain types of police or fire-protection activities. Here the question is the extent to which the reduced services entail risks. In a Kansas City, Missouri, experiment, the police department varied the amount of patrol activity in randomly

Table 11.4. Questions to Be Dealt with in Planning Controlled-
Experiment Evaluations

---

1. Can some units (for example, citizens) be given treatments different from those given others without the potential for significant danger or harm?
2. Can some citizens be given services different from those given others without violating moral or ethical standards?
3. Is the cost of the experiment likely to be less than the risk entailed in implementing the program without a formal evaluation?
4. To what extent is a decision about full-scale implementation postponable until the evaluation experiment is completed?
5. Can the essential experimental conditions be maintained throughout the experimental period?
6. Are adequate technical procedures available for assessing the program's effects?
7. Are the prospective sample sizes likely to be large enough to support definitive results?
8. Are the potential findings likely to be generalizable to the whole jurisdiction or population?
9. Are sufficient resources—staff, finances, and administrative support—available to carry the experiment through?
10. Can client privacy and confidentiality of records be maintained?
11. When it is needed, can client consent for participation be obtained without invalidating the experiment?

---

selected areas of the city. In this process, routine patrols were withdrawn from some sectors (police responded to service calls from these sectors but did not patrol them). These changes were made without major opposition. The results showed no significant difference in the amount of reported crime related to the different levels of patrol activity. One reason for the lack of opposition to undertaking the experiment was that the previously existing patrolling was sufficiently infrequent that the public did not observe a difference. The study's findings were controversial, but not on the grounds that the experiment was unsafe (Kelling et al., 1974; Larson, 1975; Larson et al., 1977).

Question 2. Can some citizens be given services different from those given others without violating moral or ethical standards? Some experimental designs, although not involving physical hazards, may call for not providing, or offering reduced, services to some groups, which may raise ethical or moral issues. For example, if an experiment to assess the effectiveness of social casework calls for no service to be rendered to a control group, this arrangement could be perceived as unethical or immoral (especially by those who believe that such casework is an essential support for its clients).

One way to reduce the likelihood of such a perception is not to withdraw services fully, but instead to use less dramatic variations.

This issue is less relevant for situations where the effectiveness of the program is in substantial doubt, and therefore where the potential deprivation is questionable.[7] Indeed, it is worth noting that the usual basic purpose of experimental evaluation is to obtain information on situations where there is considerable uncertainty about the program. Further, it can be argued that to undertake a program with no evidence of its effectiveness, or with highly questionable evidence (such as that perhaps based on equivocal trials, or merely on strongly held beliefs), is itself unethical, or at least inappropriate as a governmental activity (Boruch, 1976).

When the resources available to the organization are not sufficient to treat all clients, a random assignment to treatment and nontreatment groups seems appropriate and less controversial, provided those concerned understand what is being done. Random assignment could replace an existing first-come, first-served allocation rule. There is, however, an important limitation: The government may want to assure that at least the high-priority cases (such as those with a major need for the service) are served; such cases might have to be excluded from the experiment, with the random assignment procedure applied to the rest of the clients.

Question 3. Is the cost of the experiment likely to be less than the risk entailed in implementing the program without a formal evaluation? Although there are exceptions, evaluation by controlled experimentation is a costly venture. It requires careful planning, special data-collection activity, and much careful attention throughout the experimental period, as well as a qualified evaluation staff employed over a substantial period. Thus, to be valuable and attractive, an experimental evaluation should cost considerably less than the value of the savings or improved benefits that could possibly accrue from applying the experimental results. It is usually not too difficult to arrive at judgmental probabilities of various possible outcomes that will allow comparison of the cost of the experiment with the values of possible outcomes.

In the United States, federally-sponsored experiments have sometimes cost hundreds of thousands of dollars, and they have been for programs with many millions, if not thousands of millions, of dollars at stake. However, if a governmental unit already employs personnel with appropriate skills and responsibilities, the marginal additional costs could be substantially less. Certain small-scale experiments, such as the vehicle-maintenance experiment discussed in Section 11.3, can be undertaken at much less expense.

When a government is facing a commitment to an expensive, long-term

---

[7] For a further discussion of this point, see Boruch (1976). Even for such obvious examples of potential ethical problems as social casework, the service's inherent effectiveness has been severely questioned; see Fischer (1976), which reviews the findings of a number of past evaluations of casework, including some controlled experiments.

program, an experimental evaluation is an appropriate early step to take. New programs that, once in place, are likely to be difficult to terminate, cut back, or even modify are especially attractive possibilities. In such cases, using an experimental period to test the program on less than a full-scale basis is an attractive possibility if it is feasible, even if the experimental period is fairly long (see, for example, Lowry, 1983, for a description of a ten-year study of housing assistance for low-income families). An experiment, rather than immediate full-scale implementation, tends to reduce (but not always eliminate) the tendency for vested interest groups to build up and is likely to permit decisions about the future shape of the program without supercharged political cross-currents. For example, drug or alcohol treatment programs could be introduced in selected localities, rather than throughout the country.

To the extent that there is considerable risk or uncertainty about moving ahead with full-scale implementation, evaluating a program experimentally should be considered as a way to put the decisions about the program's future on a more reliable and less costly basis.

Question 4. To what extent is a decision about full-scale implementation postponable until the evaluation experiment is completed? Undertaking an evaluation experiment for a social program can take a significant amount of time—more than one or two years is not unusual. Thus, such an experiment should be considered only when a decision about full-scale implementation is postponable until after the experiment has been completed. It may not be possible to delay a decision when there is considerable political pressure, perhaps in response to a recent emergency, but an experiment could be used as a way to temporize, since it represents a partial implementation. It can also deal with a situation in which public officials feel that an immediate decision in favor of a full-scale implementation would be primarily an emotional response that could turn out to be wasteful.

Other factors can also make it difficult to postpone a full-scale decision while an experiment is undertaken: if even the partial implementation arising from an experimental program entails a major organizational upheaval or if to proceed at all requires a major capital investment. For example, individual local governments are not likely to find it appropriate to experiment with waste-water or water-supply treatment options that require plants to be built and operated over long periods. On the other hand, a national or provincial government could sponsor an experiment with different methods applied in different communities, provided the approach used does not have severe sample-size requirements.

An experimental approach is much more attractive if it can precede any major investments or organizational changes.

Question 5. Can the essential experimental conditions be maintained throughout the experimental period? Under the best circumstances it is difficult enough to maintain the essential experimental conditions. The problems that can occur include these:

program personnel not adhering to difficult and unusual procedures that may not be well understood and foreign to their training and experience;

program personnel who may be responsible for making the random assignments of clients to experimental groups not following the randomization procedures (perhaps because other factors appear to be more important to the program personnel);

very long experimental periods, making relaxation of essential evaluation discipline more probable.

The first difficulty can be reduced by training program personnel carefully and explaining to them the rationale behind the procedure requirements. (To maintain a realistic experiment, it will almost surely not be desirable for evaluation personnel themselves to administer treatments.) The second difficulty can be reduced by having evaluation personnel make the assignments if this is possible. The third difficulty can be alleviated by a carefully designed plan that will enable the desired results to be obtained in a reasonable time.

Experience supports the concern about program personnel by-passing the strict requirements of experimental designs. For example, social workers have been known to allow their judgments of client needs to override the imperatives of the experimental design; police dispatchers have resisted the special experimental rules that appear to them to be peculiar. Close supervision over experimental activities can help to alleviate these problems, but circumstances often make this a formidable task, and supervision cannot be so close as to destroy the natural programmatic behavior of the persons involved. Experimental designs that reduce or eliminate such problems as these are sometimes possible—and certainly desirable.

Question 6. Are adequate technical procedures available for assessing the program's effects? For many programs there are important difficulties to be overcome in order to measure program effects; developing adequate criteria to measure, and then collecting valid and reliable information on which to base the measurements. If these difficulties cannot be overcome, the experiment will be largely wasted. Thus, as part of the evaluation plan, the analysis team must identify the evaluation criteria clearly and then determine whether or not the collectable information will provide an adequate evaluation. Partial information may be usable if the cost of obtaining it is low, but the limitations should, of course, be considered when reporting the findings.

Question 7. Are the prospective sample sizes likely to be large enough to support definitive results? In general, it is desirable for enough units to be assigned to the experimental groups for the results to be statistically significant. Otherwise, the differences could reasonably be attributed to chance. This is not usually a problem in social programs with many persons available for experimental and control groups. On the other hand, the problem may be of concern when the units to be assigned are other than people, such as geographical regions, service districts, or political units. If these units are small in number, the experimental results may not be as decisive as desired—but the situation may force the evaluation to accept this fact.

Evaluators should consider using pairs of such units matched as closely as possible, with one member of each pair randomly assigned to the experimental group and the other to the control group. Even though the number of pairs is small, this process can yield more reliable information—almost certainly better information than that from not using randomization at all.

Question 8. Are the potential findings likely to be generalizable to the whole jurisdiction or population? The usual purpose of an evaluation is to discover whether an experimental program, or some revision of it, is sufficiently meritorious to be applied to a larger population. This means that the experimental groups, both treatment and control, should be representative of the population to which the program may be extended, and that the conditions of the experiment should be representative of the conditions of such a future extension. If either of these conditions is not met, then the experimental program is not likely to produce fully useful results.

Some of the difficulties in achieving generalizability have already been discussed in Section 11.4 under Actions 11 and 16; however, a few important ones are worth adding here:

The experiment is not able to cover either all of the major types of clients or all of the major conditions that will be present if the program is extended after the experiment.

Special personnel have to be used for the experiment who may influence the outcomes because they are likely to be highly skilled and motivated, but they will not be representative of the persons involved in full implementation of the program after the experiment.

The treatment and control groups, in cases where they have to be made aware that they are participating in an experiment, may respond in sufficiently unusual ways to make the outcomes unrepresentative of what would happen under the more general conditions of a widespread program. For example, the experimental activities may be accompanied by special motivational effects that would not carry over into the full program implementation.

To avoid this last difficulty, it is desirable for the experiment to be as unobtrusive as possible—certainly not to have a spotlight of publicity on it—and to continue long enough for any special attention to wear off, so that all concerned will behave as normally as possible. If it is feasible and ethical, it is preferable to conduct the experiment without its subjects knowing that an experiment is going on; this is sometimes impossible, but, where it is, it is a way to minimize the chance of unusual behavior associated with the experiment itself.

Question 9. Are sufficient resources—staff, finances, and administrative support—available to carry the experiment through? Experimental designs require skilled personnel, careful monitoring and information collection, and discriminating analysis by trained professionals. Their activities must be supported fully from beginning to end, both administratively and financially, if the experiment is to yield adequate findings. The time to spell these requirements out and to obtain an administrative commitment to meeting them is when the plan is being prepared. If there is substantial doubt about whether or not adequate resources will be available to complete the experiment, it probably should not be started.

Question 10. Can client privacy and confidentiality of records be maintained? Experiments involving treatments to individual clients generally require compiling personal information about the clients and their conditions. Human-resources programs usually require such records, whereas service programs such as solid-waste management and fire protection may not. It is accepted in many countries, including the United States, that the experiment should not be undertaken unless data can be obtained without unacceptable intrusions into individual privacy, and unless what is gathered is maintained in confidence, except for use by analysts for legitimate purposes (which must always mask individual results). The problem of confidentiality exists in any type of evaluation, not just controlled experiments, but it is likely to be a larger problem in experimental designs where both treatment and control groups need to be examined.

In general, carefully designed procedures can be identified to protect client confidentiality. And, equally important, the evaluation-experiment findings can be presented in the form of aggregate statistics with identifying information about individual clients completely concealed.

Question 11. When it is needed, can client consent for participation be obtained without invalidating the experiment? It is sometimes necessary for clients to have to give consent to be included in the experiment, or to consent to have special information about them collected as part of the experimental work. The contemplated experimental design declines in feasibility to the extent that clients are likely either to refuse consent or to alter their behavior after agreeing to participate. Thus, it is important to face this difficulty realistically in evolving the experimental plan.

Consent is not always necessary, as in experiments that do not involve providing direct services to clients. For example, in the Kansas City police patrol experiment (Larson et al., 1977), it was not deemed necessary to obtain the permissions of the areas of the community that were to obtain differential treatments. On the other hand, as we discussed under Question 1, where an experimental program may involve unusual risks to individuals, they will probably need to be apprised of such risks.

To avoid influencing participant behavior, evaluators try not to reveal to the participants whether they are in the treatment or control groups. This is possible in some experiments, but not in others. This corresponds to the common medical practice of giving some patients a placebo. Sometimes it is possible to conduct a double-blind experiment in which neither the client nor the caseworker knows whether or not the treatment is being applied.

A Baltimore experiment (Lenihan and Casey, 1977) tested the relative effects of giving financial aid or job-placement help to hard-core ex-inmates of prisons in order to reduce recidivism. It was not deemed necessary to indicate to the participants that they were part of an experiment. During the initial in-prison interview, the randomly selected inmates were not told about either the financial aid or the job-placement help that they would be receiving, only that they were to report to a certain office. The purpose of not disclosing the program's existence or its nature was to keep the effects of the first visit to the office the same for all ex-convicts, regardless of which treatment groups they would be assigned to later.

A study of the use of random assignment in twelve experiments found that half of the programs explained the randomization procedure to the clients and, in general, gained their acceptance (Conner, 1977, especially pp. 229ff.). The experiment that provided by far the greatest problem was one sponsored by the U.S. federal government that gave funds for desegregation to some randomly selected schools and no such funds (for a period of about two years) to schools in a control group. The schools were made aware of the assignment procedure, but some schools in the control group complained after they were selected for that group. The assignments were adhered to, but with the compromise of providing some funds to a few of the control schools, a procedure that weakened the contrast between the control and treatment schools but maintained the basic experimental design.

11.6. Final Comments

After a program has been implemented, it is desirable to evaluate the results to see if the program achieves the desired benefits at the predicted costs—and to identify directions for program improvement. Such a program evaluation must be approached with both logic and common sense, as well as intimate familiarity with the program and its setting. Although many successful approaches that have been used in the past are available for application, each real-world situation offers its own peculiar properties and difficulties that will challenge the ingenuity of the evaluation analysts. The purposes of this chapter were to introduce prospective participants in such program evaluations to some of the practical issues and choices they will face, and to give them an entry to the substantial literature on evaluation that has grown up in recent years and that can offer help and guidance in resolving many difficulties.

Some of the material of Sections 11.2, 11.3, and 11.4 has appeared elsewhere in a different form (Hatry, 1980). I wish to thank Carol Weiss, Sumner Clarren, Joseph Nay, and John Waller for their helpful suggestions.

References

Abramson, J. H. (1974). *Survey Methods in Community Medicine*. New York: Longman.

Armor, D. J., J. M. Polich, and H. B. Stambul (1976). *Alcoholism and Treatment*. Santa Monica, California: The Rand Corporation.

Austin, M. J., and Associates (1982). *Evaluating Your Agency's Programs*. Beverly Hills, California: Sage.

Babbie, E. R. (1973). *Survey Research Methods*. Belmont, California: Wadsworth.

Bernstein, I. N., G. W. Bohrnstedt, and E. F. Borgatta (1975). External validity and evaluation research, a codification of problems. *Sociological Methods and Research* 4 (August), 101–128.

Boruch, R. F. (1976). On common contentions about randomized field experiments. In *Evaluation Studies Review Annual* 1 (G. V. Glass, ed.), pp. 158–194.

Campbell, D. T. (1969). Reforms as experiments. *American Psychologist* 24, 409–429.

———, and H. L. Ross (1968). The Connecticut crackdown on speeding. *Law and Society Review* 8 (August), 33–53.

———, and J. C. Stanley (1966). *Experimental and Quasi-Experimental Designs for Research*. Chicago: Rand McNally.

Campbell, J. R., et al. (1969). *Household Survey Manual, 1969.* Washington, D.C.: Bureau of the Budget, Executive Office of the President.

Carter, R. K. (1983). *The Accountable Agency.* Beverly Hills, California: Sage.

Conner, R. F. (1977). Selecting a control group, an analysis of the randomization process in twelve social reform programs. *Evaluation Quarterly* 1, 195–244.

Cook, T. D., and D. T. Campbell (1979). *Quasi-Experimentation: Design and Analysis Issues for Field Settings.* Chicago: Rand McNally College Publishing Company.

Duckworth, P. A. (1973). *Construction of Questionnaires, Technical Study.* Washington, D.C.: U.S. Civil Service Commission.

Edwards, Ward, M. Guttentag, and K. Snapper (1975). A decision-theoretic approach to evaluation research. In Guttentag and Struening, vol. 1 (1975), pp. 139–148; also in Struening and Brewer (1983), pp. 139–184.

Epstein, P. D. (1984). *Using Performance Measurement in Local Government: A Guide to Improving Decisions, Performance, and Accountability.* New York: Van Nostrand Reinhold.

Fischer, Joel (1976). *The Effectiveness of Social Casework.* Springfield, Illinois: Charles C Thomas.

Fitz-Gibbon, C. T., and L. L. Morris (1978). *How to Design a Program Evaluation.* Beverly Hills, California: Sage.

Glaser, Daniel (1973). *Routinizing Evaluation: Getting Feedback on Effectiveness of Crime and Delinquency Programs.* Rockville, Maryland: Center for Studies of Crime and Delinquency, National Institute of Mental Health.

Gleeson, G. A. (1972). Interviewing methods in the health interview survey. *Vital and Health Statistics* 2 (April), 1–86.

Guttentag, Marcia, and E. L. Struening, eds. (1975). *Handbook of Evaluation Research,* volumes 1 and 2. Beverly Hills, California: Sage.

Hamrin, R. D. (1974). OEO's performance-contracting project: Evaluation bias in a social experiment. *Public Policy* 22, 467–488.

Hatry, H. P. (1980). Pitfalls of evaluation. In *Pitfalls of Analysis* (G. Majone and E. S. Quade, eds.). Chichester, England: Wiley, pp. 159–178.

——, L. Blair, D. Fisk, J. Greiner, J. Hall, and P. Schaenmann (1977). *How Effective are Your Community Services? Procedures for Monitoring the Effectiveness of Municipal Services.* Washington, D.C.: The Urban Institute.

——, R. E. Winnie, and D. M. Fisk (1981). *Practical Program Evaluation for State and Local Governments,* 2nd ed. Washington, D.C.: The Urban Institute.

Herzog, Elizabeth (1959). *Some Guide Lines for Evaluative Research.* Washington, D.C.: Children's Bureau, Welfare Administration, U.S. Department of Health, Education, and Welfare.

Horst, Pamela, J. N. Nay, and J. W. Scanlon (1974). Program management and the federal evaluator. *Public Administration Review* 34(4), 300–308.

Kelling, G. L., T. Pate, D. Dieckman, and C. E. Brown (1974). *The Kansas City Preventive Patrol Experiment.* Washington, D.C.: Police Foundation.

Larson, R. C. (1975). What happened to patrol operations in Kansas City? A review of the Kansas City preventive patrol experiment. *Journal of Criminal Justice* 3, 267–297.

————, K. W. Colton, and G. C. Larson (1977). Evaluating a police-implemented AVM system: The St. Louis experience (Phase I) *IEEE Transactions on Vehicular Technology* VT-26, 60–70.

Lenihan, K. J., and F. M. Casey (1977). *Unlocking the Second Gate,* R&D Monograph 45. Washington D.C.: U.S. Department of Labor and Employment and Training Administration.

Lowry, I. S., ed. (1983). *Experimenting with Housing Allowances: The Final Report of the Housing Assistance Supply Experiment.* Cambridge, Massachusetts: Oelschlager, Gunn, and Hain.

Magura, Stephen, and B. S. Moses (1986). *Outcome Measures for Child Welfare Services: Theory and Applications.* Washington, D.C.: Child Welfare League of America.

Majone, Giandomenico (1985). Systems analysis: A genetic approach. In Miser and Quade (1985), pp. 33–66.

Miles, M. B. (1979). Qualitative data as an attractive nuisance: The problem of analysis. *Administrative Science Quarterly* 24(4), 590–601.

————, and M. A. Huberman (1984). *Qualitative Data Analysis.* Beverly Hills, California: Sage.

Millar, A. P., M. P. Koss, and H. P. Hatry (1977). *Monitoring the Outcomes of Social Services—Volume 1: Preliminary Suggestions.* Washington, D.C.: The Urban Institute.

Miser, H. J., and E. S. Quade, eds. (1985). *Handbook of Systems Analysis: Overview of Uses, Procedures, Applications, and Practice.* New York: North-Holland. Cited in the text as the *Overview.*

Mosteller, Frederick, and J. W. Tukey (1977). *Data Analysis and Regression.* Reading, Massachusetts: Addison-Wesley.

Nunnally, J. C., and W. H. Wilson (1975). Method and theory for developing measures in evaluation research. In Guttentag and Struening, vol. 1 (1975), pp. 227–232.

Overview. See Miser and Quade (1985).

Patton, M. Q. (1980). *Qualitative Evaluation Methods.* Beverly Hills, California: Sage.

Riecken, H. W., and R. F. Boruch, eds. (1974). *Social Experimentation: A Method for Planning and Evaluating Social Intervention.* New York: Academic Press.

Roos, N. P. (1975). Contrasting social experimentation with retrospective evaluation: A health care perspective. *Public Policy* 23, 241–257.

Rossi, P. H., and H. E. Freeman (1982). *Evaluation: A Systematic Approach,* 2nd ed. Beverly Hills, California: Sage.

————, J. O. Wright, and A. B. Anderson, eds. (1985). *Handbook of Survey Research.* New York: Academic Press.

Rutman, Leonard (1977). *Evaluation Research Methods: A Basic Guide.* Beverly Hills, California: Sage.

Shaver, Phillip, and G. Staines (1972). Problems facing Campbell's "Experimenting Society." *Urban Affairs Quarterly* (December), 173–186.

Struening, E. L., and M. B. Brewer (1983). *The University Edition of the Handbook of Evaluation Research.* Beverly Hills, California: Sage.

Suchman, E. A. (1967). *Evaluative Research: Principles and Practice in Public Service and Social Action Programs*. New York: Russell Sage Foundation.

Survey Research Center (1976). *Interviewer's Manual*, revised edition. Ann Arbor: Institute for Social Research, The University of Michigan.

Tannenbaum, A. S. (1970). The group in organizations. In *Management and Motivations* (V. H. Vroom and E. L. Deci, eds.). Harmondsworth, England: Penguin, pp. 218–226.

Turner, C. F., and E. Martin, eds. (1984). *Surveying Subjective Phenomena*. New York: Russell Sage Foundation.

Webb, Kenneth, and H. P. Hatry (1973). *Obtaining Citizen Feedback: The Application of Citizen Surveys to Local Governments*. Washington, D.C.: The Urban Institute.

Weiss, C. H. (1972a). *Evaluating Action Programs: Readings in Social Action and Education*. Boston: Allyn and Bacon.

———— (1972b). *Evaluation Research: Methods of Assessing Program Effectiveness*. Englewood Cliffs, New Jersey: Prentice-Hall.

————, and H. P. Hatry (1971). *An Introduction to Sample Surveys for Government Managers*. Washington, D.C.: The Urban Institute.

Wholey, J. S. (1983). *Evaluation and Effective Public Management*. Canada: Little, Brown.

Wye, C. G., and H. P. Hatry (1987). Evaluation—its potential utility for operating managers. In *Organizational Excellence* (J. S. Wholey, ed.). Lexington, Massachusetts: Lexington Books (D. C. Heath).

Yin, R. K. (1984). *Case Study Research: Design and Methods*. Beverly Hills, California: Sage.

PART V
PROFESSIONAL ISSUES

Chapter 12

Underlying Concepts for Systems and Policy Analysis

Hugh J. Miser

12.1. Introduction

With more than three decades of experience behind them, systems and policy analysts have come to a reasonably clear and widely accepted understanding of what they are about. On operating and policy problems involving interactions among people, the natural environment, and the artifacts of man and his technology, they bring to bear the knowledge and methods of modern science and technology, together with appropriate consideration of social goals and equities, the larger contexts, and the inevitable uncertainties. In doing this, they pursue variants of a commonly understood approach (as summarized in Chapter 1) and agree generally that, if analysis is to be helpful, it must not only offer sound information about problems and their solutions but also give good advice—and persuade clients to give the information and advice proper consideration. Since systems and policy analyses have these descriptive, prescriptive, and persuasive aspects, they mingle scientific, advisory, and interactive components in ways prompted by the problem situation and the craft judgments of the analysts.

This core of generally accepted agreement has not led systems and policy analysts to pay more than modest attention to the need to develop a broader foundation of underlying concepts for their work. Nevertheless, enough experience has accumulated to point to some important elements of such a foundation, and some penetrating observers of science and professionalism have developed concepts for wider fields that can usefully be brought to bear, not only on systems analysis, but also on policy analysis as it extends its work into behavior-oriented organizations (see Section 2.4). Thus, the purpose of this chapter is to present an overview of some of the important streams of thought that I consider likely to make significant contributions to underlying concepts for systems and policy analysis as they grow and gain wider acceptance. (In order to avoid some terminological clumsiness, the phrase systems analysis will be understood generally in this chapter to refer to both systems and policy analysis; where one is meant to the exclusion of the other, this fact will be noted specifically.)

Systems analysis deals with what Ravetz (1971) calls "practical problems," that is, those for which "the goal of the task, in principle, is the serving or achievement of some human purpose" (loc. cit., p. 319). Thus, the elements of what we are discussing in this handbook fall into three categories: the problem situation and its surrounding context, the persons with interests in and responsibilities for aspects of the problem situation, and the analysts who address the problem situation. And, of course, there are the relations among elements in these three categories. It follows that a well-rounded set of foundational conceptions for systems analysis should deal with all three categories of these elements and their mutual relations.

Most of the literature on systems analysis focuses its attention on the activities of the analysts—as this handbook does—with considerably less attention to the properties of the problem situations the subject is called on to deal with, to the nature of the decisionmakers and other persons at interest, and to the mutual relations among these elements. Nor is the literature relating systems analysis to the scientific roots from which it sprang as satisfactory as one could wish. It is not the purpose of this chapter to fill all the gaps suggested by this description, nor would the existing body of literature and thought make this possible. Rather, this chapter offers overviews of three well-developed streams of thought that can deal with fundamental concepts relating to three areas:

A view of the underlying concepts of science and the activities of scientists that constitute, not only the source from which much of the inspiration for systems analysis sprang, but also useful parts of the portion of systems analysis that involves scientific work and its results; Ravetz (1971) is the dominant basis for this part.

A view of the problem-situation contexts of systems-analysis work and how the work of the analysts relates to its realities; here Boothroyd (1978) is the principal source.

An ideal—but nevertheless realistic—view of the relationship between the analysts and the decisionmakers and others at interest; this part is drawn principally from Schön (1983).

This chapter presents these conceptual elements as potential contributors to a set of underlying concepts for systems analysis as it has been described in this handbook, and as it may develop in the future. It is undoubtedly the case that these elements will evolve as the field develops and subjects its concepts to critical scrutiny, with some elements being cast away and new ones added. It is hoped, however, that these brief summaries will attract attention to the importance of developing strong and widely accepted foundations for systems-analysis work, and will play a useful role as points of departure for further development.

There is considerable literature in operations research and management science that is at least loosely related to the subject of this chapter; for critical summaries, see Checkland (1981, pp. 3–122) and Jackson and Keys (1987, especially pp. 1–25 and 133–164). Since little of the literature has generated appropriate and widely accepted conceptual bases for systems analysis, it would not be useful to summarize it here.

After three sections devoted to setting forth the views listed above, the chapter discusses the progress of systems analysis toward maturity and concludes with a plea for wider attention to underlying concepts as an important guide to the growth and influence of effective systems-analysis work.

12.2. Underlying Concepts of Science Useful for Systems Analysis

The starting point for a set of concepts of science useful for systems analysis is what Ravetz (1971, p. 1) calls "the social activity of disciplined inquiry into the natural world," work that includes an essential element of systems-analysis inquiries. Indeed, he goes on to say (loc. cit., pp. 1, 9) that "the problems of the character of scientific knowledge, of the sociology and ethics of science, and of the applications of science to technology and to human welfare, are so intimately connected that a proper study of any one of them requires an informed awareness of the others. . . . The illusion that there is a natural science standing pure and separate from all involvement with society is disappearing rapidly."

The argument proceeds in a number of steps:

A discussion of scientific inquiry as a craft

An identification of the objects of scientific work, not as natural things, but as intellectual constructs, studied through the investigation of phenomena and problems of the real world

A description of the methods of investigation, which contain important informal and tacit elements

An examination of the way science achieves what it regards as knowledge

The deduction of the character of the knowledge that is produced by the social processes that have been described

A discussion of the involvement of science with the solution of practical problems

In carrying through this argument, this section follows Ravetz (1971, pp. 69–240 and 315–363) closely; to suggest the flavor as well as the substance of his argument, it quotes him extensively.

The central concept in the argument is that of a problem, with science being considered as a special sort of problem-solving activity (Ravetz, 1971, p. 72).

> For the present, we may think of a scientific problem as analogous to a textbook 'exercise,' with the following crucial differences: a major part of the work is the formulation of the question itself; the question changes as the work progresses; there is no simple rule for distinguishing a 'correct' answer from 'incorrect' ones; and there is no guarantee that the question, as originally set or later developed, can be answered at all.

This preliminary definition serves our immediate purposes; it will be refined and discussed further later on.

Scientific Inquiry as a Craft

With a problem in mind, a natural scientist turns to the portion of nature related to his problem to see what it has to tell him. The process of interrogating nature often involves an experimental apparatus of some kind. When this is the case, the scientist must get it to work properly, an especially important issue if he has had to build it himself. When it does work, he must decide whether or not its behavior is sufficiently stable to yield reliable information about the phenomena being investigated. Finally, he must decide whether or not to accept what it is telling him as valid observations, not influenced by some sort of extraneous forces or subtle aberration inside the equipment. Although this list may seem short, any scientist with mature experimental experience can easily lengthen it considerably.

Similarly, when the problem involves people, or operations in which both people and technology are involved, considerable craftmanship may have to be brought to bear on designing observation schemes that allow stable and reproducible readings to be collected.

Such procedures produce data—and the point of the examples is to suggest that producing data in a scientific investigation involves many diverse skills that can properly be called a craft.

The situation, however, is more complicated than this. In formulating his problem, the scientist has used certain concepts, and his interpretation of the data is necessarily related to them. Working out these relations also demands craft skills. Ravetz carries the argument further in this way (loc. cit., p. 81):

> I have stressed the importance of craft skills in the production of data which is sound of itself, and useful for the later stages of work on the problem. Without such skills, the scientist will blunder into pitfalls and produce reports without soundness or significance. The craft character of the production of data is of some philosophical significance, for it is in this phase that the scientist makes new contact with the external world, or achieves a new organization of his conceptual objects. The directions and later stages of the work will depend on the results of this work. Yet . . . no set of data can be 'perfect' as a report of properties of the objects of investigation, nor can it be independent of the plans and expectations for the later stages of the work. If such reports are truly the foundations of scientific knowledge in experience of the external world, then those foundations would seem to be peculiarly insecure and complicated. And so it is; the wonder is not that our scientific knowledge is an imperfect and fallible picture of the external world, but rather that it exists at all. In the light of the history of human inquiry into the natural world, we see how difficult it is for such knowledge to come to be.

It is seldom the case that data are used in an analysis in the raw form in which they are collected. Rather, they are refined into information that is more directly related to the problem being addressed. For example, they may be plotted on a graph, fitted to a curve, or reduced in bulk by various statistical methods. Then the data in its new form has to be related to the problem in hand, a process that can go astray if the mechanisms assumed by the scientist in his problem formulation and underlying conceptions are not in fact present in the external world. Keenly aware of this, the conscientious scientist subjects his transformations of the data and their results to varied tests to seek assurance that his interpretations are not erroneous.

Scientific knowledge is not usually established, however, by data that are transformed into information. Rather, selected aspects are put to work as evidence in an analysis, and for this purpose they may well be further refined and transformed.

All of these processes—gathering data, transforming it, and putting it to work in an analysis—employ important and substantial craft skills.

Scientists use a wide variety of tools in their work: laboratory equipment, various modes of observation, methods for transforming data into information, information collected by other workers (often compiled into handbook form), various mathematical methods of analysis, the knowledge and structure of the field in which they are working, knowledge of the intellectual processes and underlying concepts of science, and so on. Using these tools calls for a wide and varied spectrum of craft skills; they may be so uniquely personal as to give a scientist's work a recognizable stamp among his colleagues. Ravetz (loc. cit., p. 90) goes on to say that "when two craftsmen with different skills are involved with the same project, they will inevitably see the work from different points of view." He then argues that a searching

> discussion of tools shows how subtle must be the craft knowledge of a scientist who is a leader in his field. He must not only be able to develop a craftsman's competence in the use of particular tools, and have a sense of their powers and limitations; but, if he is to do anything but follow fashions set by others, he must assess the sorts of problems into which the use of particular tools would lead him and his colleagues and decide whether they are appropriate for the best progress of work in his field. [Ravetz, loc. cit., p. 94.]

As a scientist leaves familiar ground to explore the unknown, he carries along expectations based on what he knows. As the new territory is revealed, however, these expectations may be fulfilled, either in whole or in part, or they may prove to have been erroneous. His expectations may throw new and unexpected findings into sharp relief, or they may lead the scientist into a concealed trap, often called a *pitfall* (see Section 15.7). When a scientist is surprised by what he finds, he has not fallen into a pitfall, provided he recognizes the novelty of what he is seeing and responds appropriately to it. He has stumbled into a pitfall when his expectations have led him into the sort of error that vitiates the solution to the problem.

A well-developed science has developed standard patterns of work that skirt its potential pitfalls, evolved after an early history that contained many pitfall experiences, even though its literature seldom records these unhappy events. Since systems analysis is young, the pitfalls into which its scientific work have fallen are reasonably fresh in mind, and to some extent they have been reported in the literature; Majone and Quade (1980) have assembled a wide and numerous collection of them, the *Overview* volume discusses some of the more important ones, and Section 15.7 contains an overview and a short list.

The nature of scientific inquiry implies that it is impossible to eliminate pitfalls entirely. A mature discipline, however, largely avoids them. In the first place, it has charted standard procedures that skirt them; in the second, its leaders are sensitive to the sorts of pitfalls likely to be encountered, and can usually plan their work around these dangers. Nevertheless, beyond the range of well-mapped territory, only previous analogous craft experience can serve as a guide around possible pitfalls, but it cannot offer a guarantee against them.

In view of the importance of the concept of pitfalls in scientific inquiry, Ravetz (1971, p. 98) and Majone and Quade (1980) argue that knowledge of them and how the existing standard procedures are designed to avoid them should be a part of the basic training of every potential scientific investigator.

Finally Ravetz (1971, p. 97) draws attention to the implications of the concept of pitfalls and their importance as phenomena in the histories of all the sciences:

> This discussion of pitfalls is of some philosophical significance, for it indicates one of the reasons why it is vain to seek for certainty, or even for a guarantee of the existence of certain knowledge, in science. There can be no absolutely certain foundations in experience, nor any absolutely certain inferences from that experience, for the achievement of knowledge of the natural world. Worse still, from this discussion of pitfalls we see that we even lack a guarantee that our errors will be revealed to us by a direct and straightforward process.

Every scientist involved in scientific inquiry brings to this activity a varied set of tools, skill in using equipment, techniques of observation, ways of reducing data to information, mathematical and statistical tools, the facts and theories of his subject, and so on. It hardly needs arguing here that along with these tools he brings a varied set of craft skills related to their use. "The scientist's craft also includes the formulation of problems, the adoption of correct strategies for the different stages of evolution of a problem, and the interpretation of general criteria of adequacy and value in particular situations" (Ravetz, 1971, p. 103).

The synoptic argument of this subsection—which could be greatly lengthened, as in Ravetz (1971, pp. 75–108)—is aimed at supporting the point that scientific inquiry is a craft activity in all its aspects.

Scientific Inquiry as Problem-Solving on Intellectual Constructs

Physical sciences such as physics or chemistry usually use the same vocabulary for the aspects of reality they treat as they do for the intellectual constructs that represent them, a fact that has misled many persons both in and out of science. Systems analysis, on the other hand, has always distinguished between the reality and its model (as Section 8.5 emphasizes). A model is an artificial object constructed intellectually; it represents a class of things possessing relevant properties.

Against this conceptual background, Ravetz (1971, p. 114) defines "science as craft work operating on intellectually constructed objects," each object defining a class. After defining the goal of scientific inquiry as "the establishment of new properties of these intellectually constructed classes," he carries the argument forward in this way (loc. cit., p. 119):

> In this, the objects themselves are altered, for they exist only as classes defined by their properties. . . . [N]ot all their properties are exhaustively

defined by the formal statements presented in the public record; each scientist must have a craftsman's intuitive, personal and partly tacit knowledge of his intellectual objects and of their physical samples, if he is to work creatively or even competently with them. But there must be a large common core of practical knowledge of the objects (. . . not all of it in the explicit record) if there is to be an effective social endeavour of their study.

The study of these objects . . . cannot progress very far without some interaction with the ultimate sources of our knowledge in that reality. But the fashion in which this interaction yields new properties of these objects is highly indirect and in principle inconclusive. We have already seen that the production of data, and its refinement into information, are craft operations which are governed by the problem at hand, and which produce less than conclusive inferences from the behaviour of particular samples in unnatural conditions. The connection between such a foundation in particular experiences, and the establishment of the properties of intellectually constructed classes of things and events, is in principle tenuous. And in practice, it has been forged only after long periods of development, extending over generations or centuries for the different fields of scientific inquiry.

If, with Ravetz, the items of information selected to support a particular argument are called evidence, then this view of scientific inquiry gives a crucial place to the evidence fashioned from the observed portions of reality and how it is used. Since each use of evidence is special and demanding, the scientist must bring to bear on it an assessment of adequacy for the use to which it is being put. In doing this he uses his own craft skills, as well as those of the field in which he is working.

In any genuinely novel work, or even in work of crucial importance for a big problem, the accepted criteria of adequacy may not extend to cover all the inferences made from the evidence in a particular argument. For the novelty of the problem will generally entail a corresponding novelty, and uniqueness, in evidence and its relation to experience and to the argument. Also, since such work is frequently done at the limit (if not beyond) of the capabilities of the existing tools for producing reliable information, the assessment of the strength and fit of the evidence can become very subtle indeed. Then there can arise disputes about the adequacy of the solution to the problem, which cannot be resolved either by a scrutiny of the data and information, nor by an appeal to accepted criteria of adequacy. Thus at such points, this aspect of the "objectivity" of scientific knowledge, which is really a result of a successful social tradition of produciing and testing the materials of that knowledge, breaks down. In the long run, to be sure, further work will decide the issue; but the decision on whether to engage in such further work, which partly depends on the assessment of the adequacy of the controversial piece, must be taken now. Thus, at such infrequent but critical junctures in the advancement of science, the assessment of the evidence adduced in an argument becomes a crucial judgment, in which the individuals are thrown back on their own personal resources. They are forced to put themselves at risk in making the judgment, and they lack the safe channels of an accepted tradition to steer them towards the correct answer. [Ravetz, 1971, pp. 124–125.]

The argument of a scientific inquiry is aimed at reaching a conclusion: the completion of its cycle and the first step toward achieving scientific knowledge. This conclusion, however, is not concerned directly with reality, but with the intellectually constructed objects that are the focus of the argument; the relation with reality is indirect. Although the conclusion is stated in terms that can be understood by other competent scientists, and the evidence and the stages of the argument are reported clearly enough to be reproduced and tested by other workers, the work as a whole has been informed by many personal judgments that depend on the scientist's personal knowledge of his craft.

At this point in developing his argument, Ravetz makes some statements that are particularly important for the descriptive aspect of systems analysis (see the introduction to this chapter and Section 1.1), and therefore also for the prescriptive and persuasive aspects that depend on it (loc. cit., p. 128):

> The artificial and fallible nature of the conclusions of solved scientific problems has an importance beyond the improved historical understanding of natural science. For the same holds true in any field of thought and action where there is an explicit body of theory in whose terms arguments are cast and conclusions are drawn. Since so much of the direction of the technical and social aspects of our existence is now done by specialists trained in a formal 'science' of their craft, an appreciation of the limitations of the conclusions to scientific problems is as important for politics as much as for epistemology.

The discussion so far has distinguished data, information, evidence, argument, and conclusion. This distinction is not intended, however, to suggest that these stages in a scientific inquiry are successive, discrete, or independent; rather, each tends to condition the others, and progress may dictate moving backward as well as forward in this list. For example, formulating the evidence to be used in an argument may dictate that more data be gathered, perhaps under new or more carefully controlled conditions; or, a surprising or inadequately supported conclusion may suggest that the entire cycle be repeated on an expanded and more critical basis. Too, as the work proceeds, the intellectual objects with which it started may undergo evolution in directions that could not have been foreseen at the beginning. This can happen when the data yield unexpected and surprising information, or when the argument evolves novel results suggesting that the objects themselves as originally formulated were inadequate.

> The work on a particular problem is completed when an argument, meeting the accepted standards of adequacy, can be framed and the conclusions drawn. But this does not occur suddenly, as with the dotting of the last *i*. Rather, the cyclic interaction of the various materials of the problem decreases in intensity, the argument is stabilized, the evidence becomes sufficiently strong and well-fitting, and fewer lots of new information and data are required. The unexpected results decrease in importance, pitfalls become negligible, and there is a sense that the whole process is 'converging' towards solution. But this does not always occur; and a warning sign of the imminent failure of a problem is when the difficulties begin to 'diverge'; when the subsidiary problems called into being by new difficulties become larger and more fundamental, and ever more extensive modifications of the argument become necessary, with increasing lots of fresh data being required for throwing into the breaches. An important part of the craft skill of a scientist is to sense whether a problem is beginning to converge as early as it should, and to detect signs of incipient divergence; and then to decide when to abandon a doomed venture. [Ravetz, 1971, p. 131.]

Against this background it is now possible to define more formally what is meant by a scientific problem and to discuss how it emerges from a problem situation. By its nature, the concept of a problem situation is difficult to define—nor would a carefully framed and restrictive definition serve our present purpose. Rather, a problem situation can be seen to exist when there is a recognition that there are questions to be asked, even though it may not yet be possible to frame them adequately. Problem situations exist when phenomena remain mysterious and unexplained, when proposed schemes of explanation seem to be in conflict, when existing explanations seem too shallow, when doubt casts a shadow over what have previously been deemed adequate explanations, when existing intellectual constructs seem inadequate as the basis for formulating scientific problems, and so on. A problem situation can also exist when there is some practical difficulty that needs overcoming, for which an improved understanding of nature is conceived to hold a key.

> Although the problem situation is in a less specified state than the problem to which it may give rise, it is already a very artificial thing. The very existence of a problem situation presupposes a matrix of technical materials: existing information (with the intellectual objects it describes), tools, and a body of methods including criteria of adequacy and value. For in the absence of such a matrix of technical materials, a genuine problem could never come into existence, and the decisions on investigating it, and on shaping it during the work, would have no foundation. [Ravetz, 1971, p. 136.]

Thus, conceiving of a problem situation demands a mixture of knowledge, judgment, and imagination, supported by the craft knowledge of what is—or may be—doable.

A scientific problem emerges from a problem situation by refinement, a process that makes the question more precise, that envisages the means that will be used, and that may develop new or revised intellectual objects as the focus of the research.

Against this background we achieve (with Ravetz, 1971, p. 134) the definition that "a scientific problem is a statement (always partial and subject to evolution) of new properties of the objects of inquiry, to be established as a conclusion to an adequate argument, in accordance with a plan (specified to an appropriate degree) for its achievement."

Methods of Investigation

Throughout the description of the process of scientific inquiry it has been clear that judgments have to be made: a basic judgment of the soundness of the data, judgments as to how to convert it to information and then into evidence related to the scientific problem being tackled, and so on.

> These individual acts of judgment do not derive solely from private in-tuitions of the scientist; rather they are based on a body of principles and precepts, social in their origin and transmission, without which no scientific work can be done. I shall use the term 'methods' for such principles and precepts, which (through their interpretation and appli-cation in particular situations) guide and control the work of scientific inquiry. Such a corpus of 'methods' has a special character, which may seem paradoxical and at variance with the objects and results it governs. For methods cannot be established 'scientifically,' through arguments resting on controlled experience; this is partly because there is no simple test of the 'correctness' of a particular method, and even more because the principles and precepts are incapable of a fully explicit, public state-ment. The body of methods associated with any field is partly informal and even tacit; it is not transmitted by the same public channel of com-munication as scientific results, but through an informal, interpersonal channel. Hence the testing, criticism, and improvement of the methods for a field must proceed by means quite different from those applicable to its scientific results; and in this aspect of scientific inquiry, the char-acter of the community engaged in the work is thus crucial for the nature and quality of its achievements. [Ravetz, 1971, pp. 146–147.]

There are a number of difficulties involved in discussing methods of scientific inquiry in a general way. The first is that it is an attempt to render explicit something that is largely tacit; a second is that, although many methods are used in several sciences, others are used in only one; another is that methods often have to be varied significantly to adapt them to the particular inquiry in hand. Nevertheless, it is possible to state some useful perspectives on the role of methods and how they should be judged.

Throughout his work a scientist must be continually aware of the issue of the adequacy of his methods in the light of the problem he is working on. This matter has already been emphasized in connection with the steps in the scientific inquiry process: gathering data, converting it to information, and using what has been found as evidence in the analysis; and conducting the analysis using the properties of the objects of inquiry.

Judgments of adequacy depend not only on the problem and its context but also on the general criteria of adequacy that the scientific community has imposed on the class of inquiries of which the scientist's work is an example. For instance, the levels of significance in statistical tests are assigned by the scientist in the light of the issues in his inquiry and standards usual in his field, not by some scientific deity (see, for example, Action 14 in Section 11.4). Similarly, the community of mathematicians may establish standards of rigor for mathematical research that are quite different from those that meet the needs of physicists for mathematical argument.

> In general, we can say that imposed criteria of adequacy are necessary for scientific work because of the inconclusiveness of the arguments used in science. This inconclusiveness follows from the peculiar character of the objects of scientific inquiry: classes of intellectually constructed things and events, the evidence for whose properties is derived from particular experiences. To move from the reported properties of particular samples, to the properties of the classes they are intended to represent, a demonstration is necessary; but a formally valid argument, yielding certainty or truth, is impossible.
>
> A scientific problem is thus incapable of having a solution that is 'true.' Rather, the solution will be assessed for adequacy; and for this every component must be so assessed. . . .
>
> The criteria of adequacy associated with a field are, however, directly relevant to the strength it attains. Indeed, they perform an essential function, and a field is matured and effective only when they perform it well. . . . [Ravetz, 1971, pp. 153, 155.]

It follows that the strength of a field grows as its criteria of adequacy become stronger and more widely recognized and used, and that one of the central issues in developing a relatively new field is to develop its accepted sense of adequacy. Thus, the descriptive (or scientific) work of systems analysis can hardly be expected to have criteria of adequacy as highly developed as the classic sciences; this is why much of the discussion throughout this handbook is aimed at strengthening this newer field's criteria of adequacy. The important point to note is that systems analysis should develop its own criteria based on its own context and work; although it can—and should—consider adapting criteria used elsewhere, it cannot simply take over those of another field.

A second well-developed set of criteria is essential for a mature field of science: those of value. Assessments of adequacy relate to work that is done, those of value not only to what has been done, but also to the choices about what new work to undertake. Thus, the criteria of value guide the future of a field, and thus constrain its prospects, for

> choices and decisions must be made at every stage of scientific inquiry, and such actions are impossible without judgments of value. The extension of the boundaries of the known into the unknown does not take place like the spreading of a wet spot on a piece of blotting paper. Without strategy and tactics, a field of scientific inquiry has as little chance of success as an army in battle which is simply told to 'advance.' The statement of a problem, to be solved is analogous to an objective to be taken; and the selection of some problem for investigation, necessarily to the exclusion of others, must be governed by competent judgments based on sound principles. The judgments which determine such decisions are those of value, feasibility, and cost. [Ravetz, 1971, p. 161.]

It is useful to distinguish three components in criteria of value: internal and external ones (both relating to the roles that can be played by finished work) and personal ones.

The internal components of value relate to how the solved problem will contribute to the field of which its objects of inquiry are a part: the extent to which knowledge will be advanced; whether or not light will be shed on an important area of concern; whether descendant problems will emerge that, if solved, will carry the field still further along; the extent to which the prospective finding will be interesting in its own right; and so on. Working scientists can easily lengthen this list—and must, *a fortiori,* agree that there is no single basis for judging internal components of value.

The external components of value relate to the contributions that the solved problem makes to problems or tasks outside its field. These contributions may be to other fields of science, to technology, to the solution of applied problems (such as those in a systems analysis study), to political or ideological problems, or even to philosophy. As in the case of the internal components of value, there can be no unitary basis for judging the external components.

Since much of science in the twentieth century has emphasized internal criteria of value while turning its back to the extent possible on the external components, some observations are in order. The external components of value exist for any scientific work, whether the scientists involved choose to pay attention to them or not; thus, the only issue is whether or not to deal with them explicitly. Since science is supported by society, which has a stake in the external components, scientists have the choice of hiding among the internal components of value, leaving society to evaluate its support without knowledgeable guidance, or of helping society to arrive at intelligent choices by giving external components of value explicit consideration, with the results communicated effectively to lay audiences. In this connection, Ravetz (1971, p. 165) has this observation dealing with the internal and external components of value:

> [I]n the distinction between these two components, we have the root of the inevitable tension between a scientific community and the lay society which supports it. Speaking very roughly, we may say that the 'internal' components of value are generally esoteric, and incomprehensible to the lay public which pays the bills; while the completely 'external' components of value in the present age are vulgar, and, if dominant, can eventually debase and destroy the activity itself. In practice, the health of a scientific community is maintained by a delicate and inherently unstable compromise between the two.

But this is not the only compromise needed; there is also one between the internal and external components of value and those relating to the purposes and desires of the individual scientist. These personal components of value range from a single-minded devotion to advancing scientific knowledge to the exclusion of personal rewards all the way to an equally single-minded devotion to power, prestige, and financial reward. These personal components of value are too well known to working scientists to call for listing here. It is worth remarking, however, that it would be as foolish as it would be unrealistic either to ignore these personal components of value or to demand that scientists ignore them. As Ravetz (loc. cit., p. 166) puts it:

> To demand that scientists abstain from including any personal components in their valuation of problems would be ridiculous and, to the extent that efforts were made to enforce such an impossible standard of behaviour, they would only breed hypocrisy and corruption.

This subsection has presented the methods of scientific inquiry, except for straightforward techniques, as a heterogeneous collection of craft skills that are communicated informally—or that may even be tacit. Although these methods are not usually specified overtly and completely, they are the common possession of the field in which they are found. They are used to produce individual conclusions based in part on data yielded by nature; however, each such conclusion is only a fragment, needing to be woven into a much larger tapestry to contribute importantly to a picture of nature's behavior. The discussion to follow must resolve the apparent paradox of how these methods, chosen and informed by so much that is in the nature of craft skill, can produce the scientific knowledge that is so useful and reliable.

Before turning to this matter, however, there is a remark to be made about "the method of science," a phrase often used in philosophical discussions. If this is intended to describe a standard pattern of craft methods used in scientific inquiry, then the preceding discussion shows that it is meaningless, as no such pattern exists; but if the phrase is intended to suggest a schematic anatomy of a complicated and varied process, then it is useful. (The *Overview* volume [pp. 18–19] presents such a schema, quoted from Kemeny (1959), and points out that it cannot be interpreted as a procedural guide for individual scientific inquiries.)

Achieving Knowledge in Science

Anyone familiar with the progress of science over the last two centuries can view the account so far as presenting some paradoxes:

> First, there is the contrast between the highly personal endeavour required for a penetrating inquiry into the natural world, always fallible and governed by mainly tacit craft methods; and the public, objective, impersonal knowledge which eventually issues from that work. Similar to this contrast is that between the ephemeral structures in which scientific inquiry is undertaken, problems succeeding each other with great rapidity and theories and concepts being ploughed under at a perceptible rate; against the steadily increasing depth and power of the knowledge which somehow remains and grows amidst the swirling currents of day-to-day work. And, finally, we have already observed the impossibility of 'proving' a single result in science, or even establishing its 'probability' of being true; and yet the stock of permanent knowledge, absorbing all reinterpretations and surviving all attempts at refutation, survives and grows.
>
> There is no magic formula to explain the success of natural science; neither some special property of natural things which makes them uniquely accessible to human reason; nor a victorious 'method' whose application has laid bare the secrets of the natural world, and which can now be turned to other fields at will. Rather, the scientific knowledge we possess is the result of a social endeavor, which over the centuries has developed an approach appropriate to its limited goals, and where the work of each individual is informed and controlled by that of his colleagues in this endeavour, of the past, the present and the future. . . .
>
> The resolution of the paradoxes in the character of achieved scientific knowledge will come from an analysis of the further development of a solved scientific problem, when it emerges from the workshop of its creator, and takes on a life of its own in the community of science. [Ravetz, 1971, pp. 181–182.].

The conclusion reached by studying a scientific problem does not yet have the status of scientific knowledge. Although the result may be presumed to have passed the tests of the investigator's criteria of adequacy and value, it must face other tests as well. Before the result is published it is reviewed by one or more referees, who bring to bear on it their criteria, which are usually those of the scientific community to which they belong.

Although the refereeing process is universal—as it should be—it has the general effect of being conservative. Really pioneering work not only produces novel results; it also recasts the criteria of adequacy and value to some extent, shifting them significantly away from the center of gravity of those the referees are used to. They are bound to find the new work lacking when judged by the older, familiar criteria. Thus, to admit revolutionary new work to the archive they control, referees and editors must not only use a venturesome imagination, they must also be able to discriminate between work with a solid basis in craftsmanship and that emerging from idiosyncracy.

> It might be thought that a research report, duly certified and published in a recognized journal, can be accepted as embodying a unit of scientific knowledge. Indeed, the belief in such a rapid and straightforward process of the achievement of scientific knowledge was dominant until recently, and is still widely held. But my analysis of the referee's work indicates at least that great things are not won so cheaply. For all the referee can assure is that in his judgment the problem is of value and the work adequately performed. The true value of the problem can be determined only through the further development of the field; and the real adequacy of the work can be properly assessed only by the more demanding tests of repetition and application. [Ravetz, 1971, p. 183.]

The tests of use are much more demanding than those of the refereeing process. The user, seeking to reproduce the results and build on them, subjects the original work to detailed scrutiny to discover its trustworthiness and strength as a contributor to a new inquiry. If the original work passes these tests, it contributes to the generation of descendant results. If it does not, it remains in the literature unused—and perhaps even representing a pitfall to other workers; however, progress leaves it behind, where, unless cited by later authors, it fades away into oblivion. Thus, the communal tests of adequacy and value by use act as a form of natural selection: Survival is achieved by the creation of descendant results.

The result of a scientific inquiry that meets the tests of publication and use is usually regarded as a scientific fact. The concept of a scientific fact, however, cannot be divorced from the process that produced it.

> Facts are the best candidates as the bearers of certainty and truth in science; they are exhibited to apprentice-scientists as the foundation and proof of scientific theories; in the short term development of science they are the court of final appeal in the judgment of theories; and in a long term retrospective survey of a matured field of science there will seem to be a few matured salient facts which stand out as survivors from

a wreckage of the erroneous theories of the past. Although the most magnificent achievements of natural science are in its great theoretical syntheses, these must be tied down to brute facts in several ways. Their foundations in experience must rest on attested facts, they must predict new experiences which prove to be real and factual, and they must survive collision with contrary facts. Finally, no matter how hard we try to illustrate the human, creative endeavor of scientific inquiry, or to show how bold theorizing and even speculation are deeply woven into any great advance in science, we must acknowledge that what counts in the long run in science is the increase in factual knowledge of the world around us. [Ravetz, 1971, p. 185.]

In spite of this fundamental importance of scientific facts, they cannot be regarded as absolute. It can be argued that any fact is intimately involved with an intellectual construct for its interpretation. But even if one sets aside this point of epistemology, the history of science shows again and again that "even the hardest facts are not impenetrable" (loc. cit., p. 186). Increasingly, searching and subtle inquiry can show that a fact is a crude summary of a complicated reality (as research has shown the 1930s concept of the atom to be), or new inquiries may show old facts to have been at least partly mistaken (as when in 1962 reports appeared identifying compounds involving gases that had for over a century been regarded as inert). Since scientific knowledge cannot be regarded as completed and static, we must return to the processes by which it is created.

We have already seen that, for the conclusion of a scientific inquiry to become a fact, it must be significant (that is, it is noticed and so regarded by other scientists) and stable (that is, capable of reproduction and use by other workers). In addition, the conclusion must be invariant. As scientific problems and their solutions succeed each other, with the earlier results contributing to the later ones, there is an evolution of the intellectual objects of inquiry. If some aspect of an early conclusion persists through this evolution, then it can be thought of as invariant. The three properties of significance, stability, and invariance are necessary and sufficient for the result of a scientific inquiry to be regarded as a scientific fact. It would be a mistake, however, to infer from this definition that a scientific fact will not undergo further evolution.

A report on a scientific inquiry contains many elements: a statement of the problem; a conclusion; the information serving as evidence in the arguments; the techniques used in producing the conclusion; and a description of the intellectual objects of inquiry, especially if they have undergone modification during the work. Such a report also contains information on the choice of the problem treated, the patterns of argument used in the process of achieving the conclusion, and the criteria of adequacy and value employed. This listing is, of course, a somewhat conventionalized description of a complicated and highly varied reality—and, of course, not all research reports contain every element. These elements provide different sorts of stimuli to later workers, serve different functions, and vary in their permanence. It is through the interweaving of the lines of descent of these various components of a succession of works that scientific knowledge develops its tough fabric, so resistant to trivial refutations and simple revolutions. [Ravetz, 1971, p. 192.]

Scientific facts, and the principal tools used to establish them, provided they continue in being long enough to be widely used, not only in their own field but also others, undergo a process of extension and standardization. This is both necessary and sufficient for survival, for otherwise the conclusion would disappear with its original problem and its immediate descendants. The standardized fact then becomes widely used.

This standardization, however, renders the fact vulnerable to new inquiries and their conclusions, if the challenge is strong enough to upset the standardized fact (as the discovery of compounds involving gases formerly believed to be inert shows).

Similar statements can also be made about the intellectual objects of inquiry, which undergo change at the ends of periods of standardization and stability.

The Character of Scientific Knowledge

The standardization of facts does not preclude a necessary differentiation as they find uses farther and farther from the context in which they were originally developed. Thus, when we speak of a fact having survived as part of the accepted canon of scientific knowledge, we also include a family of variations of the fact, related by a complex lattice of descent involving varied uses, and further, a process that is continuing to evolve.

The property of standardized facts of science that permits their differentiation is the obscurity of their intellectual objects; different uses call for varied interpretations and different shades of meaning. Contrary to the dominant traditions of the philosophy of science, which hold that such obscurity destroys the claim to worthwhile knowledge, the history of science shows that useful knowledge can be arrived at in spite of obscure intellectual objects. A scientific inquiry is not aimed at explicating the objects' meaning; although its argument deals with properties of the objects, it does so only to develop properties of aspects of the real world. As Ravetz (1971, p. 215) puts it,

the objects of a scientific argument, whose concealed obscurities contain a wealth of material for philosophical analysis, are used as the components of an argument intended to relate to something other than the objects themselves, namely the external world. Hence a rough and ready control of the properties of those objects will be adequate for the argument.

It is for this reason that the obscurity of the objects of scientific discourse is generally irrelevant to scientific work of any sort. Contact with this obscurity is made in particular ways, depending on the work being done, and (in matured fields) a practical resolution can be achieved in every case. Thus, the research scientist will usually be aware of obscurities, at the technical level, in new materials in a rapidly developing field. But as the field stabilizes, there is achieved a sufficiently clear and univocal understanding of these objects for them to be manipulated in arguments with common agreement on their meaning and an absence of pitfalls.

It might seem that a fact produced by a process such as the one just sketched would be rather casual and imperfect, but this view would fail to take account of the many and diverse tests to which the fact has been subjected, the host of competitors that have been ruthlessly ruled out, and the changes in the intellectual objects that have responded to the new needs and uses. Thus, although to observers many years later the fact may seem untidy and imperfect, it has demonstrated its robustness and its ability to flow through a lattice of descendant problems and their solutions.

On the other hand, it is fallible in the sense that new facts may eventually replace it in the collection of scientific knowledge.

How widely we should apply the term 'knowledge' is of course a matter of convention. If we cannot bear the paradox of accepting that genuine knowledge may be fallible, then we must ban the term altogether from productions of the human intellect. But if we extend it to include all that which at any moment is accepted as fact, we are left without any means of differentiating between the ephemeral and the permanent in the achievements of science. It seems best to restrict the term to those results which are so solid that they live on (in the fashion I have described) as long as does the framework of reality in whose terms they are cast. [Ravetz, 1971, p. 236.]

Science and the Solution of Practical Problems

So far, our discussion has been confined to what we have called scientific problems, that is, ones "where the goal of the work is the establishment of new properties of the objects of inquiry, and its ultimate function is the achievement of knowledge in its field" (Ravetz, 1971, p. 317). But today science and scientists are involved in a wide variety of activities lying outside the activities of solving scientific problems:

The emergence of this new sort of work has led to a variety of new descriptive terms, which reflect the confusion over its nature and its relations with science of the traditional sort. The term 'science' itself has stretched to include technology, and even any sort of inquiry whose methods are modelled on those of the experimental and mathematical natural sciences. . . .

From one point of view, all the new sorts of 'science' are entitled to that honorific appellation: their work consists of problem-solving on intellectually constructed objects; and there is no clear demarcation between the methods used in traditional pure science and those in the most speculative of social technologies. We can, however, make a useful distinction among the various sorts of problems by means of the classification of final causes involved in a task. . . . [T]he task itself has a goal, which is conditioned more or less strictly by the function which will be performed by the result of the accomplished task; and this in turn is governed by the ultimate human purposes which are expected to be served by the performance of that function. [Ravetz, 1971, pp. 317–318.]

One of the social problems of science is to maintain a harmony between the goals of the work and the private purposes of the individuals doing it, these goals and purposes being called collectively its *final causes*. To discuss this problem Ravetz distinguishes between the objective and subjective final causes. Dealing with subjective purposes and how they relate to the objective goals of scientific work will not be discussed here. It is important, however, for this discussion to distinguish three types of problems on the basis of their objective goals: scientific, technical, and practical.

For scientific problems, the goal of the task is to solve the problem, and the function to be performed by the solution is to contribute new results to its field.

For technical problems, the function to be performed sets the problem. The task is accomplished—and the problem solved—if the solution enables the function to be performed.

For practical problems, the goal of the task is to serve some human purpose, and the problem is solved when a means for serving this purpose has been devised and shown to be effective.

Practical problems and their solution are the class of principal concern to us, and Ravetz offers this overview of the way they are approached:

As in the case of the other sorts of problems, a practical problem has a phase of gestation, when it is a problem-situation: an awareness that things are not as they should be, but no clear conception of how they might be put right. Once it comes into being, its cycle of investigation may be described by five distinct phases: definition; information and argument; conclusion and decision; execution; and control. Such a division of phases may be used to describe the accomplishment of any task; once the goal is set, one proceeds to consider the possible ways of fulfilling it; then decides between alternatives; proceeds to the operation; and during the actual work, supervises it to ensure that the goal is being satisfactorily fulfilled. . . . Thus the purpose to be achieved is necessarily stated in intellectually constructed categories, and so it is necessary to have the results from an inquiry of a scientific sort on which to decide the best means of its achievement. Also, the phase of execution of the work will usually be distinct from that of decision; and finally, where there is any division of labor, the task of control cannot be neglected.

Comparing this cycle to that of scientific and technical problems, we notice certain systematic differences. Ultimate purposes, which are remote and diffuse in science, become a part of the criteria for the controlling judgments in technology, and here they determine the goal itself. Outside those limited fields where purposes are capable of being handled in terms of accepted explicit intellectual objects, the framing of a new practical problem is an essentially creative act. Whereas the setting of a problem in science involves the partial and tentative specification of a conclusion about artificial objects in a self-contained universe, and a technical problem involves imagining a device to perform a preassigned function, here the specification is of a state of affairs in human society which does not yet exist. Each of the controlling judgments of feasibility, cost, and value involves a multiplicity of factors, few of them reducible to quantitative or routine assessment. [Ravetz, 1971, pp. 339–340.]

The experienced systems analyst sees much of his experience reflected in the approach to practical problems that Ravetz describes, particularly as it relates to the descriptive aspect of his work, but misses any recognition of the prescriptive and persuasive aspects of systems-analysis work, as well as the interactions that the descriptive work has with these aspects. Although the practical-problems category is the one closest to systems analysis as it has been described in this handbook, Ravetz's description of this category would have to be expanded somewhat in order to fit systems analysis into it. Nevertheless, it is possible to use Ravetz's work as the basis for drawing some inferences about basic concepts for systems analysis. To carry out this program completely is beyond the scope of this chapter, but some points are worth making here:

The goals of the work on scientific problems and those of work on practical problems are different, as the preceding discussion brings out, and will thus call for approaches responsive to this difference in goals. For example, a scientist pursuing a program of inquiry yielding results that do not seem to be converging is free to abandon this line in order to take up another, perhaps aimed at an entirely different problem; a systems analyst finding frustration in his approach to his client's practical problem may be forced to abandon his approach, but only to take up another aimed at the same problem.

The descriptive aspect of systems analysis is the one aimed at selecting the understandings of the phenomena of reality that will be useful in the analysis, and therefore the models to represent them. When the available stock of scientific knowledge does not cover the needs of the analysis, the gaps may require that scientific problems be investigated to the extent required to meet the needs of the analysis. Thus, this aspect is the one that makes the most use of the constructs, methods, and facts of science, and that therefore most nearly partakes of the character of scientific work as Ravetz describes it. The experienced systems analyst will find the descriptive aspect of his work mirrored in many ways in Ravetz's description.

On the other hand, since practical problems pose their own schedule demands, the systems analyst is seldom allowed the leisure to test all of the results used in his work until they can be considered facts of science as Ravetz defines this term; rather, he must assemble results relevant to his analysis ranging from established facts of science to results from his own new inquiries into scientific problems, and then weave them into an analysis aimed at solving the problem on which his work is focused. Thus, the elements entering his analysis from scientific activities occupy various positions in the stream of activity that leads from a new finding to a fact of science, and the confidence that he expresses in his results must reflect a sober and realistic assessment of his level of confidence in these inputs. Perhaps paradoxically, the resulting assessment of realistic—and often modest—confidence in the findings of a systems analysis usually enhance its value in the eyes of the client decisionmaker, who is accustomed to making decisions in the face of uncertainty, and who therefore finds value in any contributions that may narrow this uncertainty or make it more understandable, even though they do not eliminate it.

A single systems analysis that melds scientific facts with the results of new inquiries must, a fortiori, occupy a bottom rung on the ladder leading to the acceptance of a result as a scientific fact, since part of its basis also occupies this position. Successful implementation of its results can move it up the ladder, as can succeeding analyses in a continuing program. The successful implementation of a blood-supply procedure in the Greater New York area (as described in Section 2.2) showed that the models used in the analysis were adequate and reliable descriptions of what went on in the blood demand and supply system. Although quite a number of continuing programs of systems analyses have existed, few have been described in the literature in their totality; for notable such descriptions dealing with a program of systems-analysis work on a university's operations, see Hopkins and Massy (1981), and on fire-protection operations, see Walker, Chaiken, and Ignall (1979).

Since scientific work aims to understand how some aspect of reality behaves, and therefore to predict how it will behave in the future, systems analysis can be seen to have analogous aims: It seeks problem solutions that, if implemented, will have the beneficial effects discovered in the analysis. Nevertheless, the patterns of work in the two cases, although occasionally showing some overlap, are quite different, owing to the differing goals.

Science even has a persuasive aspect, since its workers who have achieved what they regard as important and successful results usually try to persuade their peers that these results are worth recognizing, and that they should be given the further tests that will permit them to be incorporated into the body of scientific facts. Thus, the persuasive aspect of systems analysis has an analog in this aspect of scientific work, but with important differences in the audiences addressed and the purposes in addressing them.

Against this background of underlying concepts of science useful for systems analysis, the next section turns to another source to illuminate issues relating systems analysis as a whole to the operational and administrative context in which it works.

12.3. Concepts Underlying the Applications of Analysis

With underlying concepts now suggested for the portion of the work internal to systems analysis that uses scientific knowledge and follows patterns of scientific work, it is possible to turn to conceptual elements underlying the activity as a whole as it relates to the problem situations of reality; to discuss them is the purpose of this section. Boothroyd (1978) provides the principal basis for the argument, which proceeds by these steps:

It first defines the setting in which a systems-analysis study occurs as an *action program.*

It then conceives of systems analysis as a basis for intervening in an action program.

It calls for systematic reflection—in the case of interest here, by the analysts—as the basis for action to intervene.

The discussion in this section first deals with these three steps; and then takes up the position of scientific theories in systems analysis in the light of the underlying concepts described in the previous section; finally, it describes some problems and their implications.

Action Programs

Since the subject of this handbook is systems analysis, one might well expect that the concepts underlying it would begin with a definition of system. Indeed, at several stages in preparing the *Overview,* we tested various definitional approaches, but found them unsafisfactory as a support for the work as a whole. Rather, the concept of a problem situation was found to be more fruitful, with exploration of this situation yielding a problem definition adequate to support an ensuing analysis.

On the other hand, it can be argued that the problem situation and the attempt to solve its descendant problem define a corresponding system.

There are two ways of defining such a system (Hoag, 1957). The first—and the one usually adopted by systems analysts—is to consider its elements to consist only of the aspects of the people, nature, and technical artifacts that are involved with the problem adopted for study; this definition excludes other aspects of these elements that are included in other systems, and perhaps in other problems. The second way of defining such a system is to include in it not only all of the elements as in the first definition, but also any other properties and involvements that these elements may have, one rationale for such comprehensive inclusion being that the elements are inseparable and that an influence on one aspect may well affect others.

At first glance, the more ambitious second definition may seem to be the one that should be chosen. But each such system is a totality of people, technical artifacts, and aspects of nature that can be conceived of and described from many points of view. The people can be described from physiological, psychological, political, and demographic points of view; the technical artifacts from functional, engineering, cost, or pollution points of view; and the aspects of nature from a similarly varied number of perspectives. Only a modest amount of reflection suggests that considerations of practicality dictate that a more restricted concept of system is needed.

In seeking this concept, it is important to recognize that the problem situations and problems that systems analysis deals with involve ongoing and proposed functions or operations having at least a loose coherence.

Against this background, and prompted by the concept of a research program as discussed by philosophers of science (see, for example, Lakatos, 1978), Boothroyd introduces the concept of an *action program*: a function, operation, or response related to and given coherence by an objective, a need, or a problem, together with the known, unknown, and implicit perceptions of the actors. For the purposes of this discussion as it relates to systems analysis, I will also consider the people, equipment, organizational elements, and management structure involved.

Perhaps the best way to illustrate what this concept means is to distinguish the action programs associated with the examples of systems analysis described in Section 2.2. The analysis that showed how to improve the delivery and utilization of blood in the Greater New York area dealt with an action program related to the means of delivery together with the administrative arrangements for gathering data on needs and assigning blood units to individual hospitals. The study aimed at improving fire protection in Wilmington, Delaware, considered an action program relating to the fire-response operations of the Wilmington Bureau of Fire, together with the personnel, equipment, and stations involved. The analysis aimed at protecting a Dutch estuary from flooding considered an action program related to the operational, social, economic, and natural elements that could be affected by flooding together with a variety of possible ways of protecting the estuary. The inquiry into the world's energy supply and demand over the next half century considered an action program to respond to various possible demands; the work also investigated the technical and administrative arrangements needed.

In each of these cases there was an objective that the action program was created to attain: the Greater New York area hospitals need blood as part of their treatment programs; Wilmington, Delaware, needs a prompt and effective response to fires; the Dutch desire protection from estuary flooding; and the world's society needs an abundant supply of energy if it is to function in accordance with its desires. In each case a need existed at the boundary of the action program considered in the analysis, but it was not brought into the analysis for further scrutiny or adjustment. These arrangements were, of course, a matter of choice and not of necessity; for example, an analysis could seek to balance world energy uses with its expected possible supplies, in which case the relevant action program would include society's uses of energy as well as the demands that these uses generate.

Boothroyd (1978, pp. 21–22) offers these further views of the concept as he sees it:[1]

> I intend the notion of action program to be a point of view from which to regard human activity. Although almost any activity can be so regarded, I am *not* claiming that it is a point of view from which we can give a *comprehensive* account of the activity; I do not believe comprehensiveness is possible. . . .
>
> Action programs are characterized by the actions that happen, the theories that the actors have about their actual and potential actions, and the proposals that are made within the program and across its boundaries. . . .
>
> . . . the word 'program' . . . is not being used to denote a set of intentions such as might be included in what a company calls its program of future investments or in what a city calls its program for a week's music festival; the word 'proposal' is the general word I am using to cover plans and intentions of that kind. Rather, the word 'program' is used with the idea of a continuing process which has associations with

---

[1] In the quotations in this section I have taken the liberty of altering Boothroyd's "programme" to the American "program."

the perceptions and intentions of human actors and which also has associations of stability and adjustability as it progresses. [Italics in the original.]

The alternatives considered in each of the examples mentioned earlier can be thought of as what Boothroyd calls proposals. In the fire-protection example there were several possible locations for firehouses that were considered, together with the staffing and equipment-deployment arrangements that they implied; in the flood-protection example three principal flood-protection alternatives were studied.

It is worth noting that some elements related to an action program, such as people or equipment, may also be related to other action programs. For example, the vehicles related to a fire-protection program are often used in holiday parades to contribute to a celebratory action program; the flow-through dam became the principal element related to an action program to build it after a choice of its design was made. This is especially true of people related to action programs; each is inevitably involved with many organizational, social, and personal action programs. Thus, the advantage of the action-program concept is that it focuses attention on the relation of each element to the action program, while omitting its other concerns and activities.

Boothroyd offers this view of the behavior of an action program over time:

> The existence of [an] action program needs some degree of social agreement; indeed, . . . we may say that the continued existence of a program depends on the continued rehearsal of its meaning by those who participate in it or have transactions with it. Within that limitation, action programs can begin, expand, wither, change direction, lie dormant, modulate by stages into something completely different, and end. Although they can, and do, exhibit considerable stability, there is nothing inherent in them to prevent them becoming completely other, and it would be a mistake to invest them with the properties of organisms—individual living entities whose development is constrained by their molecular biology.
>
> . . . For the participants of an action program there is some sense of continuity through time. What gives it stability is that some of the theories or proposals or repertoire of actions are held to be central and only to be removed or relinquished under the greatest stress if at all. What gives it adaptability is that many of its theories and proposals and repertoire of actions are disposable or modifiable or replaceable by new ones: the closer they are to the core, the greater the pull to retain them and the greater the sense of partial discontinuity when they go.
>
> . . . Insofar as they are articulated, theories and proposals offer scope for reflection within a program and for advisory intervention from outside [it]. [Boothroyd, 1978, pp. 27, 29.]

There are a number of points that can usefully be made here about actions undertaken as part of action programs.

Aggregation. Actions can be conceived and examined at various levels of aggregation, depending on the ends in view. On the one hand, if actions are aggregated too much, the significance of variations in them may be masked; on the other hand, if they are divided too much, the existence of patterns may be difficult to discern. It is a common experience, however, to consider actions in aggregated form. For example, as I type the manuscript of this chapter, I do not think separately of the finger stroke for each letter, nor even of the spelling of each word; rather, I think of the aggregate sentence that I am constructing. If a finger stroke falls on the wrong letter, then my attention must turn to it as an individual; if a word is inserted wrongly, then my attention must turn to the proper word choice; if the sentence reads poorly, then I must revert to its aggregate; if the sentences do not make a smooth flow of argument, I must attend to larger aggregates of paragraphs, sections, and chapters.

Similarly, at different levels in an organization, actions are—and, indeed, must be—aggregated differently. An individual sale—attended to carefully by a clerk—becomes part of an aggregate for a department in a store, or part of a still larger aggregate for a chain of stores. Boothroyd (1978, pp. 32–33) carries the discussion further:

> Insofar as action happens within an action program which is part of one or several hierarchical programs, it happens simultaneously in all the programs but is differently regarded in each. Moreover, the action might have its own interpretation in programs to which it relates only indirectly. We might well play a game of producing the infinitely long list of action programs that a single action might affect. Many effects might be small. But there are also well known stories of the for-want-of-a-nail kind which suggest that some action can be seen in retrospect to have had a tremendously wide range of consequences, though the actions were not necessarily noticed as pivotal at the time.

Differential regard. An action is regarded differently, depending on the action program from which it is viewed. For example, related to an action program to meet a region's energy needs, a nuclear plant must be regarded as an important asset, but from the point of view of an action program aimed at promoting environmental safety it poses important problems.

> So whatever we think about the ultimate possibility of harmonizing human affairs either relative to each other or relative to some absolute, we cannot suppose that it is practical at present to regard the various significances of an action as evidently reducible to a single measure which is translatable between programs. Even if within a single action

program an action could have a single measure ascribed to it, we would need a vector of measures to represent the measures ascribed by each action program—that is, a list with a separate measure from each of the programs. In our Western culture we are used to assenting to the view that the measures from many action programs can be reduced from a vector to a single measure because we have a number of very general and longstanding action programs where it is customary for many of the actors to forget the essentially vector nature of measures. Indeed, when anyone talks of the best action to choose as though 'best' were an absolute property independent of the program from which it is declared, then he has already made a profound mistake. [Boothroyd, 1978, p. 33.]

In the Wilmington fire-protection example, the analysis was aimed at finding a new deployment of fire stations that would keep the level of protection roughly the same while achieving some economies in the operations of the fire-fighting service. Thus, from the point of view of the city administration's action program, some relocations accompanied by a reduction in the number of stations were regarded as desirable. The firefighters' union, however, perceived these changes to be undesirable from the point of view of its members and opposed them.

Since the significance of any action depends on the action program from which it is viewed, Boothroyd suggests this conclusion (loc. cit., p. 33): "No pair of actions can be specified in which one is preferable to the other in all conceivable programs."

In any case, the principle of differential regard for actions urges that systems analysts proposing actions based on their work identify clearly the action programs and objectives for which they are preferred, and more importantly, significantly relevant action programs for which the preference may be in doubt or which have not been examined carefully. The argument here makes this canon of good practice part of the essential foundation of systems analysis, and not something to be chosen on the basis of the professional taste of the analyst. Observing this canon gives the client decisionmaker information that is essential for him to consider as he works his way toward actions or policies aimed at ameliorating his problems.

Related action programs. Any action program presenting a problem situation calling for investigation by a systems-analysis team inevitably has many associated action programs; some may have a hierarchical relation to it, others may be relevant by virtue of having common elements, still others may sustain impacts from the problem situation even though they are external to its action program. It follows that any action proposed for ameliorating the problem situation can be viewed from the vantage point of any of the related action programs, often with widely disparate results. If the analyst confines his analysis to the original action program, and ignores the other associated ones, his work is likely to be seriously incomplete. On the other hand, time and resource constraints in practice invariably limit the scope of the action program that can be brought under study. These facts have important implications:

> Although it may be both undesirable and infeasible for the work to extend to all of the conceivably affected action programs, it must take up the consequences for all of those judged to be important, as well as how they may be judged from the points of view of the relevant action programs. It is not uncommon for the relevant action programs to include some important ones beyond the purview of the study's client.

> In view of the practical limitations that are always present in systems analysis work, it is important for the analysts to have them clearly in mind, and to make them equally clear to the client and decision-maker, so that they can assess the findings accordingly.

A systems-analysis study may be conducted quietly at some distance from the operations of the action program related to it, with nondisruptive cooperation from the client and his staff. In such a case, its first serious impact on the action program is likely to occur when a course of action is adopted and implementation starts. On the other hand, it is possible for such work to require involving the action program repeatedly in the search for information and data to such an extent as to make the activity obvious to all concerned, and perhaps even to cause some noticeable disruptions. Against this background, two points emerge:

> The systems-analysis work itself, whether it enters the action program being studied early or late after a course of action has been embarked on, can at some time be regarded as an action program intervening in others.

> There is no way, *a priori*, to prescribe how or when a systems-analysis study should enter an action program; rather, the approach must be evolved as one of the important early products of the inquiry.

The repertoire of actions. Actions are at the core of action programs. Hence the repertoire of available actions—that is, options or alternatives—has a central influence on the action programs that can be established pursuant to an analysis.

When systems analysis inquires into a problem situation, it may find that new actions in the corresponding action program are needed to ameliorate the situation. Indeed, progress may depend on the ability to conceive of constructive new actions. Some worth considering may be simple combinations of earlier actions, but others—and often the most important ones—may be novel in the strong sense of that word. For the latter, with

Boothroyd (1978, p. 38), "I say *conceive* them rather than *deduce* them because novel actions cannot logically follow from any theory deductively; that is, novel actions cannot be found by logic, mathematics, or computation. They must first be conceived." (Emphasis in the original). This is not to say that logic, theory, and calculations cannot suggest guiding principles for constructing new actions so that they will have the desired properties. For example, in the blood-supply analysis described in Section 2.2, economy suggested that the number of deliveries be reduced, although the imperative of having blood where it was needed when it was required suggested that the delivery process would have to be remodeled if this imperative were to be met without adding to the already large blood discard rate. However, the scheme of returning unused blood from the hospitals to the central store was conceived within the framework of acceptable policy for blood use; it was not deduced from analysis, although its properties were established by calculations.

If novel actions are important, how are they to be conceived? In particular, when systems analysis has been asked to intervene in an action program, how can it conceive of a full repertoire of actions to consider? Chapter 6 discusses this issue, the term alternatives being used there for the actions considered. There is no simple and direct answer to the question, but some good advice can be offered: Consult widely with all concerned (especially with the client's staff, which may well harbor persons with imaginative thoughts about possibilities), as well as many others with relevant knowledge and experience, and put as much imagination to work as possible; further, keep this process going as exploration proceeds— new insights emerging as the analysis matures may well suggest possibilities hitherto overlooked. An example of this last possibility occurred in the global energy demand and supply analysis described in Sections 2.2 and 9.6; the large amount of coal needed to generate the heat for a coal gasification process suggested that a nuclear plant supply the heat (*Overview*, pp. 183–184), a possibility that proved to have important advantages, including making better use of coal reserves and reducing carbon dioxide release into the atmosphere. The serious consideration of such a process depended on a technical check of its feasibility with appropriate technologists expert in the relevant energy-generation techniques.

In sum, although systems analysis can be expected to estimate the consequences of a variety of potential actions, an analyst must look as broadly as possible if he is to assemble a full repertoire for consideration.

Limits to prediction. If a proposed action can be analyzed on the basis of experience with other comparable actions and the data that they have yielded, and if the analysis finds that the action exhibits significant properties of stability under a variety of possible conditions, then the consequences of introducing it into an action program can be predicted with some confidence. On the other hand, if a proposed action is so novel as to be without precedents that yield relevant performance information, then the analysis must *a fortiori* rest on a less confident basis; the proposed action's potential stability must remain in question, and analysis results can only indicate possible effects, not estimate them with some precision. Too, novel actions may stimulate the action program into which they are introduced to have much wider interactions with other action programs, thus still further weakening the possibilities of basing predictions on results from analyses of other analogous actions.

Many action programs evolve over time, so that characteristic parameter values associated with them also change. Thus, potential predictions for the future must take account of such evolutions if it appears that they may occur.

These influences—as well as others not mentioned here—pose limits on predictions for actions within action programs, and they must be carefully considered in analyses.

These points are not intended to discourage attempts to apply systems analysis to action programs, but to caution all concerned to consider and understand the limits to prediction that this context imposes. This caution is especially important for scientists coming into systems analysis from the natural sciences, where the stabilities of the natural phenomena understood by these sciences are so common as almost to be taken for granted.

Finally, if proposals rest on predictions, then they must be presented so that their limitations, arising from the limitations of the underlying evidence and predictions, are well understood.

Intervention in Action Programs

There are many ways to intervene in action programs. An executive in the program can prescribe new policies or procedures; other persons in the program can unite, either formally or informally, in a resolve to undertake a new approach; outside forces can force important responses inside the program; higher authority may prescribe new forms of action; consultants may formulate new actions and succeed in getting them adopted by management; and so on.

The bases for such interventions may range from quick, almost thoughtless, intrusions to carefully-thought-out and meticulously researched plans. Boothroyd (1978) speaks of carefully-thought-through changes as *articulate interventions*. Using his terminology, we may then think of systems analysis as providing a basis for *articulate intervention into an action program*.

For convenience in the discussion, we shall think of the intervention as some sort of action, such as adoption of a new policy or implementation of a new program. A systems analysis and the careful consideration of its results that prompt the action are what make it articulate. And the action is thought of as entering—and altering—an already existing action program.

There are a number of points that are important to make about interventions into action programs; although I set them down here as relating to interventions based on systems-analysis work and its findings, they apply also to interventions arising from other considerations.

An intervention as an action program. Interventions of modest scope and impact can usually be considered to be part of the normal functioning of an action program. On the other hand, when an intervention has widespread influence and major impact, and involves a major segment of an action program, it is useful to consider the intervention as an action program in its own right.

> The intervening program can be regarded as staying separate at its interface with the entered program or alternatively having with the entered program jointly created a new intervention program. Each view emphasizes one or other aspect of separateness and togetherness during the intervention. It is perhaps conventional to think in contractual terms of just two programs, but for the purpose of understanding what is happening to the bundles of perceived theories and proposals it seems more useful to regard the intervention experience as a separate program. One is then likely to pay attention to the differences between the three programs. Occasionally that can be crucial. For example, the intervention program can develop a much wider range of content than the participants realize, and thus create surprises at the point where it had been thought that all that would be needed for the rest of the entered program was a brief report: it may instead need a considerable appreciation and training program in order to translate advances in understanding and repertoire from the intervention to the rest of the entered program. [Boothroyd, 1978, pp. 45–46.]

When a systems-analysis study develops an intervention program that will require major changes in the way an action program operates, then it is wise—as the *Overview* (pp. 215–216, 249–280) prescribes and Section 5.7 and Chapter 14 emphasize—to prepare a careful implementation plan for introducing the intervention; putting this plan into operation can then be considered a transitory action program aimed at producing the action program devised as the result of the analysis. The study of the distribution of blood in the Greater New York Area described in Section 2.2 developed an almost entirely new distribution system; therefore, introducing this new system was planned and carried out carefully (*Overview,* pp. 75–78). In sum, the desired action program was achieved through a transitional training and implementation program that allowed the reactions of all those involved to develop naturally in response to their experiences as they occurred.

Not all intervening action programs are so comprehensive or important in their impacts as to need a transitory program, but much systems-analysis experience warns that the issue needs explicit consideration and a decision about whether or not to develop a transitory program should be made in the light of mature consideration of the likely problems to be encountered during the transition process.

Views of an intervening program. Persons who have been part of an action program for a long time tend to view their world from its perspective. Similarly, members of an analysis team who have developed a program proposed for intervention into another have familiarized themselves with the world that the new program is expected to create. If these two worlds differ substantially there is a potential for difficulty with the intervention.

> The contrast in content between the entered program and the intervening program is a potential source of grievance. It is common for the two programs to be substantially more different than the participants of either program realize, and for the broad outlines of the likely course of an intervention to be seriously mispredicted. A part of the problem is not simply that the theories and perceptions of action repertoire are different but that the very words and images that go to make up formal statements are different: both programs are to some extent without a sufficient language for considering the other. Yet for the proposed intervention to have been seen as worth encouraging, there must have been a view in the sponsoring program that the two programs were sufficiently similar and sufficiently contrasting for there to be an interesting potential from an intervention. So it would seem that in the preliminary mutual appraisal significant differences of content can be overlooked. [Boothroyd, 1978, p. 45.]

It is vitally important for the reader to note that the difficulties foreseen in this discussion are not necessarily intrinsic, and that therefore they can be overcome. To do so, however, is the responsibility largely of the analysis team preparing the intervening program. Its members must become familiar with the vocabulary, outlooks, attitudes, and concerns of the action program being subjected to analysis and must frame the plan for the intervention so as to meet and overcome the potential difficulties. This process starts at the beginning of an analysis, and continues throughout, but becomes especially important in the reporting and persuasion phase, as well as the follow-up implementation. Chapter 14 deals with this matter further, and advice to the analysis team about how to meet this responsibility also permeates this handbook; thus, there is no need to extend the discussion here.

Professional relations. Since the persons involved in developing the intervening program must become thoroughly familiar with the entered program throughout their work, it is clear that important relationships are involved, and that the basis for these relationships should be an appropriate one. When professional knowledge is the basis for developing the intervention, the relationships should have an appropriate professional basis, and it should be a mutual one. Section 12.4 discusses this matter further.

Disciplined Reflection before Intervention

Actions can be undertaken on a variety of bases, ranging from instinct and prejudice to the most disciplined reflection. History warns us against associating success, as measured by some sort of an impression of a favorable outcome, uniquely with any one basis; good outcomes, as well as bad ones, have emerged from actions that have not been thought about before being taken, and carefully-thought-out plans have gone awry. Our natural prejudice, however, is widely shared and supported by a good deal of historical evidence: Actions based on disciplined reflection have a better chance of good outcomes than others.

When a significant body of knowledge is involved as the basis for disciplined reflection, the persons doing the reflecting are often members of a profession centering its attention on the body of knowledge that is being used. In this century, as knowledge has become segmented into more and more parts, a proposal may require consideration by a body of professionals representing a number of different bodies of knowledge (as is characteristic of most systems-analysis studies).

Joined into a team focusing its attention on the problem situation in an action program, these professionals can be considered articulate interveners (Boothroyd, 1978, p. 54):

> The role of articulate interveners then seems to be:
>
> 1. To join the [participants in the] entered program in articulating and elaborating the components of reflection before action,
> 2. To reflect on the quality of the components, and to act to improve them,
> 3. To participate in deliberative argument, and to elaborate the means for deliberative argument,
>
> though these may be limited by the perceptions or the wishes of the participants.

Analysis as a basis for articulate intervention. Although there are many possible forms of articulate reflection that can be used as the basis for action and intervention—indeed, every profession has its own catalogue of such forms—the attention here is focused on systems analysis as the basis for articulate intervention in action programs. In this connection, a number of additional comments are worth making.

Boothroyd (1978, pp. 48–49) offers three reasons supporting the use of systems analysis as the basis for articulate intervention:

> Firstly, the conventional abstract models of [systems analysis] are *in intent* fully articulated announcements of theoretical content on the basis of which action can be chosen: the fact that they are almost all structurally incomplete . . . does not vitiate the intent that they should be free of ambiguity or, at a higher level of abstraction, the intent that such ambiguity as remains shall have unambiguous statements made about it. Secondly, it is a conventional aspiration that the output of understandings from [a systems-analysis] program shall be transmissible and that they shall be able to withstand articulate well-structured critical debate. . . . Thirdly, whenever the actors in a would-be-scientific program use language about the unarticulated content of programs they raise that content to the level of language, thereby articulating what was unarticulated; in making that point I do not intend to imply that their only impact on the entered program is through articulation, since they will also have an impact through many actions that are the subject of little or no thought. [Emphasis in the original.]

Boothroyd (1978, p. 51) then offers this brief overview of what is involved in an analysis aimed at an intervention:

> During the course of articulate-reflection-before-action one can be aware of drawing on, or creating or modifying, members of three largely undeclared sets:
>
> A set of concepts of possible actions.
>
> A set of theories about what consequences follow particular actions, and perhaps even about how the actions lead to the consequences.
>
> A set of proposals about which actions and theories and consequences are to be considered and in what light.
>
> There will be a sense of incompleteness if any of these sets is unrepresented or poorly represented in what has been offered as articulate reflection to precede action.

Since practical limitations of time, knowledge, and resources place boundaries on what can be comprehended in reflection before action, the set of matters that can be considered explicitly is also limited. The omitted considerations may be ones for which intellectual constructs and knowledge would be available if called upon, or they may be ones for which such ideas have yet to be developed. In either case, there is a risk that the assumption that they can safely be ignored will prove to be false during the intervention. To eliminate this risk, it is impossible to argue as a practical matter that all possibly influential factors must be considered explicitly; rather, it is a practical necessity to choose the ones that it is important to consider and to run the risk of ignoring the others (indeed, making this choice is one of the craft skills of systems analysis).

Underlying any analysis there are factors that remain implicit; that is, they can be recognized to be present but are not considered explicitly. The most usual one is that the world surrounding an action program will continue without drastic change during and after an intervention into the action program. More generally,

> we rely every moment on the continuity of most things around us, and at a much less intense level we rely on the continuity of the social milieu we live in but only realize how extensive that reliance is when change makes us notice. Because implicit theories underpin so much of what we do, rehearsing them can be found offensive, even though they are sometimes relied on erroneously, and even though their declaration might lead to the resolution of some problem in hand. [Boothroyd, 1978, p. 56.]

In appears, then, that the prudent analyst should identify as many of the implicit factors as he can, so that he can make considered judgments about whether or not they should be given explicit consideration in his analysis. Too, in his communications with his client as part of the intervention, he may find it prudent share his judgments so that both know what has remained implicit. This process prepares all concerned for the possibility that an implicit factor can undergo a significant change not otherwise foreseen.

Many scientists have been brought up to believe that theirs is an emotionless enterprise, and this feeling often extends to their work as systems analysts. It takes very little reflection on the action programs of which we are a part to realize that actions within them—and especially interventions aimed at altering the pattern of actions—are often overlain with considerable emotional content. Thus, it would seem imprudent to ignore the emotional dimensions of an action program, particularly one subjected to intervention.

On the other hand, one approach sometimes used in systems and policy analysis is to reduce emotions to preferences, "a rather limited notion, and one which suggests somewhat passive membership of an established economic order rather than life where power is to be taken and kept by all kinds of forceful action" (Boothroyd, 1978, pp. 57–58).

This sublimation of emotion to a simple preference measure can deal with some parts of a problem, but not others. Strong emotions arise in many human contexts, as proponents of nuclear energy have discovered to their considerable regret.

> Emotions . . . are inseparably linked with mental processes, modifying

our readiness to think in certain directions and blocking the possibility of some mental sequences altogether. . . .

There are countless situations where an understanding of physiologically or psychologically describable aspects of reflection and communication will give you a greater ability to influence goings-on towards some desired consequence, or will prompt you to change your view of the sort of action to consider. . . .

So there is no question of ignoring emotions. It is a question from which point of view to regard them. The role of articulate intervenor implies that you articulate your understanding of them, but it is a role you will only be granted if your articulation generates favorable emotions in the participants of the entered program. [Boothroyd, 1978, pp. 58–59.]

Balancing strengths and limitations, how can we sum up the cases for articulate reflection as the basis for intervention?

It seems that as a species we are content with poorly articulated reflection for much of what we do. . . .

In the circumstances, the extent to which would-be-scientific intervention is welcomed is surprising. . . .

There is, however, no *guarantee* that an increase in articulate reflection will improve an outcome compared with what it might have been without the increase. By the same token, there is nothing inherent in articulate intervention that guarantees a more desirable outcome. . . .

. . . The case for articulate reflection before action is not therefore that it always gives improvements but that it does so often enough to be a major source of advance, and in default of similar innovations elsewhere to be an important factor in the natural selection of action programs.

The case for articulate intervention is similar: there is sufficient evidence of articulate intervention causing pivotal improvements for it to promise to be a significant source of long-run improvement. [Boothroyd, 1978, pp. 59–61; emphasis in the original.]

Theories

Every participant in an action program has theories about why things happen the way they do; it is usually the case that these ideas remain largely unexamined in the critical sense—and, of course, they vary in the accuracy of their predictions. The purpose of articulate reflection with respect to an action program is to make these theories explicit, to examine and test them, to construct new ones when the older ones fail, and to fit the set of theories into a logical structure. Systems analysis is a special form of articulate reflection that aims to build as fully as possible on scientific theories, uniting them with information and data yielded by the action program itself. It follows, therefore, that the trustworthiness of the results of articulate reflection, and especially those of its systems analysis form, depend importantly on the status of the theories that they contain. Boothroyd (1978, p. 73), in an overview of the situation, says that

> in an action program many theories are likely not to be particularly well founded. Nor is there any chance that all the theories in a program can be checked out. So would-be-scientific intervention is at best a way of getting things right*er*, not of getting them right. It is a tribute to the progress of our species that what we do after theorizing can in so many cases be left unchanged for years or even decades, but it would be idle to deny that much of what we do needs continual theorizing even if at relatively low levels and that we need continual recovery from our mis-theorizing. [Emphasis in the original.]

Here we must remind ourselves that even the most attractive and widely believed scientific theory has the status of a still accepted conjecture that may be damaged by new evidence that fails in some way to confirm it; indeed, every theory we have is a conjecture accompanied by a state of belief depending on how searchingly it has been tested.

The processes of testing and evolution that a theory undergoes as it finds its place in the body of scientific knowledge—as Ravetz (1971) describes and Section 12.2 summarizes—give it an appropriate status of belief. If it has just been advanced, based on an initial inquiry into the relevant portion of nature, it clearly can be trusted as a model only in a very tentative way; later, after many tests and some evolution based on the testing process and wide examination from other points of view, it can be given a much more confident status, provided it has survived these processes intact.

Thus, systems-analysis studies that must include scientific inquiries to establish properties in order to fill gaps in knowledge must regard the results of these inquiries guardedly, as they have yet to be subjected to searching tests by other workers. Similarly, knowledge used from the body of science must be accorded a status appropriate to its state of testing and review in its field. All of these judgments of status then become important inputs to the analysis team's judgment about the confidence that should be placed in the results of their work.

This discussion is not intended to deter analysts from using scientific theories in articulate reflection—or in systems analysis; to do so would deny the rich history of the last three centuries of applications of science to practical concerns, as well as the successful work that systems analysis has done in its brief history. Rather, it is intended to caution analysts to avoid exaggerated claims for their work and its results. The truly professional analyst presents realistic claims for the confidence that should be placed in his work and its findings, and in developing them he remains aware of all the practical and conceptual issues that underlie his work.

Some Problems

Articulate intervention in action programs involves many difficulties; although it would be inappropriate to provide a comprehensive catalogue of them here, a few deserve mention.

One of the cornerstones of the success of science is that its theories represent carefully selected, limited aspects of reality. "So even the largest and most coherent formal models that are constructed represent a tremendous reduction from all that could be represented" (Boothroyd, 1978, p. 99). Since these statements apply equally to models used in systems-analysis studies, it follows that the aspects of reality chosen for model representation and their resulting adequacy have a great deal to do with the effectiveness of analyses on whose basis interventions are made into action programs. Choosing what to model and how to model it so as to make systems analysis effective is one of the subject's principal craft skills, to which much of this handbook is relevant (especially Chapters 2 and 8 of this volume).

There is a sense in which each intervention into an action program is unique. This being so, is it possible to make any general statements about how to approach articulate intervention into action programs? Boothroyd discusses the matter in these terms (loc. cit., pp. 101–102):

> [T]he enhancement of intervention ability is unlikely to be best accomplished simply by articulating statements about it. . . . Indeed, statements probably have greater use for the organizing of understanding about experience after the experience rather than as a precursor of it. This has all kinds of implications for education for intervention. It more than justifies the use of case histories, case studies, and participative project experience, but is also suggests a point of view from which moment-by-monent content of educational programs might be reviewed, and the place of these programs in the career of the intervener. It also suggests that the design of abstract technology for a deliberative context may be quite seriously deficient if the designers are without first-hand experience of that deliberative context.
>
> All of which means that I am reluctant to suppose that we can write a manual of intervention that can be put into practice by any intelligent layman: a sort of algorithm of intervention. . . . On the other hand, . . . we might offer useful prescriptions against an assumed background of intervention experience provided that we do not regard the prescriptions as a set of sufficient instructions.
>
> Perhaps the message . . . is that, for all the support we have from societ[y] in general and OR [and systems-analysis] communities in particular, we are essentially on our own in each new intervention. It is in that unique situation that we try to understand and act in terms of what we already know and what we can discover. There is a place for each.

The central purpose of this handbook is to collect for the systems-analysis community the aspects of its craft that may shed light on new potential interventions that are undertaken, although understanding that the collection can neither constitute a set of sufficient instructions nor substitute for extensive professional involvement with systems-analysis work.

Since systems-analysis work draws on a variety of bodies of knowledge, it must *a fortiori* depend on facts from them of varying supportability; for example, many of the facts from the physical sciences are likely to have been subjected to a great deal more testing than facts from the newer social sciences. Thus, the state of the field from which the facts are drawn must be taken into account, as well as the more detailed support for the facts themselves.

After the problem has been formulated, information gathered, assumptions agreed on, and models built, the analyst has what Boothroyd calls a precise problem (and what Ravetz calls an intellectual construct), and he can explore it with all of the technical, mathematical, and computer tools that he has available. When these operations yield results, the problem is how to translate them into conclusions about the real problem (see Chapter 8). Boothroyd issues this warning (loc. cit., p. 119):

> In principle, the claim that precise-problem conclusions translate precisely into correct real-problem conclusions is wrong. The mapping of real problems on to precise problems is always accompanied by a considerable simplification in choosing what to map: the properties of any real system are indefinitely many. So we must maintain a clear distinction between the *theory* that an extremal precise action indicates that the corresponding real action is extremal, and the *proposal* that the entered program undertake the real action corresponding to the precise action: the theory is not justifiable, but the proposal can like all other proposals be accepted or rejected. [Emphasis in the original.]

To this warning he adds this gloomy observation of past history (loc. cit.):

> [T]he widespread provision of computing facilities has had so hypnotic an effect on the participants of intervening and entered programs alike that some very unsuitable precise problems have been used to choose actions.

Concluding Remarks

Since systems analysis can be thought of as large-scale operations research, as Boothroyd conceives it, we may conclude this short description of his conception as we have focused it on systems analysis by adapting his summary description (loc. cit., pp. 122–123):

> [Systems analysis is a form of] the practice of intervening in action pro-

grams to address any significant problem of articulate reflection before action, and by any conceivable method examining and adding to and modifying the theories, proposals, and repertoire of actions for the entered programs and other programs with which it interacts, so as to permit a clearer understanding of reality, and a more reliable and timely choice of action.

To which he adds (loc. cit.):

The intention [of this description] is that articulate intervention be seen as concerned with the full range of concerns of the participants of an entered program, insofar as we can usefully articulate views about them.

Although this description may appear superficially to include all human action-program concerns as addressable by systems analysis (such as those of public health and law), this is not the intent. Rather, Boothroyd (loc. cit.) says:

In effect I am suggesting we do away with demarcation disputes and that, if we do want to draw tighter boundaries round ourselves for the purpose of self-definition, we do so voluntarily and temporarily rather than with the idea that there is something permanently correct about the present delineation of our field of activities.

For the purpose of this section, I have drawn principally from Boothroyd's *Articulate Intervention,* because it offers a well-formulated set of concepts related to the problem situation and its context that are also directly applicable to the concerns of the systems analyst. There is, of course, a great deal of literature in widely scattered sources related to operations research and management science (ably summarized, for example, in Jackson and Keys, 1987, especially pp. 1–25 and 133–164), and some of its concepts will certainly contribute to the foundational concepts for systems analysis, but it would be difficult to make predictions in this regard.

This section describes a concept of systems analysis based on Boothroyd's formulation of it as a principal element in articulate intervention in action programs. Since this concept includes, as an essential ingredient, facts from science, it must have as one of the important elements of its foundation a mature and realistic concept of science; Section 12.2 described such a concept as developed by Ravetz. However, since the basic conception is that of intervention in action programs—not only those potentially entered but also entering programs—and all such programs contain people as essential participants, it is clear that any mature concept for systems analysis must also be rooted in a realistic and mature view of the appropriate human relationships involved. The next section addresses this subject.

12.4. Concepts of Professionalism for Analysis

The setting in which a systems analyst commonly works—as summarized in the Appendix to Chapter 1—is this: Someone with a suitable responsibility in an action program (for convenience called the client) asks the analyst to explore a problem situation looking toward the possibility that a problem identification will lead to an analysis, which may identify ways to ameliorate the problem. Thus, from the very beginning of his work, and throughout its duration, the systems analyst is involved with the client's action program; indeed, if the analyst follows the urgent and repeated advice of experienced systems analysts, he remains in close touch with the client and other persons related to the action program throughout the analysis and its follow-up activities, engaging their cooperation at many points in the work.

Since good relations between the analysis team and persons concerned with the action program are usually essential for effective systems-analysis work, it is desirable to have a mature and realistic concept to guide the analysis team's approach. The purpose of this section is to present the elements of such a conception. It is an interpretation for systems analysis of the work of Schön (1983); although he developed his ideas from quite a different perspective, they are also apt for our concerns.

Before coming to the ideas in Schön's conception of professionalism, it is desirable to look briefly at the social background and the inheritance from other science-based professions that systems analysts and their clients may be involved with. The next two subsections deal with these subjects.

The Social Background

Specialized knowledge and the crafts based on it have played important roles in the functioning of society throughout man's recorded history. In particular, the rise of modern science over the last four centuries has prompted the creation of professions that apply its knowledge to practical problems—and, indeed, the concept of professionalism has expanded to include bodies of practitioners applying other organized bodies of knowledge. The rise of professionalism in developed societies has been so rapid in the twentieth century that Schön (1983, pp. 3–4) can say, without fear of serious challenge:

> The professions have become essential to the very functioning of our society. We conduct society's principal business through professionals specially trained to carry out that business, whether it be making war and defending the nation, educating our children, diagnosing and curing disease, judging and punishing those who violate the law, settling dis-

putes, managing industry and business, designing and constructing build-
ings, helping those who for one reason or another are unable to fend for
themselves. Our principal formal institutions—schools, hospitals, government agencies, courts of law, armies—are arenas for the exercise of
professional activity. We look to professionals for the definition and
solution of our problems, and it is through them that we strive for social
progress. In all of these functions we honor what Everett Hughes [1959]
has called "the professions' claim to extraordinary knowledge in matters
of great social importance;" and, in return, we grant professionals extraordinary rights and privileges.

By the early 1960s professional involvement with the affairs of society
in developed countries was so widespread as to give the professional
community a strong sense of power and self-appreciation; in America a
1963 survey of the professional scene (Lynn, 1963) began with this statement: "Everywhere in American life, the professions are triumphant."

Since the early 1960s, however, the professions have faced a growing
crisis of confidence and legitimacy. Artifacts of science and technology
have exhibited unhappy properties about which their designers have not
forewarned the public: Large industrial and energy plants generate acid
rain at great distances; nuclear energy plants seem not as safe as the public
had thought them to be; safety in the technical workplace has become a
public issue. Professionally designed solutions to public problems have
had important unintended consequences, sometimes worse than the problems themselves. Widely publicized disagreements among professionals
on possible approaches to public problems have undermined confidence
in the knowledge bases for many of the advocated approaches. And
professional hubris has even led professionals to advocate courses of
action on public matters in which they are not expert in ways suggesting
that their views should be accorded more respect than those of nonprofessionals. And the much-admired successes in very large technical
programs, such as those of space exploration, are often contrasted with
the lamentable state of more mundane matters of more widespread importance, such as those of poverty, inadequate housing, and widespread
hunger.

To mention this crisis of confidence in professionalism is not to say
that it affects all professional work; rather, it is to suggest that systems
analysis works in a field affected by cross-currents. In spite of the evidence of crisis in the public's confidence in professionalism, it depends
on the skill and adaptability of professionals to a larger extent than ever
before—as the growth of systems analysis testifies.

This context in which today's systems analysts find themselves urges
not only that work in this field meet the highest standards possible but
also that it rest on a well-developed sense of the proper relations that
they should have with their clients.

The Inheritance from Science-Based Professions

The science-based professions have a well-known history of usefulness that is too long to be summarized here. Some of the current problems facing these professions need to be understood, however, as a background for the ensuing discussion of the professional's relations with the person he serves.

To begin, we may say, with Brooks (1967, p. 89), that

> the professional is much more than a man with knowledge, he is the middleman or intermediary between a body of knowledge and society. To the professional belongs the responsibility of using both existing and new knowledge to provide services that society wants and needs. This is an art because it demands action as well as thought, and action must always be taken on the basis of incomplete knowledge.

A professional can approach this responsibility in two contrasting ways: He can stay very close to his area of knowledge and restrict his activity to problems that fit this knowlege, or he can occupy a position much closer to society, where goal setting and problem formulation are usually initial difficulties, and where existing knowledge may have to be stretched much farther, or possibly supplemented by new research.

The first position allows the professional to devote his time and effort to applying rigorously tested and widely accepted scientific knowledge. Thus, from this position, identifying the problem means identifying in the problem situation a difficulty to which the available knowledge is relevant, and then applying it rigorously in accordance with accepted and widely sanctioned practices. Unfortunately, this narrow view also has very little place in it for larger issues of goals and welfare within the problem situation.

On the other hand, the much wider view of professional activity as taking place closer to the society being served starts with problem setting: considering various and disparate possible goals, a wide variety of means that might be adapted to achieving them, the decisions that need to be made along the way, and the spectrum of outcomes that may ensue.

The first of these positions emphasizes solving clearly identified problems, the second starts with problem setting as an essential prelude to focusing knowledge on solving the identified problem, with considerable adaptation of the knowledge usually being needed. To many, the first approach is rigorous but may not be as relevant as it should be, the second relevant but usually not so rigorous. Hence, some observers speak of the "dilemma of rigor or relevance," and it is a lively issue in many science-based professions—notably in operations research.

The first position—which has wide dominance in the science-based professions and in education for them—rests on an outdated philosophy in which the dominant epistemology of practice is what Schön (1983, pp. 21, 31) calls "Technical Rationality:"

> According to the model of Technical Rationality—the view of professional knowledge which has most powerfully shaped both our thinking about professions and the institutional relations of research, education, and practice—professional activity consists in instrumental problem solving made rigorous by the application of scientific theory and technique.
>
> . . . Technical Rationality is the heritage of Positivism, the powerful philosophical doctrine that grew up in the nineteenth century as an account of the rise of science and technology and as a social movement aimed at applying the achievements of science and technology to the well-being of mankind. Technical Rationality is the Positivist epistemology of practice. It became institutionalized in the modern university, founded in the late nineteenth century when Positivism was at its height, and in the professional schools which secured their place in the university in the early decades of the twentieth century.

The professional epistemology of technical rationality presents professional practitioners—particularly systems analysts—with difficulties. First, technical rationality presumes that the sole basis for problem solving is the knowledge of science, and that therefore a professional practitioner must be able to map a problem he faces onto this knowledge. The result is to make the arts of practice—particularly those relating to problem setting—into puzzling anomalies. Although admitting that practical, craft knowledge may exist, the positivist cannot fit it nearly into positivist categories. Thus the arts of professional craft do not have natural places in technical rationality.

Schön (1983, pp. 39–42) describes additional difficulties:

> From the perspective of Technical Rationality, professional practice is a process of problem *solving*. Problems of choice or decision are solved through the selection, from available means, of the one best suited to established ends. But with this emphasis on problem solving, we ignore problem *setting*, the process by which we define the decision to be made, the ends to be achieved, and the means which may be chosen. In real-world practice, problems do not present themselves to the practitioner as givens. They must be constructed from the materials of problematic situations which are puzzling, troubling, and uncertain. In order to convert a problematic situation to a problem, a practitioner must do a certain kind of work. He must make sense of an uncertain situation that initially makes no sense. When professionals consider what road to build, for example, they deal usually with a complex and ill-defined situation in which geographic, topological, financial, economic, and political issues are all mixed up together. Once they have somehow decided what road to build and go on to consider how best to build it, they may have a problem they can solve by the application of available techniques; but

when the road they have built leads unexpectedly to the destruction of a neighborhood, they may find themselves again in a situation of uncertainty.

It is this sort of situation that professionals are coming increasingly to see as central to their practice. They are coming to recognize that, although problem setting is a necessary condition for technical problem solving, it is not itself a technical problem. When we set the problem, we select what we will treat as the "things" of the situation, we set the boundaries of our attention to it, and we impose upon it a coherence which allows us to say what is wrong and in what directions the situation needs to be changed. Problem setting is a process in which, interactively, we *name* the things to which we will attend and *frame* the context in which we will attend to them.

Even when a problem has been constructed, it may escape the categories of applied science because it presents itself as unique or unstable. In order to solve a problem by the application of existing theory or technique, a practitioner must be able to map those categories onto features of the practice situation. . . .

Technical Rationality depends on agreement about ends. When ends are fixed and clear, then the decision to act can present itself as an instrumental problem. But when ends are confused and conflicting, there is as yet no "problem" to solve. A conflict of ends cannot be resolved by the use of techniques derived from applied research. It is rather through the nontechnical process of framing the problematic situation that we may organize and clarify both the ends to be achieved and the possible means of achieving them. . . .

We can readily understand, therefore, not only why uncertainty, uniqueness, instability, and value conflict are so troublesome to the Positivist epistemology of practice, but also why practitioners bound by this epistemology find themselves caught in a dilemma. Their definition of rigorous professional knowledge excludes phenomena they have learned to see as central to their practice. And artistic ways of coping with these phenomena do not qualify, for them, as rigorous professional knowledge. [Emphasis in the original.]

Finally, technical rationality poses the dilemma of rigor or relevance: Should the practitioner stay close to the scientific knowledge of his profession or move closer to the reality that contains so many elements not treated by his stock of scientific knowledge? Schön (1983, p. 42) poses the dilemma in this way:

In the varied topography of professional practice, there is a high, hard ground where practitioners can make effective use of research-based theory and technique, and there is a swampy lowland where situations are confusing "messes" incapable of technical solution. The difficulty is that the problems of the high ground, however great their technical interest, are often relatively unimportant to clients or to the larger society, while in the swamp are the problems of greatest human concern.

> Shall the practitioner stay on the high, hard ground where he can practice rigorously, as he understands rigor, but where he is constrained to deal with problems of relatively little social importance? Or shall he descend to the swamp where he can engage the most important and challenging problems if he is willing to forsake technical rigor?

Computer modeling in operations research and management science offers an example of overemphasis on rigor at the expense of relevance. Early models of modest size have been used successfully as tools in solving many operating problems, a fact that has led academic courses in these fields to give such models a very great emphasis—so much so that many regard the fields as devoted almost entirely to modeling. These successes have led to the creation of much larger computer models, which, however, have been much less satisfactory in yielding results about large-scale operating and policy problems, where the surrounding situation is much more complex. However, the strong impulse—particularly in academia—behind the positivist ideal of rigor has given the construction of such large models a life of its own, increasingly divergent from real-world problems of practice, that has found its way into research institutes as well. Meadows and Robinson (1985) review an important segment of this field, and provide a trenchant critique of it; similarly, Meadows, Richardson, and Bruckmann (1982) offer a critical overview of the field of global modeling.

This discussion is not intended to suggest that the working professional should reject the benefits of rigorous analysis and careful thought about the problems he faces. Rather, its intent is to suggest that analysts should move toward the goal of relevance, thus moving toward the "swamp" wherein lie the problems of major social importance. How is the professional to do this? Can he carry along ideals of careful and rigorous thought, appropriately adjusted to the newer context? Or must he learn an entirely new set of approaches? The ensuing subsections address these questions.

The Concept of Reflection-in-Action

Although technical rationality is a strong force in the training and practice of the science-based professions, it does not follow that all practitioners hew exclusively to practices expressive of its tenets; many have moved successfully into the swamp of relevance. Thus, Schön suggests that we look at what these latter practitioners do: "Let us search . . . for an epistemology of practice implicit in the artistic, intuitive processes which some practitioners . . . bring to situations of uncertainty, instability, uniqueness, and value conflict" (loc. cit., p. 49). In sum, let us examine the craft practices of successful practitioners to see what principles may be inferred from them.

The examination begins with the observation that practitioners with substantial experience in actual problem situations develop patterns of response that remain consistent but with variations suited to each case, these responses being a mixture of conscious and unconscious elements. These elements respond to scientific knowledge, craft practices learned from master practitioners in apprenticeship experiences, and other elements emerging from personal experience. Schön (1983) supports these statements with detailed examinations of professional behavior in a variety of contexts and involving a number of quite disparate professions. His principal conclusion is this (loc. cit., p. 51):

> Although we sometimes think before acting, it is also true that in much of the spontaneous behavior of skillful practice we reveal a kind of knowing which does not stem from a prior intellectual operation. . . . Over the years, several writers on the epistemology of practice have been struck by the fact that skillful action often reveals a "knowing more than we can say."

This conclusion is the basis for the first step in developing a realistic epistemology of practice: Reflect explicitly on—and then record—what is done in the interactions of practice and what the bases are for it. Schön calls this process reflection-in-action.

> When a practitioner reflects in and on his practice, the possible objects of his reflection are as varied as the kinds of phenomena before him and the systems of knowing-in-practice which he brings to them. He may reflect on the tacit norms and appreciations which underlie a judgment, or on the strategies and theories implicit in a pattern of behavior. He may reflect on the feeling for a situation which had led him to adopt a particular course of action, on the way in which he has framed the problem he is trying to solve, or on the role he has constructed for himself within a larger institutional context.
>
> Reflection-in-action, in these several modes, is central to the art [or craft] through which practitioners sometimes cope with the troublesome "divergent" situations of practice.
>
> When the phenomenon at hand eludes the ordinary categories of knowledge-in-practice, presenting itself as unique or unstable, the practitioner may surface and criticize his initial understanding of the phenomenon, construct a new description of it, and test the new description by an on-the-spot experiment. Sometimes he arrives at a new theory of the phenomenon by articulating a feeling he has about it.
>
> When he finds himself stuck in a problematic situation which he cannot readily convert to a manageable problem, he may construct a new way of setting the problem—a new frame which, in what I shall call a "frame experiment," he tries to impose on the situation.
>
> When he is confronted with demands that seem incompatible or inconsistent, he may respond by reflecting on the appreciations which he and others have brought to the situation. Conscious of a dilemma, he may attribute it to the way in which he has set his problem, or even to

the way in which he has framed his role. He may then find a way of integrating, or choosing among, the values at stake in the situation. [Schön, 1983, pp. 62–63.]

Finally, Schön (1983, pp. 68–69) gives us this view of how reflection-in-action can lead to an epistemology of practice that can dissolve the dilemma of rigor or relevance:

> When someone reflects-in-action, he becomes a researcher in the practice context. He is not dependent on the categories of established theory and technique, but constructs a new theory of the unique case. His inquiry is not limited to a deliberation about means which depends on a prior agreement about ends. He does not keep means and ends separate, but defines them interactively as he frames a problematic situation. He does not separate thinking from doing, ratiocinating his way to a decision which he must later convert to action. Because his experimenting is a kind of action, implementation is built into his inquiry. Thus reflection-in-action can proceed, even in situations of uncertainty or uniqueness, because it is not bound by the dichotomies of Technical Rationality.
>
> Although reflection-in-action is an extraordinary process, it is not a rare event. Indeed, for some reflective practitioners it is the core of practice. Nevertheless, because professionalism is still mainly identified with technical expertise, reflection-in-action is not generally accepted— even by those who do it—as a legitimate form of professional knowing.
>
> Many practitioners, locked into a view of themselves as technical experts, find nothing in the world of practice to occasion reflection. They have become too skillful at techniques of selective inattention, junk categories, and situational control techniques which they use to preserve the constancy of their knowledge-in-practice. For them, uncertainty is a threat; its admission a sign of weakness. Others, more inclined toward and adept at reflection-in-action, nevertheless feel profoundly uneasy because they cannot say what they know how to do, cannot justify its quality or rigor.
>
> For these reasons, the study of reflection-in-action is critically important. The dilemma of rigor or relevance may be dissolved if we can develop an epistemology of practice which places technical problem solving within a broader context of reflective inquiry, shows how reflection-in-action may be rigorous in its own right, and links the art of practice in uncertainty and uniqueness to the scientist's art of research. We may thereby increase the legitimacy of reflection-in-action and encourage its broader, deeper, and more rigorous use.

The Concept of Reflective Practice

Although the outlines of an epistemology of practice based on reflection-in-action have only been suggested in a very general way in what has been said so far, the concept of reflective practice can be seen to have important implications for the relations between a practitioner and his client.

These implications are thrown into relief when contrasted with the professional-client relationship that is the classical heritage of technical rationality:

> In the traditional professional-client contract, the professional acts as though he agreed to deliver his services to the client to the limits of his special competence, to respect the confidences granted him, and not to misuse for his own benefit the special powers given him within the boundaries of the relationship. The client acts as though he agreed, in turn, to accept the professional's authority in his special field, to submit to the professional's ministrations, and to pay for services rendered. In a familiar psychological extension of the informal contract, the client agrees to show deference to the professional. He agrees not to challenge the professional's judgment or to demand explanations beyond the professional's willingness to give them. In short, he agrees to behave as though he respected the professional's autonomy as an expert. [Schön, 1983, p. 292.]

On the other hand, the concept of reflective practice requires that the client be reflective in a sense that is complementary to the way the professional is reflective, the difference being that the client reflects on the material of his everyday activities, the professional on his experience and knowledge as they relate to the client's difficulties. Each then contributes to a new form of partnership that is in marked contrast with the subservience of client to practitioner that is the hallmark of the traditional relationship; Schön calls this new relationship a reflective contract.

> Just as reflective practice takes the form of a reflective conversation with the situation, so the reflective practitioner's relation with his client takes the form of a literally reflective conversation. Here the professional recognizes that his professional expertise is embedded in a context of meanings. He attributes to his clients, as well as to himself, a capacity to mean, know, and plan. He recognizes that his actions may have different meanings for his client than he intends them to have, and he gives himself the task of discovering what these are. He recognizes an obligation to make his own understandings accessible to his client, which means that he needs often to reflect anew on what he knows. . . . The reflective practitioner tries to discover the limits of his expertise through reflective conversation with his client.

> Although the reflective practitioner should be credentialled and technically competent, his claim to authority is substantially based on his ability to manifest his special knowledge in his interactions with his clients. He does not ask the client to have blind faith in a "black box," but to remain open to the evidence of the practitioner's competence as it emerges. For this relationship to work, however, serious impediments must be overcome. Both client and professional bring to their encounter

a body of understandings which they can only very partially communicate to one another and much of which they cannot describe to themselves. Hence the process of communication which is supposed to lead
to a fuller grasp of one another's meanings and, on the client's part, to
an acceptance of the manifest evidence of the professional's authority
can only begin with nonunderstanding and nonacceptance—but with a
willing suspension of disbelief.

Thus, in a reflective contract between practitioner and client, the client
does not agree to accept the practitioner's authority but to suspend disbelief in it. He agrees to join the practitioner in inquiring into the situation
for which the client seeks help; to try to understand what he is experiencing and to make that understanding accessible to the practitioner;
to confront the practitioner when he does not understand or agree; to
test the practitioner's competence by observing his effectiveness and to
make public his questions over what should be counted as effectiveness;
to pay for services rendered and to appreciate competence demonstrated. The practitioner agrees to deliver competent performance to the
limits of his capacity; to help the client understand the meaning of the
professional's advice and the rationale for his actions, while at the same
time he tries to learn the meanings his actions have for his client; to make
himself readily confrontable by his client; and to reflect on his own tacit
understandings when he needs to do so in order to play his part in fulfilling the contract. [Schön, 1983, pp. 295–297.]

Schön does not advance this concept of the reflective contract as the
universal goal for all professional work, since there are obvious varieties
of situations where more limited relationships may be desired by the parties involved—and to establish and work under a reflective contract is
expensive in time and resources. Nevertheless, for broad contexts in
which the client is a large organization, the problem situation involves
many conflicting interests, and whatever response to the problem situation is adopted will have far-reaching effects, the reflective contract offers
many advantages, especially if there is a need to reconcile adversarial
conflicts about what should be done.

There seems . . . to be some growing recognition of the need for cooperative inquiry within adversarial contexts. The idea of reflective practice leads to a vision of professionals as agents of society's reflective
conversation with its situation, agents who engage in cooperative inquiry
within a framework of institutionalized contention. The question remains, however, whether it is utopian, in the pejorative sense, to suppose
that professionals who occupy key roles in the public policy process can
learn on a broad basis to engage in reciprocal reflection-in-action. . . .
What kind of a question are we posing when we ask whether such a
vision is merely utopian? The existence of a widespread capacity for
reciprocal reflection-in-action is unlikely to be discovered by an ordinary
social science which tends to detect, and treat as reality, the patterns
of institutionalized contention and limited learning which individuals

transcend, if at all, only on rare occasions. The extent of our capacity for reciprocal reflection-in-action can be discovered only through an action science which seeks to make what some of us do on rare occasions into a dominant pattern of practice. [Schön, 1983, pp. 353–354.]

The Position of Systems Analysis

Historically, systems analyses have occupied many positions on the line between the classical professional-client relation of client subservience and the full realization of a reflective contract; and, no doubt, this will continue to be the case for some time to come, depending on the circumstances of each study. The examples described in Section 2.2 support this point. Nevertheless, the emphasis on craft—and particularly on the importance of client participation in the work—in the *Overview* and throughout this volume support the contention that the profession of systems analysis has much to gain from working with its clients in accord with the reflective-contract concept. To make this point, it is worth taking up several steps in the systems-analysis process in its light.

Figure 1.1 shows the principal activities in the systems-analysis process, and Table 1.1 lists where each is discussed in the *Overview*; the reader may want refer to them as he reads the brief discussions of some of the key activities viewed from the perspective of the reflective-contract concept. Too, the case descriptions in Sections 2.2, 9.6, and 10.4 can usefully be reviewed in the light of this discussion.

Formulating the problem. The process of formulating the problem begins when a client asks an analyst to explore a problem situation. Since the situation is usually new to the analyst, the client must assume the burden of making everything available to the analyst that might possibly shed light on the situation's difficulties, and how solvable problems might be formulated as a step toward ameliorating the difficulties. Although much of the work of investigating the situation must be done by the analysis team, the burden on the client and his staff to cooperate in this work is usually substantial—and essential if the analysis team is to gain a true picture of the situation. Chapter 5 in the *Overview* details a generic approach to the work of problem definition. Checkland (1981) describes a successful approach that has been used in problem situations dominated more by human relationships than by technology, but it contains ideas and approaches that could well be considered by analysts facing problem situations of the type envisaged in this handbook. Finally, the problem formulation adopted as the basis for the analysis work must be adopted by the client as well as the analysis team, if the later work and its findings are to gain acceptance. One of the best ways to be sure that the client claims ownership for the problem formulation is to have his active participation in the work of deriving it. All of these arguments urge the relevance of the reflective-contract concept for this stage of the analysis process.

Formulating alternatives. Since the client's problem situation and its technology are seldom likely to be matters of the analysis team's intimate acquaintance, the team is not likely to be the best and most imaginative source of possible alternatives. This is not to say that the team's analysis will not have a great deal to do with shaping the alternatives after they have been conceived; rather, it is to say that intimate knowledge of the situation and the relevant technology, aided by a lively and creative imagination, is more likely to generate alternative ideas. The client and his staff are one good source of such ideas that should be explored, along with others (see Section 6.2). If an alternative proposed by the client and his staff emerges from the analysis as preferred, then—human nature being what it is—it has an advantage in gaining acceptance that an idea from an outside source does not have.

Forecasting future contexts. Although much of the context surrounding a problem situation may lie outside the direct purview of the client, and therefore needs to be marshaled by the analysis team, much of the future within the client's purview is known to him, or is at least subject to his estimates, however crude. Therefore, his knowledge of this future must be elicited from him and factored into the larger contextual picture.

Comparing and ranking alternatives. In early operations-research and systems-analysis work, this step in the process was often carried out by the analysts. As problems became more complex, and particularly as indices of impacts became more numerous, and usually incommensurable, it became necessary to bring the client's knowledge and intuition into this process. Evaluating the three alternatives for protecting the Oosterschelde estuary from flooding, a case in which dozens of potential impacts had to be considered, is an outstanding example of the successful use of the client as an evaluator of the relative desirabilities of alternatives (see Section 2.2 and the *Overview*, Section 3.4).

Communicating results. Unless this essential step is carried out effectively, everything that had been done before has been wasted; thus, the analysis team must plan for this phase of the analysis carefully and carry it out with meticulous care (see Sections 2.7 and 13.6, and Chapter 14). Most of the burden falls on the analysts, of course, but participation of the client and members of his staff can be of great help in choosing the communication approaches and instruments, and what their content should be. Too, in a large organization, a simple communication by a report or briefing is likely to have to be supplemented with many additional briefings and consultations; it is highly desirable for the client—or at least members of his staff—to take a significant part in this work, especially in helping to develop how best to persuade others who may have a say—or a veto—on what is to be done.

Implementation. The burden of this step in the process lies with the client and other implementors (Chapter 14), but experience shows that, if it is not to go awry, the continuing help of the analysis team is needed. Thus, the partnership initiated with the exploration of the problem situation needs to continue through the implementation process, although the scale of analysis involvement may be considerably reduced.

The relevance of the reflective-contract concept. These brief reminders of some of the craft practices in systems analysis that have been advocated in this handbook serve to underscore the relevance of the reflective-contract concept as the basis for the relationship between analyst and client. The precise form that this relationship should take will vary from case to case, but its principles are likely to be helpful in formulating the details. Similarly, case-by-case analysis will serve to flesh out the form that the reflective-contract concept should take for systems analysis.

This case-by-case analysis, however, must recognize that, for systems-analysis work to lead to successful outcomes, the client must make important contributions throughout, as the preceding paragraphs suggest. Thus, although Schön's reflective-contract concept was developed on the basis of his observations of classic professions such as law, medicine, engineering, and architecture, the setting in which systems-analysis work typically takes place makes it even more important. Indeed, without the appropriate participation of the client, the systems-analysis work has little chance of being useful. In sum, the authoritarian postures based on technical rationality of many practitioners in the classic professions offer an inappropriate analogy for systems analysis.

12.5. Systems Analysis: Mature or Immature?

In his essay on the history of systems analysis (*Overview*, pp. 33–66), Majone describes it as a field striving to emerge from an immature past into a stage of early maturity, and he points out that adequate and widely accepted criteria of quality and effectiveness will be the necessary hallmarks of a state of maturity. There is a sense in which this handbook works in a variety of ways toward the goals of quality and effectiveness. Nevertheless, it is worth saying a bit more about the matter in general.

In his discussion of mature and immature branches of science, Ravetz (1971, pp. 364–402) makes a number of points that can be extended to systems analysis, even though only a portion of its work can be thought of as following scientific patterns. Four of these points will be mentioned here.

The establishment of facts. Mature sciences have a body of established facts, where this term has Ravetz's meaning (as described in Section 12.2). By analogy, systems-analysis work should establish a factual basis for its descriptive work (see Section 13.6) that is validated when analysis findings make their way into implementation. They may be somewhat universal facts drawn from established sciences, or they may be facts primarily associated with the context of the analysis, but they must be adequate descriptors of the relevant aspects to the phenomena of the problem situation if the findings of the work are to find their way to successful application. A continuing stream of analysis and application offers a particularly strong test of such facts; the fire department deployment work described in Walker, Chaiken, and Ignall (1979) arrived at such a body of underlying facts, as did the seven-year program of analysis of university operations reported by Hopkins and Massy (1981).

The recognition and avoidance of pitfalls. Majone and Quade (1980) offer evidence that systems analysis has identified a great many pitfalls— and stumbled at one time or another into most of them; but it also shows that they have been recognized and are avoided today by skilled analysts. Section 15.7 discusses pitfalls and provides a short list.

The control of quality. To control the quality of systems-analysis work demands that criteria must exist to act as guides for workers as they proceed, as well as a basis for judging findings and their implications. There is a sense in which much of this handbook addresses this issue, at least indirectly, but Chapter 15 focuses its entire attention on it.

The effectiveness of the work. Systems analyses may be judged to be effective on a variety of grounds: if the work helps the parties at interest understand the problem situation better than before, if decisionmakers factor its findings into their considerations of possible courses of action, if implemented programs bear out the validity of the findings, and so on. In general, the effectiveness of systems analysis may be judged in terms of its adequacy in meeting its goals. (See Section 14.4.)

A net judgment. It would be imprudent here to make more than a transitory judgment of the maturity or immaturity of systems analysis. On the one hand, there is much evidence of a movement toward maturity (a record of effective work, consciousness of pitfalls, evidence of widespread recognition of patterns of good work, a recognition of a need for standards of quality in both product and process), but on the other there is evidence of immaturity (analysts still stumble into pitfalls, standards of quality are not yet as widely accepted as they should be in a mature discipline, and there is as yet no widely accepted conceptual basis for the work). Thus, an observer must make his own judgment of where the field is on this ladder from immaturity to maturity—and one can hope that the progress in the future will be upward.

Ravetz (1971, pp. 400–402) offers an instructive view of the dangers, challenges, and rewards of working in a socially relevant scientific field that the reader can easily extend to systems analysis:

> It is only to be expected that the application of scientific inquiry to new practical problems should be even more hazardous than the management of deeply novel results within science itself. To the extent that the investigation of problems loses its protective framework of accepted and successful methods, it becomes exposed to pitfalls of every sort. On being associated with an influential field, an immature field, in chaos internally, experiences the additional strains of hypertrophy, and its leaders and practitioners are exposed to the temptations of being accepted as consultants and experts for the rapid solution of urgent practical problems. The field can soon become identical in outward appearance to an established physical technology, but in reality be a gigantic confidence-game, combining the worst features of entrepreneurial and shoddy science. The dangers of such corruption are at present more acute for some of the social sciences and technologies (especially those using mathematical and computational tools) than for the natural sciences, since they are related to the most urgent practical problems and they lack a base of fully matured disciplines; pseudo-research is one of the symptoms of the diseased state, and sophisticated criteria of adequacy of results (as 'falsifiability of theories') are irrelevant to situations where conflicting purposes and ideologies are central to the problems, and the discipline's function as a folk-science cannot be eliminated. To thread one's way through these pitfalls, making a genuine contribution both to scientific knowledge and to the welfare of society, requires a combination of knowledge and understanding in so many different areas of experience, that its only correct title is wisdom.
>
>   . . . To be involved in a field just entering maturity is the most rewarding career for a scientist; for then one can make great achievements at relatively little risk. But estimating the points of transition between phases is a very delicate task; a field or area of science which is approaching senescence is a dreary place; and immature fields with the hope of imminent maturation are, with all their attendant hazards, the place where the greatest challenge is found.

12.6. Conclusion

Normally one expects a handbook to present summaries of bodies of knowledge and concepts of craftmanship that have achieved reasonably standard form and wide acceptance in its professional community—and this handbook does this in some areas. But since the systems-analysis profession does not yet have a well-developed and widely accepted set of underlying concepts for its work, an authoritative chapter presenting such concepts cannot be written. Nevertheless, this subject is so vitally important to the future of systems analysis that it should not be overlooked in this handbook. Therefore, this chapter has taken a step in the desired direction: It has summarized three well-developed streams of thought that I feel will have important influences on the conceptual basis that systems analysis will eventually develop. The first of these is Ravetz's developmental view of the growth of scientific knowledge, the second is Boothroyd's description of operations research as articulate intervention into action programs extended to systems analysis, and the third is Schön's view of the relations between professionals and their clients and how they should be understood systematically. These elements—plus others that emerge as important—need to be combined into a synthesis focused on systems analysis, a task that urgently needs to be pushed forward.

To review what has been achieved, and to project what the systems-analysis community still needs to attend to, it is useful to consider the three principal categories of elements involved in systems-analysis work, as represented by the circles in Figure 12.1, and the three principal in-

Figure 12.1. The three principal categories of elements involved in systems analysis work. The circles represent the categories and the lines represent the principal interactions between the pairs.

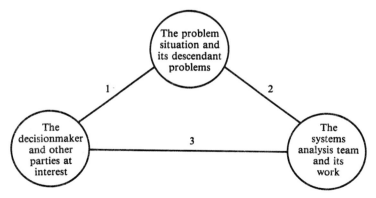

teractions among these categories, represented by the lines in the figure. Most of the rest of this handbook concentrates its attention on the work of the systems-analysis team (that is, the right-hand circle in the figure), with some attention to how it emerges from the problem situation (interaction 2) and relates to the decisionmaker and other parties at interest (interaction 3). As related to this framework, this chapter's contributions are these:

The discussion of science and how its facts and methods enter systems-analysis processes relates to the work and how the team should regard its results (the right-hand circle).

The discussion of action programs presents a view of the sort of setting in which the problem situations occur that systems analysis deals with (the top circle), and into which it intervenes (interaction 2).

The discussion of professionalism deals with the relations between systems analysts and their clients (interaction 3).

With respect to matters not covered by this handbook, it is useful to make the following observations: The handbook does not cover the behavior of decisionmakers and other parties at interest in the problem situation (the left-hand circle), or the interactions between these persons and the problem situation (interaction 1), partly because other specialties and their literature deal with them (see, for example, Drucker, 1974). Although the handbook offers some insights into the sorts of problem situations dealt with by systems analysts (the top circle) by describing some examples, it does not present any sort of descriptive categorizations. As the fundamental concepts underlying systems analysis develop, and as the structure built on them grows, we can expect appropriately standardized and realistic descriptions to be extended to include these matters.

It is hoped that this chapter, by presenting some thought-provoking candidate concepts, will serve to attract attention to the need for a conceptual synthesis on which to found systems-analysis work, and will persuade workers in this field to develop it.

References

Boothroyd, Hylton (1978). *Articulate Intervention.* London: Taylor and Francis.

Brooks, Harvey (1967). Dilemmas of engineering education. *IEEE Spectrum* 4, 89–91.

Checkland, Peter (1981). *Systems Thinking, Systems Practice.* Chichester, England: Wiley.

Drucker, P. F. (1974). *Management: Tasks, Responsibilities, Practices.* New York: Harper and Row.

Hoag, M. W. (1957). What is a system? *Operations Research 5*, 445–447.

Hopkins, D. S. P., and W. F. Massy (1981). *Planning Models for Colleges and Universities*. Stanford, California: Stanford University Press.

Hughes, Everett (1959). The study of occupations. In *Sociology Today* (R. K. Merton, L. Brown, and L. S. Cottrell, Jr., eds.). New York: Basic Books.

Jackson, M. C., and Paul Keys, eds. (1987). *New Directions in Management Science*. Aldershot, Hampshire, England: Gower.

Kemeny, J. G. (1959). *A Philosopher Looks at Science*. New York: Van Nostrand Reinhold.

Lakatos, Imre (1978). *The Methodology of Scientific Research Programs*. Volume I of *Philosophical Papers* (J. Worrall and G. Currie, eds.). Cambridge, England: Cambridge University Press.

Lynn, Kenneth (1963). Introduction. *Daedalus 92*, 649.

Majone, G., and E. S. Quade (1980). *Pitfalls of Analysis*. Chichester, England: Wiley.

Meadows, Donella, J. Richardson, and G. Bruckmann (1982). *Groping in the Dark: The First Decade of Global Modeling*. Chichester, England: Wiley.

———, and J. M. Robinson (1985). *The Electronic Oracle: Computer Models and Social Decisions*. Chichester, England: Wiley.

Miser, H. J., and E. S. Quade, eds. (1985). *Handbook of Systems Analysis: Overview of Uses, Procedures, Applications, and Practice*. New York: North-Holland. Cited in the text as the *Overview*.

Overview. See Miser and Quade (1985).

Ravetz, J. R. (1971). *Scientific Knowledge and its Social Problems*. Oxford, England: Oxford University Press.

Schön, D. A. (1983). *The Reflective Practitioner: How Professionals Think in Action*. New York: Basic Books.

Walker, W. E., J. M. Chaiken, and E. J. Ignall, eds. (1979). *Fire Department Deployment Analysis: A Public Policy Analysis Case Study*. New York: North-Holland.

Chapter 13
Validation

Hugh J. Miser and Edward S. Quade

13.1. Introduction

Models are used throughout the systems- and policy-analysis process; they range from the mental models and assumptions that analysts and their clients use to diagnose and frame the problem—and to guide its solution—to the formal mathematical and computer models that the analysts often use to estimate what would ensue from adopting various alternatives (see Figure 1.1, Chapter 2, and Chapter 8). In fact, it can be argued that the logical structure created by the analysis process is itself a model of the features of essential interest in the situation that it purports to describe.

Since the findings emerging from a systems- or policy-analysis process are intended to improve the understanding of the problem and the situation surrounding it—and are often used as guides for decision and action—it is important for the models used in this process, and the logical structure that it creates, to represent the relevant phenomena of reality closely enough to serve these purposes adequately.

It is intrinsic that any model is an intellectual construct that represents only a portion of reality, and represents it only partially (Chapter 8 and Section 12.2). Therefore, an analyst must judge whether or not this representation is adequate for his purposes, a judgment based on assessing two relations: the relation between the model and the relevant phenomena of reality and the relation between the model and the purposes for which it is being used. In sum, for systems and policy analysis a judgment of model adequacy is aimed at answering this question: Are the model results close enough to the outcomes of the phenomena of reality being modeled to improve understanding, and, when aided by judgment, to support decisions about policies and actions?

To reach this standard of adequacy, analysts use two processes, validation and verification (the latter actually a step in the validation process, but worth distinguishing here); the purpose of this chapter is to discuss them and how they are used in systems and policy analysis. The next section defines these two processes and reminds the reader briefly of the setting in which they take place. The third section takes up model verification, explaining it in detail and describing the approaches to carrying it out. The fourth section deals with some of the problems and difficulties associated with validation, the fifth discusses it on the basis of fundamental concepts of science and systems analysis, and the sixth extends the concept of validation to analyses as a whole. The seventh section discusses various validation approaches and procedures. A final section sums up the importance of verification and validation in the analysis of real-world problems.

Although all of the models used in assembling a systems- or policy-analysis study need to be verified and validated, the most complicated ones—and therefore the ones most in need of such attention—are usually those used in connection with the central process of predicting the consequences of various alternatives (see Figure 1.1). Therefore, this chapter focuses its attention on these models, it being understood that the discussion applies, with appropriate modifications, to the other models as well. Unless otherwise noted, when the rest of the chapter speaks of systems analysis, this term—as in the rest of this handbook—is understood to represent policy analysis, as well as operations research, management science, and other disciplines that employ models in a similar way.

The terms *verification* and *validation* do not have clear and standard meanings in the literature, and indeed, they are sometimes used in overlapping and almost interchangeable meanings. Nevertheless, in order to sharpen the discussions, this chapter adopts as standard terminology what we believe is by far the most common usage. Although it may differ from some of the uses in the literature, the contexts and discussions elsewhere usually make what is intended clear enough to enable one to translate what is said here into these other contexts.

13.2. The Setting

As this chapter uses them, the terms *verification* and *validation* have the following meanings:

Verification is the process by which the analyst assures himself and others that the actual model that has been constructed is indeed the one he intended to build.

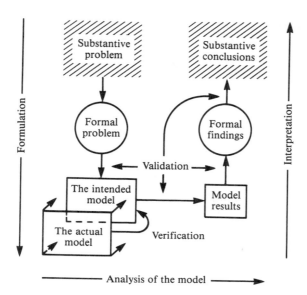

Figure 13.1. The positions of verification and validation with respect to the central model in applying analysis to practical problems. (*Adapted from Figure 8.4, which is based on Strauch, 1974, p. 14.*)

Validation is the process by which the analyst assures himself and others that a model is a representation of the phenomena being modeled that is *adequate for the purposes* of the study of which it is a part.

Systems analysis begins with a problem situation, and the early work is aimed at identifying a substantive problem in it that can be formalized, and thus made the basis for a study. The phenomena associated with the formal problem are then translated into an intellectual construct consisting of assumptions, variables, parameters, submodels, and so on that is sometimes called the central model, or more often, just the model. The problem-solving analysis then deals with this intellectual construct, or model, with its formal findings translated back into substantive conclusions by means of interpretations that take account of the original translation of the substantive problem into a formal problem and then its associated phenomena into an intellectual construct.

Figure 13.1 shows a somewhat simplified view of verification and validation as part of what in actuality is a rather circuitous process containing many feedbacks; Chapters 2, 8, and 12 discuss these matters, except for verification and validation, in detail.

This chapter is also concerned with extending the process of validation to the entire analytic structure, as well as with validating the models contained in it; Chapter 15 shares these concerns from the point of view of quality control. Validating the analytic process as a whole is particularly difficult, because the links, as Figure 13.1 shows, between the substantive problem and the formal problem, and among the model results, formal findings, and substantive conclusions depend heavily on judgments by the analysts and others.

Any realistic analysis makes use, of course, of many models, and each must undergo an appropriate set of verification and validation tests. Since each is involved in a situation at least roughly analogous to that of the central model, as shown in Figure 13.1 and discussed here, it is sufficient to deal with only the central case, leaving the others to the reader.

13.3. Verification

Verification is the process of testing the extent to which a model "has been faithful to its conception, whether or not it and its conception are valid" (Greenberger, Crenson, and Crissey, 1976, p. 70).

Regardless of whether they are simple or complex, the careful analyst verifies all of the models he uses. The verification procedures depend on the nature of the model involved, and range from very simple and intuitive ones for simple models to very complicated and subtle procedures for large mathematical and computer models. Since the analyst uses his own scientific knowledge—or invokes the specialized expertise of his colleagues—for simple models drawn from established fields, little needs to be said about this case. However, since the models used in systems and related analyses are constructed to represent real-world situations and are therefore often extremely large and complicated, how they are to be verified is an important issue.

There is no standard procedure for verifying models; rather, the analyst must bring to bear on them such knowledge and processes as are available in the related scientific and technical specialties. Nevertheless, there are some commonly used practices that may be mentioned: the first and most obvious way to ensure that a model can later be verified is to check each step carefully as the model is constructed to verify its logic and relevance; during the construction process, it may be possible to carry out some of the later steps listed here on the partially constructed model; when the model is complete, it should be checked carefully, a step best done by someone who did not participate in its original construction, and who is therefore in a good position to identify hidden assumptions or difficulties that the original builder may have overlooked. The completed model can be tested in such partial ways as checking what happens on the boundaries or for values of the variables where the correct results are known, considering degenerate cases where the formulas for such cases are known, or comparing results from the model with ones from other models in which there is high confidence. None of these procedures is adequate taken alone, but, if all of them—together with others adapted to the verification process in hand—yield results established as correct, or what the analyst intended, he may have a high confidence that his model has been verified.

Simple mathematical models can be verified to an extent that may be regarded as complete; however, for large computer models, such a complete verification may be very difficult, or even impossible, so that an analyst may have to settle for a verification process that is somewhat less than complete, but that gives him a high confidence that the model has been constructed and programmed as planned.

It is possible to have verification by validation; that is, if an implemented program or policy shows that a study's findings—and therefore the models that yielded them—are substantially correct, then it may fairly be assumed that the models are correct (even though from a strict logical point of view this may be deemed doubtful). Similarly, if the analysts and the users have confidence in the results because the form of analysis has been used in other similar cases, in all of which the predictions from the analysis have agreed with the actual outcomes within acceptable limits, then a form of verification by validation can be deemed to be present. Thus, if the results of an analysis as a whole are accepted as valid, then the component models can be accepted in practice as having been verified.

Clearly, the issue of verifiability is one that should be considered carefully when the analysis team is deciding on what modeling strategy to adopt (see Chapter 2). Admission that the central model can be only partially verified undermines the confidence that can be stated in the study results, and this fact must be balanced against the possibility that a series of more approximate, but verifiable, models could be used (see Section 2.8).

The discussion so far applies to any form of modeling. Computer modeling, however, is so ubiquitous in virtually every form of analysis as to call for some additional consideration.

Computer Models

To verify a computer model is to demonstrate that it produces the results intended. Gass (1983, p. 609) divides this verification process into three parts:

> (1) ensuring that the program, as written, accurately describes the model as designed, (2) ensuring that the program is properly mechanized on the computer, and (3) ensuring that the program as mechanized runs as intended.

He goes on to say that:

> Verification tests fall into two indistinct categories: experiments to debug the logic of the computer program, and checks to demonstrate the correctness of the numerical and data procedures as carried out by the computer program. . . . Verification tends to be the domain of the programmers, with some assistance from the analysts in devising the numerical tests.

Simple computer programs can be verified quickly and unambiguously, but, as they increase in size, the difficulties of verification multiply. The analyst who is bending his efforts to get his arms around a complex problem may find himself adding complexities to his models, and thus being driven toward larger and larger computer programs to perform their calculations. But Greenberger, Crenson, and Crissey (1976, pp. 72–73) offer him good advice:

> A simple model may have better performance characteristics over the long run than a more complicated model, even though it provides less information. As more detail is added, a model can become more opaque to the user, harder to understand, and more expensive to operate. Even models of simple systems can be complicated if they include unnecessary items or attempt to be overly precise. Garry Brewer [1973, p. 4] makes a useful distinction between complexity and complication. "Complexity refers to substantive logico-mathematical interrelations and difficulties; complication can arise in almost any arrangement of facts, concepts, or thoughts. Complication is an undesirable characteristic of any construct; complexity may be an inherent feature. Complication often expresses lack of effort to give the construct its appropriate form." Complication is to be avoided, but complexity may be unavoidable (if the reference system [that is, the set of phenomena being modeled] is complex).
>
> One of the most important advantages of a model is its ability to illuminate causal relationships that are obscured by complexity in the reference system. A modeler who feels that his simple but understandable model is not "rich" enough in behavior may fall victim to the tendency to add layer upon layer of detail to the model in an attempt to make it more realistic.

They go on to warn against what Carlson (1970, p. 713) calls the *artichoke effect*:

> We have a proclivity to add features, add functions, and add interfaces— layer upon layer—onto existing systems. Each succeeding layer has less and less useful or tasty substance on it, until the outside layers merely add weight, complexity, and a prickly hindrance to reaching the core of the problem.

In effect, this advice—which is echoed in Section 2.8—says that, if the verification process is to be feasible and reasonable, the models must be chosen with this essential process in mind.

Despite this urgent—and by now classical—advice, many large-scale computer models have been built in recent years with a view to policy influence, so that substantial experience with them has accumulated (for overall reviews, see, for example: Arthur and McNicoll, 1975; Gass, 1983, which contains an extensive bibliography; Greenberger, Crenson, and Crissey, 1976; House and McLeod, 1977; Meadows, Richardson, and Bruckmann, 1982; and Meadows and Robinson, 1985). This experience, however, has not yielded widespread knowledge about how to verify such large models; indeed, Gass (1983, p. 609) offers this gloomy summary of the situation: "There is general agreement that for large and complex computer programs (and most models of public systems are of this type) it is impossible to demonstrate that a program has been completely verified." He goes on to describe how some workers have tried to bandage this analytical wound by introducing concepts of confidence and reliability that depend on factors external to the program (such as the reputation of the programmers) and on somewhat superficial experience (such as how long the program has been running). None of these concepts can be conceived properly as logical replacements for verification; rather, they are expedients aimed at making the best of a somewhat unsatisfactory situation.

Thus, the current state of the art presents the analyst with a choice between a set of simpler computer models (each dealing with a modest number of variables) that can be verified, or more complex models where verification may become a problem, with the knowledge that, in the latter case, the ability to persuade the client to have confidence in the findings of the analysis may be undermined by the knowledge that they emerge from models that are only partially verified.

Summary

In theory, it should be possible to verify any model completely, and such a basic procedure can be regarded as a fundamental of professionalism in systems and policy analysis. The prudent analyst will do well to presume that this fact is understood by his client. There may be cases, however, in which the reasons for using a reasonably comprehensive model are so compelling that the analyst may feel it necessary to sacrifice some verifiability to attain it. In such a case, a balance must be struck between increased comprehensiveness and the loss of a measure of confidence; if the model must be so complex as to defy complete verification, the gains should be sufficiently compelling to offset the loss of confidence in the results.

In sum, verifiability is one of the important factors to consider when a systems or policy analyst is choosing an analytic strategy and the modeling tactics to pursue it (see Chapter 2), and when he is making a judgment about a model's adequacy.

The difficulties in verifying large-scale models form one of the reasons for urging systems and policy analysts to use coordinated collections of smaller verifiable models rather than a single very-large-scale model (see, for example, Section 2.8 and the advice from Goeller in the *Overview*, pp. 214–216).

An analyst may be able to verify a model that he has constructed on the basis of his memory of the assumptions underlying it and the steps taken in its construction, but simple prudence dictates that he set down a complete record of them, not only to guard against memory lapse, but also to act as an overt check against conflict, ambiguity, and oversight. Such a complete documentation becomes absolutely essential if another person is to conduct the verification process.

13.4. Validation of Models

A model is an intellectual construct that represents a set of phenomena in the real world; however, it represents only carefully selected aspects of these phenomena, aspects chosen for their special interest. Holling (1978, p. 96) has put the matter well:

> Any model is a caricature of reality. A caricature achieves its effectiveness by leaving out all but the essential; the model achieves its utility by ignoring irrelevant detail. There is always some level of detail that an effective model will not seek to predict, just as there are aspects of realism that no forceful caricature would attempt to depict. Selective focus on the essentials is the key to good modeling.

Validation is the process by which analysts and other users judge the extent to which outputs obtained from a model represent the behavior of the selected aspects of the phenomena being modeled, and thus estimate the degree of confidence that should be placed in the model. There are two important facets that must be noticed immediately.

> The comparison process is theory-laden in two ways: (1) the observations that are to be the basis of the comparison are based on the analysts' theories of the phenomena and how to measure them and (2) the comparison process itself is based on a—perhaps implicit— theory of comparison.

> The comparison is not made against an absolute standard. Rather, it is relative in two ways: (1) it depends on the nature of the model and the phenomena being imitated and (2) it depends on what the model will be used for in the systems-analysis work.

As an example, consider a model devised to estimate the probability that a pilot will be able to see a given linear object (such as an airport runway) on the ground through partial cloud cover involving discrete clouds of cumulus, fractocumulus, scattered stratocumulus, and altocumulus types. Such a model must relate this probability to the cloud cover (measured, as is usual in meteorology, in tenths of the sky covered), the sizes and shapes of the clouds, and the length of the linear object—as well as possibly to other factors.

Since the situation of concern involves discrete clouds, one such theory assumes that they are circular discs, obviously a considerable smoothing of reality (adopted, of course, largely for analytic convenience in working with the model). This choice, however, is not without some basis in reality, since clouds of the four types represented in the model tend toward some compactness, but this statement is itself theory-laden, and any measurements of actual clouds aimed at checking it or the circular-disc assumption are equally theory-laden, both as to the shapes of the clouds (they are compact enough to be representable by simple geometrical shapes) and the method of measurement (which may involve schemes of observation and statistical procedures for handling the data and the comparisons).

This model is clearly intended to be a very crude representation of the phenomena of cloudiness aimed at suggesting how the probabilities of seeing the full length of a linear object decline with increasing cloudiness and the length of the object, not to produce precise numerical estimates of this probability. Further, it is intended to be simple enough to be readily usable.

How should an analyst then proceed to validate such a model? As to the shapes of the types of discrete clouds of concern, he could observe them visually or check photographs in cloud atlases, or perhaps take photographs in such a way as to facilitate measurement. Then he could check to see if the results from these measurements combined with the amounts of total cloudiness are accurately reflected by the model. Although most analysts would probably be content with the assumption of random location of the cloud centers, a meticulous validation procedure could even explore this issue with actual photographs of partially cloudy skies.

Throughout the validation process, the crude nature of the model and its rather imprecise purposes are kept in mind, a background that is especially relevant to the sort of claims that may be made for the model results. If all that is claimed for the results is a general indication of what happens, then quite informal and intuitive approaches to validation suffice; if more is intended to be claimed, then more vigorous and disciplined attention to validation will be needed (in such a case the circular-disc assumption could well be brought under scrutiny).

Finally, perhaps the most ambitious and persuasive way to validate the model would be to conduct an operational experiment in which airborne observers report whether or not they can see a given length (perhaps of a road or an airport runway) through partial cloud cover and compare these results with those of the model.

Most models used in systems analysis are more complicated than this example, but it is not uncommon for a somewhat crude assumption to be adopted for a variety of reasons. The model for the expected distance from a firehouse to a fire described in the *Overview* (pp. 203–205) incorporates a number of such assumptions. Since it was intended for use as a building block in other models aimed at obtaining reasonably accurate estimates to support important decisions and policies, considerable effort was expended in validating it (see Kolesar, 1979, pp. 185–199); since these tests showed that it gave good estimates of actual average times, the model has been extensively and satisfactorily used.

The variety of issues that are considered in validating models for systems analysis makes this process differ from one that might be used to test whether or not a new model represents progress in developing scientific theories, where a model that is shown to be a closer fit to the phenomena is considered to be an improvement. The systems analyst, on the other hand, balances goodness of fit with the uses to which the model and his findings will be put, and with the complications the model will introduce into the analysis. For example, he prefers a simple model, even though it is quite approximate, unless a more complex model can be shown to improve the findings enough to make the additional computational complexity worth its costs in time and money.

For instance, an analysis team developing a plan for water resources in the Netherlands adopted, as one premise in their modeling philosophy, the intention to "build both simple and complex versions of some models," and they discuss this premise in this way (Goeller and the PAWN Team, 1985, p. 9):

> [This] premise helps us to analyze the problem adaptively. For a particular part of the problem, we build the least detailed model possible that still has sufficient breadth and tolerable accuracy; even with the detail limited, this is nevertheless a complex model. This complex model is then used to explore the problem space, educating the analyst about both the problem and the model—their sensitivities and insensitivities. With this knowledge, the analyst creates a simpler version of the same model. . . .
>
> We use the simple model in most of the analysis, either as a component of a more comprehensive model, such as the water distribution model, or as a free-standing model for extensive trade-off analyses. We might use the original complex model in the final rounds of the analysis or whenever greater accuracy or more detailed outputs are desired.

In sum, in systems analysis the model chosen may not be the one that is the best fit to the phenomena if a simpler and more approximate one offers significant advantages to the analysis without the prospect of damaging the usefulness of the results.

13.5. The Conceptual Basis for Validation

The conceptual basis of science has undergone a substantial evolution in recent decades that is important for systems analysis (see Section 12.2 and Majone, 1980). This evolution has arrived at a view that is important for systems analysis (Majone, 1984, p. 144):

> Few scientists and philosophers of science still believe that scientific knowledge is, or can be, proven knowledge. If there is one point on which all major schools of thought agree today, it is that scientific knowledge is always tentative and open to refutation.

This view implies that a model—or a theory—cannot be validated in the sense of being proved correct. Rather, how the model survives repeated use determines the level of confidence that the scientific community places in it; this use involves two components, continuing comparisons between the model and the relevant aspects of reality, and evolution of the model's uses in shifting new contexts. For a brief discussion of this process, see Section 12.2; Ravetz (1971, pp. 181–208) describes it more fully.

Thus, perhaps to oversimplify Ravetz's description somewhat for our present purpose, progress in science involves continuing tests of its models against aspects of reality and by further use, the ones surviving these tests with high degrees of confidence being incorporated into the body of knowledge thought of as scientific, those with low degrees being discarded. This process can be thought of as giving the appropriate meaning to the term validation: Models that survive the tests to achieve high degrees of confidence are valid, whereas those that do not survive are invalid. On the other hand, validation is not so decisive a process as this last statement would suggest, because high credibility depends on the tests to which the model has been subjected. Thus, an accepted model has only been validated relative to the tests that have given it a high degree of confidence among scientists. Validation must therefore be thought of, not as an absolutely decisive process, but as one related to the tests that associate with a model the credibility that scientists hold in it.

This description of the general process of validation in science can have practical value for the systems analyst in several ways.

When the analyst uses a model from science, this view reminds him to consider the position of the model in the stream of activity in the field from which it is drawn, and judge it accordingly; this judgment then contributes to the level of confidence that he should adopt with respect to the findings of his analysis.

When the analyst must conduct an original scientific inquiry as part of his work in order to establish a conclusion about some aspect of reality not previously studied by scientists—as is not infrequently the case—this view of scientific validation urges that his claim for the new conclusion be appropriately qualified, since it has not yet been subjected to the demanding tests of the larger scientific community.

The conditions under which any model has been developed, and the tests to which it has been subjected, create a set of circumstances in which its degree of confidence can be held. Therefore, in contemplating using a model, an analyst must consider the context in which it is deemed sufficiently valid for his purposes.

Finally, the testing process may invalidate some or all of a model, that is, show that in some way it does not portray the phenomena of reality adequately. When this happens, the analyst may amend the model or develop a more realistic alternative (Holling, 1978, p. 97, describes such an example briefly). Sometimes, however, the opposite happens:

> Another example occurred in our budworm work . . . , where the model predicted that forest volume would decline independently of insect damage, while it was "common knowledge" that volume was high and would remain so if insects were controlled. We spent 2 months checking the model for errors when we should have spent 2 days looking at the available data on forest volume. When we belatedly took this obvious step, the model was vindicated, and "common knowledge" was shown to be at variance with the data on which it should have been based. We suspect that this is not a rare occurrence. [Holling, 1978, p. 97.]

If a carefully prepared test dealing with centrally relevant phenomena fails significantly to verify a model's predictions, analysts lose confidence in it, usually so much as to abandon it. They must then turn to efforts either to repair the model or replace it, although they must keep in mind the remote possibility that the test itself is at fault (as the example quoted above illustrates). After a new or repaired model has been constructed, the cycle continues with new testing. For an example of the testing cycle required with an ecological model, see Holling (loc. cit., pp. 95–106).

To recapitulate, no tests can validate a model or analysis in some absolute sense. The purpose of testing is not to establish the ultimate truth of a model, but to raise or lower the confidence that should be placed in it, and thus shed light on whether or not it is adequate for its prospective uses. One test can destroy all confidence in a model, but it may take many to establish enough confidence in a model to make it adequately usable.

It follows that demands for "valid" models are impossible to satisfy—and claims of validity for models are based on a faulty conception. Acceptance of a model does not imply that its predictions are certain, but merely that the potential user has enough confidence in them to consider them adequate for his prospective use. To achieve this confidence, he subjects the model to tests in which he has high confidence—such as checking it against other established models, carefully gathered data, outcomes in known circumstances, field experiments, and so on—designed to reveal where the model is supported and where it fails.

13.6. Validation of Analyses

What we have said so far relates to models intended solely to describe phenomena in the world of reality. As we say at the beginning of this book (Section 1.1), systems analysis (here as well as there construed to represent policy analysis) is concerned with much more than modeling and describing phenomena in a scientific way:

> Systems analysis is concerned with theorizing, choosing, and acting. Hence, its character is three-fold: descriptive (scientific), prescriptive (advisory), and persuasive (argumentative-interactive). [Majone, 1985, p. 63.]

Thus, high confidence in the descriptive aspect and its scientific models is not enough; if the analysis as a whole is to have credibility for the users, they must also have high confidence that the prescriptions are soundly based and the persuasion is not based on fiction.

As we have already seen, the adequacy of a model of the sort we have been discussing might be judged on the basis of two relations: the relation between the model and the relevant phenomena of reality and the relation between the model and the purposes for which it is being used. The adequacy—or validity—of the analysis as a whole as the basis for action must be judged on the basis of the same two relations. In addition, since the user's confidence in the analysis as a whole is affected by his confidence in the three aspects of analysis, we shall first discuss validation as it relates to each of the descriptive, prescriptive, and persuasive aspects of analysis, and then return to the validation of the analysis as a whole.

Validation Relative to the Descriptive Aspect of Systems Analysis

What this chapter has said so far about validating models applies to models used in the descriptive aspect of systems analysis, provided the validation effort is extended to include consideration of how the models are being used in the analysis. For example, a model can represent in a simple way a phenomenon that is peripheral to the concerns of one analysis, whereas a much more detailed model will be needed in another analysis that is centrally concerned with the same phenomenon; and, of course, the validation criteria for the differing models will differ in the two cases, being quite loose in the first case, and much tighter in the second.

Similarly, if the models in an analytic structure are intended to act as surrogates for the phenomena associated with the problem being studied (Section 8.5), then a demanding validation process is called for; but, if the analytic structure is intended to act as a perspective on the problem (Section 8.6), then a less demanding validation procedure is acceptable, for it need only respond to the perspective chosen for investigation.

In either case, the aim of the validation process is to establish the level of credibility of the model in the light of its expected uses. For an example involving the validation of a complicated model dealing with phenomena central to a study, but to be used in connection with strategic-level planning issues, see Holling (1978, pp. 156–166).

The fact that the adequacy of a model to be used in a systems-analysis study is a relative concept suggests that it is a pitfall to believe that there is a universal model for a phenomenon that can be used, no matter what study may be contemplated (Quade, 1968, pp. 357–359). In the early days of systems analysis, clients sometimes asked for—and perhaps analysts new to the field dreamed of—universal models that would solve very large classes of problems; however, experience has shown that there is no basis for the hope that the wild frontier of systems analysis will ever see the lone analyst with a white hat and a calling card with the legend "Have model, will travel." The study and its models must be tailored to the problem situation that the client faces. Thus, model validation must also relate to this situation.

This last point does not mean, however, that a model designed for one situation might not work well in another similar one; for example, the model for the average travel distance between a firehouse and a fire mentioned in Section 13.4 has been widely used.

Although the descriptive, or scientific, aspect of a systems-analysis study must consider both the past and the present, it must also relate, much more importantly, to the future and to what alternative should be chosen in order to create a different and more desirable future. Thus, validating the descriptive aspect of systems analysis must pay serious attention to how the work deals with the future.

How the study deals with the future. Any complete systems analysis must contain—or at least imply—models about the future, and the validity of these models must be explored.

Owing to the widespread awareness of dramatic and unexpected changes that have occurred in recent years in nature and in political, economic, and industrial arenas, it is common to think of the future as highly changeable—and, of course, it cannot intrinsically be checked fully until it arrives. On the other hand, as many successful analyses offer ample testimony, the future can be dealt with in such work.

To begin, many aspects of any future are stable continuations of the past; others may be adequately represented by continuing existing trends; and uncertainties in others can sometimes be adequately bracketed. Some phenomena exhibit random fluctuations, and therefore must be represented by probabilistic models; although yielding uncertain results, such models can be highly valid. Other uncertain elements in the future can be represented by considering them separately, or uniting them in coherent scenarios (see Chapter 9).

It may seem, at first glance, unfortunate that many problem situations—and therefore the analyses that deal with them—are importantly affected by uncertain future elements, notably those involved with social and economic phenomena. On the other hand, if the future could confidently be expected to be an exact mirror of the past, or a simple continuation of its trends, the most important reasons for analyzing the future would not be present. Future uncertainties are virtually always an important part of the concerns of a systems-analysis study, and its findings must embrace them in a reasonable way, even if the logical structure of the analysis can deal with them only in part.

Thus, validation processes for systems analysis must take account of this future perspective. Some studies include a base case in which the future mirrors the past, such a case forming a valuable basis for evaluating the changes that the future may bring when changed circumstances occur. Such a base case can offer a cornerstone for validation to support the confidence in the results of the more venturesome cases. Most studies, of course, must contemplate significantly altered futures, and models and procedures to represent them will be needed. Chapter 9 on forecasting and scenarios deals with some approaches to modeling the future when it cannot be assumed to be a simple extrapolation of the past.

Validation Relative to the Prescriptive Aspect of Systems Analysis

Deciding what should be done on the basis of the analysis is not—and should not be made—the responsibility of the analysts. The findings of the analysis may prescribe one or more desirable courses of action, but the decision about what to do is for the client or user to make. The analysts may, however, particularly if requested to do so, make recommendations or otherwise give advice; if they do so, the prescription is part of the analysis and should be tested—or validated—to the extent possible.

There are three cases to consider.

The problem is so framed that any prescription, advice, or recommendation as to what should be done is developed in the analytic process as part of the descriptive aspect of the work, is incorporated in the findings of the analysis, and is accepted by the user. Thus, when the analysis is validated, the prescriptive aspect is included in the validation, and no separate evaluation is needed—or may be possible.

The prescription—advice or recommendations—is not a finding of the central analysis work but is developed by the analysts on the basis of the findings of the analysis, their judgments and those of consultants, and possibly, further analysis.

The user prescribes for himself, but makes his decisions as to what to do in the light of the findings of the analysis.

In the second and third cases, further validity testing should be required, in the second case by both analysts and user to increase their confidence that the analysts' prescription is an adequate basis for action and has a reasonable probability of leading to a successful outcome, and in the third case by the analysts to protect the user from making a poor decision, apparently—but perhaps not adequately—based on the findings of the analysis.

Assuming that the validity of the findings is not in question, how can the validity of the prescriptions in the second and third cases be tested? For the second case, the potential user, his staff, and other interested parties need to probe the analysis with questions (see Section 15.10 for lists of possibilities) to uncover the reasoning, the arguments, and the judgments that led to the recommendations; the nature of the analysis; the biases, backgrounds, and track records of those involved; what consultants contributed important inputs; and so on. The goal of this process is to reach an opinion about the professional competence of the analysis team and the competence of their work. In their own validity testing, the analysts go through much the same process, but ask questions of themselves and their colleagues; their concerns are that the findings support the recommendations, that no significant factors are omitted, that the elements of the analysis are open and clear, and that the prescription is not likely to be misinterpreted.

For the third case, it may be impossible to involve the user in the discussion and debate needed to uncover his chain of reasoning, his hidden assumptions, his unconscious biases, and so on. But in such a case, the analysts can test the decision as if it were their own. If these tests fail to indicate a reasonable level of confidence in a favorable outcome, the analysts should find a way to make their opinions, and the reasons behind them, known to the user in the hope that this will stimulate further review.

Validation Relative to the Persuasive Aspect of Systems Analysis

A potential user may be persuaded to accept the findings and recommendations of an analysis solely because he trusts the analysis team, but this halcyon possibility seldom occurs—and is probably not even desirable from the analysts' point of view. Far more often, effective persuasion depends on two facets of the analysis. The first—the stronger and more important one—consists of the structure of the analysis and the evidence and arguments by which the results and recommendations are reached. The other is the part of the communication process by which the analysts try to convince the users that the analysis is adequate for their purposes and should be used in evolving courses of action. One hopes that the analysis itself will do the persuading and that a vigorous second effort will not be needed, or is called on only for completeness.

The need to persuade the client to accept the results of the work urges that the analysis team choose elements for inclusion that will facilitate this process (insofar as it can be done without detriment to the analysis): clearly identified assumptions, facts that are easily understood, data that are summarized in simple ways, models that can be described in such a way as to convince the user that they are adequate representations of the phenomena involved, views of the future that can be supported with confidence (or that need exploring as valuable extreme cases), and so on. This process should help the client judge the validity of the analysis for himself, and thus lessen the need for a later effort to convince him that the results are well suited to his purposes. Further discussion of persuasive validity of this type appears in a subsection with this heading in Section 13.7.

In spite of such efforts, not all analyses—and sometimes even those that appear the most transparent and valid to the analysts—leave the client and others convinced that the findings should be accepted and used. Additional information and evidence, designed specifically to address the doubts and presented persuasively, may change dubious minds. If such a persuasion effort leads to full acceptance of the study's results, it can be said to be adequate, but we should not call it valid. For one reason, an enthusiastic analyst may convince an impressionable client by appearing to make a stronger case than is justified by what the analysis has actually shown; for another, it can happen during the effort to persuade that the client (who may know far more about his operations and their context than the analysts have been able to learn) may convince the analysts to change their conclusions. In sum, it is inappropriate to speak of validity in this connection, but merely of the quality and adequacy of the persuasion.

The need to persuade the client of the validity of the work forces the members of the analysis team to review their own assessment of its validity and refine the estimates of confidence that they have made of its findings. It also forces the client to bring his own point of view to bear on the work and its results, and forces the analysis team to recognize the client's point of view explicitly. The hope is that this process will yield a common view and strengthen the sense of validity. It could, of course, merely strengthen a confidence in misleading results, but the diversity of points of view brought to bear offers a considerable safeguard against this serious pitfall.

Validation Relative to the Analysis as a Whole

One of the central purposes of any systems-analysis study is to erect a structure representing reality with sufficient validity to persuade clients and decisionmakers that this view is worth incorporating into their understanding, thus contributing authoritatively to policies and actions. Testing the validity of the various parts and aspects of the work can raise the potential user's confidence in the adequacy of the analysis for his purposes, but, when it can be done, testing the validity of the analysis as a whole is more effective.

The ultimate test of the validity of a study is to implement its findings and then compare them with what actually happens. Although such a procedure is not unknown, it may not be possible (for example, when the study looks so far into the future as to place measurable results out of reach within a reasonable time frame), and the prudent decisionmaker is seldom willing to gamble so much. Rather, he almost always wants an estimate of the study's validity as one of the inputs to his process of deciding what to do about its findings. A perceptive and intelligent client need not depend completely on the analysts or on his staff for such an estimate. He can test the analysis and its results in a variety of ways to assure himself of its validity for him. In this process he reviews what factors the analysis has included and which ones it has not considered, whether its view of the future is one that he shares, whether its assumptions are reasonable, whether the factual basis for the work is believable from his point of view, whether the variables interact in a believable way, and so on.

Throughout an analyst's work, he aims to achieve a structure and results in which he has high confidence, or at least a degree of confidence appropriate to the goals of his work. Although this state of mind is necessary for successful work, it is far from sufficient; the client must also arrive at an appropriately high degree of belief in the validity of the work and its findings if he is to use it effectively in understanding his problem and deciding what to do about it. Thus, behind all that is said here is the presumption that an overriding goal is to achieve a situation in which both analysts and client share a common—and appropriately high—degree of confidence in the adequacy of the analysis and its findings.

A client or other potential user will often ask his staff to evaluate the validity of an analysis in order to help him decide whether or not it is adequate for use as a basis for decision. But to do so, the staff members assigned the task should be chosen from a peer group with experience and training either equivalent to or superior to that of the members of the analysis team who did the work. The judgment of such a peer group with respect to the study as a whole is usually made up of a suitable amalgam of many subsidiary validity judgments, a number of which are discussed in the next section. With no standard procedure to ensure correct results—and, indeed, no simple criteria of correctness to act as guides—the evaluators use their knowledge of analysis to probe for errors, omissions, and weak links in the argument by asking many questions (for some possibilities, see Tables 15.3 and 15.4), testing correlations for the effects of chance, examining data for reliability and relevance, looking at judgments underlying the analysis to see how sensitive the findings are to them, searching for new alternatives that might have been overlooked but which appear attractive, and so on. In some ways, what they do may be close to a brief recapitulation of the original analysis. They are also likely to be strongly influenced by their opinion of the professional quality of the work and the members of the analysis team. In the end, unless they uncover a serious error or an important omitted factor, or unless a validity test fails, the net estimate of adequacy for the study as a whole must be based on their pooled judgment.

13.7. Validation Approaches and Procedures

This section presents some approaches to testing validity that have been used in systems analysis. Since no single test can offer adequate assurance of the validity of a model or an analytic structure, a number of successful tests are needed to offer reasonable assurance that further testing will yield similar results, thus assuring that the model or structure is valid in the sense of being adequate for the purposes of the analysis. Validity testing is thus an interactive process involving many elements, and the approaches described here are not mutually exclusive; they are distinguished merely to facilitate discussion.

Since studies vary, as do their purposes and the models used in them, the validation procedures must also vary accordingly. Of the appropriate mix of validation procedures required, some may be from the list described in this section, but others may have to be invented and specially adapted to the study and its models. Consequently, it is not possible to specify in advance, or in any standard way, how validation should be approached. Efficiency in testing, however, suggests that, if cost is not an important consideration, the test that seems most likely to put validity at serious risk should be applied first. It is possible to state categorically, however, that serious attention should be paid to validation in any systems-analysis study and to recommend that an appropriate complementary mix of procedures be carried out (since any validation procedure itself carries its own biases of theory and practice with it).

The Face-Validity Test

In exceptional cases, the structure and details of the analysis may be such a clear representation of the obvious properties of the situation being analyzed that both analysts and their clients can regard it as valid: in sum, they can regard the analysis and its findings as valid on their face. This test can be extended by polling other knowledgeable people and relating their views to those of the analysts and their clients.

Although passing this test may appear to be the most desirable situation, the test involved is not very strong—and, indeed, the analysis may founder when a chosen option is implemented if all those consulted share a misconception about reality. Thus, in spite of its desirability, it should never constitute the only test of validity that the analysts make; the objective component of their professionalism demands that they subject their work to other tests. In most cases, they will, in fact, use some of the procedures described in the remainder of this section.

On the other hand, it sometimes happens that an adequate analysis or model may fail to pass the face-validity test because the knowledge or intuition of the analysts as to what the findings should be is faulty (for an example, see the quotation from Holling, 1978, in Section 13.5). Thus, failure to pass the face-validity test may not necessarily be fatal; rather, it may indicate that more information and other tests are needed before a decision can be made.

The Pilot-Experiment Test

When both analysts and clients have high degrees of belief in the validity of an analysis, it may be desirable—if the situation makes it possible—to conduct an experiment, the most convincing of all validity tests. (This test, however, is sometimes expensive; if costs in time, money, and public relations were not factors to be weighed, it would undoubtedly be used in many more cases than it is.) Such an experiment can take either of two forms: a specially designed and limited test, after which, if successful, the validity-assessment program is terminated, or a small-scale initial implementation made in the hope that, if it verifies the accuracy of the analysis, it can be expanded into a fully implemented program. As an example of this possibility, the Greater New York Blood Program undertook an early, limited implementation of a new blood-supply system recommended by a systems analysis study (see the *Overview*, pp. 68–79); as the test verified the accuracy of the analysis, the system was gradually extended from the 4 hospitals in the pilot program to over 200 in the entire network.

In addition to supplying very convincing evidence relating to the validity—or lack of it—of the analysis results, depending on whether or not the pilot experiment results agree—or disagree—with the analysis findings, a pilot test has a number of advantages: It can often be conducted at moderate cost compared to a full-scale implementation; it allows procedures and variables to be calibrated; it offers important lessons about implementation that may be useful later on if a complete program is undertaken; and it controls the costs and impacts of failure, if this unhappy outcome should ensue.

Thus, if the situation permits a pilot experiment, it is a very desirable step to take between an analysis and implementing its findings fully.

There is a wide variety of situations in which verifying the validity of the findings of a study directly is not possible on a limited basis, such as in a pilot experiment. Such limited testing can be denied by:

Equity considerations, such as may occur when a proposed system or policy confers benefits on one class of persons or enterprises but denies them to others (and it is not possible to compensate the losers adequately).

The impossibility of partial implementation, as, for example, in closing a Dutch estuary to prevent flooding (see Section 2.2), where a partial closure would be valueless in shedding light on the properties of a full closure.

The impossibility of bringing any significant portion of the future into the present, as, for example, in a study of world energy supply and demand looking 50 years into the future (see Sections 2.2 and 9.6), where no segment of the economic and operational systems using energy are under the control of a decisionmaker with the power to bring this segment rapidly to a new state that might characterize what the future would be like on a much wider basis.

The interactive nature of a problem's associated phenomena, which may defy some sort of sectorization that would enable a pilot experiment to be undertaken.

Conflicting views of validity sufficiently strong to deter undertaking a pilot experiment.

Political issues that may make it undesirable to telegraph the possibility of a full implementation until a very confident public sales campaign can be conducted based on significant confidence in the proposed course of action.

When limited testing is not possible, other less direct forms of validation can be undertaken, as described below. None by itself can lead to the high degree of confidence that could result from a successful pilot test, but a combination of several can create a climate that will yield a degree of confidence high enough to prompt suitable consideration and action by decisionmakers.

Using Established Knowledge

If the models used in an analysis are taken from well-established scientific knowledge, this fact usually supports a high degree of belief in the systems analysis itself, provided this knowledge is assembled in a way that supports this confidence. Such a situation can occur, for example, when some phenomena involved in the analysis are combined in a physical or engineering system whose properties are well established. More generally, such well-established knowledge may include items from accepted theories of relevant phenomena, inferences drawn from stable bodies of performance data, or structural components that have been used successfully in other studies. Many problems, however, involve questioning established knowledge and sometimes drawing conclusions at variance with the established view. Consequently, one must be certain that the knowledge is indeed well established, and not merely commonly or widely accepted, for using knowledge with only the latter support can cast undeserved doubt on the validity of a model being tested if the common belief is erroneous (for an example, see Holling, 1978, p. 97, quoted in Section 13.5).

In order for the well-established knowledge to support a high degree of belief in the structure of the analysis, the analysts must conduct a theoretical validation with good results. This form of validation requires that analysts evaluate how their models have been assembled from existing theories and the assumptions adopted for the study and whether or not the models represent faithfully the conception of reality that the analysts have brought to their work. As Gass and Thompson (1980, p. 436) put it, the process of theoretical validation

> requires the evaluators to review the theories underlying the model and the major stated and implied assumptions which are embodied in that set of theories or which have been made to develop or adapt a theory to a problem. The applicability and restrictiveness of these assumptions

in relation to the internal and external problem environments as viewed by decision-makers must be examined.

The evaluators must also verify that the transition from the theoretical model of reality (or perceived reality) to the mathematical model has been made correctly. This process will involve identifying and assessing the reasonableness of the most important assumptions made by the modeler in formulating the mathematical model. This is not as easy as it might seem, for assumptions come in many different forms. Explicitly stated assumptions are easy to verify. The difficulty lies in isolating the most important unstated or implied assumptions. Such implied assumptions may be present in underlying theory or in the methodology chosen to apply the theory to the problem. Sometimes they are affected by the implicit and/or unintended biases of the model developers and computer programmers.

When existing knowledge is used in systems analysis, it is usually stretched somewhat to cover situations that are at least partially new. Thus, the problem of testing the validity of existing knowledge as it is used in analysis includes justifying the validity of these extensions.

Inquiries such as these may assist in validating models from well-established sciences used in a systems or policy analysis:

Are the hypotheses on which the analysis is founded comparable to ones that are widely accepted in other work?

Do the variables in the analysis exhibit behavior that can be checked with phenomena in the problem situation, and do these behaviors agree?

When the models in the analysis are simplified at their margins, do they agree with accepted theories elsewhere, and do they represent behaviors that are well known and accepted as valid?

Do the assumptions in the analysis agree with reality as it is known to the analysts and decisionmakers?

Are the elements in the analysis assembled in accordance with relationships that enjoy a high confidence as to their validity in other contexts?

Do the ways the models from other sciences have been stretched represent reasonable extrapolations from the existing knowledge in those fields?

In addition to these questions, each analysis will generate others from the context of its work that will need investigation.

The usual analysis does not make use of the most fine-grained results from other sciences, but rather adopts a reasonable summary that fits well into its analytic structure, a choice that raises this question: When a variable or subtheory is introduced to summarize a field about which a great deal is known, what form is most suitable for this condensation? For example, in the systems analysis dealing with the global energy supply and demand for the next 50 years, how can the myriad economic activities generating the demand be summarized (Section 9.6)? Clearly, the summarizing subtheory should capture the essential properties needed for the analysis, but at the same time be economical enough not to dominate the work on the central issues in the analysis of supply. In this case, the analysis team adopted a simple function of the expected growth of the gross domestic product to represent the growth of energy demand (Energy Systems Program Group, 1981, pp. 173–174). In an analysis covering a wide spectrum of impacts, such summarizing subtheories may become a substantial portion of the analysis, as in the case of the Dutch estuary closure aimed at preventing flooding (Section 2.2), where dozens of such models were needed and used, with their results displayed explicitly and separately in the findings reported to the client.

When a model is chosen from the standard scientific literature for use in a systems analysis, it is a pitfall to ignore the need to validate the relevance of this model to the situation being studied. Lest this remark be seen as stating the obvious, consider a case in which two analysts chose a queuing model that had been developed earlier by one of them for use in connection with planning police operations in a large metropolitan area.

> The seriousness of the actual and potential applications of the . . . model led us to an extensive validation effort for which we have not found parallels in the queuing literature. There have been few reports of actual applications of analytic queuing models and even among these, validation is not usually treated in depth. It is noteworthy that the best discussions of the validity of a queuing model are found in the early classic papers of Edie (1954) and Brigham (1955). There is no treatment in the operations research literature of the validity of queuing models of systems in which human behavior plays an important role. Such validations are clearly harder to perform and are less likely to be definitive than those of purely mechanistic systems such as computers, or automated production systems. While the question of what constitutes adequate validation of such models is ambiguous, we strongly agree with Gass (1983) that knowledgeable and confident use of them can only be based on more thorough validation efforts than have been reported to date. [Green and Kolesar, 1984, pp. 1–2.]

In sum, Green and Kolesar are reporting that, even though the mathematical literature of queuing can certainly be regarded as standard, there is little literature reporting the validation of its models for use in applied studies. It would appear that the community of appliers has not listened to Lee, who wrote at the end of his 1966 review of applications:

> The theory of queues . . . is a theory almost without empirical foun-

dation, largely based upon experiences in the single—albeit important and respected—field of telecommunications. We need to know more about the real behavior of people in queues: until such time as we have that information, much of the theory that appears in the journals will remain no more than a collection of charming, mathematical acrostics. Mathematics, however ingenious, is not a proper substitute for knowledge. [Lee, 1966, p. 211.]

Green and Kolesar went to considerable trouble to validate their model, dividing their validation work into three parts:

> *Output Validity:* A determination of whether, given reasonable estimates of system parameters, the model produces reasonable estimates of system performance. We do not question here *why or how* the model works—it is viewed as a black box and judged only on its performance.
> *Structural Validity:* An examination of the validity of its key structural assumptions—asking, in effect, why it works.
> *Sensitivity Testing:* Testing the sensitivity of the model to violations of its key assumptions. Of particular concern are those assumptions which were identified as being suspect in [the previous] stage . . . , and those assumptions that might be violated in future applications. [Green and Kolesar, 1984, p. 2; emphasis in the original.]

In carrying out this work, the analysts used the somewhat limited data that they could obtain from the experience of an active police department. Their judgment that the model was adequate for its intended uses represented a balance between how it fitted the experience data and keeping the model simple enough to be useful in a variety of future circumstances.

In further work in this context, Green and Kolesar (1985) examined "the issues of the validity and utility of queuing models of service systems in which adaptive behavior by the (human) customers or servers is likely." They found that "in addition to the technical accuracy of its assumptions, the accuracy of such a model will depend on the level of managerial control of the system and the adequacy of resources." They concluded further that the validation process itself could provide important information to a management trying to establish a stable service system and they recommended that "queuing models of human service systems be used in a normative fashion and incorporated in the management feedback loop." They summarize the situation this way (loc. cit., pp. 17–19):

> Our validation study of the . . . queuing model of police patrol operations revealed several potential sources of discrepancy between the model's predictions and actual system behavior. While some of these discrepancies can be attributed to technical limitations of the model, others are indicative of problems in the management and operation of the system itself. Both issues must be addressed in order to improve system performance. We believe that this will be true in modeling many other human service systems. . . .

We have found that the key factor that distinguishes "human" from "mechanical" service systems is adaptive behavior. And our major conclusion on adaptive behavior is that a queuing model cannot be used as a passive oracle that states once and for all time what should be done, but rather it should be employed interactively as a prescriptive tool. In effect, the model must become an integral part of the management process. We see this process as a feedback loop similar to that often used to describe the scientific method. The manager/management scientist employing a queuing model to assist in operating a human service system must, we believe, follow a cycle of collecting data, building a model that reflects the data *and* management's prescription of how the system should operate, run the model, modify the resource allocations, etc., on the basis of the model's outputs, collect more data to monitor actual performance and to recalibrate the model, then rerun the model, etc. It is crucial that the comparisons of actual data and the model's predictions be used to make judgments about *both* the model and the actual system.

It thus follows that field management be capable of bringing actual performance into correspondence with specifications. It has been our experience that, in the past, field management of the police patrol forces in New York City was very loose. In such a situation, inaccurate predictions or poor allocations cannot be simply attributed to problems with the model itself. A generalization of this point is that accurate models cannot be built of ill-managed systems.

The concept of implementation of the queuing model within a feedback loop also strongly suggests that resource modifications to the actual system be undertaken *incrementally* and that changes in actual performance be monitored and evaluated after each addition. This allows management to zero in on a good solution and avoid making big changes that might prove to be big mistakes. Active and integral use of the model in the process of management places a burden of understanding the model on the manager. But, it is only the logical structure of the model, and the physical/operational meaning of all inputs and outputs that need to be understood by management, not the mathematics.

Our other fundamental conclusion is that a queuing model of a human service system will be useless when congestion is severe. The adaptive behavior of servers and customers alike becomes so dominant when long delays develop that any predictions from a model become nearly meaningless. There is just one prediction that can be made: system performance will be unsatisfactory. Thus, our last proviso is that queuing models of human service systems will be most effective when they are used to keep delays at tolerable levels.

On balance, although there is clearly more potential for discrepancies between model predictions and actual performance in human service systems than in mechanistic service systems, we feel that, when used in the fashion outlined above, queuing models can contribute significantly to improved management and performance. [Emphasis in the original.]

Thus, in situations like the one examined by Green and Kolesar, the validation process can become an important link in an adaptive process of implementing a study's findings.

Before leaving the subject of validating models, it is well to return to the point that Gass and Thompson make about the difficulty of isolating the important unstated or implied assumptions that inevitably accompany the models in systems and policy analysis. The uninitiated, or analysts of modest experience, may be tempted to regard this matter as something easily coped with by sweeping carefully through the work toward its conclusion with pencil in hand, but experience does not support this view. In a conscientious and very searching review of nine large-scale models developed to represent social systems and their problems, Meadows and Robinson (1985, p. 367) reach this unhappy conclusion:

> A final obstacle to critical understanding in nearly all these models is that their most important assumptions are implicit. Problem definition, choice of method, boundary, and selective omission of facts about the system being modeled are the most essential assumptions of any model. These assumptions are almost never documented, not because modelers wish to hide them, but because they are largely unconscious of them. As we have discovered in compiling this book, considerable training and effort are required to ferret them out. There are undoubtedly crucial implicit assumptions within each of the nine models that we have not yet discovered, in spite of the fact that we were actively looking for them.
> . . . important parts of all these models are in fact not explicitly stated.

The moral is clear: Analysts must be clear and explicit about the assumptions implied by their models and by the structure of their analysis, and they should take considerable pains in this regard, both to understand these assumptions and their implications and to share them with their clients. Failure of diligence here is to risk serious degradation of the client's degree of belief in the analysis.

Data and the Validation Process

Data play two key roles in the analysis process: suitably converted, they are used as inputs to the analysis and as the basis for comparison with outputs from models in the analysis.

The issue of the validity of data for use in systems and policy analyses is a major one on which much has been said and written. To summarize all of the points in this literature is beyond the scope of this chapter; however, Altman (1980) provides a useful entry to it.

Data used in systems and policy analysis studies can come from two sources:

They may be gathered by the analysis team in a program of collection designed specifically to support the analysis. Such a program must be very carefully designed and controlled if the needs of the analysis are to be fully met.

They have been gathered by others for other purposes and compiled in handbooks, census reports, and other publications. Here it is essential to know how they were compiled, what controls on the compilation process were exerted, what purposes animated the collection, and so on, so that the analyst can have some assurance that the relevance they purport to have with his analysis does in fact exist.

A great deal could be said about data gathered in both of these ways, but it suffices here to say that validity with respect to the analysis in hand is a vitally important issue that must be dealt with constructively. Further brief discussion here might suggest—wrongly—that there are a few simple principles that will carry amateurs in data gathering and analysis easily past the ubiquitous shoals: this is a place for deep professionalism, and it should be well represented on the analysis team.

As one of the important steps in judging the adequacy of a model, it is usually desirable to compare some of its outcomes with a set of related phenomena (what has earlier been called its reference system), and such a comparison calls for appropriate links. As Greenberger, Crenson, and Crissey (1976, p. 70) put it:

> Data provide a tangible link between a model and its reference system, and a means for gaining confidence in the model and its results. A model that closely reproduces data on observed past behavior of the reference system gains credibility and wins the acceptance and trust of potential users.

Although the goal suggested here is highly desirable, to achieve it is not usually as simple and direct as the quotation might suggest. Each model reflects only certain aspects of the reference system; similarly, data usually emerge either from selected aspects of this system or from controlled operations of it. Thus, the analyst must assure himself that the data represent the same aspects of the reference system as the model does, unperturbed by disturbing phenomena; otherwise, the comparison is not warranted. This issue becomes especially important when, as is often necessary, the comparison must involve data collected somewhere else, by other analysts, and for other purposes.

Two further points about the comparison of models with reality are worth making:

Such comparisons may be made with individual models in the analysis, or with the analysis structure as a whole. The actual situation usually determines what can be done: comparisons with both the whole structure and some of its models, or with just one possibility—or even none.

The path to such comparisons is strewn with pitfalls, but there is a vast literature on what they are and how to avoid them, so that it is important for the analysis team to have on it someone with a knowledge of these pitfalls and willing to play the devil's advocate.

Beyond the technical issues of data validity in analysis, the analysts must keep clearly in mind the essential limitation of data with respect to the validity of the analysis as a whole. They have, *a fortiori*, been collected in the past, and, although having a model that agrees with past behavior is desirable—even in most cases essential—such a model may or may not predict the future. Almost any complex model with a few adjustable parameters can be made to fit a given pattern of historical data. Thus, since the findings of a systems-analysis study can only affect how things are done in the future, the question of whether or not the data validity of the past can be extended into this future needs additional inquiry. The only case where this last statement would not apply would be the unlikely one in which the analysts could assume that the future will be exactly like the past, as far as the reference system is concerned.

Predictive Validity

Although bringing the future down to the present remains beyond our powers, gaining some confidence about the ability of a model to predict the future behavior of its reference system is not entirely foreclosed. In particular, three schemes can be employed to shed some light on this issue:

Comparing the outcomes predicted by the systems analysis with new data, that is, data not used in the analysis and drawn from a source independent of the sources of the analysis, but representing a reference system like the one on which the analysis was based.

Establishing, if possible, that the future will depart at most to a limited extent from the reality incorporated in the model of the reference system.

Exploring the extent to which changed conditions in the future will affect the predictions of the analysis, thus conducting a sensitivity analysis focused on the future context in which an implementation could take place.

The first of these three schemes is a common way to move toward validation. For instance, to validate the ecologic model in the Oosterschelde estuary study (described briefly in Section 2.2), the analysts looked to data from Grevelingen, a similar estuary that had been closed recently, and for which data taken both before and after closure were available. The parameters of the model were first calibrated to reflect the situation observed when the Grevelingen estuary was open; then the model's predictions for the closed Grevelingen estuary were compared with the data taken after closure. This comparison, together with other factors present in the situation, convinced the analysts that the model had been validated to the extent that the available data would permit, and adequately enough for their purposes. Their confidence in the model was increased later, when, in one instance, the model's prediction appeared to be better than the original estimate made directly from the data taken soon after closure, based on data from a more recent experiment that became available. For a description of this work, see Goeller et al. (1977, pp. 69–70).

At best, predictive validity must remain somewhat less than certain. On the other hand, one should not let this fact give rise to undue pessimism about analysis. The future is always uncertain, and some doubt will always be present about what changes it will bring. Rather, it is the uncertainty of this future that prompts us to explore how forecasting can be improved; as we proceed into this uncertain future, it is well to have the best estimate we can develop of how the futurity of our present posture casts its shadow into this future, even though the outlines of the shadow may be fuzzy (Drucker, 1959). And it is even more important to have at least some idea of how changes in what exists in the present might shift the position of this shadow, even if this information is also somewhat less than certain.

Extending Technical Validity

The expression technical validity can be thought of as comprehending all of the matters discussed so far in this section, which are largely focused on the more technical concerns of the analyst. On the one hand, for the results of a systems-analysis study to be adopted, the clients and decisionmakers must come to share with the analysts a high degree of belief in the technical validity of the findings, but, on the other hand, there may be essential elements of the analysis that have not—or perhaps cannot—be subjected to technical validity tests of the sort discussed so far.

Two such possibilities may usefully be distinguished here:

The possibility that the analysis may point to a new way of organizing an operation that has never been tried before, with a pilot test of the possible new system being either undesirable or impossible.

The possibility that administrative or operational tactics may form an essential part of a new approach, and it is impossible to model these tactics in the analysis.

The first possibility—that of an untried system—can be subjected to a validity test by a means of a simulation, but this device brings in validity questions of its own (Van Horn, 1971). On the other hand, if the untried system has been recommended by the analysis, it likely has been represented there by means of models that have been devised to represent it by extending or revising other models that are believed to have validity. These new models, which cannot be tested with reference to an existing system, can sometimes be assessed by means of a simulation; Ignall, Kolesar, and Walker (1978) report four such examples in connection with New York City police and fire operations (see also Walker, Chaiken, and Ignall, 1979, pp. 545–548).

The second possibility—that tactics may be an essential part of a new system—may pose difficulties that prevent analytical modeling. These difficulties can be overcome in two ways:

By relying on the judgment of the clients about how the tactics would work in practice.

By exploring the tactics in a gaming exercise.

The first of these approaches, although perhaps not very strong, may be acceptable to the clients, who may be willing to try it based on their judgment—hardly a new situation for decisionmakers, who act on their judgments daily. The approach can be made somewhat more tentative by a trial period aimed at validating it. For a discussion of the second approach (which, like computer simulation, has validity problems of its own), see Chapters 2, 3, and 4.

Both simulation and gaming involve considerable time and effort, factors that are important considerations in whether or not to adopt them; for light on these issues, see Shubik and Brewer (1972), and in connection with gaming, see also Chapter 4.

In any case, for the results of a systems-analysis study to be transformed into practice, the clients and decisionmakers must be able to generate a high degree of belief in the validity of the findings, a process in which the analysts play an important role. The perceptions of technical validity that the analysts can convey are an important ingredient in this process, but cannot be expected to cover all issues; the clients must also bring to bear their judgments, both in what the analysis covers and the relevance of what has of necessity been omitted from the technical work. These matters are discussed further in what follows.

Persuasive Validity

For the analyst to be persuaded of the validity of his work is far from enough; the clients and decisionmakers must also have high confidence in it if they are to be persuaded to act on its findings. Much of the evidence emerges from the exploration of the findings and their technical validity, but the clients and decisionmakers cannot be expected to dig into all the detail that the analyst considers. Rather, the analyst has the task of summarizing and focusing this evidence on key issues as they relate to the client's concerns. This is a vitally important task for the analyst, and must be performed with both sensitivity and precision, as well as with the fact and appearance of reasonable objectivity with respect to the decisionmaker's concerns. Not all issues of validity will yield uniformly favorable answers, and it is important for the less favorable ones to be reported as candidly as the more favorable. In fact, perhaps curiously, candor about the less favorable comparisons may do a great deal to bolster the client's confidence in the analytical work.

Although there is a great deal of discussion in the literature about getting clients to understand models, experience shows that much of this arises from both inexperience and misperception. Very few clients want to accompany the analysis team on a detailed tour of the models used in an analysis (although some may want a technically trained staff member or consultant to review what was done). On the other hand, a perceptive and intelligent client will want to examine key aspects of the modeling work, such as these:

The assumptions lying behind the analytic structure and the models in it, to see if they correspond with his view of reality.

The data and evidence used in constructing the models and testing their validity, again to see if they correspond to his view of reality.

The key factors in the structure and models, as well as the essential features of their interactions.

In all three of these categories the analyst must be able to summarize the information concisely and effectively.

In all three categories, it is usually necessary to be selective, suppressing purely technical matters in which the analyst has high confidence of correctness in order to emphasize issues of significant concern for the client. In the third category, unless the client is a scientist, the analyst must construct a simple version of the analytic structure—sometimes called a model of the model, or a semitransparent model—in order to have a basis for discussing the structure with the client, if the client is interested in such a discussion. In some past cases the client trusted the analysis team to construct an appropriate analytic structure, and did not inquire into it—being satisfied with his checks of the data and assumptions. On the other hand, it often happens that interactions occurring within the analytic structure are essential bases for understanding the shape of the findings; here the analyst may have compelling reasons for wanting his client to understand the essential interactions within this structure.

In addition, the analyst needs to explain in detail, and support with logical arguments, the judgments that link the various stages of the analysis (as pictured in Figure 13.1), particularly those by which the model results are converted to formal findings and substantive conclusions about the problem. In particular, if the analyst provides models to aid the final decision, they should be designed in close collaboration with the decisionmaker.

Indeed, persuasive validity always implies a full and detailed set of discussions with the client. Experience shows that these discussions are best spread over the life of the project, rather than crammed into limited time at the end; assumptions can usually be discussed early, data and evidence as they are gathered, processed, and summarized for the project. The best situation is one in which the analysts and client have a close working relation as the project moves along, what Schön (1983) calls a "reflective contract," as described in Section 12.4.

A study will not always speak for itself; indeed, a complex one can hardly be expected to do so. Some persuasive communication by the analyst is usually needed; without it, the client may not understand or accept—or even hear—what he needs to know. The question of professional ethics here is what form of persuasion is acceptable, not whether it is necessary. Surely, if the analyst leaves the content and message of his work intact, it is acceptable to change the style and form of presentation.

> No doubt many analysts will be uncomfortable with the notion that they may have to manipulate a client or audience. They will argue that it is impossible to change form without changing substance, and they will consider attempts to do so expedient and unethical. They may well be right, but one thing is certain: if they do not use the manipulative arts of persuasion, their communications will be effective only in situations in which the values, attitudes, and beliefs of the participants are reflected in the analysis. [Meltsner, 1980, p. 129.]

The goal is for the client to have before him enough evidence to make up his own mind about the degree of belief to place in the analysis. It is this confidence that will have to inform his thinking as he considers other surrounding matters, as well as the findings of the analysis, while working toward decisions about what to do.

Bowen (1985) puts this challenging summary before us:

> Ideally, if analysts and client are a team, validation in all its aspects becomes a continuing aspect of the total decision process and is not a separable issue. All members of the team are committed, in their various roles, to ensuring that what they do is as professionally sound as they can make it, within whatever constraints apply.

Validation may have become a separated issue because the analysis effort and the decision effort have themselves been separated rather than being integrated in a decision *process*, in which all participants are concerned with the whole rather than having demarcated roles.

However, an ideal situation is only realizable in theory. Depending on the observed reality, analysts have to decide, sometimes unilaterally, what steps *they* must take to meet their professional standards and to ensure that their work is found relevant and not, ultimately, wasted. [Emphasis in the original.]

Operational Validity

If an alternative recommended by a systems analysis study is—or can be—implemented successfully, and if it achieves the desired objective, it is said to be operationally valid. Since the analysis is concerned, *a fortiori*, with only selected aspects of reality, whereas any implemented program involves all of the relevant aspects, checking this form of validity is concerned with the gap between any analytic structure (and its models) and the full reality of their reference systems.

Operational validity can only be checked finally and fully by implementing the proposed program and observing its results. On the other hand, early indications that can increase the degree of belief in operational validity can be obtained in several ways:

If the analysis examines the organizational changes that may be required to manage a new system, develops an appreciation of the limitations of an implementing organization that cannot be significantly changed, or examines how any new equipment or procedures can be introduced effectively, then confidence in any alternative that invokes such possibilities will be increased (Bowen, 1985).

If a well-developed, detailed, realistic, and practical implementation plan can be formulated, then confidence in the proposed alternative will increase. In other words, confidence in how to achieve the alternative will increase the confidence in the alternative itself.

If a pilot experiment has been successful, then the confidence in the operational validity of the alternative will increase.

If the early stages of an implementation yield desired results, or if these early stages can be adjusted to do so, then belief in the operational validity of the full program will increase.

If an evaluation of an early phase of a complete implementation shows that desired results have been achieved, then confidence in continuing will be enhanced (see Chapter 11).

In sum, the degree of belief in the operational validity of a proposed program can be enhanced before the program is undertaken, and increased during its earlier stages, but full confidence must await full implementation. Although this prospect may appear undesirable to analysts, it is something that experienced administrators live with daily.

What we have said so far about operational validity presumes an analysis that has been able to project a fairly complete view of the future that each alternative offers, and in which both analysts and clients share a reasonably high degree of belief. One might be tempted to jump to the conclusion that this is the only sort of situation in which systems analysis can be helpful. However, experience shows that this is not the case. Even when the problem situation does not allow the analysis to cover it completely, or when important uncertainties remain about both the desirability of courses of action and their operational validity, an important contribution can be made.

Some systems-analysis work either is not aimed at implementable changes, or does not result in them. In such cases, we may realistically adopt the view of an analyst who surveyed the work of a large organization that had completed many important systems-analysis studies (Smith, 1966, p. 230):

> The end product of most [systems-analysis work] . . . is not an agenda of mechanical policy moves for every contingency—plainly an impossible task—but rather a more sophisticated map of reality carried in the minds of policy makers.

Section 8.7 carries the discussion of this point of view further.

Considerable evidence exists (see Majone and Wildavsky, 1978, and Wittrock and deLeon, 1986) that, if a decisionmaker decides to implement a program, he will do best to proceed adaptively with a decision to begin a cautious and partial implementation, carefully monitored and analyzed, followed by an altered implementation, again carefully monitored and analyzed, and so on. Such an adaptive approach, both to management and analysis, has been used when irreducible uncertainties were present in connection with ecological problems (see the *Overview,* p. 12, and, for details, Holling, 1978). This adaptive approach, like the one employing a pilot experiment, not only allows the analyst and decisionmaker to manage and analyze iteratively, it also permits them to build their confidence iteratively in the operational validity of their work.

Dynamic Validity

It is one thing to initiate a program as a result of an analytic study; it is another to keep it running effectively over a long period. A program is dynamically valid if it can be maintained—and adjusted if need be—to remain effective throughout its life cycle. This aspect of validity has been little discussed in the literature, but should not be neglected. Experienced analysts and decisionmakers will be able to make preliminary judgments on this matter as plans for implementation are prepared, and early performance outcomes should also shed light on the issue. In any case, it seems likely that judgments of dynamic validity must be based on intrinsic properties of the emerging program as the work moves from design through planning to implementation. An implementation plan prepared before a decision to implement is made final can be an important early harbinger of the dynamic validity of a program, and this is one of the virtues of preparing such a plan.

Summary Remarks

Efforts at validation must inevitably respond to the special features of the study and its setting. Thus, validation work can be expected to differ from case to case. Although some of the types of validation procedures discussed in this section may be used, others may also be called into play, and the mix and emphasis must be adapted to the case in hand.

It is not uncommon for the various validation procedures to yield differing results—perhaps even contradictory ones. This is not always the disaster that it might appear to be; rather, it may merely indicate that the analysis needs adjustment, or that the overall evaluation of adequacy must be guarded, with exceptions noted, or both. The need to make many such adjustments does nothing to increase confidence in the analysis, however.

There are two key points for every analysis effort:

It is essential for the analysis team to approach the issue of validation explicitly, conscientiously, energetically, and candidly.

The results of the validation work must be shared with the client, this sharing being best done on a continuing basis as the work of the study proceeds.

13.8. Conclusion

The verification process is an essential part of building analytic structures, as well as mathematical and computer models, and of insuring that they may be used with reasonable confidence to support operating and policy decisions. Verification is an essential component of validation, particularly when computer models are involved.

The validation process is an important step in systems analysis, although it cannot be absolutely or logically conclusive. To summarize the situation:

[T]he aim of the validation . . . attempts is to increase the degree of confidence that the events inferred from the model will, in fact, occur under the conditions assumed. When you have tried all the reasonable [validation and] invalidation procedures you can think of, you will not, of course, have a valid model (and you may not have a model at all). You will, however, have a good understanding of the strengths and weaknesses of the model, and you will be able to meet criticisms of omissions by being able to say why something was left out and what difference including it would have made. Knowing the limits of the model's predictive capabilities will enable you to express proper confidence in the results obtained from it. [Quade, 1980, p. 34.]

And, of course, the phrase "analytic structure" could replace the word model throughout this quotation.

The validation process is not designed to establish the truth or ultimate validity of a model or an analysis. It is aimed instead at establishing limits of credibility and determining if the degree of belief in both the analysis teams and the users is high enough for the findings to support their intended uses adequately. Although no particular numbers or types of tests can be stated *a priori* as establishing such a confidence, a single failure of a significant test can destroy it.

Unfortunately, the writings on the subjects of verification and validation are scattered and nowhere near as complete and systematic as one would like; however, Gass (1983) gives a useful overview that includes 137 references to the literature. Thus, one of the important opportunities facing systems and policy analysts is to contribute to the further growth and maturity of this aspect of their profession.

Finally, it must be said that conscientious and candid attention to the processes of verification and validation is one of the hallmarks of professionalism in systems and policy analysis.

References

Altman, S. M. (1980). Pitfalls of data analysis. In Majone and Quade (1980), pp. 44–56.

Arthur, W. B., and G. McNicoll (1975). Large-scale simulation models in population and development: What use to planners? *Population and Development Review* 1, 251–265.

Bowen, K. C. (1985). Personal communication, September 25, 1985.

Brewer, Garry (1973). *Politicians, Bureaucrats, and the Consultant.* New York: Basic Books.

Brigham, Georges (1955). On a congestion problem in an aircraft factory. *Operations Research* 3, 412–428.

Carlson, W. M. (1970). A decade for dialogue. *Communications of the Association for Computing Machinery* 13, 713.

Drucker, P. F. (1959). Long-range planning: Challenge to management science. *Management Science* 5, 238–249.

Edie, L. C. (1954). Traffic delays at toll booths. *Operations Research* 2, 107–138.

Energy Systems Program Group of the International Institute for Applied Systems Analysis, Wolf Haefele, Program Leader (1981). *Energy in a Finite World: Vol. I. Paths to a Sustainable Future*. Cambridge, Massachusetts: Ballinger.

Freeman, H. E., ed. (1978). *Policy Studies Review Annual*, Volume 2. Beverly Hills, California, Sage.

Gass, S. I. (1983). Decision-aiding models: Validation, assessment, and related issues for policy analysis. *Operations Research* 31, 603–631. (Contains a bibliography of 137 items.)

———, and B. W. Thompson (1980). Guidelines for model evaluation: An abridged version of the U.S. General Accounting Office exposure draft. *Operations Research* 28, 431–439.

Goeller, B. F., A. F. Abrahamse, J. H. Bigelow, J. G. Bolten, D. M. de Ferranti, J. C. DeHaven, T. F. Kirkwood, and R. L. Petruschell (1977). *Protecting an Estuary from Floods—A Policy Analysis of the Oosterschelde. Vol. I: Summary Report*, R-2121/1-NETH. Santa Monica, California: The Rand Corporation.

———, and the PAWN Team (1985). Planning the Netherlands' water resources. *Interfaces* 15(1), 3–33.

Green, Linda, and Peter Kolesar (1984). *Testing the Validity of a Queuing Model of Police Control*, Research Working Paper No. 512A. New York: Graduate School of Business, Columbia University.

———, and ——— (1985). *On the Validity and Utility of Queuing Models of Human Service Systems*, Research Working Paper No. 85-5. New York: Graduate School of Business, Columbia University.

Greenberger, Martin, M. A. Crenson, and B. L. Crissey (1976). *Models in the Policy Process: Public Decision Making in the Computer Era*. New York: Russell Sage Foundation.

Holling, C. S., ed. (1978). *Adaptive Environmental Assessment and Management*. Chichester, England: Wiley.

House, P. W., and John McLeod (1977). *Large-Scale Models for Policy Evaluation*. New York: Wiley.

Ignall, E. J., Peter Kolesar, and W. E. Walker (1978). Using simulation to develop and validate analytic studies. *Operations Research* 26, 237–254.

Kolesar, P. J. (1979). Travel time and travel distance models. In Walker, Chaiken, and Ignall (1979), pp. 157–201.

Lee, A. M. (1966). *Applied Queuing Theory*. London: Macmillan.

Majone, Giandomenico (1980). Policies as theories. *OMEGA: The International Journal of Management Science* 8, 151–162.

——— (1984). The craft of applied systems analysis. In *Rethinking the Process of Operational Research and Systems Analysis* (Rolfe Tomlinson and István Kiss, eds.). Oxford, England: Pergamon, pp. 143–157. This paper was prepared in 1974 as an internal document in the project at the International Institute for Applied Systems Analysis, Laxenburg, Austria, that was beginning work on this handbook; it was issued there in 1980 as WP-80-73.

——— (1985). Systems analysis: A genetic approach. In Miser and Quade (1985), pp. 33–66.

————, and E. S. Quade, eds. (1980). *Pitfalls of Analysis*. Chichester, England: Wiley.

————, and A. Wildavsky (1978). Implementation as evolution. In Freeman (1978), pp. 103–117.

Meadows, Donella, J. Richardson, and G. Bruckmann (1982). *Groping in the Dark: The First Decade of Global Modeling*. Chichester, England: Wiley.

————, and J. M. Robinson (1985). *The Electronic Oracle: Computer Models and Social Decisions*. Chichester, England: Wiley.

Meltsner, A. J. (1980). Don't slight communication: Some problems of analytical practice. In Majone and Quade (1980), pp. 116–137.

Miser, H. J., and E. S. Quade, eds. (1985). *Handbook of Systems Analysis: Overview of Uses, Procedures, Applications, and Practice*. New York: North-Holland. Cited in the text as the *Overview*.

Overview. See Miser and Quade (1985).

Quade, E. S. (1968). Pitfalls and limitations. In *Systems Analysis and Policy Planning* (E. S. Quade and W. I. Boucher, eds.). New York: American Elsevier, pp. 345–363.

———— (1980). Pitfalls in formulation and modeling. In Majone and Quade (1980), pp. 23–43.

Ravetz, J. R. (1971). *Scientific Knowledge and its Social Problems*. New York: Oxford University Press.

Schön, D. A. (1983). *The Reflective Practitioner: How Professionals Think in Action*. New York: Basic Books.

Shubik, Martin, and G. D. Brewer (1972). *Models, Simulations, and Games—A Survey*, R-1060-ARPA/RC. Santa Monica, California: The Rand Corporation.

Smith, B. L. R. (1966). *The Rand Corporation: Case Study of a Nonprofit Corporation*. Cambridge, Massachusetts: Harvard University Press.

Strauch, R. E. (1974). *A Critical Assessment of Quantitative Methodology as a Policy Analysis Tool*, P-5282. Santa Monica, California: The Rand Corporation.

Van Horn, R. L. (1971). Validation and simulation results. *Management Science* 17, 247–258.

Walker, W. E., J. M. Chaiken, and E. J. Ignall, eds. (1979). *Fire Department Deployment Analysis—A Public Policy Analysis Case Study (The Rand Fire Project)*. New York: North-Holland.

Wittrock, B., and P. deLeon (1986). Policy as a moving target. *Policy Studies Review* 6(4), 44–60.

Chapter 14
A Framework for Evaluating Success in Systems Analysis

Bruce F. Goeller

14.1. Introduction

The central purpose of systems analysis is to help parties at interest in problem situations to understand what is at work in them and to respond effectively to their problems. In achieving this purpose, an analysis team must deal skillfully and imaginatively with many technical matters, as discussed in this handbook. There are, however, some additional non-technical matters that must be handled successfully if the analysis is to be effective, such as:

Identifying the parties at interest in the problem situation and developing appropriate relations with them.

Understanding the kinds of success a study may achieve, so as to help the parties at interest as much as possible.

This chapter has two related purposes. The first is to appreciate and identify the roles that various parties at interest may play in a systems analysis. This discussion can help analysts in identifying the parties at interest in a particular problem situation, in developing appropriate relations with them, and in communicating the study's results to them effectively.

The second purpose is to present a framework that describes the different kinds of success that may be achieved by a systems analysis and the different parties' possible criteria for evaluating success. This framework for evaluating success can aid analysts in setting realistic success goals for a study and in pursuing them effectively so as to avoid pitfalls. In addition, understanding gained from the framework can aid analysts in setting up and managing studies. For example, the framework has implications for recognizing circumstances that are such harbingers of failure as to raise doubts whether a study should be undertaken, for anticipating and limiting the handicaps that might hamper the success of an analysis, and for staffing the study in ways that may promote the use as well as the technical quality of its results.

The next section takes up three examples that will be used to illustrate many of the points. Then Section 14.3 discusses the parties at interest and Section 14.4 describes the framework, identifying the kinds of success and various parties' criteria for measuring them. Finally, Section 14.5 offers some conclusions concerning success in systems analysis.

14.2. Illustrative Cases

In order to clarify and illustrate the general points made, this chapter makes use of three systems-analysis studies in which the author was a project leader. Although these three studies do not exemplify the full range of what may be involved in systems-analysis work, they were all complex and various enough to support the points of concern in this chapter.

The purpose of this section is to offer a brief introduction to each case and list where it is discussed further in this volume and elsewhere. The reader would be well advised to refresh his memory of these cases before proceeding to the rest of the chapter.

The San Diego Clean Air Project (SANCAP)

For some time air pollution has been a significant problem in the United States, particularly in the air basins surrounding urban areas, which accrue pollution from industry, electric power generation, and transportation systems. In 1970 the U.S. Congress passed the Clean Air Act Amendments setting various air-quality standards for air basins in the United States and requiring that the states prepare and promulgate air-quality implementation plans for each region within their boundaries to meet the standards by 1975. The federal Environmental Protection Administration (EPA) was given major responsibilities to review these plans, evaluate their ability to meet the air-quality standards, and penalize states for shortfalls.

In 1970, when the amendments passed, San Diego County in southern California violated the standard for oxidant, its most difficult species, on 226 days. In 1971 the county's Office of Environmental Management concluded that, even after the new controls legislated for new cars and other sources had been implemented, the region would still be violating the standards substantially in 1975, the deadline year. Recognizing that additional controls were needed and that the overall impact on San Diego

County could vary considerably with the control strategy selected, the county in late 1971 submitted a grant application to the EPA to support a thorough evaluation of alternative strategies for meeting air-quality standards in 1975 and subsequent years. Concurrently, The RAND Corporation proposed to the EPA a research design for building upon past impact-assessment work to create a new methodology for evaluating urban transportation and environmental strategies. The EPA then authorized a grant to the County with the understanding that the principal research activity would be subcontracted to RAND, funded by approximately two-thirds of the grant. After negotiations between the county and RAND to determine the work assignments for both organizations, a research contract with RAND for a one-year-long study was approved by the county in late May, 1972.

The resulting San Diego Clean Air Project (SANCAP) was undertaken to analyze alternative strategies in terms of a comprehensive set of impacts on the quality of life in San Diego and to help identify the most promising strategy for implementation. The project was a joint venture of the county's Office of Environmental Management and the RAND Corporation, and involved nearly five and a half person-years of effort by RAND and about three by the county.

The county's role in the research and analysis phase of the project was to (1) manage the joint project, (2) develop much of the regional data base, (3) prepare initial sets of alternative emission-reducing tactics, and (4) analyze institutional and implementation issues.

The RAND Corporation's role was to (1) develop analysis methodology; (2) define and combine the alternative pollution-reducing tactics into cost-effective mixes (strategies); (3) evaluate quantitatively the environmental, service, economic, and distributional impacts of the strategies, and (4) develop and apply presentation techniques for the findings.

After analysis results were presented to the public and to policymakers for selection of the preferred alternative, the county's role in the implementation-planning phase of the project was to be the revision of the San Diego Air Quality Implementation Plans to reflect the preferred alternative.

The alternative strategies considered in the study were mixes of: *fixed-source controls,* which use technological and managerial controls on emissions from fixed sources, including not only smokestacks and organic solvent users, but also aircraft, because their significant emissions occur within airfields; *retrofit devices and inspection-maintenance policies* for vehicles; and *transportation management,* which includes bus-system improvements, carpooling incentives, and gasoline surtaxes or rationing to reduce the person-trips and vehicle-miles that generate vehicle emissions.

The project's analysis results were briefed to county policymakers in May 1973, and the final report (Goeller et al., 1973a) was circulated as a draft in early July and published in December, 1973. Section 10.4 gives a fairly extended summary of the SANCAP analysis and its findings. Sections 14.3 and 14.4 discuss what the various parties at interest did with the results and the consequences for San Diego County.

The Policy Analysis of the Oosterschelde (POLANO)

In 1953 a severe storm flooded much of the Delta region of the Netherlands, killing several thousand people and inundating 130,000 hectares of land. In 1954 the Dutch government embarked on a massive construction program for flood protection. By 1975 the new dams, dikes, and other works were nearly complete for all Delta estuaries except the largest, the Oosterschelde. There the program was interrupted by controversy, as goals other than security began to compete for attention.

The original plan had been to construct an impermeable dam across the nearly 9-kilometer-wide mouth of the Oosterschelde, thereby closing it off from the sea completely, a step that would have entirely changed the rare ecology, wiped out the local oyster and mussel fishing industries, and produced many other effects. People with a special interest in protecting the fishing industry or preserving the natural environment voiced strong opposition. Those primarily concerned with safety, however, supported the original plan.

As a response to enormous controversy, in November 1974 the Dutch Cabinet directed the Rijkswaterstaat, the government agency responsible for water control and public works, to prepare a report within 18 months on the technical feasibility, financial costs, and construction time of an alternative approach: constructing a *storm-surge barrier* in the mouth of the Oosterschelde. The barrier was to be a flow-through dam with large gates that would be closed during severe storms, but that would be open in normal weather to allow a reduced tide to pass into the basin, the size of the tide being governed by the aggregate size of the opening in the barrier.

At the same time, the Cabinet, after heated debates in Parliament, also declared that the storm-surge barrier would be built—provided that the study showed that the barrier could meet a set of specified conditions: it must be technically feasible, must be completed by 1985, must cost no more than a stipulated amount, and must provide protection against a storm so severe that it might be expected to occur only once in 4,000 years. Unless these conditions were met, the original plan would supposedly be carried out. This "rather unusual political maneuver," Goemans (1986, p. 6) points out, temporarily "cleared the air but at the same time set the stage for stormy weather one and a half years later."

The Rijkswaterstaat found itself in a dilemma. Although it found the barrier feasibility study a stimulating challenge, it had previously "defended the original plan and warned that a storm-surge barrier . . . might be beyond the limits of technology . . ." (Goemans, 1986, p. 6). Thus it feared the conclusions of its report would be greeted with suspicion, regardless of what they were. If the report concluded that all conditions could be met, there might be suspicion that the Rijkswaterstaat had been too optimistic because it was eager for the challenge of the barrier, which would produce cost and schedule overruns if it was started. On the other hand, if the report concluded that some condition(s) could not be met, there might be suspicion that the Rijkswaterstaat had been too pessimistic because it still favored the old plan.

To avoid this dilemma, the Rijkswaterstaat decided to start, in parallel with the technical feasibility study, a policy-analysis study that would compare the consequences of the different approaches to the fullest possible extent. The *Policy Analysis of the Oosterschelde* (POLANO) project was established in April 1975 as a joint research venture between The RAND Corporation and the Rijkswaterstaat. The Rijkswaterstaat asked RAND to help because it had extensive experience with similar kinds of analysis (such as SANCAP) and for several years it had been working with the Rijkswaterstaat on other problems. Also, the involvement of a nonprofit corporation from the United States was expected to enhance the study's credibility because the United States was then perceived as the world leader in dealing with environmental issues.

The project began with one year of analysis, during which each organization spent about eight person-years of effort on joint research, concentrating on different but complementary tasks. RAND's primary task was to develop and then apply a methodological framework for predicting and comparing the many possible consequences of the alternatives; RAND's other tasks were to help the Rijkswaterstaat staff coordinate their various study activities on the Oosterschelde (including the feasibility study) by showing interrelations and identifying data problems, and to make them familiar with policy-analysis techniques by participating in joint research. The Rijkswaterstaat's primary tasks were, on the basis of special engineering and scientific studies, to develop a specific design for each alternative approach, to analyze the consequences of the designs in which it had special expertise (such as the effects on salinity), and to provide data, as well as assistance, for the methodology being developed with RAND.

The study considered three major alternatives (each with variations within its major design concept):

A dam across the mouth of the Oosterschelde, closing it off from the sea completely.

A storm-surge barrier at the mouth of the Oosterschelde that would be closed during severe storms but that would be open under normal weather conditions to allow a reduced tide to flow into the estuary.

A system of large dikes around the estuary's perimeter that would leave the estuary open in order to maintain the original tidal conditions.

There was an obvious fourth alternative—to do nothing, except possibly providing for enhanced storm prediction complemented by disciplined evacuation procedures and indemnification funds to compensate those experiencing storm losses—but such a possibility was not politically acceptable to the Dutch public, and so was not given serious consideration.

Seven categories of consequences were considered for each alternative: financial costs, ecology, fishing, shipping, recreation, national economy, and regional effects. Within each category there were several types of consequences to consider.

In April 1976, RAND presented an all-day briefing to the Rijkswaterstaat describing the methodological framework that had been developed and summarizing the results of the POLANO analysis. The Rijkswaterstaat combined this work with several special studies of its own and, in May 1976, submitted its report (Rijkswaterstaat, 1976) to the Cabinet. Sections 14.3 and 14.4 describe what the Cabinet and Parliament did with the report and what has happened in the Oosterschelde.

The POLANO results and approach are discussed briefly in Section 2.2, and more fully in the *Overview* (pp. 89–109). Goeller et al. (1977) provide a much more comprehensive summary, together with an overview of the extensive supporting analyses.

The Policy Analysis for Water Management of the Netherlands (PAWN)

Historically, the water-management problem for the Netherlands has been too much water, not only during storms but also in normal circumstances, as much of the land lies below the level of the North Sea. But in recent decades the Dutch have faced the less dramatic, but no less urgent, problem of too little fresh water and too much pollution brought on by increased industrialization and a burgeoning population with a high standard of living.

Much of the Netherlands' wealth is derived from crops grown on irrigated land. Agriculture is by far the largest user of fresh water in the Netherlands, so water shortages can cause large economic losses.

The Rhine River, which enters the Netherlands from West Germany and flows through the country, is the Netherlands' major source of surface water for agricultural irrigation and other purposes. Unfortunately, it also brings with it a substantial amount of pollution. Salinity, which has both foreign and domestic causes, is the Netherlands' most serious water-quality problem, and eutrophication—heavy growths of algae in relatively stagnant waters of lakes and reservoirs—is the next most pressing and widespread.

Along with several other rivers and numerous Dutch canals, the Rhine is also a major artery for the inland shipping fleet of Western Europe, on its way to Rotterdam Harbor, the world's busiest port. Low water levels in the rivers and canals can cause serious shipping delays and economic losses because only partially laden ships can navigate the inland waterways.

In 1976 a severe drought cost the Netherlands more than $2,500 million in agricultural losses alone, over 4 percent of its gross domestic product, while worsening water-quality problems. Although at this time the supply of fresh surface water was adequate except in dry years, the country faced the prospect that the supply would be inadequate in normal conditions by the late 1980s. Moreover, the supply of groundwater, which was very popular because of its quality, was already inadequate.

Facing such water-management problems, and others not mentioned here, the Rijkswaterstaat—which, of course, had had a successful experience with the POLANO project—commissioned an analysis that it hoped would provide a basis for a new national water-management policy. Begun in April 1977, the *Policy Analysis for the Water Management of the Netherlands* (PAWN) project was conducted jointly by The RAND Corporation, the Rijkswaterstaat, and the Delft Hydraulics Laboratory (a leading Dutch research organization).

The PAWN project was a major undertaking. Considering both research and documentation, it directly involved over 125 person-years of effort (more than one-third in data gathering), about 48 by RAND.

PAWN's primary tasks were to:

Develop a methodology for assessing the multiple consequences of possible water-management policies.

Apply it to generate alternative water-management policies and to assess and compare their consequences.

Create a Dutch capability to conduct further analyses of this kind by training Dutch analysts and by documenting and transferring the methodology to the Netherlands.

After more than two and a half years of research, the PAWN final briefing described the methodology and joint results in December 1979 to an audience of several hundred senior representatives of governmental and private organizations concerned with water resources, environmental quality, and the economy.

Although this final briefing marked the end of the analysis phase for RAND, the documentation phase continued for several years. The Rijkswaterstaat wanted unusually thorough and extensive documentation, both to support the new policy and to use in training new analysts and in performing new studies.

The Rijkswaterstaat combined results of the joint PAWN analysis with results of its own (performed with the PAWN methodology by Dutch analysts trained in PAWN) to draft its highly detailed new national policy document on water management. This document went to Parliament for formal approval in November 1984.

Section 6.3 offers an overview of the work with particular attention to the issue of screening the many alternatives that could have been considered in order to arrive at a small set for detailed analysis. Goeller and the PAWN Team (1985) offer a more extensive overview; Goeller et al. (1983), a volume of almost 500 pages, describes comprehensively the methodology and results and offers an entry to the large literature supporting them.

The next two sections identify many of the parties at interest in these three cases, describe the different kinds of success that may be achieved by systems analysis, and discuss how successful the cases were by various criteria.

14.3. Parties at Interest

The analysis of an important problem situation may be of interest to many parties, including study sponsors, analysts, policymakers, and organizations and individuals potentially affected by the situation or what is done about it. Together, these parties largely determine the success of an analysis, for their actions and interactions govern how the analysis is conducted, what is done with the findings, who is affected, and how the effects are perceived.

There is no general prescription for identifying the parties at interest for a particular study other than to make their discovery an important early goal. This section's purpose is to define the potential parties in terms of the roles they perform. Understanding these roles will help discover the parties at interest in a particular study and provide context for the framework described in the next section.

This section now presents and discusses two lists, one from the perspective of the problem situation and the other from the perspective of the analysis, showing the roles of potential parties at interest. A particular individual or group may play more than one role on both lists. The lists are intended to be suggestive rather than exhaustive.

Parties Related to the Problem Situation

Some parties may be affected by the problem situation or what is done about it whereas others may affect the decisions about what is to be done. We can describe these parties by their roles in relation to the problem situation; they include:

Policymaker for the problem situation

Implementor for the policy or program

Operator of the implemented policy or program

Decisionmaker for the problem situation

Responsibility taker for the problem situation (Shubik, 1984, p. 22)

Staff member for any of these types of persons

Persons affected by the problem situation

Lobbyer

Advisor on the decision

Evaluator of the implemented policy or program

Enforcer for the implemented policy or program

For some of these roles the definition is obvious, but others deserve some discussion, with examples from the illustrative cases.

Policymaker. Although the definition of decisionmaker given below is consistent with typical systems-analysis definitions (see, for example, the *Overview, p. 167*), this chapter's definition of policymaker, which is generally a synonym, is narrower. This is because I find it essential to distinguish between policymaking and implementation: they have innately different concerns, they occur at different stages in the attempt to improve a problem situation, and they are usually done by different organizations.[1]

In my view, policymaking is concerned primarily with selecting among options (policies) that are relatively general descriptions of *what* is to be done. Implementation, by contrast, is more concrete and specific, for it concerns *how* to do something; that is, what actions by what institutions will bring the selected policy into being. For example, if the problem situation involves improving a regional bus system, policymaking might specify that the route spacing should be two blocks rather than four blocks whereas implementation might specify that the local transit company, financed by a special subsidy, should run buses on First and Third streets rather than on Second and Fourth streets. (See the discussion by Goeller in the *Overview,* pp. 214–216.)

This chapter defines a *policymaker* as an individual, group, or organization that can establish or modify policies (and programs), which, through implementation, affect problem situations. Although policymaking authority or responsibility is concentrated within a single individual

---

[1] Implementation and its differences from policy selection are discussed more fully in the *Overview* (Chapter 9) and in Brewer and deLeon (1983, Chapters 2 and 7–10).

in some situations, this is rare; it is usually distributed among multiple individuals or organizations.[2]

Thus we often find multiple actors playing three distinct policymaking subroles:

The *nominator*, who recommends a particular option or presents a short list with several promising options.

The *selector*, who chooses the preferred option, but not necessarily from those offered by the nominator.

The *ratifier*, who may veto, approve, or modify the selector's choice.

As examples, consider these subroles in the illustrative cases. For POLANO, the nominator was the Rijkswaterstaat, which submitted a report to the Cabinet (Rijkswaterstaat, 1976) that compared the three disparate options considered in the analysis but, by intention, did not recommend a particular alternative. The selector was the Cabinet, which decided on the storm-surge-barrier alternative. The ratifier was the Dutch Parliament, which approved the Cabinet decision to proceed with the storm-surge-barrier approach, but raised the issue of possible modifications to the barrier design that, in normal weather, would permit a larger flow of water into the basin, benefiting the ecology and the fishing industry. In response, the Minister of Transport and Public Works commissioned a Rijkswaterstaat study on the implications of a larger opening in the barrier, from which the Cabinet decided, and Parliament subsequently accepted, that a modest increase (to an area of 14,000 sq. meters) "would be a reasonable balance between additional costs and benefits" (Goemans, 1986, pp. 9, 10).

For PAWN, the nominator was also the Rijkswaterstaat, which used the PAWN analysis (Pulles, 1985) to draft the national policy document on water management (Rijkswaterstaat, 1985) that presented one coherent policy. However, the policy was shaped in part by other government ministries, provincial agencies, water boards, industrial organizations, environmental groups, and so on, which were involved in the extensive, consensus-building discussion process typical of Dutch-style democracy. The Minister of Transport and Public Works, acting as the selector, readily accepted the policy, which was near consensus, and Parliament ratified it soon after.

For SANCAP, the analysis results were the primary analytical inputs to the deliberations of the San Diego County Task Force on EPA regulations, which served as the nominator: The Task Force report recom-

---

[2] Indeed, House (1982, pp. 40–41) and others contend that the solitary policymaker-decisionmaker may be a "myth" and question whether, at the national governmental level, "the monarchical or powerful individual paradigm is operable. . . ."

mended a particular strategy, developed in the analysis, as an alternative to an EPA plan. The San Diego County Board of Supervisors endorsed the report and selected the recommended strategy to be presented at EPA hearings. The Region IX office of the EPA, acting as the ratifier, subsequently promulgated the Final Implementation Plan for the San Diego Air Quality Control Region (October 15, 1973), a plan that was similar in many respects to the San Diego strategy, but had some major differences in the transportation management controls (Goeller et al., 1973a, pp. 71, 94).[3] Subsequent revisions and compromises occurred that will not be discussed here.

Implementor. The *implementor* attempts to execute the policy chosen by the policymaker. In this process, decisions and actions must be taken to flesh out the relatively general description of the selected policy; concrete details must be added and open issues resolved. Sometimes the result is consistent with the policymaker's intent, but it often differs substantially from what the policymaker intended. Some differences result from adapting the policy to make it more practical or to accommodate new political concerns. Others occur because the policymakers and the implementors are usually different—different individuals and organizations, each with different values, perceptions, and incentives.

Brewer and deLeon (1983, p. 20) suggest these activities as characteristic of implementation:

Developing rules, regulations, and guidelines to carry out [policy] decision

Modifying [policy] decision to reflect operational constraints, including incentives and resources

Translating [policy] decision into operational terms

Setting up program goals and standards, including a schedule of operations

To these we may add:

Developing a complete and detailed design for the policy

Preparing the implementation plan

Constructing or procuring needed facilities and equipment

Hiring and training needed personnel

---

[3] Note that the 1970 Amendments to the Clean Air Act required the EPA to determine the efficacy of implementation plans submitted by the states. If the State of California had submitted a plan acceptable to the EPA in time to meet its various deadlines, the state would have been the selector (and the county the nominator).

Drafting and letting contracts for other organizations to perform activities such as the previous two

Administering contracts

Establishing organizations or associations of existing organizations

Operating the implemented policy or program

Of course the particular mix of activities will depend on the nature of the policy and the problem situation.

Implementation often finds different actors playing three distinct subroles:

The *installer*, who creates the facilities or assembles the resources necessary to get the (new or modified) policy into operation.

The *operator*, who is responsible for the day-to-day operation of the implemented policy.

The *implementation manager*, who develops guidelines, prepares plans, administers contracts, and so on, and may select and manage the installers and operators or perform some of their functions.

For POLANO, the contractors who built the barrier and related facilities such as compartmentation dams and locks can be considered installers. But the Rijkswaterstaat was the implementation manager: Its staff selected the contractors from several competing groups; it drafted, negotiated, and administered the contracts under which the contractors and various research institutes conducted their work; it established the schedule and the implementation plan; and it worked with the contractors to develop the detailed design for the barrier, which underwent several revisions while adapting to practical problems.

The Rijkswaterstaat also had several open issues to confront during implementation, which it did by initiating several smaller policy studies with the POLANO approach and then taking appropriate actions. These open issues included deciding on the exact location of the compartmentation dams and determining the type and dimensions for the shipping locks to go in the dams.

However, the largest managerial decision, and supporting study, considered an issue of operational policy rather than installation: selecting the control strategy for operating the barrier's many large gates. Such a strategy includes (1) the actions that govern the times and rates at which gates close and open, (2) the rules underlying the decisions for these actions, and (3) gathering and processing the required information. In choosing a control strategy, the familiar POLANO conflict between security and ecology was confronted again: From the standpoint of flood security the barrier should be closed as long as necessary, although from the standpoint of ecology it should be closed as briefly as possible. And the choice was complicated by uncertainties in forecasting storm behavior and resulting water levels. The strategy choice was made largely on the basis of analysis by the BARCON (barrier control) project, a joint effort of the Rijkswaterstaat and The RAND Corporation (Catlett et al., 1979).[4]

For the national water-management policy developed through PAWN, the Rijkswaterstaat is probably the major implementation manager and operator. Among other things, it is responsible for constructing the Brielse Meer pipeline, for adopting a more stringent thermal standard to limit heat discharge into canals, and for introducing and operating a new flushing policy to reduce the salinity of the Markermeer (an enormous freshwater reservoir three-fourths the size of San Francisco Bay). But the provinces and water boards (local water-management authorities, similar to school boards) are also important implementors. Under the regional approach to eutrophication recommended by PAWN, provinces are using improved versions of the PAWN eutrophication models to develop tailored combinations of control tactics for their lakes and reservoirs (Goeller and the PAWN Team, 1985, pp. 30, 31). Moreover, about half the provinces and numerous water boards have cut back their plans for new facilities on the basis of PAWN recommendations (Pulles, 1987).

For SANCAP, fragmentation of power complicates implementation, particularly if the County tries to implement its preferred strategy. Goeller et al. (1973a, p. 73) note:

> When one examines the various tactics for improving air quality, he discovers how many different organizations are potentially involved in implementation. The local APCD [Air Pollution Control District] and the Navy may influence fixed source control. County government may indicate the need for a particular retrofit strategy, but only the state and Federal governments have the power to require installation. The San Diego Transit Corporation and the city of Oceanside manage the bus service in the San Diego region, but the various penalities against the automobile that might make the bus more attractive are beyond their powers; additional parking fees require action by each of the thirteen cities plus the county, while additional gasoline charges require state or Federal action. And, of course, there are helpful actions by citizen groups and individuals.

Operator. On the one hand, operators are considered independent actors who play a major role after implementation is complete; on the other, they should often play a subrole during implementation, as mentioned

---

[4] Another important operational policy decision during implementation involved developing an ecosystem management plan for each water basin being created by the construction. This required extensive negotiation between the different levels of government (national, provincial, and municipal) because the administrative structure did not correspond with the ecosystem boundaries.

above. The *operator* is responsible for managing and maintaining the implemented policy, whatever form it finally takes, in day-to-day operation. If the operators are different organizations or individuals than the implementation managers or installers, the result may differ from what the managers intended. The difference will depend on the adequacy of the operational guidelines (their concreteness and comprehensiveness), the character of the operators, and the nature of their incentives. It will also depend on the extent to which the operators' concerns and constraints are considered during implementation planning and installation.

For POLANO, the operator is the Rijkswaterstaat Directorate for the Province of Zeeland, the location of the storm-surge barrier. In an artful attempt to minimize differences between implementation and operation, as the barrier neared completion the man who had headed the Rijkswaterstaat organization primarily responsible for constructing the barrier (the Delta Service) was appointed to head the organization that would be responsible for operating it.

For PAWN, the major implementors—the Rijkswaterstaat, the provinces, and the water boards—are operators as well as managers. And the same is true for most of the implementors in SANCAP.

But many problem situations include operators not commonly considered to be implementors. This generally occurs when the policy being implemented by one organization employs pricing measures (charges and surtaxes) or regulation measures (administrative and legal restrictions) to influence the behavior of others. Consider some examples from SANCAP. When the county and the various cities implement parking surcharges, or when the state or federal government implement gasoline surtaxes, they are trying to influence automobile operators to drive less and, in the long run, to buy automobiles that are simultaneously more fuel efficient and less polluting. When the Air Pollution Control District sets quotas on the permissible emissions by individual firms, it is trying to influence the operators of the firms; in the short run operators may reduce the production of goods to reduce associated emissions, although in the long run they may substitute less-polluting equipment or production processes.

Price and regulation measures usually afford the affected operator considerable flexibility in choosing his response. But monitoring and enforcement activities are often necessary to ensure that the operator's response has the desired effects.

Decisionmaker. In this chapter's usage, a *decisionmaker* may be either a policymaker or an implementor. This is equivalent to typical system-analysis definitions, as mentioned before (although not all definitions seem to include the operator subrole).

Persons affected by the problem situation. This role includes persons directly affected; for example, those threated by floods and those employed constructing the barrier in POLANO, and those using the water supply in PAWN. But the role also includes those indirectly affected—in future as well as present generations. Future generations may suffer deaths or deformities from the present generation's discharges of toxic metals into the water. And expenditures on the barrier or water-management facilities create jobs not only in the industries directly involved in construction but also indirectly in other interrelated industries, such as steel and concrete production.

Lobbyer. A *lobbyer* is an individual or group that seeks to influence policymakers toward a particular viewpoint. Lobbyers may be persons affected by the problem situation, persons in organizations affected by the problem situation but not directly involved in it themselves, or concerned persons outside the problem situation.

In POLANO, for example, formal and informal lobbyers included the Provincial Government of Zeeland, which wanted primarily to assure adequate flood protection for its residents; oyster and mussel fishing-industry groups who wanted to maintain their livelihood; environmental groups that wanted to preserve the Oosterschelde's ecology; and Comite Samenwerking Oosterschelde (SOS), a consortium of various groups that wanted to keep the mouth of the Oosterschelde open. (Even several years after Parliament ratified the decision to build a storm-surge barrier in the mouth of the Oosterschelde, SOS was actively criticizing the decision and lobbying for an open Oosterschelde.)

In pursuing their goals, lobbyers may apply political pressure, present arguments, provide information, or offer political support on other issues. Many are ardent volunteers rather than professional advocates paid to represent a viewpoint (that is, professional lobbyists).

Advisor on the decision. When a decisionmaker consults an advisor about the problem situation, he is seeking help. The *advisor* may supply information, conduct analysis, recommend action, suggest political strategy, or provide emotional support.[5] And it is in this advisory role that policy analysts usually appear in relation to a pending decision.

The main difference between analysts and other advisors is the degree to which their recommendations are based on problem-specific analysis, their assumptions are made explicit, and their personal preferences are set aside. (If an analyst allows his preferences to shape materially the

---

[5] Goldhamer (1978) discusses the role of the advisor from the perspective of advising the leader of a nation. Benveniste (1977, p. x) discusses "how experts influence public . . . policy," asserting that experts such as policy analysts fill a "new social role combining political and technical dimensions."

analysis or recommendations, he is functioning instead as a different kind of advisor—or even a lobbyer—and should make this role shift, and its implications, clear to the clients, as discussed in the *Overview*, pp. 320–325.)

There are many kinds of advisors. They may be inside a decisionmaker's organization or outside. They may be private consultants or academic researchers who are considered experts on the problem situation, or staff members who are considered nonexperts. And advisors may be individuals or groups, such as a council, committee, commission, or task force.

For example, the POLANO analysts were preceded by several other advisors. According to Leemans (1986, p. 50), a study group of the Technical University of Delft developed the initial idea of a tidal dam, the forerunner of the storm-surge barrier. A special advisory committee to the Dutch Cabinet, the Oosterschelde Committee, considered hundreds of different plans for protecting the Oosterschelde and recommended an approach (Goeller et al., 1977, p. 15; Goemans, 1986, p. 5). Then an important standing committee of the Dutch government, the Spatial Planning Committee, which included top officials of most ministries, was responsible for evaluating these recommendations and reaching its own conclusions (Leemans, 1986, p. 51).

As another example, the SANCAP analysis results were used by another advisor, the San Diego County Task Force on EPA regulations, as the source of the strategy recommended to the County Board of Supervisors. And Dutch water-management policymaking received advice not only from the PAWN study, but also from the ICWA, the Interdepartmental Committee for Water Management, which included top-level representatives of all government ministries concerned with water policy.

Evaluator. After the implementation of a policy or program, an *evaluator* compares the actual effects on the problem situation with the expected effects, using "established criteria."[6] An evaluator may belong to an organization involved in the policymaking, implementation or operation, or be a totally distinct and autonomous entity.

Enforcer. An *enforcer* is an evaluator with teeth; that is, when an enforcer observes an implemented policy performing unacceptably, he has power to induce policy changes. For POLANO, the Delta Institute at Yerseke is enforcer for the ecology; it can induce changes in the barrier-control strategy if it adversely affects the ecology. For PAWN, different

---

[6] Brewer and deLeon (1983, pp. 19, 20) pose the comparison in terms of performance levels. I feel this is too narrow, for a policy or program may have major effects—including some that are indirect, some that are spillovers, and some that are unanticipated—far beyond its performance levels.

parts of the Rijkswaterstaat, under Parliamentary oversight, monitor and enforce compliance with water-quality standards. For SANCAP, the EPA monitors regional compliance with air-quality standards and control plans, and can impose sanctions that withhold federal funds or restrict regional growth.

Parties Related to the Analysis

Thus far this section has discussed parties at interest from the perspective of the problem situation. Now it takes the perspective of the system analysis: Some parties may shape the analysis whereas others may affect how it is perceived or used. The roles of the parties at interest in relation to the systems analysis include:

Problem poser (*Overview*, p. 167)

Sponsor

Client

User

Member of the staff of any of these four types of persons

Systems-analysis team

Systems-analysis peer group

Research program director

Advisor on the analysis

Formal reviewer

Implementation planner

The categories where the definition is not obvious are discussed below.

Sponsor. This chapter refers to the individual who commissions the work and sees to its support as the *sponsor*. For POLANO this was H. Engel, Chief Engineer-Director of the Rijkswaterstaat's Delta Service, and his predecessor, H. A. Ferguson. For PAWN it was H. M. Oudshoorn, Chief-Engineer Director of the Rijkswaterstaat's Directorate of Water Management and Water Movement. And for SANCAP it was L. Edwin Coate, Director of the Office of Environmental Management during most of the project, and his successor David Nielson, who served as Acting Director.[7]

---

[7] Stanley Greenfield, EPA Assistant Administrator for Research and Development, was the godfather, if not the sponsor, of the project. He realized the potentialities of a San Diego County/RAND collaboration, helped frame the project objectives and guide its evolution, and provided the EPA grant to San Diego County that funded most of the project's activities.

The sponsor plays a major role in shaping the analysis. He commonly selects the research organization to perform the analysis, determines the funding level, and may influence the staffing. He also participates in problem definition and reviews progress and findings.

Note that a study sponsor is not necessarily a policymaker for the problem situation. In POLANO and PAWN, the sponsoring organization (the Rijkswaterstaat) played the nominator's role in policymaking, but did not select the policy. In SANCAP, the study sponsor was not a policymaker, although the study results were an essential element in the policymaking process.

Client. Much of this handbook refers to the client as meaning various things, including the sponsor. This chapter uses the term *client* to mean a potential user of the study findings, such as a decisionmaker, his staff, or an interest group trying to influence the decisionmaking.

Through their actions, clients determine directly how an analysis is used. But they may also influence indirectly how the analysis is performed, as analysts emphasize issues they know or expect to be important to clients.

Client examples appear earlier in this section, as parties related to the problem situation.

Research program director. A policy or systems analysis is typically organized as a project, with a small research team managed by a project leader. When multiple analysis projects are under way in an organization, similar projects are often organized as a research program, managed by a program director. The program director usually selects a project leader, helps negotiate the research agenda with the sponsor, and reviews progress and findings. Based on his strategic plan for the program, the program manager may allocate resources among competing projects and determine which prospective projects will start. The strategic plan reflects such goals as building intellectual capital for the analysts, promoting synergy among projects, developing a center of expertise, and creating a foundation for growth into new areas.

Advisor on the analysis. As it designs and conducts a systems analysis, the analysis team may receive guidance from selected advisors. Although such advisors may be solitary individuals, in complex problems they are often formal advisory groups, either study advisory groups—sometimes called steering committees—or technical advisory groups.

Study advisory groups provide guidance on the scope and emphasis of the research, that is, the issues to be considered and the relative importance of each. Because of its complexity and size, PAWN benefited from three such advisory groups. One was the Rijkswaterstaat advisory group, which consisted of senior managers from various Rijkswaterstaat departments concerned with different aspects of water management (such as shipping or water quality). Another was the IWW, a special working group of the ICWA (the Interdepartmental Committee for Water Management) established specially for discussions about PAWN and the national policy document on water management. The third was the RAND advisory group, which consisted of a senior RAND expert on hydrology who was born and educated in the Netherlands (J. J. Leenderstse), the head of the System Sciences Department (G. H. Fisher), and RAND's Senior Vice President (G. H. Shubert) (Goeller et al., 1983, pp. xxviii, xxix).

By contrast, technical advisory groups provide data and expertise on the more concrete and specialized aspects of the problem. They may suggest options for solving the policy problem, identify potential pitfalls for the analysis, propose analytic strategies, and critique methodology and recommend improvements. In PAWN there were technical advisory groups on pollution, nature, shipping, and technical and managerial options for improving water management (Goeller et al., 1983, p. xxviii). There were also individual technical advisors for particular topics in various study areas. For example, a senior researcher of the Ministry of Agriculture and Fisheries (H. Ton) spent a week at RAND reviewing PAWN's agricultural "research plan and preliminary versions of several . . . models" (Petruschell et al., 1982, p. xiii).

Implementation planners. Those who perform analysis specifically to assist implementation planning or decisionmaking are referred to here as *implementation planners,* a term I believe reflects the detailed and prescriptive nature of such activity better than its synonym "implementation analysts."

Depending on circumstances, the group of implementation planners may or may not include analysts from the team that supported the policymaking process. Such inclusion is certainly desirable, for the analysis team understands the problem situation, has experience assessing the implementability of proposed solutions, and should be well equipped to devise ways to resolve implementation difficulties. Only in this way can the policy-analysis team influence implementation directly. Otherwise its influence on implementation, if any, will be indirect, arising through its influence on policy selection or the subsequent use of the study results by implementation planners and decisionmakers.

14.4. Kinds of Success

For policy and systems-analysis studies, "success" is difficult to define and measure, partly because different parties at interest have different goals and perspectives. Definitions and criteria for success exist, but they all have weaknesses as well as strengths, and some, if used alone, might mask serious defects or ignore important contributions. This section describes various definitions and attempts to synthesize them into a reasonably coherent framework for describing success in systems analysis.

Three general kinds of success can be distinguished:

Analytic success considers how the study was performed and presented.

Utilization success considers how the study was used in policymaking or implementation.

Outcome success considers what happened to the problem situation (and those affected by it) as a consequence of the study.

On a *direct* basis, one assesses these three kinds of success with respect to the decision and the problem for which the analysis was commissioned. On an *indirect* basis, one assesses them with respect to other decisions and situations.

There is a hierarchy of dependence and difficulty among the kinds of success. Analytic success is the easiest to achieve, although it requires careful work to produce a high-quality study, one that is valid, credible, and pertinent to policymaking. Analytic success is also the foundation for the other kinds of success. For utilization success, study results must not only be used in the policymaking process but should also be of high quality; utilization of poor-quality results is no credit to a study. Outcome success is the most difficult to achieve, for it requires establishing that the study results were an important factor in policy selection and that implementation of the selected policy helped the problem situation.

Each kind of success clearly pertains to a different domain; the first pertains to the study itself, the second to the policymaking or implementation process, and the third to the problem situation. And each kind may apply to a different time period; for example, outcome success, if it occurs, may not be discernible until long after the analysis is completed.

The success of a study may appear different from the perspectives of the various parties at interest, for they often have different objectives. One interest group may use the results of the study in the policymaking process, although another may not. One interest group may find a particular implementation outcome to be beneficial, although another considers it harmful.

With these general comments as background, this section now examines, first on a direct basis and then on an indirect basis, the kinds of success and the criteria for measuring each. The criteria are incommensurable, and their relative importance varies with the analysis and problem situation. Nevertheless, oversimplifying somewhat, the more criteria by which an analysis is successful, the greater its success.

Analytic Success

There are two general ways to determine the extent to which a study is an analytical success: formal quality control and approval of selected parties.

Formal quality control. Formal quality control attempts to apply a clearly defined process and explicit standards to evaluate the quality of a study and its findings. In my view, study quality has these main components:

Technical validity: The study methodology appears sound, the analysis considers the relevant policy alternatives and the important uncertainties, and the findings follow explicitly from the analysis. (This is discussed in Chapters 13 and 15.)

Persuasive validity: The analysis and its findings are presented cogently and clearly, without exaggeration or oversimplification. (See Chapters 13 and 15.)

Availability: The study content is readily available, through reports or briefings, to its clients and reviewers.

Credibility: The study content appears believable or trustworthy to the study clients.

Timeliness: The study content is available in sufficient time to be digested and used for policymaking.

Pertinence: The study content applies to clients' substantive concerns and spheres of responsibility.[8]

Usefulness: The study content is potentially valuable to its clients; for example, the study provides a helpful new perspective or devises feasible ways to improve the problem situation. (Chapter 15 considers the "worth" of a study as a composite of its pertinence and usefulness.)

Unfortunately, the field of policy analysis lacks the agreed-upon standards and evaluation procedures necessary for formal quality control to be applied and accepted. (Chapter 15 discusses quality control issues in detail.) Thus, until more progress is made toward formal quality control, there must still be heavy reliance on the approval of selected parties as the means of gauging analytical success.

---

[8] I refer to this component as "pertinence" rather than "relevance" because the latter term can have a variety of meanings in systems and policy analysis. Lynn (1978, pp. 18–19) provides an excellent discussion of these meanings.

Approval of selected parties. Approval is innately subjective. So one who contemplates using the approval of another party to measure a study's analytical success should consider the party's perspective and possible biases, as well as the specific context in which the judgment was made. Consider these different marks of analytical success:

1. The first mark of analytic success is that the work and its findings satisfy the sponsor. SANCAP, POLANO, and PAWN all satisfied their sponsors. It seems obvious that any analysis team should want a satisfied sponsor, but closer examination suggests that the matter is not so clear. It must, unfortunately, be admitted that there has been work aimed frankly at confirming the sponsor's prior prejudices; if such work ignores relevant disturbing facts and bends the analysis to this motivation, then it clearly is bad analysis, and the satisfied sponsor is not a mark of success. It is far commoner for a meticulous and searching analysis to produce findings that disturb the sponsor, at least as they first reach him; here the unsatisfied sponsor would seem to be the mark of greater success. Of course, there are discriminating sponsors who realize that the news emerging from a searching examination of a problem situation is quite unlikely to be all good, and who respond to it by examining its basis; if they find that the basis is solid and the interpretations emerge from sound analysis, then they may be unhappy about the troubling findings but satisfied with the analysis—and thankful that the analysis has brought the troubling matters to light. In sum, for the criterion of a satisfied sponsor to be judged a mark of success demands that the context from which the satisfaction emerged be known in some detail.

2. Another mark of analytic success is that the work and its findings satisfy the analysts. Here again knowledge of the context conditions how such a mark should be viewed. On the one hand, a study that gets only a few rudimentary steps beyond the information-gathering stage may turn up an insight that causes the sponsor to view his central question in an entirely new light, enabling him to approach its resolution from a new direction; yet the analysts may remain somewhat unsatisfied with such a rudimentary analysis, even though its findings helped the sponsor. Similarly, since few systems analyses escape some sort of time constraint on their completion, a quite competent study may be forced by this limitation to draw to a close earlier than a meticulous analysis team might wish. On the other hand, an analysis may be so complete and presented in such detail as to swamp the sponsor with information, thus perhaps satisfying the analysts but leaving the sponsor less so. Too, if the drive for analytic completeness and satisfaction overrides the time constraint so far as to yield findings after the related decisions have been made, the analysts may be satisfied with the technical quality of their work but have a disappointed sponsor. I believe, however, that competent analysts should not be fully satisfied with their work if the findings are late or presented

ineffectively, as in these two examples. Thus, satisfied analysts are a mark of success in systems analysis only if the surrounding circumstances support this judgment and the analysts are competent.

3. Yet another mark of analytic success is that the project satisfies the research program director. One might suppose a program director would be satisfied if a project was of high quality and pleased the sponsor. But matters are not always so simple. The program director may be dissatisfied with such a project if it has a high opportunity cost: if, for example, its findings antagonize another sponsor, or its resource consumption (money or staff) delays or undermines a more important project. And he may be disappointed if the project evolves in directions that produce little of the synergy with other projects envisaged in his strategic plan.

A research program director judges the success of a project in the context of his portfolio of projects and in terms of its contribution to his program's overall goals. As an example, consider a project that failed to satisfy the sponsor because serious conceptual or methodological difficulties kept it from producing any findings. Yet the program director might consider the project successful if it led to a follow-on project for the same sponsor, built intellectual capital for other projects, or identified a promising topic for future research.

SANCAP not only pleased its sponsors and reviewers, it was also particularly pleasing to its program director. It led directly to two other studies—studies of air pollution and transportation management strategies for Los Angeles—and indirectly to POLANO. POLANO led directly both to BARCON and PAWN.

4. A systems-analysis study may be considered an analytic success if peer reviewers judge it to have high quality (as discussed, for example, in Chapter 15) and if the basis of the review is appropriate to the study. Although peer review is universally acknowledged to be an essential component in judgments of quality for science and technology, the process is not without its faults, as a number of critics have made clear (see, for example, Armstrong, 1982). To detail possible faults in the peer review process is unnecessary here, but several of its potential limitations for systems-analysis studies should be discussed.

One limitation concerns novelty. A reviewer necessarily uses his knowledge of the craft of his subject as the background for his review. Thus, if the study being reviewed falls within this envelope of craft knowledge, then the existing standards of quality with which he is familiar apply, and are an appropriate basis for judgment. If the work is novel in important respects, however, the existing criteria may not be appropriate, and, if used, will inevitably find the work to be below standard in some ways. It follows that the peer review system tends to be conservative in its effects, unless the reviewers can generate and apply the new criteria of quality that pioneering work may demand (see Ravetz, 1971, pp. 182–

184). Since an important systems analysis almost inevitably plows new ground in its analysis, it is especially important for its review to be based on criteria that are appropriate to the work and its relation to its problem-situation context. It is unfortunately the case that this principle has not always been observed; the demand for rigor rather than relevance has even been observed (Section 12.4).

Another limitation concerns multidisciplinarity. A systems analysis commonly involves multiple disciplines. As the number of disciplines (and perhaps the synthesis among disciplines) increases, it becomes more difficult to obtain adequate peer review. Finding individual reviewers who are simultaneously expert in systems analysis and all relevant disciplines is next to impossible. And using multiple reviewers with complementary expertise is often insufficient: An impractical number of reviews may be required to cover all the disciplines involved, and the reviews may well be piecemeal judgments rather than comprehensive and holistic appraisals.

Perhaps the most serious limitation concerns competence. Because systems analysis is more art than science, it can be difficult to get reviewers with the special competence required. The reviewers must not only be fairminded and unbiased, but their expertise in the disciplines involved and in the craft of systems analysis must also be equal, and preferably superior, to that of the analysis team; otherwise one risks obtaining reviews of questionable worth, as when a Salieri critiques a Mozart or a Warhol critiques a Picasso. The reviewers also must be sufficiently knowledgeable about the problem situation that they can properly judge a study's policy findings as well as its methodological basis.

The methodology and results of PAWN and SANCAP were approved by the appropriate technical advisory groups, and the RAND reports on all three studies received a careful technical examination from at least one reviewer unaffiliated with the project.

PAWN provides two unusual illustrations of approval by reviewers. First, in a 1984 international prize competition, PAWN won The Institute of Management Sciences Award for Management Science Achievement (the Edelman Prize).

Second, the Minister of the Netherlands Ministry of Transport and Public Works, Mrs. N. Smit-Kroes (1984) made the following statement about PAWN:

> [T]he various cost and benefit numbers have been carefully examined without challenge by various other government ministries, provincial and local government agencies, private industries and associations, and public interest groups.
>   . . . On the basis of extensive examination [the PAWN methodology] appears to give credible results and to represent a substantial advance in the state of the art.

Despite the potential limitations discussed above, peer review is clearly valuable for judging a study's technical validity. And peer reviewers can usually detect when statements of study findings are so unclear or exaggerated as to undermine persuasive validity. But they are not well equipped to predict whether the various study clients will deem the statements credible or will consider the findings pertinent and useful. For these components of quality I believe the study's clients are generally much better judges than the analysts' peers.

5. Another mark of analytic success is that the study satisfies its clients. But the meaning of this mark is blurred by client multiplicity. A study's multiple clients may reach different judgments as to analytical success because of their different goals and perspectives. Of the main components of study quality, usefulness probably produces the most disagreement. Clients who agree that a study is valid (perhaps on the basis of the same peer review), credible, and pertinent may nevertheless disagree on whether the study is useful and thus successful. Client A may consider the study successful because it devises a new solution to the problem that he finds attractive, although client B may do so because its findings support a solution he already favors, as was the case with SOS and POLANO.

Goemans (1986, p. 8) notes the following client reaction to POLANO: "Although nobody [except the sponsor] had asked for POLANO, there was widespread appreciation for the [Rijkswaterstaat (1976)] report which showed the issue in its broader context."

There are even instances where clients' biases dominate their judgments of analytical success. When valid study findings, through distortion or presentation out of context, can be made to appear to support his position, a client may consider the study successful. And a client may consider a study successful if its findings support his position, even though they are of dubious validity.

Utilization Success

Although analytic success concerns how the study was performed and presented, utilization success concerns how it was used. For utilization success, the study must be *used* in the policymaking process.[9] Otherwise, its influence can only be academic and its monuments merely journal citations and reports gathering dust on bookshelves.

Various authors conclude, on the basis of empirical investigation, that policy research utilization has been seriously underestimated because

---

[9] Recall that the study must also be of high quality (an analytic success), as utilization of poor-quality results is no credit to a study.

"use" is too narrowly defined.[10] Numerous researchers have defined and refined concepts of knowledge use in policymaking.[11] I have synthesized several ideas (some from the literature and some my own) into a framework for characterizing a policy research study's utilization. This framework addresses three component questions: (1) who uses (2) which elements of study (3) for what purpose?

Study users. A policy study cannot be considered a utilization success if it is used only by users who do not participate in the policymaking process (such as students and research managers). Rather, for a study to be a utilization success, at least one of its users must participate in policymaking. Such users may be policymakers (and sometimes implementors), their staff, advisors on the decision, or lobbyers.

One might contend that a study's utilization success increases with the number of users. But matters are not that simple. Some users are more important than others; for example, use by one policymaker probably has greater significance than use by several staff members or lobbyers. Also, collective (group) users are generally more important than individual users, even when the same people are involved. For example, a study would probably be considered a greater utilization success if it led a group of policymakers to a common position than if it led the individual members to divergent positions. The distinction between individual and collective users focuses on mutuality of use and shaping of consensus, whether for problem formulation or policy selection. (Dunn [1983] distinguishes individual and collective use, but with a somewhat different slant.)

Elements used. A policy study typically contains many elements of knowledge that might be used in the policymaking process. The types of elements include terminology, issues, arguments, concepts, models, alternatives, estimates, findings, and recommendations.

A user may elect to use some study elements and disregard others. Usually elements are used or disregarded on the basis of whether they appear pertinent and useful. Sometimes, however, elements that appear useful are deliberately disregarded. This may happen because they raise political difficulties, as Larsen (1980, p. 430) suggests, because they lack credibility, or because they are less compelling than knowledge from other sources.

Mere consideration of an element of knowledge in the policymaking

---

[10] For example, Pelz (1978, p. 346), Weiss and Bucuvales (1980, pp. 10–11), Larsen (1985, p. 145), and Whiteman (1985b, p. 204).

[11] They include Rich (1975, 1977, 1981), Caplan et al. (1975), Weiss (1977a, 1977b, 1980), Pelz (1978), Larsen (1980, 1985), Dunn (1983), Beyer and Trice (1982), and Whiteman (1985a, 1985b).

process does not constitute use. Rather, the element must affect a user's thinking about the problem and its solution (conceptual use)[12] or his observable behavior toward them (instrumental use).[13]

One might contend that a study's utilization success increases with the number of elements used in policymaking. But, as with users, some elements are more important than others; the use of a recommendation, for example, is probably more significant than the use of particular terminology. In principle, a study could be considered a utilization success if only one individual element is used. In practice, more than this would usually be needed, unless the element is quite important. However, given the usual number and variety of elements, it is unrealistic to expect that every element in a study would be used.

It is also unrealistic to expect that a particular element would be used exactly as it was presented in the study. Adaptation occurs naturally during the utilization process. Users, sometimes with assistance from the analysts, often modify study elements to fit their needs (Berman and McLaughlin, 1977; Larsen, 1980, p. 428). Ideally, utilization is an interactive process that involves analysts as well as users.

Purpose of use. Knowledge from a systems or policy analysis may be used for various purposes. Each purpose represents a different kind of utilization in the policymaking process. A particular study thus may be a utilization success for some purposes and a failure for others. We shall now examine seven purposes and the meaning of utilization success for each. Note that the purposes correspond roughly to phases of the policymaking process.

1. The first purpose is *problem formulation*. Here utilization success occurs when participants in the policymaking process use study knowledge to shape or elaborate their conception of the policy problem: the

---

[12] Of course the inclusion of effects on thinking greatly complicates the empirical measurement of knowledge use. (One cannot accurately measure such effects by counting citations of a study or by administering a simple questionnaire.) However, I shall not discuss this topic further here because my focus is not on measuring knowledge use but on understanding it so as to learn how to increase utilization success. Larsen (1980, p. 425) discusses problems with direct observation. Dunn (1983) reviews various measurement procedures, including naturalistic observation (such as ethnography), content analysis, and questionnaires and interviews.

[13] Rich (1975) and Caplan et al. (1975) distinguish conceptual use from instrumental use. Instrumental use has been defined as "knowledge for action"; it refers to instances where "respondents . . . could document the specific way in which . . . information was being used for decisionmaking or problem-solving purposes." By contrast, conceptual use has been defined as "knowledge for understanding"; it refers to instances where information "influenced a policymaker's thinking about an issue without putting information to any specific, documentable use" (Rich, 1977, p. 200). Whiteman (1985b) suggests that instrumental use should be called concrete use, to provide a more intuitive antithesis to conceptual use, whereas Dunn (1983) distinguishes conceptual and behavioral effects of knowledge.

objectives, the issues, the boundaries and constraints, the relevant consequences of policies and the ways to measure them, the important uncertainties, the relations among problem components, and so forth. (Section 4.3 and Chapter 5 of the *Overview* discuss problem formulation, but from the perspective of the analysts.)

Although a systems analysis generally provides a problem formulation, participants in the policymaking process may not choose to use it; they may cling to a prior conception of the problem or may prefer one developed among themselves. On the other hand, they may derive part of their formulation from a systems-analysis finding or some other aspect of study content that is not part of the analysis's formulation of the problem; for example, participants in policymaking might create a problem formulation that deliberately neglects a particular uncertainty because a prior systems analysis has shown through sensitivity analysis that the uncertainty had negligible effects.

A systems analysis may be judged successful if it influences the participants' problem formulation, individually or collectively. Usually this influence is partial, as when the problem statement or several other elements from the systems analysis are adopted. Consider three, quite different, examples from SANCAP. First, the SANCAP approach and terminology for describing alternatives hierarchically in terms of tactics, pure strategies, and mixed strategies (discussed in Section 10.4) were used in the policymaking process to identify points of leverage. Second, the SANCAP analysis helped the policymakers pull in the problem boundaries to focus on strategies for controlling reactive hydrocarbons rather than other species of pollutants. The analysis demonstrated, for San Diego, the dominance of strategies that control reactive hydrocarbon emissions for controlling other species of emissions as well; that is, a strategy that reduces reactive hydrocarbon emissions sufficiently to meet the oxidant standard will have already reduced all other species sufficiently to meet their respective standards. Third, the SANCAP analysis helped the policymakers to relax a major constraint in their formulation. The necessary reduction in pollutants to meet the oxidant air-quality standard was determined originally on the basis of a certain historical oxidant concentration. When the analysis led to this concentration being rejected as anomalous by San Diego County and EPA Region IX, the necessary reduction became substantially smaller, making control strategies feasible that were less stringent and thus less costly (as much as $50 million per year less costly). (Goeller et al., 1973a, pp. 62, 63.)

An analysis's influence on problem formulation is not always partial, however. Sometimes essentially the complete problem formulation from a systems analysis is adopted in the policymaking process. This happened with POLANO and PAWN, and with the nominators and selectors—but not the EPA ratifiers—in SANCAP.

2. The second purpose is *generation of alternatives*. Here a systems analysis may be judged successful if it influences the nature or specification of alternatives considered by participants in the policymaking process. An analysis need not influence all the alternatives being considered, or all the participants considering them, to be deemed successful; it should usually be sufficient to influence a few alternatives, or components of alternatives, considered by a few participants.

This influence can happen in several ways. First, the participants may consider existing ideas for alternatives that the analysis team has systematically collected and presented. Examples include the technical and managerial tactics considered in PAWN (see Section 6.3) and the retrofit emission-control devices for used vehicles considered in SANCAP (see Section 10.4). An unusual example comes from POLANO. About six months before the study began, the Cabinet and Parliament had declared in principle that they favored a storm-surge barrier—provided the Rijkswaterstaat could devise one that met certain conditions (such as acceptable levels of cost and flood security). Unless the conditions could be met, the mouth of the Oosterschelde would supposedly be closed with an impermeable dam. Besides these two general alternatives, the analysis team investigated a third—leaving the mouth of the Oosterschelde open and constructing a system of large dikes around the perimeter of the estuary. They did this on their own initiative because, as Goemans (1986, p. 7) notes,[14] they believed that "the knowledge base for decision-making should be as complete as possible." The analysts expected this alternative to be omitted from the Rijkswaterstaat's report to the Cabinet and Parliament because it went beyond the issue posed by Parliament. However, after seeing the three alternatives compared in a common framework, the Rijkswaterstaat boldly elected to include the open alternative in their report, where it was considered by both the Cabinet and the Parliament.

Second, the participants may consider new ideas for alternatives that were either invented by the analysis team or synthesized by them from existing ideas. For example, PAWN policymakers considered a new managerial policy for flushing the Markermeer to reduce its salinity, several new technical tactics for treating water-supply problems, and a new managerial strategy for operating the network of major water-management facilities, all of which were invented by the PAWN analysis team (Goeller et al., 1983, pp. 217, 231, 287). And SANCAP policymakers considered candidate retrofit strategies (promising combinations of emission-control devices for retrofit on used vehicles) and promising mixed strategies (that

---

[14] Goemans, who was then with the Rijkswaterstaat, served as the POLANO project leader in the Netherlands (B. F. Goeller was the project leader at RAND), helped write the Rijkswaterstaat report, heard the Parliamentary debate, and later interviewed members about their reaction to the study.

is, overall strategies) that were synthesized by the SANCAP analysis team.

Finally, the participants' own creation of alternatives may be stimulated by study content. They may create alternatives that are hybrids or descendents of alternatives presented in the systems analysis. They may create alternatives that are intended to fill needs or holes identified by the analysis, as when PAWN found the previous Dutch national policy of phosphate control would be ineffective by itself in solving the eutrophication problem. And they may create alternatives that reflect systems-analysis findings or other results, as when they concentrate on the kind of alternatives that the analysis has shown to be cost-effective or slight issues that the analysis has shown to be unimportant or intractable. For example, after the SANCAP analysis found that fixed-source controls were much more cost-effective than retrofit emission-control devices and both were much more cost-effective than transportation management measures, strategies devised for San Diego by EPA Region IX appeared to reflect this finding. As another example, PAWN found that "dilution was no solution to pollution"; that is, no national policy to redistribute the water would appreciably improve water quality nationwide, although there might be highly localized changes. This finding led participants in policymaking to slight such policies and to emphasize instead regional approaches to pollution problems.

3. The third purpose is *estimation of consequences*. Before they can properly compare and select among alternatives, participants in the policymaking process need relatively comprehensive estimates of the alternatives' potential consequences—consequences for the problem situation, for persons affected by the situation, and for other parties at interest.

A systems analysis may be judged successful for this purpose if (some) participants use its results or methodology to estimate alternatives' consequences. This need not happen for all, or nearly all, alternatives or consequences for an analysis to be deemed successful. Yet an analysis that supplies a large fraction of the estimates participants believe they need will probably be considered more successful than one that supplies a small fraction.

Participants may use an analysis in several ways for the purpose of estimation. The most obvious way is to adopt its estimates. POLANO, PAWN, and SANCAP provided most of the estimates used in the associated policymaking, although some came from complementary studies such as the Rijkswaterstaat's technical feasibility study of the storm-surge barrier. Goemans (1986, p. 8) notes that POLANO's systems analysis "report [Rijkswaterstaat, 1976] was actually used in both the Cabinet and Parliament; the [values of the] impacts [consequences] were hardly questioned and the discussion focused on preferences."

Another way is to use the methodology developed in the analysis to estimate the consequences of alternatives developed in the policymaking process. For example, the SANCAP methodology was used to estimate, in a common framework, the consequences of the EPA Region IX Final Implementation Plan (October 15, 1973) for the San Diego Air Quality Region.

Still another way is for participants to use findings or results to understand the issues or appreciate the influence of assumptions. For example, SANCAP sensitivity analysis showed that shifting the technical assumptions used to calculate emissions from the reference values considered realistic for San Diego to the more pessimistic values defined by EPA Region IX would greatly increase the cost and extent of strategies to meet the air-quality standards. As an additional example, PAWN found that farmers, in the absence of government restrictions, might triple the number of sprinklers—and the corresponding demand for water—while seeking to maximize their profits. And, as a last example, POLANO's "disaggregate way of presentation using scorecards enhanced understanding of the issues and did much to get the idea accepted that there [was] no 'best' solution; the explicit treatment of uncertainties . . . and . . . sensitivity analysis (e.g., for various . . . [size openings in] . . . the barrier . . .) was useful to give a feeling for the relative influence of assumptions" (Goemans, 1986, pp. 8–9).[15]

And the final way is for participants to recognize from the study the need for certain information that it lacks, and then to obtain and use this information. The information might be consequence estimates that the participants are better equipped to produce (such as the reactions of other actors), or information that must be obtained before particular consequences can be estimated or the corresponding estimation methodology completed. There is also a possibility that the information's potential usefulness might not have been appreciated by the analysts; the participants may have recognized this only after examining the information the analysts did provide.

4. The fourth purpose is *rejection of alternatives*. A systems analysis may be judged to be successful if it helps policymakers[16] to reject inferior alternatives, particularly if they had been strongly supported by parties who expected to gain from their adoption. To someone not experienced in systems analysis, this purpose may seem less important than some of

---

[15] For examples of scorecards, see Tables 10.4 and 10.8 and the accompanying discussions on pages 385–388 and 413–416, respectively. The *Overview* (pp. 96–99) also discusses scorecards and their rationale and presents examples from POLANO in Section 3.4.

[16] The careful reader may have noticed that I focus on "policymakers" here, rather than on "participants in the policymaking process," as with previous purposes. What is important here is whether the analysis helps policymakers collectively decide to reject one or more alternatives, not whether the analysis influences a minority of policymakers or a majority of other participants (such as, staff members or lobbyers) to argue for rejection.

the others, but experience tells us that it is quite common in large organizations for courses of action to be strongly advocated—and even sometimes adopted—that will not have the desired effects, with a consequent waste of energy and resources. Thus, it is fairly important to help reject the really inferior alternatives, and much successful systems analysis work has done not much more than this.

Although the SANCAP and PAWN analyses did considerably more, they were quite successful at getting inferior alternatives rejected. Much of this success resulted from extensive screening analysis, described in Sections 10.4 and 6.3, which identified alternatives that were either unpromising (not cost-effective) or dominated by other promising alternatives. PAWN's screening of technical and managerial tactics led to the rejection of most of the 57 national and regional tactics proposed, including several with construction costs over $100 million.

Rejection of inferior alternatives does not always result from screening, however. It also results from comparing alternatives' estimated consequences in detail. In SANCAP, for example, the San Diego County policymakers rejected alternatives D and E because their consequences were similar to but dominated by alternatives B and C respectively (see Table 10.8).

5. The fifth purpose is *selection of a preferred alternative*. Out of the many alternatives considered in the policymaking process, one must be selected for eventual implementation.[17] A systems analysis may be judged a success here if it has a significant influence on the selection of the preferred alternative by policymakers.

Such success has several dimensions. The first is the significance of the study's influence; that is, did it have an essential influence on the selection or merely an important one? This can be established by the answer to the question: Would the selection have been different without the study? If the answer is "probably so," then the study should be considered an essential influence on the selection. If the answer is "possibly so," then the study should be considered an important influence.

This dimension is relevant whether there is one policymaker or more. The other dimensions are relevant only when there are multiple policymakers, as with a committee, the U.S. Congress, the San Diego Board of Supervisors, and the Dutch Cabinet or Parliament.

---

[17] Occasionally, where the implementability or the potential implementation costs of the preferred alternative remain a matter of concern after the analysis, two or three different alternatives may be selected provisionally and ranked by preference. If implementation planning subsequently shows unacceptably large implementation difficulties or costs for the alternative with highest ranking, then the next-highest ranked alternative that seems acceptable probably will be implemented (unless the implementation planning results appear likely to change the rankings and the policymakers are willing and able to reconsider their preference).

The second dimension is the extent of the influence; that is, what proportion of the selections by individual policymakers did the study influence? A majority? A few? Of course the policymakers need not agree on their selections for the study's influence to be considered extensive; even if the study influenced all the policymakers, equal numbers might come to prefer different alternatives.

The final dimension is the decisiveness of the influence; that is, did the study influence the policymakers' collective selection? Although this may happen when the study influences a majority of policymakers to select the same alternative, it may also happen when the study influences only a few policymakers. The study might influence the few deciding votes in circumstances where most votes were divided equally between two alternatives for reasons other than the study. Or it might influence a few leaders among the policymakers to cause the group to agree on a selection where most members were motivated by the leaders rather than the study.

Multiple policymakers introduce an additional complication. Success in these dimensions may differ with the policymaker's subrole (nominator, selector, or ratifier).

The PAWN study's success was clear-cut. The study's influence on selection was essential, extensive, and decisive for policymakers in all subroles. The Minister of the Netherlands Ministry of Transport and Public Works, Mrs. N. Smit-Kroes (1984), stated that

> the new national water management policy for the Netherlands is based largely on the PAWN project. Without the PAWN methodology and analysis, or something equivalent, to assess the cost and benefits of alternative policy actions in a credible way, many of the changes to the previous policy would not have been made.

As illustrations, consider several components of the national water-management policy that were selected on the basis of PAWN (Goeller and the PAWN Team, 1985). First, although bids had gone out for constructing the Waddinxveen-Voorburg Canal, the Brielse Meer pipeline is being built instead, on PAWN's recommendation. This will yield a $38 million investment saving plus $15 million per year average net benefit in reduced salinity damage to agriculture.

Second, the new flushing policy in PAWN for the Markermeer has been implemented, yielding expected net benefits of between $1.2 and $5.4 million annually.

Third, there has been a drastic change in the Dutch approach to eutrophication, the next most serious water-quality problem. Because PAWN concluded that the previous Dutch national policy of phosphate control would be ineffective by itself, the Netherlands adopted a regional approach to pollution control. Complementary tactics are to be added to the continuing national policy of phosphate control.

And finally, the national policy recommends that water-board plans and the regional technical and managerial tactics identified as promising in PAWN's screening analysis be seriously considered during the development of provincial water-management plans. Implementation of all recommendations would produce expected profits of between $53 million and $128 million per year.

For SANCAP, success was almost as clear-cut. The study's influence on selection was essential, extensive, and decisive for the nominators and selectors; on the basis of the SANCAP analysis they chose an overall strategy (and technical assumptions for evaluating its emissions) developed in SANCAP in preference to those developed by EPA Region IX. For the ratifier, EPA Region IX, SANCAP had an important influence on part of the strategy and some of the assumptions selected; but it did not appear to have much effect on the ratifier's choice of transportation management controls, where there were some major differences from San Diego's selection.

For POLANO, the reception of the study was auspicious. The results were summarized by RAND in an all-day briefing to the Rijkswaterstaat on April 5, 1976, one year after the study began. Events then moved quickly to a conclusion. The Rijkswaterstaat (1976) report, based largely on POLANO, was presented to the Cabinet one month later, and then to Parliament, along with the Cabinet's selection of the storm-surge barrier alternative, which Parliament approved in June 1976.

But success was not as clear-cut for POLANO as for the other illustrative cases. On the one hand, the study had an essential, extensive, and decisive influence on the selection of a key storm-surge-barrier design component: the gates. The storm-surge barrier has sixty enormous gates that are normally open but can be closed to prevent large storm surges from entering the Oosterschelde. Each gate consists of a slot (the gate opening) that can be closed with either one or two slides. Whether single or double slides should be used was a crucial design question for the barrier. Double slides had been standard practice for other barriers in the Netherlands so as to minimize the risk of flooding if a slide failed to close on command. But single slides were used in the storm-surge barrier, saving about $200 million, because POLANO showed they offered a very high level of safety and that it was doubtful whether double slides could increase this sufficiently to justify their much larger costs.

The study also had an essential, extensive, and decisive influence on the decision by the nominator—the Rijkswaterstaat—as to which alternative to recommend to the selector (the Cabinet) and how to pose the recommendation. The Rijkswaterstaat chose to mirror the POLANO study conclusions in its report (Rijkswaterstaat, 1976). By intention, the study did not conclude by recommending a particular alternative. Rather, it attempted to clarify the issues by comparing, in a common framework, the many different impacts of the alternatives, but left the choice among alternatives to the political process, where the responsibility properly resides. Significantly, there was no dominant alternative—one that was best for all the impacts. Indeed, each case had a major disadvantage that might be considered sufficiently serious as to render the alternative politically unacceptable: The storm-surge-barrier case was by far the worst for cost, the closed case for ecology, and the open case for flood security.

On the other hand, it is hard to appraise the study's influence on the selection of the storm-surge-barrier alternative by the Cabinet and the Parliament (the ratifier), for several reasons. First, the Rijkswaterstaat's report to them did not recommend a particular alternative. Second, the privacy of the Cabinet and the size and complexity of Parliament handicap those who try to understand the process leading to a particular decision. It is clear that the POLANO results, embodied in the Rijkswaterstaat report, were an essential factor in Cabinet and Parliamentary deliberations. But it is not clear whether this was also true for their selections. Some observers believe it was, although others do not. Finally, since the Cabinet and Parliament favored the storm-surge barrier before POLANO began, provided certain conditions could be met, it is difficult to determine whether POLANO influenced the selection more than it helped confirm or justify a selection that had already been made on other grounds.

Even when it is of high quality and potentially useful, a systems analysis may fail to have a significant influence on the selection of the preferred alternative. One reason is the political nature of policymaking. As Springer (1985, p. 490) points out, policymakers

> have obligations to honor past commitments, to balance opposing values, and to respond to the political give and take that is part of the institutional "cement" of our pluralist system. If specific analytic recommendations do not translate directly to action in this setting, it does not necessarily mean that the analysis "failed." It may simply have lost to some more pressing, or deserving, claim.

Another reason is that, compared to other sources, systems analysis may be at a disadvantage in supplying the type of information that turns out to be decisive, or especially important, in a particular selection process. Sabatier, quoted in Springer (1985, p. 491), contends that there are four other "types of information that a politically rational agency can be expected to pursue" besides the type of substantive information on policy problems and impacts typically generated by systems analysis. These competing types are (1) legal rules and regulations that affect agencies' decisions, (2) the history of agency decisions, (3) the "preferences of important actors," and (4) "the probable reactions" of these actors to the estimated "consequences of each policy alternative."

Sabatier's list has some communality with one by Brewer and deLeon (1983, pp. 190, 191), who suggest that a policymaker "probably considers, either consciously or unconsciously, at least five factors before making a choice:" the context of the problem,[18] the points of leverage (that is, variables manipulable by policymakers), the availability (and trustworthiness) of information, the personalities of the participants, and the importance of the problem (which "governs how much of the decisionmakers's limited time, attention, and political resources will be expended on a given policy"). Again, systems analysis may be at a comparative disadvantage in supplying information on some of these factors.

6. The sixth purpose is *promotion of a preferred alternative*. A systems analysis may be judged a success for this purpose if information from the study is used to "advocate, [justify or] reaffirm policy positions after they have been determined."[19] It does not matter whether a position was influenced by the study or predetermined.[20]

For a study to be successful for this purpose, it must be used without distortion, which can occur through misrepresentation or selective use of arguments out of context. Distorted use is no credit to a study.

Consider some examples of promotion. Promotion occurs when an individual policymaker uses information from the study to advocate his preferred alternative to another policymaker.[21] And it may occur when he uses it to confirm or reinforce his original preference.

Promotion also occurs when policymakers collectively use information from the study to advocate or justify their position to another policymaking body. In SANCAP and POLANO, for example, the selectors used dy results in advocating and justifying their preferred alternative to the ratifiers. In POLANO, after Parliament had ratified the Cabinet's selection of the storm-surge barrier, members of both groups used study

---

[18] The context, as they define it, includes such questions as, "How is the problem defined and bounded? . . . When or how soon must the decisions be made? What are the rules of the game and the players? . . . [And] on what precedents is the decisions based, and how do they relate to present conditions?"

[19] Although Whiteman (1985a, p. 200) calls this type of knowledge use "strategic," we feel it is better described as "promotion."

[20] Weiss (1977a, p. 15) argues persuasively that using research "to support a predetermined position" is "neither an unimportant or an improper use . . . just because sides have already been taken is no reason to discount the effects of research." However, Pelz (1978, p. 351) warns of "symbolic" use, where performing a study is a substitute for a decision or where study knowledge is distorted to support publicly a decision that was predetermined or made on other grounds.

[21] If the other policymaker uses information conveyed by such advocacy in formulating the problem, estimating consequences, or rejecting or selecting alternatives, then information from the study is also being used for these purposes—and may contribute to corresponding successes.

results to justify their position to the general public and interest groups such as SOS.

7. The final purpose is *implementation*. A system analysis may be judged successful if it helps to implement the selected alternative.

This is almost automatic when the analysis has a significant influence on the alternative's selection and implementation ultimately occurs. Such was the case for POLANO, PAWN, and SANCAP. The Queen of the Netherlands formally dedicated the fully operational storm-surge barrier on October 4, 1986, in a ceremony attended by 25,000 people, including foreign dignitaries such as the presidents of France and West Germany, and televised throughout the Netherlands. Nearly all the national water-management policy developed through PAWN has been implemented, although some implementation is still in process. And, after multiple rounds of negotiation and revision, taking several years, a clean-air strategy derived partly from SANCAP has been instituted in San Diego.

Even when it was not a significant influence on the alternative's selection, a system analysis may help implementation, and particularly implementation planning, in several ways. One way is through carryover of knowledge. An analysis generally explores the implementability of various alternatives, as well as their estimated consequences, and this knowledge may be useful in implementation.

Another way is by identifying issues for resolution in the final detailed design or plan for the selected alternative. Some POLANO examples include identifying the need for a storm-surge-barrier control strategy and raising the possibility of dike redesign as an inexpensive means of hedging against uncertainty in certain assumptions.

Yet another way is by identifying data or research needed to facilitate implementation. For example, POLANO recommended additional research on scour processes (which potentially could undermine the barrier) so that barrier design modifications could be made, if necessary, with greater understanding. POLANO also recommended measuring the amount of nutrients being imported into the Oosterschelde by the North Sea. This quantity would affect the amount of biological life the Oosterschelde could support with different-size openings in the barrier; also, if much of the import occurred during storms, it could restrict the choice of an operating policy for the barrier.

And the final way is by providing analysts knowledgeable about the problem who may participate in implementation planning or become implementation managers. For example, H. N. J. Smits, who was one of the Dutch analysts working on POLANO, subsequently worked on planning for the barrier and then became head of the Rijkswaterstaat department responsible for its planning and financial management. The analysis team's background makes it well suited to help constructively in devising ways of overcoming the difficulties that are bound to arise in an implementation.

Conclusions on utilization success. Utilization success is complex; a study may succeed in some purposes and fail in others, be used by some clients and not by others. And different users may use different elements of a study for different purposes. Utilization success, if it occurs, generally involves partial use of a study's content rather than complete use. Utilization success is thus a matter of degree.

The framework for utilization success offered here has important implications for those who would empirically measure knowledge utilization: It is difficult to measure the degree of success of one study. And it is more difficult to compare the success of several studies. The framework also suggests that empirical investigations of utilization must cast a wide net, lest they miss uses or users. Fortunately, the concern in this chapter is not with measuring knowledge use but rather with understanding it sufficiently to learn how to increase utilization success. However, I believe that this framework can help those concerned with utilization measurement.

Outcome Success

Although utilization success considers how the study was used in policymaking, outcome success considers what happened to the problem situation and those affected by it as a result of this use. Specifically, a systems analysis may be judged an outcome success if two conditions are satisfied: (1) the implemented alternative improves the problem situation, and (2) the collective selection of that alternative was significantly influenced by the analysis.

The second condition is identical with utilization success for the purpose of selection, which was discussed above. The first condition is the subject of evaluation research. Since Chapter 11 offers a useful overview, there is no need to summarize the burgeoning evaluation-research literature here (for two excellent but different overviews, see Brewer and deLeon, 1983, Part V, and Patton and Sawicki, 1986, Chapter 9). It is useful, however, to discuss briefly several evaluation-research complications that make it very difficult to determine whether the first condition has been satisfied—and thus whether outcome success has been achieved.

The first complication is attribution of cause. Simple comparisons of the problem situation before and after policy implementation often cannot establish whether the policy actually caused any or all of the observed differences; the differences might result instead from changes in the environment or other factors. This complication has sometimes fostered experimental-design approaches to evaluation where the policy is applied to one situation or group and withheld from another (the control). Unfortunately, such experimental approaches are impractical for many policy problems.

The second complication is the measurement of multidimensional outcomes. The outcome of an implemented policy generally involves many and varied consequences, which differ in their ease, expense, and accuracy of measurement. Resource limitations may force unfortunate compromises in outcome evaluation. If an evaluation measures only a few potential consequences, it is limited and may miss important shortcomings of the policy. Suppose, for example, that an evaluation of SANCAP concentrated on the policy's performance in improving air quality. It might miss the fact that the implemented policy led to a subtle but serious degradation in transportation service for a particular segment of the population. On the other hand, if an evaluation tries to measure all the potential consequences of a policy, the evaluation costs would probably be overwhelming. And if one tries to economize on measurement processes, the measurements may be so inaccurate as to be misleading about a policy's actual outcomes.

The third complication is balancing multiple evaluation criteria. "Considering the number and diversity of possible outcomes and effects for even the simplest program, one nearly always must use multiple criteria to evaluate" (Brewer and deLeon, 1983, p. 331). But the existence of multiple evaluation criteria leads to the problem of how to balance or weigh diverse and often conflicting criteria. (This is conceptually similar to the multicriteria problem that policymakers face when selecting among alternatives, as discussed in Section 8.4.) The problem is aggravated by the fact that the implemented policy may have both favorable and unfavorable consequences, and some consequences may be larger and others smaller than expected or desired.

The fourth complication is the standard of comparison. Evaluations generally have a standard of comparison, explicit or implicit, that can affect the conclusions. Many evaluations compare actual outcomes to expected outcomes (Brewer and deLeon, 1983, p. 20), which often reflect the analysis estimates on which the policy was selected. This standard primarily evaluates the accuracy of the expectations—and the underlying estimates.[22]

Problem situations evolve with time, of course, so that, by the time a systems analysis is complete and a resulting implementation is in place, the situation may have changed enough for the outcome to be somewhat different from that predicted by the analysis. In such circumstances, the analysis team can sometimes adjust the analysis results for the changed situation so the comparison with the actual outcome will be a fair one.

A different standard is whether the actual outcome is considered at-

---

[22] This comment usually applies to a related standard that compares actual outcomes to policy objectives. The policy objectives for the selected policy are usually established on the basis of systems-analysis estimates. Sometimes, however, they are general objectives that were established before the analysis.

tractive (desirable or even acceptable) by the affected people. This attractiveness standard is developed after the policy has been implemented and taken effect (that is, during evaluation) whereas the expectation standard is developed before implementation. Thus the expectation standard can evaluate the attractiveness of a policy's outcome to the affected people only to the extent that the policymakers considered the people's preferences (preference functions) during selection and these preferences have remained unchanged. The longer the period between policy selection and evaluation, the greater the possibility that these preferences will change.

The fifth complication is the multiplicity of audiences. Evaluations have multiple audiences with different concerns. Policymakers are concerned with the implemented policy's performance in the problem situation. Is it effective or should it be terminated because of ineffectiveness or unacceptable side effects? Implementation managers are concerned with where the policy is having problems. Where and how can policy implementation and operation be improved? Interest groups of affected persons are concerned with how the policy is affecting their group in comparison to others. How are the actual consequences distributed among different groups? Is my group getting its fair share or bearing an unfair burden? (Interest groups are much more interested in policies' distributional effects than other audiences.)

Given limited resources, each evaluation must limit itself as to intended audience and included concerns. Depending on this choice, an evaluation may not be helpful for judging outcome success. Evaluations focused on implementation managers are a prime example. Such evaluations emphasize the issue of how to improve the policy rather than the issue of how the policy, as originally implemented or modified, has improved the problem situation.

The sixth complication is the timing of the evaluation. If an evaluation is performed too soon after implementation, there may not have been sufficient time for the policy to take effect and its resultant consequences to become apparent. For POLANO, as an example, the Rijkswaterstaat (1986) performed an evaluation 10 years after the selection of the storm-surge-barrier alternative. Although implementation had proceeded with all deliberate speed, the barrier was still a few months short of full operation. The Rijkswaterstaat evaluation report (1986) evaluated "the extent to which the predictions . . . conform[ed] to reality" by comparing the observed outcomes with the predictions in the "White Note," the Rijkswaterstaat report (1976) to the Cabinet and Parliament that was largely based on POLANO.

The evaluation report found that observations showed (essentially) no difference from predictions for most consequences, improvements for several consequences, and declines for two consequences: financial cost and completion date. The cost estimates for the overall project were overshot by 10 percent, and those for the storm-surge barrier by 30 percent. (The completion date was overshot by a year, partly because of a political decision to delay completion to reduce budgetary impacts.)

The report skipped the evaluation of two important aspects of the problem—the marine ecology and the fishing industry—because the policy's consequences for these aspects would not be apparent until perhaps a half dozen years after completion of the construction.

And although the report attempted to evaluate flood security, it lacked the necessary data for a complete job. During the first fifteen months of its operation, according to Dye (1988), the barrier "has closed off the sea at least three times." But the security of the barrier has not been severely tested. The barrier, after all, was designed to provide protection against a storm so severe that it might be expected to occur only once in 4,000 years. Flood security thus provides an extreme illustration of how some consequences cannot be fully evaluated within any reasonable time horizon.

Of course an evaluation can also be performed too long after implementation. Long delays complicate attribution of cause, and make some improvements in consequences unmeasurable. And, as mentioned earlier, the longer the period between policy selection and evaluation, the greater the possibility that the preference functions of policymakers or interest groups may change.

The final complication is the cost of evaluation. Good evaluation is costly. It takes sufficient resources—time, money, and talented researchers—that many implemented policies will not be deemed important enough to warrant detailed evaluation. And, as Brewer and deLeon (1983, p. 360) point out, evaluations may have unacceptable political costs; they "disrupt, reveal institutional weaknesses and limitations, and threaten individual policies and their sponsors."

Because of the complications we have discussed, an implemented policy will often receive no evaluation, or an evaluation that cannot establish whether or not it improved the problem situation. Thus, it is difficult for a systems analysis to satisfy the first condition for outcome success mentioned above. When this difficulty is compounded with the difficulty of satisfying the second condition, the clear conclusion is that it is extremely difficult for a study to be judged an outcome success—even when it truly is one.

On the other hand, if a policy clearly fails to achieve its goals or improve the situation, it does not necessarily mean that the reason for this outcome failure was either a defective analysis (theory or assumptions) or a defective selection. Among other reasons, it may be, as Schneider (1986,

p. 357) suggests, because "the agencies responsible for implementation failed to properly translate the policy theory into an operational plan or [because] inadequate resources were available to those responsible for implementation."

Indirect Success

Thus far this section has discussed success on a direct basis, where one assesses the three kinds of success with respect to the decision and problem situation for which the analysis was commissioned. It addressed such questions as: How were the POLANO methodology and results used in selecting a flood protection alternative for the Oosterschelde? And how were the SANCAP methodology and results used in selecting a strategy to clean up the air in San Diego?

Now this section will discuss success on an indirect basis, where one assesses success with respect to *other* decisions or situations. The discussion will address such questions as: How were the POLANO methodology and results used in selecting a strategy to control the storm-surge barrier in the Oosterschelde? And how were the SANCAP methodology and results used in selecting a strategy to clean up the air or improve transportation in Los Angeles? Because this section has previously considered the three kinds of success in detail, the discussion of them on an indirect basis will be fairly brief, mainly providing examples and describing unusual features.

Utilization success. For utilization success, a study must be used in the policymaking process. When study methodology or results are used for some purpose in the policymaking process of another decision or problem situation, it counts toward indirect success for that purpose.[23] This is easily clarified with examples.

The POLANO methodology was used for several purposes in other decisions and problem situations. A refined version of POLANO's SIMPLIC model for estimating water levels was used in the BARCON project to generate alternative control strategies for the storm-surge barrier and estimate their consequences. It was also used by civil engineers developing the detailed design for the barrier.

The POLANO model of eutrophication (algae blooms) in salt water was extensively modified for PAWN, where it became the primary tool for developing and estimating the consequences of fresh-water eutrophication control strategies.

---

[23] Collateral success is another term for indirect success.

All the models in the SANCAP methodology, except the air-quality model and the economic input-output model, were recalibrated and used for Los Angeles. (Some were extended to account for new problem features.) They were applied to generate alternatives and estimate their consequences in two major studies of Los Angeles; one study was concerned with alternative strategies for oxidant control (Goeller et al., 1973b) and the other was concerned with near-term transportation alternatives for simultaneously treating air pollution, energy conservation, and transportation service problems (Mikolowsky et al., 1974). Both studies were utilization successes. The oxidant-control study led to several inferior alternatives being rejected; the transportation study produced a strategy (strategy E) that became the basis of the short-range transportation plan adopted for Los Angeles (Southern California Association of Governments, 1974).

But methodology was not the only SANCAP product carried over to these Los Angeles studies. In constructing alternatives, they used the fixed-source and retrofit strategies developed in SANCAP, along with the SANCAP finding about the relative cost-effectiveness of these strategies and transportation-management measures in controlling emissions.

Much of the SANCAP problem formulation (for example, issues, relevant consequences of policies and how to measure them, important uncertainties, and relations among problem components) also carried over to the Los Angeles studies, although some region- and problem-specific items did not. The SANCAP approach and terminology for describing alternatives hierarchically in terms of tactics, pure strategies, and mixed strategies was not only used in the Los Angeles studies, but also in PAWN. In this way, and others, PAWN also counts toward SANCAP's indirect success in problem formulation.

PAWN identified a number of water-board plans and regional technical and managerial tactics as promising during screening (see Section 6.3), all of which were recommended by the national water-management policy. Their implementation, which is beyond the authority of national policymakers, is being seriously considered during the development of provincial water-management plans.

For many lakes in the Netherlands, the provinces are also using improved versions of the PAWN eutrophication models to develop a tailored combination of eutrophication control measures.

The comprehensive methodology developed by PAWN has been adopted for national water-resources planning by the Rijkswaterstaat and by the Delft Hydraulics Laboratory and has been used in several major studies. Parts of the methodology are being used by the Ministry of Agriculture and Fisheries, and by 6 of the 11 Dutch provinces.

As the previous example suggests, a study's methodology or results can produce "ripples" of use that spread more and more widely among decisions and problem situations.[24] This may happen haphazardly, by word of mouth or by migration of the analysts or policymakers; or it may happen more systematically as the methodology and results become part of the state of the art or are propagated through the literature or courses. Tracing such ripples is difficult, and often impossible. Nevertheless I can offer a few examples.

The use of findings is probably hardest to trace. This is partly because they often arrive at a new situation as part of general knowledge, so their source is unclear, and partly because they are often adapted to fit the new situation. For example, I believe, but cannot prove, that several SANCAP findings were used in developing other regions' strategies to clean up their air. These include: (1) the finding that fixed-source controls were generally most cost-effective, (2) the finding that tractor towing of jet aircraft within airports and the use of adsorption devices and substitution of nonreactive coatings in surface-coating (such as painting) operations were promising measures to reduce emissions, and (3) the finding that some technical assumptions promulgated by EPA were unrealistic for certain regions and others were flawed. I also believe that EPA modified, or permitted the modification of, several assumptions on the basis of SANCAP and the Los Angeles oxidant-control-study findings.

Although the scorecard technique was first used in another study in 1971, it was refined and propagated through SANCAP. It has now been used in a wide variety of problems throughout the world.

In SANCAP, screening received an early, and perhaps the earliest, application in a major systems analysis. Here it also became the first stage of a multistage process to generate and evaluate alternatives. (This process performs what has previously been referred to as "explicit hierarchical design and evaluation of alternatives"; see the discussion in Chapter 6.) I believe that much of the subsequent use of screening and this multistage process in other studies stems from SANCAP and its progeny.

According to Goemans (1986, p. 12–13), the POLANO approach "attracted much attention. Somewhat reluctantly at first, but more enthusiastically in later years, the scorecard was included in courses for government officials. [Furthermore,] after the Oosterschelde controversy, the Ministry of Transport and Public Works and other Ministries felt the need for an independent institute . . . to do policy studies. . . ." This led

---

[24] For a particular study result, Dunn (1983) characterizes the scope of knowledge use as either specific (for example, the use of a study recommendation for a particular decision) or general (for example, ideas in wide use). Weiss, as cited by Larsen (1980, p. 425), employs the term "enlightenment" to describe the broad influence of knowledge, where "it is not the findings of a single study nor even a body of related studies that directly affect policy [but] . . . the concepts and theoretical perspectives that . . . research has engendered . . ." (see also Weiss, 1977b and 1980).

to the establishment in 1981 of the SIBAS Joint Institute for Policy Analysis, with Goemans, the POLANO project leader in the Netherlands, chosen to be its director.

The Delft Hydraulics Laboratory has applied PAWN's general approach in master-planning studies for Taiwan and several other countries, and in studies of shore protection for the Italian coast near Ravenna and the Dutch island of Texel. SIBAS has applied the approach in several studies, including an ongoing study of strategies to manage the Netherlands' section of the North Sea and its associated resources.

PAWN provides a thoroughly documented case study to educate decisionmakers and train analysts in the analysis of complex natural-resource and environmental questions. It has already been used to train water-resource planners in the Rijkswaterstaat and all 11 Dutch provinces, as well as planners from Bangladesh in a three-month course sponsored by the World Bank. It was also the basis of a two-week training course given in November 1984 for member nations of the United Nation's Economic and Social Commission for Asia and the Pacific. With changes to the data base, many models are transferable to other countries.

The illustrative case studies we have employed in this chapter have appeared in various policy-analysis books; for example, POLANO has appeared in the *Overview* and in Quade (1982); SANCAP has appeared in Mood (1983) and in Quade (1975) as well as Chapter 10 of this volume; and PAWN appears in Chapter 6 of this book and will be discussed further in the *Cases* volume of this handbook.

Outcome success. It has already been noted that it is extremely difficult for a study to be judged an outcome success on a direct basis—even when it truly is one. This difficulty is enormously greater on an indirect basis.

There are two general ways for a study that was commissioned for a particular decision and problem situation to become an outcome success on an indirect basis. One way is for one or more items (such as, findings, estimates, pieces of problem formulation or methodology) from the study to significantly influence the policymaking for another decision or problem situation, where that policymaking results in outcome success (as defined earlier). The other way is for one or more items from the study to sufficiently influence the results of a subsequent study for another decision or problem situation so as to be an important factor in the subsequent study's outcome success (on a direct basis).

I have no examples of the first way arising out of our illustrative case studies, partly because I have not tried to trace their influence carefully and partly because knowledge propagation can be diffuse and tenuous.

The best example of the second way is the study of near-term transportation alternatives for Los Angeles, mentioned above. Viewed on a direct basis, that study is an outcome success; the implementation of a strategy developed and selected largely on the basis of the study has apprarently improved the problem situation. Furthermore, the approach and results of the study were strongly influenced by SANCAP. Most of the problem formulation and methodology carried over. And in constructing alternatives, the Los Angeles study used fixed-source and retrofit strategies developed in SANCAP, along with the SANCAP finding about their relative cost-effectiveness in comparison to transportation measures for controlling emissions. Viewed on an indirect basis, it therefore appears quite reasonable to count the Los Angeles study as an outcome success for SANCAP.

Another, less definitive example of the second way is the BARCON study. POLANO's SIMPLIC model for estimating water levels was the tool BARCON used to develop and estimate the consequences of control strategies for the storm-surge barrier, including the strategies that were subsequently selected and implemented. These strategies appear to work well and meet their goals, which means that BARCON would probably be considered an outcome success when viewed on a direct basis, although evaluation has been limited by the fact that the barrier began operation only about a year and a half ago. Viewed on an indirect basis, it therefore appears plausible to count the BARCON study as an outcome success for POLANO. By contrast, it is not clear whether POLANO itself was an outcome success when viewed on a direct basis; this is because of the previously discussed uncertainty as to the extent POLANO influenced the selection of the storm-surge barrier alternative by the Cabinet and Parliament.

Analytic success. Recall that analytic success is concerned with the quality of a study's methodology and results. It is also the foundation for the other kinds of success. The utilization of poor-quality results or the implementation of a policy whose selection was based on poor-quality results is no credit to a study.

On a direct basis one assesses a study's quality of methodology and results with respect to the decision and problem situation for which the analysis was commissioned. But on an indirect basis one assesses the quality of methodology and results with respect to their use in *another* decision or problem situation. That is, one asks whether such use for a particular purpose is invalid or inappropriate. For example, is this SANCAP finding invalid for the Los Angeles near-term transportation study? Is this POLANO estimate invalid or inappropriate for BARCON? Is this POLANO model inappropriate for PAWN? Is this SANCAP model, suitably modified and recalibrated, invalid or inappropriate for the Los Angeles oxidant-control study? Findings and estimates have underlying assumptions that may be violated if they are translated to different circumstances. Models have ranges of validity that may be exceeded and calibration factors that should be superseded if they are applied to sufficiently different circumstances.

I believe that the studies mentioned above as utilization and outcome successes on an indirect basis are also analytic successes on an indirect basis. When a subsequent study was borrowed from an earlier study, considerable care was devoted to ensuring that what was borrowed was not invalid or inappropriate for the new circumstances. For example, models were generally recalibrated and revalidated. And findings were only translated to circumstances where the underlying assumptions were fairly well satisfied. Furthermore, the analysts on the earlier study often tried to anticipate other applications or circumstances where estimates, findings, or models might be applied and then to suggest which might be inappropriate.

Although it is important to assess analytic success on an indirect basis to avoid misapplying a study's products, it is also important to learn how to apply the elements of one study to other decisions or problem situations whenever possible, so as to avoid reinventing the wheel.

14.5. Conclusions about Success

In the light of the framework for describing success set forth in the preceding section, it is appropriate to close this chapter with some general conclusions about success in systems analysis.

In managing an analysis to achieve success, analysts should give primary emphasis to analytic success and secondary emphasis to utilization success viewed on a direct basis, largely ignoring outcome success and slighting success viewed on an indirect basis. This recommendation reflects conclusions concerning the relative importance and the analysts' degree of control of each. There are several reasons for these conclusions.

Analytic success is the foundation of other kinds of success, and also the one over which analysts have the most control.

Measures of utilization and outcome success for a policy study necessarily confound the effects of study quality with the effects of political incentives and constraints. Even a study of "perfect" quality may not succeed in influencing policy, or even in being used, because some results conflict with existing political constraints or incentives. To the extent that such conflicts can be anticipated, the "perfect" study attempts to minimize them; it may try to undermine or remove constraints, and it may try to weaken conflicting incentives or to bolster supportive ones. Inevitably, some study findings will have conflicts that either could not be anticipated or about which the analysts can do nothing. Thus, a "perfect" study can only maximize its own potential; it cannot assure its own utilization (Fisher, 1987).

Nagel (1984) nicely supports the recommendation about the relative emphasis on analytic success and utilization success: "The primary obligation of a policy researcher is to do valid research, rather than research that is utilized. Greater sensitivity to the factors that facilitate research utilization can, however, be helpful in suggesting legitimate ways to increase the probability that valid research will be utilized." The wisdom of this view becomes clearer when one considers the implications if analysts viewed their primary obligation as utilization. Analysts would tend to tolerate or condone distorted use of their study results, either through misrepresentation or selective use of arguments out of context. And they would be tempted to perform analyses that pandered to policymakers' prejudices or *a priori* preferences among alternatives rather than analyses that sought the truth.

Outcome success is highly problematic. It is hard to establish whether the alternative's selection was significantly influenced by the analysis. Many things outside the analysts' control can go wrong in translating the policy alternative into an operational plan and in implementing the plan. And evaluation's many complications make it extremely difficult to establish that a study is an outcome success even when it truly is one.

Finally, it is impossible to foresee most of the other decisions or problem situations where a study's methodology or results might be applied. Thus a study should concentrate on the decision and problem situation for which it was commissioned and seek direct success there; of course, in developing a study's methodology and results, it is worthwhile to give some attention to identifying other decisions and problem situations where the approach and findings might be useful and to suggesting how they might be generalized, in order to facilitate indirect success.

One can explain the recommendation for giving primary emphasis to analytic success and secondary emphasis to utilization success by drawing an analogy to the saying: "The proof of the pudding is in the eating." In performing analysis we are primarily concerned with successfully cooking and presenting the pudding (analytic success)—how to equip and staff a kitchen, how to choose prime ingredients, and how to prepare, bake, and serve the pudding. If the pudding is not tastily prepared and attractively presented, it probably will not, and should not, be eaten. We are secondarily concerned with successfully getting customers to eat the pudding (utilization success)—how to establish a trustful relationship with the customer, have a congenial serving staff, choose a popular flavor for the pudding, and ensure that the pudding complements the rest of the menu. If the pudding is not eaten, it can neither please the customer nor nourish him. We are not really concerned with suggestions for successfully using the pudding to improve the customers' nourishment (outcome success); too many factors besides pudding affect his nourishment for us to evaluate it properly. Nor are we particularly concerned with successfully enhancing what we or the customers do with leftover pudding (indirect success).

This chapter is largely an extract and partially a condensation of a draft Note that will be published as *Managing a Successful Policy Analysis*, N-2640-A (Santa Monica, California: The RAND Corporation). The writing of this Note has been sponsored by the Arroyo Center, The U.S. Army's Federally funded Research and Development Center operated by The RAND Corporation. The views expressed are those of the author and do not necessarily represent those of the sponsor. Small portions of this chapter have previously appeared in B. F. Goeller and the PAWN Team, Planning the Netherlands' water resources, *Interfaces* 15(1), 3–33, copyright © 1985 The Institute of Management Sciences and the Operations Research Society of America, 290 Westminster Street, Providence, Rhode Island 02903.

References

Armstrong, J. S. (1982). The ombudsman: Is review by peers as fair as it appears? *Interfaces* 12(5), 62–74.

Benveniste, G. (1977). *The Politics of Expertise*. San Francisco, California: Boyd Fraser.

Berman, P., and M. W. McLaughlin (1977). *Federal Programs Supporting Educational Change, Vol. VII: Factors Affecting Implementation and Continuation*, R-1589/7-HEW. Santa Monica, California: The Rand Corporation.

Beyer, J. M., and H. M. Trice (1982). The utilization process: A conceptual framework and synthesis of empirical findings. *Administrative Science Quarterly* 27(4), 591–622.

Brewer, G. D., and P. deLeon (1983). *The Foundations of Policy Analysis*. Homewood, Illinois: The Dorsey Press.

Caplan, N., A. Morrison, and R. J. Stambaugh (1975). *The Use of Social Science Knowledge in Policy Decisions at the National Level*. Ann Arbor, Michigan: Institute for Social Research, University of Michigan.

Cases. See Miser and Quade (to appear).

Catlett, L., et al. (1979). *Controlling the Oosterschelde Storm-Surge Barrier—A Policy Analysis of Alternative Strategies: Vol. I. Summary Report*, R-2444/1-NETH. Santa Monica, California, The Rand Corporation.

Dunn, W. N. (1983). Measuring knowledge use. *Knowledge* 5(1), 120–133.

Dye, L. (1988). Barrier against the sea. *Los Angeles Times*, January 25, 1988, Part II, p. 4.

Fisher, K. T. (1987). Personal communication.

Goeller, B. F., A. F. Abrahamse, J. H. Bigelow, J. G. Bolten, D. M. de Ferranti, J. C. DeHaven, T. F. Kirkwood, and R. L. Petruschell (1977). *Protecting an Estuary from Floods—A Policy Analysis of the Oosterschelde: Vol. I. Summary Report*, R-2121/1-NETH. Santa Monica, California: The Rand Corporation.

——— et al. (1973a). *San Diego Clean Air Project: Summary Report*, R-1362-SD. Santa Monica, California: The Rand Corporation.

——— et al. (1973b). *Strategy Alternatives for Oxidant Control in the Los Angeles Region*, R-1368-EPH. Santa Monica, California: The Rand Corporation.

——— et al. (1983). *Policy Analysis of Water Management for the Netherlands:*

Vol. I. Summary Report, R-2500/1-NETH. Santa Monica, California: The Rand Corporation.

———— and the PAWN Team (1985). Planning the Netherlands' water resources. *Interfaces* 15(1), 3–33.

Goemans, T. (1986). *The Impact of the POLANO Study,* M-86165. Delft, The Netherlands: SIBAS.

Goldhamer, H. (1978). *The Advisor.* New York: North-Holland.

House, P. W. (1982). *The Art of Public Policy Analysis.* Beverly Hills, California: Sage.

Larsen, J. K. (1980). Knowledge utilization: What is it? *Knowledge* 3(1), 412–442.

———— (1985). Effect of time on knowledge utilization. *Knowledge* 7(2), 143–159.

Leemans, A. (1986). Information as a factor of power and influence: Policymaking on the Delta/Oosterschelde Sea Defense Works. *Knowledge* 8(1), 39–58.

Lynn, L. E., Jr. (1978). The question of relevance. In *Knowledge and Policy: The Uncertain Connection* (L. E. Lynn, Jr., ed.). Washington, D.C.: National Academy of Sciences, pp. 12–22.

Mikolowsky, W. T., et al. (1974). *The Regional Impacts of Near-Term Transportation Alternatives: A Case Study of Los Angeles,* R-1524-SCAG. Santa Monica, California: The Rand Corporation.

Miser, H. J., and E. S. Quade (1985). *Handbook of Systems Analysis: Overview of Uses, Procedures, Applications, and Practice.* New York: North-Holland. Cited as the *Overview.*

———— and ———— (to appear). *Handbook of Systems Analysis: Cases.* New York: North-Holland. Cited as *Cases.*

Mood, A. M. (1983). *Introduction to Policy Analysis.* New York: North-Holland. See pp. 257–268.

Nagel, S. S. (1984). MS/OR utilization in public policy making. *Interfaces* 14(4), 69–75.

Overview. See Miser and Quade (1985).

Patton, C. V., and D. S. Sawicki (1986). *Basic Methods of Policy Analysis and Planning.* Englewood Cliffs, New Jersey: Prentice-Hall.

Pelz, D. C. (1978). Some expanded perspectives on use of social science in public policy. In *Major Social Issues: A Multidisciplinary View* (M. Yinger and S. Cutler, eds.). New York: Free Press.

Petruschell, R. L., et al. (1982). *Policy Analysis of Water Management for the Netherlands: Vol. XIII, Models for Sprinkler Irrigation System Design, Cost, and Operation,* N-1500/13-NETH. Santa Monica, California: The Rand Corporation.

Pulles, J. W. (1985). *Beleidsanalyse voor de Waterhuishouding in Nederland/ PAWN.* The Hague: Rijkswaterstaat.

———— (1987). Personal communication.

Quade, E. S. (1975). *Analysis for Public Decisions.* New York: American Elsevier.

———— (1982). *Analysis for Public Decisions,* 2nd ed. New York: North-Holland.

Ravetz, J. R. (1971). *Scientific Knowledge and its Social Problems*. Oxford, England: Oxford University Press.

Rich, R. F. (1975). Selective utilization of social science related information by federal policymakers. *Inquiry* 13(3), 72–81.

——— (1977). Uses of social science information by federal bureaucrats: Knowledge for action versus knowledge for understanding. In *Using Social Research in Public Policymaking* (C. Weiss, ed.). Lexington, Massachusetts: D. C. Heath, pp. 197–212.

——— (1981). *Social Science Information and Public Policymaking*. San Francisco, California: Jossey-Bass.

Rijkswaterstaat (1976). *Analysis of Oosterschelde Alternatives*. (An English translation is available: *Policy Analysis of Eastern Scheldt Alternatives*, Nota DDRF-77.158. The Hague, Netherlands: Rijkswaterstaat.)

——— (1985). *De Waterhuishouding van Nederland*. The Hague, Netherlands: Rijkswaterstaat.

——— (1986). En evaluatie van de witte nota. *Oosterscheldewerken/Voortgangsrapportage* 20 (January–June 1986). (An English translation of this article entitled "An evaluation of the White Note" was prepared by M. Jas.)

Schneider, A. L. (1986). The evolution of a policy orientation for evaluation research: A guide to practice. *Public Administration Review*, July/August 1986, 356–361.

Shubik, M. (1984). Gaming: A state-of-the-art review. In *Operational Gaming: An International Approach* (I. Ståhl, ed.). Oxford, England: Pergamon, pp. 15–22.

Smit-Kroes, N. (1984). Letter from the Minister of the Netherlands Ministry of Transport and Public Works. The Hague, Netherlands.

Southern California Association of Governments (1974). *SCAG Short Range Transportation Plan*. Los Angeles, California.

Springer, J. F. (1985). Policy analysis and organizational decisions: Toward a conceptual revision. *Administration and Society* 16(4), 475–508.

Weiss, C. H. (1977a). Introduction. In *Using Social Science Research in Public Policy Making* (C. H. Weiss, ed.). Lexington, Massachusetts: D. C. Heath, pp. 1–22.

——— (1977b). Research for policy's sake: The enlightenment function of social science research. *Policy Analysis* 3(4), 531–546.

——— (1980). Knowledge creep and decision accretion. *Knowledge* 3(1), 381–404.

——— and M. J. Bucuvales (1980). *Social Science Research and Decision-Making*. New York: Columbia University Press.

Whiteman, D. (1985a). The fate of policy analysis in Congressional decision-making: Three types of use in committees. *Western Political Quarterly* 38(2), 294–311.

——— (1985b). Reaffirming the importance of strategic use: A two-dimensional perspective on policy analysis in congress. *Knowledge* 6(3), 203–224.

Chapter 15

Toward Quality Control

Hugh J. Miser and Edward S. Quade

15.1. Introduction

The aims of systems and policy analysis are practical: to improve decision- and policymaking by helping those responsible for making changes. To achieve these aims effectively, analysts must do high-quality work, a goal central to this handbook's purposes. The primary focus of this chapter is on how to control the quality of analysis work as it is carried out; that is, on what the analysts should do at each stage to ensure quality adequate to the purposes of their study. A secondary focus is on how a potential user is to evaluate a systems or policy analysis as effective professional work before accepting the results as a basis for action. This chapter is also concerned with evaluation after the eventual outcomes are known. (On the other hand, it does not deal with evaluating a program—that may or may not have resulted from decisions based on analysis—with a view to terminating, altering, or improving it; Chapter 11 takes up this process.)

In analysis, as in law and medicine, the eventual outcome is frequently not a satisfactory—and never a comprehensive—guide to the quality of the professional work involved; rather, any full assessment of quality must consider not only outcomes but also the process that yielded them, its inputs, procedures, and choices.

For an analysis team to control the quality of its work and for others to judge it competently, both should have not only relevant professional experience but also criteria—or standards—of quality to serve as guides. Although no literature, no matter how extensive or comprehensive, can substitute for significant practical experience in analytic work, neverthe- less one of the chief aims of this handbook is to offer a great deal of information about achieving quality in the craft of systems and policy analysis. In particular, the purpose of this chapter is to deal explicitly with criteria and standards as they might apply both to a study taken as a whole and to what analysts do at various stages in it.

To simplify terminology, throughout this chapter we mean the term systems analysis to represent policy analysis, operations research, and similar advisory disciplines, and what we say here to be construed as applying to all.

Regrettably, as this chapter is being written, the community of producers and users of systems analysis has not yet evolved a mature and coherent consensus about how such work can and should be evaluated from a professional point of view. Thus, systems analysis shares this failing with the evaluation of scientific advice to policymakers in general; as Clark and Majone (1985, p. 7) put it,

> no comprehensive appreciation of what constitutes "good work" underlies the critical tradition now being applied to science in policy contexts. Instead, parochial sniping picks off one study for not being sufficiently rigorous to be real science, another for not being sufficiently open to be socially legitimate. Missing is any indication that good scientific inquiry in policy contexts might have more appropriate objectives than the emulation of either pure science or pure democracy.
>
> More effective evaluation of the implications of ignorance and conflict for science in policy contexts will surely require that the critical standards currently arrogated from both classical physics and Jeffersonian democracy be replaced by standards more consciously and intelligently tailored to the task in hand. In light of this need, surprisingly little attention has been devoted to the development of an appropriate theory of criticism for scientific inquiry conducted in policy contexts.

Nevertheless, there are many indicators of the directions that such evaluations should take, and there is a body of relevant experience. Hence, this chapter, although it does not provide an authoritative and comprehensive consensual overview of how systems-analysis work should be evaluated, takes steps in this direction. It provides a structure for considering quality, and it assembles some of the criteria and standards that have appeared in the literature and that seem at this time to have lasting value.

As in any professional craft, to judge the quality of current work, there can be no wholly satisfactory substitute for experience and knowledge of the elements of craft established by earlier workers. This handbook can only supplement experience, but it can and does provide an entry to the body of high-quality work that is described in the literature; the *Overview* (especially Chapter 3) gives brief descriptions and references; Sections 2.2, 5.2, 6.3, 9.6, and 10.4 in this volume pay particular attention to examples of good work; and the *Cases* volume devotes its entire attention to describing outstanding work. The last section of this chapter also refers to the literature that records a segment of high-quality work in the United States since 1975 as judged by a jury of peers.

This chapter suggests the outlines of a framework for controlling and evaluating the quality of systems-analysis work, and fleshes out this framework where experience so far makes it possible. The point of view taken is similar to the one stated by Majone (1982, p. 2) with respect to a theory of policy evaluation.

> Such a theory must deal with policy in terms of a specific conceptual framework. The framework is not that of particular policies, for this would limit the evaluator's critical perspective and would transform him into a program manager or policymaker manqué, but neither is it something outside of policy, for in that case the whole subject of evaluation would be assimilated to some particular methodology, like statistics or cost-benefit analysis, or some academic discipline like economics or sociology. General principles of policy evaluation cannot be taken over ready-made from economics, sociology, or statistics, or any combination of these, but have to grow out of the activities evaluation deals with. Systematic policy evaluation must be distinguished from the critical assessments inspired by any particular discipline, political ideology, or ethical conviction, though such partial assessments may represent important inputs to the process of systematic evaluation.

Also with Majone (1982, p. 2), we view evaluation and quality control "as different faces of the same coin, since not only do we generally evaluate in order to control, but in order to control any activity it is necessary to evaluate and monitor it." Quality control is prospective in outlook, evaluation retrospective. We can distinguish the actors and the times: during an analysis, the analysis team controls the quality of its work in the light of appropriate criteria, standards, and relevant prior experience; with the analysis complete, others may review it, but not necessarily against the same backdrop of evaluative principles.

Although the analyst is obviously interested in the professional quality of his work, the potential user of its findings is primarily interested in its validity and pertinence, that is, its adequacy as a contribution to his purposes. Unless he is considering using the same analysis team again, he is interested in the quality of the work mainly because he believes it is positively and closely correlated with validity. Evaluation of the professional quality of a study after the eventual relevant outcomes are known may be of interest to professional systems analysts, scholars, educators, and historians, but to few others.

Validity, as defined and discussed in Chapter 13, is the most important ingredient of quality; we cannot have a high estimate of quality for an analysis without a correspondingly high confidence in its validity.

After pointing out the variety of groups that may be interested in a systems-analysis study, this chapter discusses the need for evaluating individually each of the three aspects of systems analysis: descriptive, prescriptive, and persuasive. It next turns to methods, considering appraisal of the scientific aspects based on the inputs and outputs and on the process of the analysis, an approach necessary for quality control and when evaluation precedes decisions by users and is intended as a contribution to the decision process. After considering appraisal of the prescriptive and persuasive aspects, the chapter sets forth some checklists for evaluators, discusses quality control as an element of professional practice, and describes a perspective for future attention to quality.

15.2. The Various Parties at Interest

When a systems-analysis study addresses an important issue, there are inevitably many parties at interest. The analysts, the sponsor, and the decisionmakers will all have special parochial interests stemming from their responsibilities. In addition, everyone who might be affected by any action based on the findings will have a stake in their outcomes, public-interest groups may have relevant views, the general public may be concerned with what ensues, and news media may find the work and what emerges from it something to exploit. (For definitions and additional discussion of parties at interest, see Section 14.3.)

The findings of the work may be used in a variety of ways by the different interests: analysts for further work; decisionmakers for decisions and actions; other groups to support relevant—even partisan—views; or reporters as material for news media. Work that may have a positive value for one constituency may have negative value for another, and estimates of worth may differ. An analysis exhibiting meticulous technical care may fail to address the primary concerns of a decisionmaker; another deemed highly appropriate by a special-interest group may not come up to standards deemed important by analysts; a third may focus so intently on technical matters of primary interest to the client as to be badly misinterpreted by the media and the general public. What needs to be recognized is that all of the relevant constituencies may have legitimate bases for making judgments of worth and they may, in fact, conflict—but, in judgments of quality, worth to the designated user is the worth that dominates.

A systems-analysis study has specific objectives. For it to be of high quality, the intended user must find it of value, as well as be convinced of its validity. Other parties at interest may find the analysis worthless to them, but that is irrelevant to its quality. Analysts do not select a problem formulation because, as in science, it will advance the field, but because, if solved, it will help the intended user with his problem.

As other chapters, particularly Chapter 8, make clear, the findings of an analysis—and any conclusions and recommendations that may emerge from it—depend heavily on the judgments of the analysts and those they consult as well as on the craftsmanship of the work; too, if the client is playing the partnership role we would like him to play (as described in Section 12.4), his judgments will also be importantly involved in the inputs, in problem formulation, and in the prescriptions resulting from the analysis results—and hence in the quality of the finished analysis. No matter how careful these two parties may be to achieve a balanced view, their judgments are biased to some extent; although deliberate biases may have been avoided, other factors may introduce unconscious ones. The most frequent causes of bias arise from organizational goals and strong desires for favorable outcomes, both of which are intensely human—and even desirable—motivations. One can easily be partially blinded by a preconceived idea of how a favorable result should ensue from a program, or so immersed in the goals of an organization or a cause as to be unable to see contrary indications or possibilities as clearly as one should.

Biases are especially likely to play a role in analysis when interested parties who may be involved in the programs—or opposed to them—conduct the analyses:

> After having looked at the results of countless social science evaluations of public policy programs, I have formulated two general laws which cover all cases with which I am familiar:
>
> *First Law:* All policy interventions in social problems produce the intended effect—if the research is carried out by those implementing the policy or their friends.
>
> *Second Law:* No policy intervention in social problems produces the intended effect—if the research is carried out by independent third parties, especially those skeptical of the policy.
>
> These laws may strike the reader as a bit cynical, but they are not meant to be. Rarely does anyone deliberately fudge the results of a study to conform to preexisting conditions. What is frequently done is to apply very different standards of evidence and method. Studies that conform to the First Law will accept an agency's own data about what it is doing and with what effect, adopt a time frame (long or short) that maximizes the probability of observing the desired effect, and minimize the search for other variables that might account for the effect observed. Studies that conform to the Second Law will gather data independently of the agency; adopt a short time frame that either minimizes the chance for the desired effect to appear or, if it does appear, permits one to argue that the results are "temporary" and probably due to the operation of the "Hawthorne effect" (i.e., the reaction of the subjects to the fact that they are part of an experiment); and maximize the search for other variables that might explain the effects observed. [Wilson, 1973, p. 133.]

Perhaps the most important steps toward eliminating bias from analysis are to recognize that it may exist (for example, to realize that neutrality is an ideal never achieved), and to take vigorous steps to bring divergent points of view to bear during the work. Another positive step is to admit one's biases as candidly and completely as possible (for an interesting example, see Meadows, Richardson, and Bruckmann, 1982, pp. xxv–xxvii). Nevertheless, some hidden biases may remain to be lived with.

Differing judgments, perhaps emerging from different biases, are to be found in judging worth and validity and are not to be resolved by appealing to higher principles or by more precision in stating findings.

> Required instead is a mutual comprehension of the different critical perspectives being employed.
>
> That different critical appraisals are arrived at by people in different roles is not a bad thing as such. It may simply reflect different needs and concerns of different segments of society, or different degrees of freedom in making certain key methodological choices. So long as the judgments leveled from the perspective of one particular role are not presented or misinterpreted as judgments relevant to or speaking for all possible roles, we have a healthy state of pluralistic criticism. Difficulties begin to arise when this neat partitioning of roles and the criticism voiced from them begins to break down. . . .
>
> Is there a cure for . . . common tendencies to confound critical roles? Probably not. At a minimum, however, efforts to build a critical capacity for judging scientific inquiry in policy contexts should explicitly recognize that multiple roles ("peer" groups) exist, each with a legitimate claim to set critical criteria. Further, such efforts should appreciate the complex pulls and pushes that the resulting diversity of critical criteria will exert on the inquiry itself.
>
> For the appraisal of science [and, by extension, systems analysis] conducted in policy contexts, the minimum set of roles to consider would probably include the individual scientists performing the inquiry, their disciplinary peer groups, the sponsor or manager of the research program, the client or decisionmaking group for whose use the results of the inquiry are intended, and, in most cases, some version of the interest groups that could be expected to have a stake in decisions being contemplated.
>
> In principle, it might be possible to envision a grand scheme that would combine all the role perspectives into one common critical standard, thus providing a weighted evaluation of any scientific [or systems-analysis] inquiry in its appropriate policy context. Some decision analyst has probably already proposed such a scheme, but we are grateful not to have seen it. Technical difficulties aside, we suspect that a common standard, like any other aggregational procedure, would solve none of the important problems and would create new ones of its own. Our own suspicion is that efforts to develop better critical skills for science [and systems analysis] with policy implications should aim, not for a unique evaluation, but rather for an enhanced understanding of different evaluative criteria on the part of all role players. We most need, in other words, a

more sophisticated and sympathetic understanding of the multiple perspectives involved. [Clark and Majone, 1985, pp. 7–9.]

These views imply that a systems-analysis team conducting or evaluating a major study should consider criteria and standards of quality emerging from the interests of all of the parties likely to be significantly affected by their findings and any actions or policies that might emerge from them, as well as any biases that they might have introduced from their own points of view. This is obviously an exacting requirement—but experience shows that it can be met.

The quality of strictly scientific work is commonly evaluated by disciplinary peer review. It is used by journal editors as a tool in deciding what to publish and by the U.S. National Science Foundation and other research-sponsoring organizations for ranking projects and scientists within disciplinary fields, and sometimes for allocating resources among fields. Nevertheless, the belief that it works well in judging quality in science—and there is considerable doubt that it does (Chubin and Jasanoff, 1985)—does not imply that it will serve well as a quality control mechanism for systems and policy analysis, where politics and other nonscientific factors play major roles. Yet, for judging the quality of professional work in systems analysis, we have uncovered no alternative that appears to be superior to a jury of analysts with qualifications and experience equal or superior to those of members of the analysis team. At present, such a jury lacks a set of explicit and widely accepted—and thus defensible—standards for comparison and judgment, but, if supplemented by representatives from the other disciplines involved in the systems-analysis work, might also help to formulate such standards.

15.3. Quality Control in the Three Aspects of Systems Analysis

As Section 1.1 describes, systems analysis can have three aspects: descriptive (scientific), prescriptive (advisory), and persuasive (argumentative-interactive). Since these aspects are often inextricably intertwined, it is reasonable to expect that evaluating them will be similarly related, and it will be difficult to discuss separately. Nevertheless, we will do so, although recognizing that there are links and that the evaluation results must be brought together into a final synthesis. Too, since systems analysis borrows and adapts its evaluation approaches from other disciplines and professions to a large extent, it is convenient to discuss these borrowings aspect by aspect. This is particularly true for the descriptive aspect, which borrows many concepts of evaluation from the social and physical sciences.

15.4. Considerations and Approaches in Maintaining Quality

The various parties interested in a systems-analysis study tend to regard it from different points of view. Nevertheless, analysts, clients, and other potential users should have an interest in the overall quality of the analysis, particularly its validity; other parties at interest could well have more limited concerns.

For the sponsor, the results are the bottom line. He asks himself: Was the study useful? If the recommendations could not be accepted, was the work helpful in thinking about the problem area that the study considered? Enough to justify the cost, delay, and criticisms that it may have generated? Would a follow-on study be worth supporting? By the same analysis team? If the recommendations were acceptable to the client, could his fellow decisionmakers be convinced that they should be implemented? If implemented, did the resulting program achieve what was wanted and predicted? At a cost somewhere near the one calculated by the analysis? If the chosen course of action brought about what was wanted, was the problem situation substantially ameliorated? What new problems did the new program cause? Were they foreseen by the analysis? Should these new problems be subjected to further analysis? And so on.

On the other hand, although the working analyst must be interested in his client's concerns, his immersion in the details of the work is bound to focus his attention on these details, and on the quality of his choices and procedures relevant to data, tools, and argument.

For someone not familiar with systems analysis, it may appear that the way to judge the correctness of the analysis and its findings would be to compare the recommendations and predictions with what actually happens when a new program or policy based on the work is accepted and implemented. In some cases, such as the blood-supply case described in Section 2.2, this can be done, but this situation is rarer than one would expect—and hope for. Much work is done with no thought of later implementation, but rather to help a problem diagnosis, to suggest important considerations for further work, or merely to produce background information. Other work may not produce decisions to implement a program or policy because the client judges that it would not be wise to do so; he may make such a judgment because he does not believe the study's results, because factors it did not consider may have become dominant in his mind, because he cannot persuade other decisionmakers to cooperate, because he no longer agrees with the study's input data and assumptions, or for other reasons. Or it may be because, as in some strategic studies, implementation is not intended unless certain contingencies arise.

Moreover, even when a course of action is chosen and implemented, a comparison of the actual results with what was predicted may be either impossible or so partial as to be of little value to the parties at interest. The study of how best to protect the Oosterschelde estuary from flooding (see Section 2.2) led to the construction of a storm-surge barrier designed to protect the estuary against a storm so severe that its probability of occurrence is less than once in 4000 years. Thus, although the effects of this barrier on the many ecological, economic, operational, and social factors considered in the systems analysis can be observed in a few years, the extent to which the barrier serves its main purpose of protection from flooding may not be tested for centuries.

Other difficulties can attend an attempt to judge the quality of a systems-analysis study by comparing actual events to its predictions. For example, uncertainty always plays a role in policy problems, and unexpected events can sometimes alter predicted results significantly. For instance, one suspects that few pre-1970 analyses of energy futures foresaw the sharp increases in oil prices that took place in 1973 and 1974, and the scenarios for the next 50 years that were the basis for the study of the world's energy supply and demand summarized in Section 9.6 deliberately excluded by assumption such unexpected events as sharp price changes, in spite of the notable earlier occurrence of such an event. Such a choice of assumptions does not necessarily suggest bad analysis. Every analysis is limited by its assumptions; to limit them by excluding sharp changes merely implies that, if such a change occurs after an analysis is completed, comparisons between the findings and actuality will have to be adjusted accordingly.

It is also notorious that programs being implemented are subjected to many influences aimed at deflecting them from their goals, diverting their resources, and adding activities not foreseen in the studies that led to their creation (Bardach, 1977). If these influences have important effects on what ensues, actual results can differ markedly from what analysis might have predicted. A medical analogy suggests other problems: If a physician experiences a higher-than-normal percentage of deaths among his patients, it may mean not that his medical skills are questionable but that he is so firm in his craftsmanship and courage that he can, without damage to his reputation, accept a very risky class of patients. Or it may mean that his patients do not follow with sufficient care the after-treatment health-care regime that he has recommended (Freymann, 1974, p. 6), although as in systems analysis successful persuasion should be considered part of the practitioner's craftsmanship.

It is not uncommon for the client of a major systems-analysis study, particularly if its findings make him uneasy, to seek an independent review of its validity before proceeding to make decisions based on it that may involve major expense and risk. Here again the evaluation must be based on considerations other than outcomes of actual programs (see Section 13.6).

Success in real-world programmatic terms and high-quality professional work in systems analysis are neither identical nor synonymous, although they often occur together. For the user, an analysis is a success if its findings lead to an action or policy that fulfills his purposes; it is a failure if his decision, based on the analysis, cannot be implemented, or, if implemented, does not bring about the desired outcome—or even if the outcome turns out later to be undesirable. But such a failure may occur because the world changed in ways that could not be anticipated, not because the analysis was not of high quality.

The foregoing discussion is not intended to imply that systems-analysis studies cannot be evaluated, or that every such evaluation must be based completely on considerations other than program or policy outcomes; rather, it is intended to undergird these points:

To be useful to a user facing decisions, an evaluation cannot be based on outcomes unless they have already occurred.

Because the processes of a systems-analysis study must be examined carefully and critically, evaluating the quality of the work is as demanding in meticulous care, penetrating analysis, factual knowledge, and skilled craftsmanship as doing the analysis itself, although the effort involved may be considerably less.

An evaluation of a systems-analysis study must consider both the study itself and its context, with particular attention to its potential effects on the parties at interest if its findings lead to adopting policies or programs.

The analysis team, as well as the later evaluators, needs criteria of quality to guide its work as it proceeds toward effective results, and these standards should be those established by the profession at large, based on its best work and the experience that work represents.

These standards must relate, not only to the finished products emerging from systems-analysis work, but also to the many kinds of steps required to achieve these products.

Since systems analysis has a three-fold character (descriptive, prescriptive, and persuasive), any comprehensive evaluation must consider all three of these characteristics. We begin with the descriptive aspect, and employ three different modes of appraisal (Majone, 1982; Clark and Majone, 1985): input (Section 15.5), output (Section 15.6), and process (Section 15.7).

The input mode considers the people involved and the wide variety of material that enters the study: data, assumptions, models, mathematical procedures, possible biases, professional specialists on the analysis teams and their track records, and so on.

The output mode focuses on the findings and how they relate to reality and to the process of analysis, on the prescriptions emerging from the findings, and, if implemented, possibly on the eventual outcomes.

The process mode considers the appropriateness of all of the steps taken in carrying out the analysis, the basis on which they were chosen, how effectively this process and the material entering it have been communicated to the various parties at interest, and the relationship of all this to accepted criteria of quality in systems analysis.

Although any complete evaluation should consider all three modes to some extent, in some cases one mode may dominate. For example, in a study for which appraisal of the output mode is handicapped by relations to reality that are difficult to assess, the input and process modes may have to constitute the principal basis for the evaluation. If the processes in such a case are simple and reasonably standard but the inputs are so difficult and venturesome as to attract important attention, then the appraisal of the input mode may dominate the evaluation. In any case, the circumstances of the analysis will force the evaluator to select an emphasis for each mode in his evaluation.

After Sections 15.5, 15.6, and 15.7 complete the discussion of the descriptive aspect, Sections 15.8 and 15.9 consider quality control and evaluation for the prescriptive and persuasive aspects. Then, since many appraisal matters overlap more than one mode or aspect, the later sections bring the separate discussions together. Section 14.4 considered evaluating the analysis in terms of "success" from the varying points of view of the various parties at interest.

15.5. Evaluation of the Descriptive Aspect: Appraisal by Input

Systems analysis demands a great number and variety of inputs, ranging from information gathered for the problem formulation to the judgments that suggest actions based on the findings. Assumptions, data collected or processed by others, facts established by science or earlier systems analyses, constraints imposed by the sponsor, models chosen as means for uniting and processing the inputs, the many choices made as the analysis proceeds, and the professional qualifications and track records of the analysts, all are inputs.

An obvious way to begin to evaluate an analysis is to assess the quality, accuracy, and relevance of the inputs; if they appear biased or below a reasonable standard, then the evaluator has identified a difficulty that must be traced: how the doubtful inputs have influenced the analysis and its findings.

To oversimplify, one might argue that, if the inputs are not satisfactory, then the findings are not likely to be satisfactory either. As the cliché goes: garbage in, garbage out (GIGO). In fact, a knowledgeable client with experience in using the results of systems-analysis work often turns first to the inputs in assessing the work's worth to him, particularly if he has found the findings surprising or unsettling. If the inputs do not represent reality as he sees it, he is not surprised if the findings are not to his liking—and he can then ask for a study of the implications of changed inputs. Indeed, if the inputs are unsatisfactory enough, he can—and should—postpone acting on the findings until new results based on revised inputs are available. The analysis team should have avoided such a difficulty by keeping the inputs that are likely to be of interest to its client before him as the work proceeds; in this process they try to anticipate last-minute changes in the state of the context, or prospective changes of special interest to the client that may dictate a revised analysis.

In the early days of systems-analysis practice, when largely technical systems were being analyzed and when, therefore, the models used to translate the inputs to outputs were relatively true and unambiguous surrogates, dictated largely by the technologies involved, clients could feel with some confidence that, if the inputs were all that they should be, the outputs would have to be accepted as valid, whether or not they were welcome. Now, however, with the problem situations much more complex, and with the analysis itself requiring so many judgments and choices, such a simple picture of evaluation is no longer realistic. Indeed, to represent such an approach as adequate should be recognized as a pitfall of evaluation, just as it is a pitfall to represent output performance by the resources used to achieve it (see Section 8.4).

One matter relevant to inputs is so important as to deserve mention: the data collection and analysis procedures (*Overview*, pp. 297–298; Altman, 1980), where many pitfalls await the unwary analyst.

15.6. Evaluation of the Descriptive Aspect: Appraisal by Output

Here it is important to distinguish between the *outputs* of an analysis and the *outcomes* in the world of reality that may ensue from actions taken on the basis of the outputs or the findings of an analysis. Outcomes emerge from implementing programs and policies. Outputs are the findings of the analysis.

Since the outcomes relevant to an appraisal of the quality of a systems-analysis study are those that emerge from carrying its findings into some form of reality, it is clear that one form of appraisal of the output is to observe that the outcomes are what are expected and desired. This approach to appraisal, although too late to help the initial user, is from many points of view the most desirable one, but, even if available, it can be difficult to apply.

> The common-sense view of evaluation . . . may be summarized by the slogan "only results count." In fact, for many people evaluation and outcome evaluation are synonymous. . . .
> This explains the strong intuitive appeal of this mode of evaluation. Indeed, if a tree is judged by its fruits, one may well wonder why any other form of evaluation is at all needed. What common sense overlooks is that outcome evaluation can be successfully performed only under rather special conditions. The most basic condition is that it must be possible to measure, with reasonable precision, the quality or level of the desired output or performance. If the indicator of performance is expressed by the distance between goals and outcomes, goals have to be clearly defined, outcomes must be unambiguously measurable, and the measuring instrument should be reliable. [Majone, 1982, p. 3.]

Evaluation by outcome is not feasible unless the outcome is known at the time the evaluation is carried out. Thus, for the evaluation that often occurs when a decisionmaker, on being presented with a study, asks his staff, and possibly other analysts, to evaluate the work as an input to deciding what action to take, no outcome is available—and the evaluation must be based on other grounds. In some cases, a partial outcome can be helpful; for example, in the flood-protection example described in Section 2.2, even though no major storm may have arrived to test safety, the outcomes with respect to the ecology, fishing, shipping, and so on, will help to evaluate quality. In some long-range strategic studies, full implementation may depend on some contingency that does not occur, but nonoccurrence may itself be a partial outcome useful for evaluation.

It is important to remember also that, although an outcome—if a relevant one is available—is usually a good indicator of the quality of the analysis, this is not necessarily always the case. If an outcome evaluation shows that a policy or program adopted on the basis of analysis findings is a failure, it does not necessarily follow that the analysis was faulty; the failure may have arisen from administrative ineptitude during the implementation, from countervailing political forces, from events in the surrounding environment that could not have been foreseen, or from other causes. The opposite can also happen: a policy or action can succeed in spite of a faulty analysis when there is a happy conjunction of favorable factors—although this is hardly an expectation on which a prudent decisionmaker should rely.

When an appropriate outcome is not available, appraisal by output depends on individual or group judgment applied to such things as the nature and relevance of the findings, the clarity and logic of the arguments on which they are based, and the adequacy of the prescriptions with respect to the objectives.

In systems analysis, as in law and medicine (where craftsmanship is so important), input and output are inadequate guides to quality. Hence, as in those professions, we turn to process as a better basis for judging quality.

15.7. Evaluation of the Descriptive Aspect: Appraisal by Process

Since most systems analysts come originally from well established and mature fields of science, they bring with them the critical and evaluative outlooks of those sciences, and apply them, insofar as possible, to the processes of their new work (Ravetz, 1971, especially Chapter 10). The criteria and procedures of evaluation cannot, however, be taken over slavishly from other disciplines; rather, they must emerge from the material being dealt with (Ravetz, 1971; Majone, 1982). In other words, although systems analysis can—and should—find both example and inspiration for its criteria of quality from classical fields of science and technology, it must develop its own based on the problems it treats and the processes that it uses to treat them. Although this point applies equally to all three modes of appraisal, it is especially important for process appraisal.

Although an evaluator will inspect the records of the process by which a systems-analysis study achieved its results, and assess the extent to which these processes reflect high standards of quality, the most important use of such standards is made by the analysts themselves from day to day as they decide how to pursue their work. Thus, the main value of criteria and standards of quality for the process of systems analysis is to act as guides to craftsmanship in carrying out the work.

Although an investigator carries with him the craft knowledge and its associated criteria of quality that have accrued from what has been learned in the past, his accumulated experience cannot guarantee that new ventures will not face new difficulties calling for the creation of new criteria to help subsequent workers avoid the new difficulties. Among these difficulties are new pitfalls, and, because they tend to prompt the more important criteria of good craftsmanship, they deserve some discussion here.

Pitfalls of Analysis

Ravetz (1971, pp. 94–97) discusses the concept of pitfalls in scientific work, and his ideas are equally applicable to systems analysis:

> The craft character of scientific work is exhibited most systematically

through the concept of "pitfall." . . . Whenever we extend from the known to the unknown, we do so on the basis of expectations of what we will find; these are necessary to give direction to our moves, and to supply interpretations of what we encounter. These expectations are always incorrect in some measure, and so we learn through the discovery of our errors. However, not all our errors are so considerate as to announce themselves as soon as they are made. It can happen that we follow an erroneous path of investigation for some time, investing great resources into its pursuit, and only much later discover that we are mistaken. . . .

Thus the path of discovery is beset by concealed traps for the unwary, which we can call "pitfalls." In some ways these are more dangerous than the physical hazards from which they take their name, for one learns only in retrospect that one has stumbled into a pitfall at some earlier point in the work. . . .

From the nature of the work, it is impossible to eliminate pitfalls from scientific inquiry. The experience of matured scientific disciplines is that they can largely be avoided. This is done in two ways: by the charting of standard paths which skirt them, and by each investigator becoming sensitive to the clues which indicate the presence of special sorts of pitfalls he is likely to encounter in his own work. The first of these requires a tradition of successful work in the subject where there is a body of standard techniques which can successfully be applied as a routine. . . . [T]hese techniques are an embodiment of successful craft practice, built up over generations. Then, when an individual scientist explores beyond the range of the well-established techniques, his craft knowledge must necessarily be more subtle and personal, for the pitfalls he is likely to encounter are peculiar to his particular materials and tools. The clues to the presence of pitfalls are all he has, since he is beyond the range of the charted paths. Thus the accumulated social experience of his field must be supplemented by his personal craft experience of his portion of it.

This discussion implies that, to a considerable extent, the maturity of a technical field can be measured by the extent to which its techniques and procedures have become widely standardized so as to avoid pitfalls, and the extent to which the field has leaders whose work exemplifies high craft skills that guide their excursions into new territory.

By the middle of the 1950s a few such systems-analysis leaders had emerged, but the rigors of the field and the shifting patterns of support for its work have kept the number small, with earlier workers dropping out almost as fast as new leaders emerged. Nevertheless, as this handbook implies, there is a small body of such leaders whose work can be turned to as exemplications of craft skill.

With the emergence of these leaders, some attention began to be paid to pitfalls, with scattered lists being circulated informally, one paper devoted to them being published in 1956 (Koopman, 1956), another in 1957

(Kahn and Mann, 1957). The first major survey, however, did not appear until 1980 (Majone and Quade, 1980); it is a volume with which every serious systems analyst should be thoroughly acquainted, both for what it says and for the literature that it refers to. It is a valuable step toward establishing a profession-wide knowledge of pitfalls and the criteria that should help avoid them. It provides substantial support for Majone's 1980 judgment that

> there is enough experience by now to suggest at least minimal criteria of adequacy. Such criteria can be derived by studying the most serious conceptual errors into which analysts occasionally fall (in collecting and analyzing data, in choosing tools, in drawing conclusions, and in communicating them). Thus, by identifying the pitfalls of analysis and charting safe paths around them, we are slowly building a solid foundation on which subtler criteria of quality can be based. [Majone, 1980, p. 27.]

On the other hand, although recognizing pitfalls and avoiding them is an important step toward controlling the quality of systems analysis work, it has limitations that must be recognized:

> Avoidance of pitfalls guarantees minimal standards of quality, nothing more. It does not imply originality, depth, or any other intangible qualities that distinguish the brilliant from the merely competent study; nor can it ensure the success of the proposed solution. And yet, even meeting these minimal standards is not an easy task in a world of increasing complexity, pervasive uncertainty, and fallible knowledge. [Majone and Quade, 1980, p. 5.]

Table 15.1 exhibits a listing of some common pitfalls.

The Quality of the Analysis Process

Although avoiding pitfalls is an important step in controlling the quality of the analysis, it is by no means all that must be done. There are many other matters that must be attended to properly; indeed, one of the main purposes of this handbook is to offer both advice and examples of how high-quality work is done.

Nevertheless, there are some areas that are so important for appraising quality as to call for special mention here:

How the problem is formulated, that is, the process of understanding a problem situation and transforming a suitable portion of it into a clear-cut problem with goals to be achieved and possible approaches to achieving them.

How data are collected, and the processes that are used to convert the data to information, and how this information is used as evidence in

Table 15.1. Some Common Pitfalls of Analysis

Insufficient attention to problem formulation.
Adherence to the belief that one already knows the problem and its solution.
Modeling more of reality in more detail than the problem requires.
Seeking academic rather than policy goals.
Underestimating the large margin of error found in most socioeconomic data and statistics.
Using ratios to measure effectiveness while disregarding absolute magnitudes.
Treating data and the results of statistical and mathematical calculations as conclusive rather than as evidence.
Confusing the future with the past; not emphasizing the recent data.
Being concerned with sunk rather than with opportunity costs.
Measuring total rather than marginal utilities when only marginal valuations can provide guidance.
Treating statistical significance as necessarily implying operational significance.
In costs, confusing the question of relevance with that of concern.
Fitting the problem to familiar mathematical tools rather than the tools to the problem.
Assuming the client is an analyst and telling him what was done and how rather than what was learned.
Neglecting the qualitative uncertainties while dealing with just those that can be treated mathematically or computationally.
Building overambitious models; modeling what one can, not what is relevant.

---

developing the arguments of the analysis (Ravetz, 1971, pp. 76–88; *Overview*, passim).

Verifying and validating the models used in the analysis, and therefore the assumptions and evidence used in constructing them (Chapter 13). This, and their documentation, are particularly important when large-scale computer models are used.

Documenting the work done, the choices made, and the reasons (Chapter 14; *Overview*, passim).

All of the matters on this list are important, but of far greater importance is the core of the analytic process: the argument and its basis in analysis and judgment, the evidence that links the inputs to the findings, the judgments that turn these findings into the final output (often a tabulation of the superior alternatives and, if warranted, recommendations for action).

15.8. Evaluating the Prescriptive Aspect of Systems Analysis

In the minds of sponsors, users, and other parties interested in systems-analysis studies, the prescriptive or advisory aspect is usually the most significant one, for it brings to light the implications for action that are the purpose of the analysis: to assist those responsible for decision or policy to select what should be done to improve the problem situation. The analysts may not make formal recommendations or offer an explicit statement of their opinion as to what should be done, leaving this matter to the client, but nevertheless the estimates, arguments, and predictions yielded by their work provide a basis for policy or action. The other two aspects of systems analysis are, in a sense, subservient to this role; the scientific (descriptive) aspect provides the foundation for what is prescribed, and the persuasive aspect offers the additional clarification and explanation that is almost always needed before new analytic work can win acceptance.

As has already been said, it is not the responsibility of the analysts— nor should it be—to decide what action or policy should be adopted. Rather, the purpose of the analysis is to help with this decision, and advice from the analysts, steeped in the problem, possibly more fully informed than the person who must act, and without an axe to grind, can be very helpful. Recognizing this, the potential user often asks specifically for conclusions and recommendations. Although he intends to make up his own mind, he expects the analysts to offer new and useful perspectives on his problem. For this and other reasons, it is clear that the influence of the client on what is produced is much greater in this aspect of the analysis than in the other two.

In offering advice—whether requested or not—that extends beyond the findings of his analysis, and thus requires the exercise of his judgment and intuition, the analyst must be careful about making suggestions that may be biased by personal objectives and values. In such a case, it is important for the user (and for the analyst himself) to distinguish what the work actually implies from any suggestions as to what should be done that might rest in part on what the analyst thinks or feels about a context larger than what was actually considered in the systems-analysis study.

Some experienced users of analysis feel that even conclusions, let alone recommendations by the analyst, are out of place.

> Simply said, the purpose of an analysis is to provide illumination and visibility—to expose some problem in terms that are as simple as possible. This exposé is used as one of a number of inputs by some "decision-maker." Contrary to popular practice, the primary output of an analysis is not conclusions and recommendations. Most studies by analysts do have conclusions and recommendations even though they should not, since invariably whether or not some particular course of action should be followed depends on factors quite beyond those that have been quantified by the analyst. A "summary" is fine and allowable, but "conclusions" and "recommendations" by the analysts are, for the most part, neither appropriate nor useful. Drawing conclusions and making recommendations (regarding these types of decisions) are the re-

sponsibility of the decision-maker and should not be pre-empted by the analyst. [Kent, 1967, p. 50.]

Nevertheless, on occasion the problem and the investigation of it are such that the findings from the descriptive aspect of the analysis suggest clearly what should be done to improve the situation (as in the blood-supply case described in Section 2.2). When this happens no additional judgment by analyst or client to arrive at advice is needed, and conclusions and recommendations are warranted—and usually should be made.

More often, however, the problem, the complications of the situation, or the portion of it dealt with in the analysis will not permit such a firm and unique result. The findings are more likely to be a list of alternatives that will accomplish what the client wants, each with associated costs and other consequences (possibly varying for different contingencies that might arise) that may or may not be acceptable or desirable. In such cases, the analysts, supplementing their findings with judgment, may offer helpful advice by pointing out cases of near dominance, or that alternative A would be preferable to B if a way could be found to placate group G so that it could be implemented without excessive cost, or that, if the client thinks that contingencies C and D are so unlikely as to be neglectable, then alternative A' would be the most attractive.

In other cases, particularly in the public sector where human behavior plays a dominant role, inquiry can rarely be anywhere near exhaustive; time and money are limited, public pressure for early answers is great, and usable data are scarce. The analysts can report findings, mostly about parts of the problem, but suggestions for action or policy depend so much on judgment covering a wider span of concerns that the responsible decisionmaker must take the lead in supplementing the prescriptive aspect of the analysis.

Thus, for most analyses, the advice or prescription is based largely on judgments by both analysts and users. Hence, in this case the quality of the prescriptive aspect must be evaluated by the standards used to judge the validity of judgments, as discussed in the second part of Section 13.6.

At the risk of oversimplifying a complicated situation, we may summarize what we have been saying by distinguishing two cases:

If the analysis rests on clearly valid judgments and surrogate models (see Section 8.5), then its findings may be adequate to support strong conclusions and clear recommendations for policy or action.

If the analysis rests largely on judgments of less confidence and/or offers only a perspective on the problem situation (see Section 8.6), then analysts are seldom likely to have an adequate basis for unambiguous conclusions and recommendations, and the main burden for even suggesting what to do shifts to the decisionmaker.

15.9. Evaluating the Persuasive Aspect of Systems Analysis

To investigate the persuasive aspects of a systems-analysis study is to answer two questions: Did the analysis deal effectively and ethically with the sponsor's and user's concerns, and those of other parties at interest (including adversary parties), insofar as these other interests could be known to the analysts and could be included in the effort? Were the findings and their limitations communicated effectively and fairly?

Most of the material aimed at answering the first question will emerge from the investigation discussed in the previous sections, but the tie to effectiveness of communication is so close that the matter must be raised here, for it is often in the communication process that the final elements leading to the decision to accept or reject the findings will emerge.

When systems-analysis studies—as is often the case—prescribe a policy or course of action, even one formulated by the client or with his help, its general acceptance and implementation are strongly influenced by the other parties at interest. Therefore, an important factor in evaluating the quality of the work, and particularly its persuasive aspect, is how it handles the concerns of the other interested parties.

To help identify these parties, we turn to a survey by Clark and Majone (1985) of a variety of historical efforts to evaluate scientific studies in policy contexts in which they tabulated the kinds of factors that they encountered as a function of the input, output, and process components of analysis, and of five types of interested parties: scientists, peer groups, program managers and sponsors, policymakers, and public interest groups. Table 15.2 shows their results. (See also Section 14.3.)

Since any systems-analysis study must be tailored to its context, the problem situation, and the parties at interest, this table must not be considered as definitive; rather, it suggests a wide variety of issues that can arise in such an enterprise—and it emphasizes the importance of considering the concerns of the interested parties throughout the analysis, and especially during the work of preparing and reporting the findings. Thus, a key element in the control and evaluation of the persuasive aspect is assessing how well the analysis has met these concerns.

Since each context that engenders a systems-analysis study has unique properties of its own—and especially its own parties at interest—it is impossible to construct a generic and globally relevant set of categories of matters to be considered in evaluating the persuasive aspect, as exemplified by Table 15.2. Rather, the analysis team must evolve its own set on the basis of its work and knowledge of the context in which it takes place; needless to say, this is best done in close consultation with the sponsor of the work and the decisionmakers who appear to be most likely

Table 15.2. Some Critical Criteria Used in Evaluations of Scientific Studies in Policy Contexts

| Party at interest | Critical mode | | |
	Input	Output	Process
Scientist	Resource and time constraints; available theory; institutional support; assumptions; quality of available data; state of the art.	Validation; sensitivity analyses; technical sophistication; degree of acceptance of conclusions; impact on policy debate; imitation; professional recognition.	Choice of methodology (e.g., estimation procedures); communication; implementation; promotion; degree of formalization of analytic activities within the organization.
Peer group	Quality of data; model and/or theory used; adequacy of tools; problem formulation; input variables well chosen? Measure of success specified in advance?	Purpose of the study; conclusions supported by evidence? Does model offend common sense? Robustness of conclusions; adequate coverage of issues.	Standards of scientific and professional practice; documentation; review of validation techniques; style; interdisciplinarity.
Program manager or sponsor	Cost; institutional support within user organization; quality of analytic team; type of financing (e.g., grant vs. contract).	Rate of use; type of use (general education, program evaluation, decision-making, etc.); contribution to methodology and state of the art; prestige; can results be generalized, applied elsewhere?	Dissemination; collaboration with users; has study been reviewed?
Policy maker	Quality of analysts; cost of study; technical tools used (hardware and software); does problem formulation make sense?	Is output familiar and intelligible? Did study generate new ideas? Are policy indications conclusive? Are they consonant with accepted ethical standards?	Ease of use; documentation; are analysts helping with implementation? Did they interact with agency personnel? With interest groups?
Public interest groups	Competence and intellectual integrity of analysts; are value systems compatible? Problem formulation acceptable? Normative implications of technical choices (e.g., choices of data).	Nature of conclusions; equity; analysis used as rationalization or to postpone decision? All view points taken into consideration? Value issues.	Participation; communication of data and other information; adherence to strict rules of procedure.

Source: Clark and Majone, 1985, p. 11. © 1985 Massachusetts Institute of Technology and the President and Fellows of Harvard College. Used by permission.

affected by the findings. In doing this, the team may find the matters listed in Table 15.2 and Section 15.10 helpful as reminders of things to be considered.

It is a regrettable, but absolutely necessary, fact of systems-analysis work that not everything in a problem situation can be included as part of an analysis: The imperatives of relevance to the client's objectives and of time and resource constraints always force the team to select the portions of reality to include in their work. Therefore, it is not surprising that occasionally a factor considered important by an interested party has not been included. Sometimes an estimate of its importance and effect can be made informally after the analysis has been concluded; sometimes the judgment of relevance and importance must be left to the various parties at interest. In any case, it is of overwhelming importance for the analysts to include in their reports clear statements of what has been considered in the study and what has not. Paradoxically, most clients and critics do not consider the statement that certain factors have not been treated to be a sign of weakness, but rather one of strength, since they are likely to be aware of the fact that any analysis completed in a reasonable length of time must restrict itself to centrally important issues, or even to a suboptimization. And any experienced decisionmaker is accustomed to weighing evidence dealing with part of his problem situation and exercising his judgment and intuition on the parts for which less explicit information is available.

To answer the question about effective communication is to do more than ascertain whether or not good communication techniques such as those discussed in the *Overview* (pp. 313–315) have been used and the pitfalls avoided (Meltsner, 1980); it also means answering questions such as these:

Were the communication efforts tailored carefully to the various parties addressed?

Were they timely?

Did the communication efforts identify clearly questions and concerns of the parties at interest and deal with them effectively?

Did the communication efforts actually create a meeting of the minds in which all the parties concerned understood one another?

As a result of the communication effort, did the potential user understand fully the implications of the analysis?

In dealing with the first of these questions, it is important to realize that different persons respond differently to various arguments and ways of presenting them. For example, to visualize the behavior of related variables most people prefer graphical presentations, but this is by no means universal; some people, particularly those with financial or accounting backgrounds, see more in a rather complete table of data. Some people would rather focus on the quantitative facts, others on how these facts might affect various persons and organizations. Some persons prefer a tightly knit argument so that they can see how the findings emerged; some prefer just to review the results in order to explore their potential implications; and so on. In any case, the ways to communicate effectively cannot be specified universally; rather, they depend on a careful assessment of how the persons to be communicated with best receive messages.

If the analysis team has the opportunity of presenting its ideas, and eventually its findings in oral briefings, its task is usually to raise the confidence of the client and the decisionmaker (possibly the same person) and their staffs in the adequacy of the work for their purposes. Other interested parties may be unknown—or may not even exist—until later, when rumors of action on the issue presented or the beginnings of implementation bring them forth. To succeed, of course, each presentation must be adapted to its audience. Since the analysis team may not have an opportunity to communicate with the representatives of the other interests that may develop, it must anticipate their possible concerns and objections and the arguments supporting them, and, where possible, supply the client and decisionmaker with the means to meet them. In sum, a key person must have full appreciations of both the pros and the cons of what is being considered, especially the cons:

> [T]o be a persuasive advocate he . . . needs to know all about the cons and the counters to these cons. Skeptics have a very nasty habit and a diabolical instinct to focus on the poorer aspects of any proposition you may make, as distinct from the better aspects. . . . [E]ven in the dirty business of advocacy it pays to be honest. [Kent, 1969.]

Answers to all the questions about effective communication will emerge from careful observation of the communication process. Since it is a dynamic process of interaction, another question can be asked and answered: How well did the analysis team respond to new questions and unexpected issues as the communication process moved along?

Candor in self-evaluation strengthens, rather than weakens, the impacts of an analysis team's communications. Candid assessments of the limitations and weaknesses of an analysis—limitations of time and effort are bound to have forced some omissions or shortcuts on the analysis team—create the impression that the team knows very clearly what it is talking about, and thus strengthen the rest of the communication effort.

Similarly, particularly if hostile audiences are involved, a candid statement of the role being played by the analysis (see the *Overview*, pp. 320–322) is helpful: If the analysis is aimed at giving as balanced and realistic a picture of reality as possible, taking the bad with the good, then the audiences expect the analysts to exhibit openness throughout the discussion, regardless of whether or not the findings support preconceived positions or not. On the other hand, if the analysis is a technical brief (*Overview*, pp. 321–322) aimed at supporting certain conclusions (usually favorable to the client's parochial interests), then an open statement of this fact helps all concerned and commands respect for the analysis team and its work—but also, of course, warns all concerned of the limitations of the analysis.

Plain communication, however, is frequently not enough; it must also be convincing (Meltsner, 1980, pp. 128–129). To be effective, systems-analysis findings must often serve to modify the attitudes and beliefs of potential users. To achieve such results, it is usually not enough for analysts merely to present reasons and facts. Facts do not speak for themselves, nor do systems-analysis findings, which can seldom be regarded confidently to be facts (see Section 12.2). Systems analysis cannot prove its conclusions as a mathematical theorem can be proved, and, even if it could, few users would be persuaded by such means that the results were correct, for users are seldom trained or able to accept such forms of argumentation. What it can do is provide evidence, which, with corroborating evidence from other sources, can be used to support the conclusion. Such evidence must be selected and shaped so as to be understandable to the user, and persuasive to him. This should be done, of course, in such a way as to represent the study's findings fairly.

It is clearly helpful for the team's best briefer to make the oral presentations, using material from the work that is most relevant and convincing to the audience being addressed. So long as the findings are not misrepresented, it is both ethical and desirable to make the best case possible for the findings, especially since any oral presentation is, *a fortiori*, limited to a tiny portion of the full story developed by the analysis. To do this, the speaker selects the combination of facts, data, methods, arguments, and modes of presentation that seems most likely to convince his audience, recognizing that any issue of interest that is not covered in the talk will most likely be explored during ensuing discussions—and can always be looked into in detail by individuals through expert team members and the analysis documentation.

What will persuade a given audience deserves considerable thought and exploration. Kahn and Mann (1956, p. 148) suggest that what to present be approached in two stages: "a first stage to find out what one wants to recommend, and a second stage that makes the recommendations convincing even to a hostile and disbelieving, but intelligent, audience." This implies that one should explore the properties of the audience: what its beliefs, prejudices, and biases are; what its objections are likely to be; and what forms of communication it is comfortable with.

The point is that, in presenting the findings of an analysis, an analysis team is not trying to prove that they are correct—such support for them as has been adduced in the study will be embodied in the supporting documentation and can be checked there—but to convince the audience that here is an adequate means to obtain the client's objectives, and what its consequences are likely to be. The question is not whether to use persuasion, but when and how to use it, and how to make it as effective as possible. After all, a valid study with a user who remains unconvinced of its worth is clearly a waste of effort and resources.

Thus, in evaluating the persuasive aspects of systems analysis, the way the analysis team has handled the elements discussed in this section is a central issue. How candor with respect to the strengths, weaknesses, and limitations of the work is handled should get particular scrutiny.

15.10. Evaluating the Analysis as a Whole: Some Useful Checklists

As said earlier, any evaluation of the quality of a systems-analysis study must be tailored closely to its context: the problem situation, the resources and data available, the parties at interest and their concerns, the responsibilities and authorities of the various decisionmakers, and so on. The purpose of the evaluation and when it is conducted are also important; for example, one conducted to bolster a potential user's confidence in the adequacy of the findings before he takes action must be quite different from one undertaken for academic purposes after implementation has been completed and the outcomes are known.

Thus, it is impossible to stipulate a single form of evaluation; rather, it depends on time and purpose and on bringing to bear appropriate elements of the craft of systems analysis. The traditional way for the potential users of the analysis to increase their understanding of its implications and their confidence in its adequacy is to ask the analysts questions aimed at exploring strengths and weaknesses, and uncovering biases, errors, and omissions. Evaluators who come later, after implementation, also depend on questioning. Although they may now have the outcomes for guidance, they are unlikely to have the original analysts available to answer questions or do additional calculations; thus, they may have to depend on documentation and themselves for answers.

In addition, all evaluators should trace through the central arguments of the analysis to make sure they understand them, compare the results with those of other analyses and well-based opinions, and ask the analysis team for additional calculations to examine intermediate and final results for plausibility and robustness—and they may do small "quick and dirty" analyses themselves. If the arguments are not clear, or if the results appear to differ too much from ones based on other sources, further questions aimed at clearing up these problems will be needed.

To help in the evaluation process, various workers have prepared lists of questions or statements that may be used as checklists to remind analyst, client, and critic what to search for; for examples, see Table 15.2, Wynne (1984, p. 310), and Quade (1964, pp. 323–325). The most important role for such checklists, however, may not be for evaluation by others but for quality control by the analysts themselves: to remind them of considerations that they might otherwise overlook or neglect during their work.

Questions Related to Standards of Quality

Although no list of questions can be prepared, even one designed for a particular analysis, with the expectation that favorable answers will guarantee high quality, such a list can at least remind all concerned of the wide range of considerations, some of which might otherwise be overlooked, that enter a judgment of quality. Nor can a list of questions guard against bias, error, or poor craftsmanship, but it may alert questioners to the possibility that such defects exist. Unlike a polished briefing or an impressive written report, which may direct attention away from weaknesses, questions are designed to reveal them, provided the questions are numerous, detailed, and searching.

This section presents two lists of questions that appear to have general value as aids to quality control and evaluation. Although there is some overlapping, those in Table 15.3 concern the internal processes of a systems-analysis study, the various procedures, assumptions, judgments, decisions, data sources, and so on having to do with the analysis itself or with what is done during the work that affects its findings. In the main, the questions in Table 15.4 concern the relations and interactions with the client and other parties at interest, and therefore deal more directly with the advisory and persuasive aspects of the systems-analysis activities. In neither case can it be claimed that these lists are complete in any sense; every experienced analyst should be able to add other questions. They are presented here to indicate the sorts of questions to be asked, not to specify the ones that should be used in a specific situation.

These questions, moreover, are designed to be answered *by the questioners themselves*: asked and answered by the analysts in their attempt to uncover error, neglect, or misplaced emphasis and by would-be evaluators to clarify what was done and to generate more specific questions to ask the analysts, or sometimes the client, about the analysis being evaluated.

Evaluators first try to answer these questions by using the documentation and briefing information available. There is little to gain from asking the briefer a question such as "Were the analytic methods used well suited to the problem and the analysis?" The answer is bound to be positive, for, if the analysts doing the work did not think so, they would have tried

Table 15.3. Questions Related to the Internal Processes of a Systems Analysis Study

1. Was a significant portion of the study effort devoted to exploring the problem situation and to formulating the problem to be addressed? Was the problem formulated the most important one in the problem situation for the client to deal with? If not, what alternative problem should have been considered and why? Was the problem bounded appropriately so that the study could be completed within the time and resources allotted? Were the problem and its boundaries satisfactory to the sponsor and/or client? Was the problem deficient in any respect? Did the analysts discuss possible objections to the goals selected, especially with persons possibly opposed to them?

2. Were the assumptions underlying the problem formulation and other actions made specifically? Are there any assumptions that could conceivably affect the results that seem so ordinary, so usual, or so obvious that they were overlooked or not deemed necessary to state explicitly? Are there alternative assumptions not considered that might be just as reasonable? If so, why were they rejected?

3. Are any of the assumptions unusual in any important way? If so, what was the justification for their use? Were any assumptions adopted for analytical or calculational convenience? If so, what effects do they have on the findings?

4. Were any limitations imposed by the client on the types of alternatives that might be considered? Did the analysis examine a full range of possible alternatives? Did this search include an attempt to design new alternatives, and in particular, ones combining desirable features of other alternatives?

5. What was the basis for the preliminary elimination of alternatives from detailed consideration? On what grounds were they judged to be inferior? Was this screening based on formal rules, preliminary calculations, or judgment—or all three? If a formal screening device was used, what are its features? If judgment, who exercised it and what aspects of the alternatives and the context entered the judgment? Are there other alternatives that should have been investigated?

6. Were the alternatives being presented examined for feasibility? How? Could some of them be infeasible because of political, policy, or cost considerations not taken up in the analysis?

7. Were any variables that might have affected the outcomes omitted from the analysis because the effects they might have had were judged to be insignificant? How was this judgment arrived at? Was it supported by analysis?

8. Were any proxy variables used? If so, was this because the more direct variable could not be measured, or because the cost or the effort to do so was prohibitive? What sort of error does this substitution introduce, and how was it estimated?

9. If data and information gathered for other purposes or by other investigators were used, what inquiry and what assumptions were made regarding the resulting information and any biases or errors that might ensue?

10. How were outliers—that is, extreme cases that appeared to contradict the general pattern of data—handled? Did the existence of these outliers affect the results? How?

11. Were all relevant costs included in the analysis? Or were some nonmonetary, hard-to-quantify costs omitted? If so, was any attempt made to estimate the effects their omission would have on the results? Were any costs included that might be considered irrelevant or not of concern? If so, did they have important effects on the results?

12. If a discount rate was used in comparing costs or benefits incurred at different times, what rate was used, and what was the justification for setting it at the chosen level or levels?

(continued)

Table 15.3. Questions Related to the Internal Processes of a Systems Analysis
Study *(continued)*

13. Were the costs considered only those associated with resources consumed, or were benefits from alternatives purchasable by the same expenditures considered?

14. Do the analytic methods employed appear to have been chosen because they were well suited to the problem and its analysis or because of the analyst's training or background? Before commitments were made to elaborate modeling exercises, was an extensive exploration of the policy field with far simpler, provisional, and interactive modeling (including real policy actors) conducted?

15. Should more up-to-date or state-of-the-art technical methods (from, for example, mathematics, operations research, or econometrics) have been used? Were adequate tools of analysis used? Could better choices of methods have been made?

16. What contingencies or scenarios were considered? Were any important or likely possibilities ignored?

17. If probabilities are assigned to various contingencies, are they intended to represent objective or subjective probabilities? If subjective, whose? Is there any question as to the legitimacy of any of these probabilities?

18. Does the analysis allow for uncertainties about the correct forms of relations in the models, about future environments, or about probabilistic uncertainties, as well as about uncertainties in the parameter values? If so, how are these uncertainties handled, and how important are they for the findings?

19. What validation processes were undertaken to establish the credibility of the models used?

20. Were the findings tested for sensitivity to simultaneous changes in the critical parameters, or was the sensitivity testing devoted merely to separate changes in single parameters? Were the findings also tested for their sensitivity to uncertainties other than those of parameter values? If so, how was this done? Do the findings appear to be robust?

21. Does the analysis support the stated findings fully? If not, to what extent do the findings depend on the intuition of the analysts? If the findings appeal to the intuition as being eminently correct, could they have been reached intuitively without a major study effort? If they do not, does this appear to be a case where intuition is unreliable, and does the analysis show this? Or, on the other hand, does lack of agreement between findings and intuition indicate that the analysis is untrustworthy, perhaps because it ignored some subtle and analytically intractable considerations that an intuitive approach would have included?

22. If there are any special cases for which the correct results are known or are obvious, are they consistent with the broader findings of the study?

23. Does the analysis ignore any consequence of a choice of alternatives that should be considered in making a decision about what action to take?

24. Are the criteria used in screening and ranking the alternatives reasonable? Did they eliminate only inferior alternatives, or were some desirable ones lost in the process? Are the criteria used consistent with higher-level criteria? Are there other criteria that should be brought to bear?

25. Are the recommendations, if any, made in full recognition of the uncertainties involved in the analysis? Are the implications of these uncertainties fully explained in the statement of the findings and recommendations?

26. Did the study generate any new and significant ideas or any new methods?

27. Does it appear that any of the results can be generalized? Or applied elsewhere?

28. Has the analysis considered the difficulties that might be encountered if the preferred alternatives are implemented? Has the outline of an implementation plan been prepared for each attractive alternative?

29. If informal craft judgment was accepted as a legitimate part of modeling and analysis, was each occurrence clearly labeled for what it was?

30. Did the study follow accepted standards of professional practice, as understood by all concerned, based on the current standards of the profession?

Table 15.4. Questions Related to Interactions with the Client and other Parties at Interest

1. Did the analysis team work closely with the sponsor and client during the problem formulation and later during the analysis? Did they agree that the assumptions were reasonable? Were the principals and their staff kept fully informed?
2. Were the findings of the analysis presented in a useful form, with the conclusions and recommendations spelled out so as to be fully meaningful to the client and to other parties at interest? Was a scorecard provided? That is, were the relevant consequences or outcomes associated with each alternative shown for each contingency considered? Could improvements have been made in these presentations?
3. Are the limitations of the analysis, as well as its strengths, pointed out fully and clearly? In particular, is it clear what the study did not cover and why, as well as what it considered?
4. Was the quality and interdisciplinary mix of analysts on the analysis team appropriate to the investigation?
5. Does the analysis team disclose where it made subjective judgments during the analysis? Is the information and logic behind them made explicit?
6. If the analysis falls short of giving the expected help to the decisionmakers, does it at least eliminate conclusively a number of inferior alternatives? Does it tabulate the consequences of some that merit consideration?
7. Were the findings presented early enough to enable the decisionmakers to consider them in depth before their deadline for action arrived?
8. Did the study take into account the institutional and political context in which the preferred alternatives would have to be implemented? If so, was this done before the conclusions were formulated? In considering political feasibility, did the analysis team search for alternative political pathways to policy approval and implementation, including investigating changes in the policymaking system needed to make otherwise clearly preferable policies feasible? If the implementation lies in the future, do the analysts plan to help with it? If it is ongoing, are they helping with it?
9. Is the general scheme of the analysis and its findings appropriate to the decisionmaking situation?
10. Do the findings of the work appear to follow logically and directly from the analysis work without any suspicion that they might have been modified to fit what the client would like to hear?
11. Is the analysis documented fully and clearly? If not, why? Does the report stress the client's concerns and what was learned about them, or is it merely a detailed documentation of the technical path pursued in the analysis? Do the oral and written reports take care to separate opinion from facts, evidence, and findings?
12. Is it clear that the analysis team thought about its communication problems early in the analysis and fashioned its presentations so as to reach its various audiences effectively? Were the various reports well adapted to their audiences? Does the analysis yield some relatively simple logical arguments, computation rules, or other schemes that a decisionmaker can use to eliminate inferior alternatives by himself? Are they clearly explained in the reports? Does the analysis yield a way of answering quickly questions of this form: How would the results be affected if such-and-such an assumption were changed?
13. Was the organizational context given consideration during the work? Was the analysis team sensitive to such factors in the organizational context as timing, structure, competition among organizational elements, the environment, and the functions of various organizational elements that might be affected by the findings?

(continued)

Table 15.4. Questions Related to Interactions with the Client and other Parties at Interest *(continued)*

14. If the analysis failed to win acceptance, was this owing to inadequate presentation? To a defect in the study itself? To changing conditions that could or could not have been foreseen? Was it because the team, in its reports and presentations, discounted or neglected factors not included in their quantitative models?
15. Was the study reviewed before its findings were presented to the client? If so, by whom?
16. If failure was due to lack of persuasiveness, was it because the arguments were poorly selected to convince the audience? Was the personality of the head of the analysis team a factor?
17. Was the analysis used properly by the client? Or improperly, such as, for example, solely as an excuse to postpone decision?
18. Did the analysis have any impact on the problem situation that is so far apparent?
19. Did the analysis team attract any professional recognition for its work on the study?
20. Were the decisions or policies adopted ostensibly based on the analysis adequately justified by the analytic work? Or did the decisionmakers have to bring to bear a significant amount of additional information and judgment?

a different approach. But if, after checking the oral and written material, the evaluator suspects that another approach might have been better, he can phrase much sharper questions about what methods might have been considered, and, if rejected, why.

As remarked in the introduction to this chapter, the purpose of the evaluation and the qualifications of the evaluators are importantly related; for example, for an evaluation intended to rate the professional qualifications of the analysts, one would be satisfied with a jury of evaluators with professional and analytic skills at least as strong as those represented on the analysis team. For an evaluation aimed at helping a decisionmaker set policy, one would like the group of evaluators to include experts in the field of the problem situation, as well as other decisionmakers with relevant experience—something, fortunately, that is hard to avoid.

The Marks of a Quality Analysis

Despite what has been said about how an evaluation of the quality of a systems-analysis study must be adapted closely to the study itself, to the context within which it was carried out, and to the purpose of the evaluation, there are nevertheless some marks that are generally associated with quality work. Most can be expressed as short statements related to longer discussions elsewhere in this handbook; Table 15.5 gathers them together for convenient reference (see also Enthoven, 1975). Possession of these marks does not guarantee quality, nor does it guard against hidden error, but experience suggests that there is a fairly high correlation be-

Table 15.5. Some Marks of Quality in Systems Analysis

---

1. A substantial effort devoted to formulating the problem.
2. An exhaustive search for new ideas and alternatives.
3. Explicit recognition and careful treatment of uncertainties.
4. Substantial testing for sensitivity, including sensitivity to simultaneous variation of groups of parameters and other factors.
5. Clear statement of the assumptions, the boundaries, and the constraints.
6. Input data scrutinized for accuracy and relevance before being transformed into information and applied as evidence.
7. Models selected and developed so as to be appropriate to the problem.
8. Models verified, with reasonable efforts to test their validity.
9. Subjective judgments made explicit and supported by the reasons behind them.
10. Extensive documentation of the work, including not only what was done, but also why, with backup material included.
11. Findings and recommendations, although focused on the interests of the decisionmakers, giving adequate attention to other interests, including those of the general public.
12. Reports, both written and oral, focused on the audiences, to the end that they can use the findings in their further thinking about the problem situation.
13. The ability to answer searching questions promptly about the work and its implications.
14. At least a preliminary plan for how to undertake an implementation, in case a decision to act is taken.
15. Explicit recognition given to the environment, to future generations, and to interest groups that might be negatively affected.
16. Attention to questions of equity that might arise, and if they are significant, ways to compensate losers.
17. The client's objectives explored for consistency with moral standards and the public welfare.
18. The leading alternatives investigated for feasibility, political, organizational, and otherwise.
19. In addition to the monetary and other obvious costs, an effort to discover hidden costs that might appear later to plague the implementers.
20. Communication between the analysis team and the client/sponsor and their staffs so frequent as to be considered continuous.

---

tween analysis quality and the possession of these marks. They are, in our view, incipient standards.

Documentation for a Computer-Based Model

Comprehensive documentation is one of the marks of a quality analysis. One of the most difficult—and usually least satisfactory—elements of such documentation is that of its computer models, particularly if large-scale ones are involved. Therefore, this subject deserves some additional attention here.

Gass (1984) has proposed standards for the form and content of the documentation for large-scale computer models; however, his approach can easily be adapted to computer models on any scale. He describes the goals of such documentation in these terms (loc. cit., pp. 85–86):

> The documentation of computer-based models should provide specific and detailed information organized and presented in a manner that will satisfy the needs of each segment of a model's audience. This audience consists of the model's sponsors and users (possibly non-technically oriented); the model's analysts, programmers, and computer operators; other users, analysts, programmers, and computer operators; and independent model evaluators. Although I am concerned mainly with the documentation requirements for large-scale decision models, my approach is applicable to all computer-based models. It is predicated on making documentation achieve the following purposes:
>
> > to enable systems analysts and programmers, other than the originators, to use the model and programs,
> >
> > to assist the user in understanding what has been done and why,
> >
> > to record technical information that enables system and program changes to be made quickly and effectively,
> >
> > to facilitate auditing and verification of the model and the program operations, that is, model evaluation,
> >
> > to provide managers with information so they may determine whether requirements have been met,
> >
> > to improve "organizational memory" and thus reduce the effects of personnel turnover,
> >
> > to provide information about maintenance, training, changes, and operations, and
> >
> > to enable potential users to determine whether the model and programs will serve their needs.
>
> Accomplishing these purposes rests on the following assumptions:
>
> Model documentation must describe all historical, technical, developmental, maintenance, and implementation aspects of the model, including assumptions, implications, and impact of using the model in a decision situation.
>
> Computer program and software documentation of a model should follow the guidelines of FIPS PUB 38. These guidelines have been promulgated and disseminated by the [U.S.] National Bureau of Standards (1976).
>
> The organization of a modeling project must include a formal documentation activity with stated objectives and assignment of resources (personnel, funds, time) for their accomplishment.
>
> As a means of managing the documentation activity of a modeling project, documentation must be produced that corresponds to the

phases of the model life cycle, and the production and maintenance of the documentation must be concurrent with the time span of each phase.

Gass proposes that the documentation consist of four manuals, *Analyst's Manual, User's Manual, Programmer's Manual*, and *Manager's Manual*, and he gives detailed specifications of the purposes and contents of each manual (loc. cit., pp. 86–93).

In presenting his prescriptions for computer model documentation, Gass stresses that they are based on his review of actual cases and proposals that others have made, and that they have yet to be tested thoroughly in actual practice. Nevertheless, his proposals emerge from considerable experience, and, as the most mature ones available when this chapter is being written, they deserve careful consideration in connection with systems analyses that may use large-scale computer models.

As is the case for any checklists aimed at eventual standards, these prescriptions will no doubt have to be modified in the light of any actual case.

15.11. Evaluating the Relationship between Analyst and Client

As Section 12.4 brings out, systems analysts are professionals who can—and should—seek a cooperative form of partnership with their clients that Schön (1983, pp. 296–297) calls a reflective contract:

> [I]n a reflective contract between practitioner and client, the client does not agree to accept the practitioner's authority but to suspend disbelief in it. He agrees to join the practitioner in inquiring into the situation for which the client seeks help; to try to understand what he is experiencing and to make that understanding accessible to the practitioner; to confront the practitioner when he does not understand or agree; to test the practitioner's competence by observing his effectiveness; to pay for services rendered and to appreciate competence demonstrated. The practitioner agrees to deliver competent performance to the limits of his capacity; to help the client understand the meaning of the professional's advice and the rationale for his actions, while at the same time he tries to learn the meanings his actions have for his client; to make himself confrontable by his client; and to reflect on his own tacit understandings when he needs to do so in order to play his part in fulfilling the contract.

To the extent that the systems-analysis profession has evolved criteria of practice for its role in this partnership, as we have interpreted them, the earlier sections of this chapter deal with them. To evaluate professionalism fully, however, requires one not only to consider the work of the practitioner as it relates to his client, but also the nature of the partnership between them; further, it also implies taking into account the contribution that the client brings to the mutual enterprise.

The client for systems analysis almost always knows far more about the problem situation for which he seeks help, and far more about ways to improve it, than the clients for such other professions as law, medicine, or engineering. Moreover, he contributes much more to the results that ensue from his association with the professionals. In other disciplines, the practitioner depends on his client for certain information not obtainable elsewhere and possibly for general guidance, but very seldom for helpful advice, ideas, and even criticism as in systems analysis. This is partly because systems analysis is as yet a young profession without the traditions and prestige of law and medicine, but more so because of the nature of its problems and subject matter. Even in engineering the relationship is different:

> If a city hires an engineer to design a bridge, it may perhaps have his work checked by another engineer, but the city fathers will not presume to study his report with a view to seeing for themselves whether the proposed bridge is likely to collapse. They believe, with more or less reason, that the field of civil engineering is sufficiently well developed and a licensed engineer is so likely to be firm in his science that his judgment in this matter is overwhelmingly better than their own. Similarly, they will trust the authority of their engineer that the clearance, carrying capacity, safety, and durability of the proposed bridge cannot be increased without an increase in cost. The trade-offs among these values might in principle concern the city fathers, and in special cases they will. But, by and large, there will be none among them capable of or feeling the responsibility for going deeply into these matters. [Savage, 1961.]

Where, however, questions of fire protection, water supply, flood protection, energy, or the environment are concerned, typical situations involving many disciplines in which systems analysts are called on for help, the tradeoffs between the various values involved are felt to be the responsibility of those in authority, and those in authority must involve themselves in helping analysts to determine a satisfactory and equitable outcome.

As to the nature of the partnership, systems analysts have learned that it is essential for the client to allow open access to all aspects of the problem situation under his control, and for him, not only to cooperate fully throughout the study, but also to make active contributions to the work, but it cannot be alleged that actual or potential clients uniformly appreciate these points. Nor has there been any sort of wide appreciation of the criteria that should apply to the other facets of the partnership, as Schön's description of it should be adapted to systems analysis.

Although most experienced systems analysts have developed intuitive feelings for when a client is likely to be a good customer for their work, this experience remains to be systematized, and the literature does not contain any generally accepted criteria for how the client should play his role in the partnership beyond what is implied in what Schön says in the quotation above.

In these regards, systems analysis has much in common with other professions (Schön, 1983) and shares with them the need to develop an epistemology of practice solidly based on actual successful experience that will not only describe the detailed nature of the reflective contract between practitioner and client and how the activities under it should be carried out but also provide the foundation for criteria of quality for the contributions of both partners to the problem-solving work. Schön argues—and we agree—that the activities of professionals and their clients are susceptible to scholarly inquiry, and that the resulting findings will yield the elements that can be combined into the needed epistemology of practice. The challenge to the systems-analysis profession is to undertake such inquiries in connection with its work and to weave them into the fabric that can be recognized as generally accepted ideals.

15.12. Conclusion

Although this chapter stresses the need for generally accepted standards as a necessary foundation for controlling the professional quality of systems-analysis work, it also argues that no rigid code can be applied successfully; rather, the criteria must be adapted individually to each study and its context.

The key forces in developing generally accepted standards of quality will be—as they are now—the examples of work carried out or criticized by leaders of the profession and cited for their excellence. As one example, the College on the Practice of Management Science of The Institute of Management Sciences holds an annual competition for outstanding examples of practice, and the leading papers in each year's competition are published;[1] although these papers do not usually pay principal attention to craft issues, they offer much relevant experience and are well worth consulting in this regard. Similarly, the examples discussed in the other volumes of this handbook also serve this purpose, and they may be explored further in their extensive background literature.

The key transmitters of high standards of craftsmanship in systems

---

[1] See *Interfaces* 5 (February 1975, Part 2), 6 (November 1975, Part 2), 7 (November 1976, Part 2), 8 (November 1977, Part 2), 9 (February 1979, Part 2), 9 (November 1979), 10 (November 1980), 11 (December 1981), 12 (December 1982), 13 (December 1983), 15 (January–February 1985), 16 (January–February 1986), 17 (January–February 1987), and at roughly annual intervals thereafter.

analysis are now—and will remain—its professional leaders, and the most important experience for anyone entering the profession may well be an apprenticeship under one or more of these leaders. Nevertheless, a handbook such as this one can make an important contribution to this process by systematizing key elements of the profession's experience as a whole and making them readily available. This is a central purpose of this handbook, and its contributors and editors will be more than amply repaid for their efforts if it comes to occupy such a place in the activities of the systems-analysis profession.

The ability to do work of high quality is important to success in systems analysis, but it is not enough. The analyst must have another quality that a handbook can do little to supply:

> The systems analyst must have high standards for the quality of the technical work that goes into the study but the standards should not be so high that they are self-defeating. If he insists on checking every fact with every possible person who could have any opinion on the subject, then he would never finish the study. He must do enough cross-checking to convince himself that, in all probability, he has the correct facts, and then he takes his chances. This means that once in a while he will be misled and will look foolish, but one cannot do effective work in this field unless one is willing to take this risk. [Kahn and Mann, 1957, p. 51.]

Although the analyst's imperative of achieving timely and helpful findings must subject him to some risks of looking foolish, his client's risks are much greater and more important; thus, the responsible analyst must take his risks with due caution and in the light of full consideration of the consequences of error for his client.

Systems analysis provides more than information and advice that, along with the associated decisions taken on their basis by the responsible parties, can seldom be established unambiguously as correct. Open and explicit analysis, however, can—and often does—provide the material and framework for constructive consideration and debate (Enthoven, 1975, p. 457). Brooks (1976, p. 115) goes so far as to say:

> In this sense, policy or systems analysis perform a function with respect to political-technological decisons similar to that performed by a judicial process with respect to conflicts between individuals. A court decision is accepted by the disputing parties largely because it is based on a set of rules both parties accept applied through a procedure which both parties are prepared, before knowing its outcome, to accept as unbiased.

The design of options and systematic inquiry into, and arguments to estimate, their costs, consequences, constraints, and risks, as provided by systems analysis, furnish material and structure for policy debate. In our society, the choice of policy and action depends far more on public debate than on science and analysis, even though some might wish it otherwise. Nevertheless, the quality of the arguments and information developed by systems analysis can contribute much to this debate. Thus, we need to strengthen its foundations by evolving standards to ensure its adequacy, pertinence, and credibility.

References

Altman, S. M. (1980). Pitfalls in data analysis. In Majone and Quade (1980), pp. 44–56.

Bardach, Eugene (1977). *The Implementation Game: What Happens After a Bill Becomes Law*. Cambridge, Massachusetts: The MIT Press.

Brooks, Harvey (1976). Environmental decision making: Analysis and values. In *When Values Conflict* (L. H. Tribe, C. S. Schelling, and John Voss, eds.). Cambridge, Massachusetts: Ballinger, pp. 115–136.

Cases. See Miser and Quade (to appear).

Chubin, D., and S. Jasanoff, eds. (1985). Peer review and public policy. *Science, Technology, and Human Values* 10(3).

Clark, W. C., and Giandomenico Majone (1985). The critical appraisal of scientific inquiries with policy implications. *Science, Technology, and Human Values* 10, 6–19.

Enthoven, A. C. (1975). Ten practical principles for policy and program analysis. In *Benefit-Cost and Policy Analysis* (Richard Zeckhauser et al., eds.). Chicago: Aldine, 1975, pp. 456–465.

Freymann, J. G. (1974). *The American Health Care System: Its Genesis and Trajectory*. New York: Medcom Press.

Gass, S. I. (1984). Documenting a computer-based model. *Interfaces* 14(3), 84–93.

Kahn, Herman, and I. Mann (1956). *Techniques of Systems Analysis*, RM-1829. Santa Monica, California: The Rand Corporation.

———, and ——— (1957). *Ten Common Pitfalls*, RM-1937. Santa Monica, California: The Rand Corporation.

Kent, G. A. (1967). On analysis. *Air University Review* 18(4), 50.

——— (1969). *The Role of Analysis in Decision Making*. Keynote speech before the 24th meeting of the Military Operations Research Society.

Koopman, B. O. (1956). Fallacies in operations research. *Operations Research* 4, 422–426.

Majone, Giandomenico (1980). *The Craft of Applied Systems Analysis*, WP-80-73. Laxenburg, Austria: International Institute for Applied Systems Analysis.

——— (1982). Evaluation. Unpublished manuscript.

———, and E. S. Quade, eds. (1980). *Pitfalls of Analysis*. Chichester, England: Wiley.

Meadows, Donella, J. Richardson, and G. Bruckmann (1982). *Groping in the Dark: The First Decade of Global Modeling*. Chichester, England: Wiley.

Meltsner, A. J. (1980). Don't slight communication: Some problems of analytical practice. In Majone and Quade (1980), pp. 116–137.

Miser, H. J., and E. S. Quade, eds. (1985). *Handbook of Systems Analysis: Overview of Uses, Procedures, Applications, and Practice*. New York: North-Holland. Also cited as *Overview*.

————, and ———— (to appear). *Handbook of Systems Analysis: Cases*. New York: North-Holland. Also cited as *Cases*.

National Bureau of Standards (1976). *Guidelines for Documentation of Computer Programs and Automated Systems*, Federal Information Processing Standards Publication FIPS PUB 38. Washington, D.C.: U.S. Government Printing Office.

Overview. See Miser and Quade (1985).

Quade, E. S., ed. (1964). *Analysis for Military Decisions*. Amsterdam: North-Holland.

Ravetz, J. R. (1971). *Scientific Knowledge and its Social Problems*. Oxford, England: Oxford University Press.

Savage, L. J. (1961). Personal communication to E. S. Quade, June 21, 1961.

Schön, D. A. (1983). *The Reflective Practitioner: How Professionals Think in Action*. New York: Basic Books.

Wilson, J. Q. (1973). On Pettigrew and Armor: An afterword. *Public Interest* 30, 132–135.

Wynne, Brian (1984). The institutional context of science, models, and policy: The IIASA energy study. *Policy Sciences* 17, 277–320.

Author Index

Subject Index